Patrick Moore's Practical Astronomy Series

For other titles published in this series, go to
www.springer.com/series/3192

3,000 Deep-Sky Objects

An Annotated Catalogue

Ted Aranda

Springer

Ted Aranda
Chicago, USA
taranda@uicalumni.org

ISSN 1431-9756
ISBN 978-1-4419-9418-9   e-ISBN 978-1-4419-9419-6
DOI 10.1007/978-1-4419-9419-6
Springer New York Dordrecht Heidelberg London

Library of Congress Control Number: 2011932225

© Springer Science+Business Media, LLC 2012, Corrected at 2nd printing 2012
All rights reserved. This work may not be translated or copied in whole or in part without the written permission of the publisher (Springer Science+Business Media, LLC, 233 Spring Street, New York, NY 10013, USA), except for brief excerpts in connection with reviews or scholarly analysis. Use in connection with any form of information storage and retrieval, electronic adaptation, computer software, or by similar or dissimilar methodology now known or hereafter developed is forbidden.
The use in this publication of trade names, trademarks, service marks, and similar terms, even if they are not identified as such, is not to be taken as an expression of opinion as to whether or not they are subject to proprietary rights.

Printed on acid-free paper

Springer is part of Springer Science+Business Media (www.springer.com)

ACC LIBRARY SERVICES AUSTIN, TX

For

*Our Mother, our Home, our Garden of Eden –
Our little Blue Marble in the depths of space –
And all who love her.*

Preface

An amateur astronomer is observing galaxies in the constellation Cetus on a clear night at her favorite dark site. She notices that the galaxy currently in the eyepiece of her 12-inch telescope is considerably brighter than the previous one. It is also much more concentrated. In general she finds the large variation in the brightness and concentration of galaxies fascinating. But there must be something wrong. She has observed dozens of galaxies so far but has yet to see one with a distinct core or nucleus within its halo. Furthermore, in only a handful of galaxies has she seen any variation in the light other than a smooth, continuous concentration toward the center. And in these cases the irregularities have been quite vague; certainly they have not resembled spiral arms.

Dawn is approaching. Our observer has had great fun tonight. But, once again, as in past observing sessions, she has failed to see in any but a few galaxies even a trace of the things that photographs show so spectacularly and routinely, that professional astronomers study so diligently, and that observing guidebooks urge their readers to look for so optimistically. She will try harder next time.

Clearly there is something amiss in the above scenario. But the fault lies not with our intrepid observer, who no doubt saw many fine sights during her enchanted night under the stars. It lies instead with a body of astronomical literature that does not sufficiently acknowledge the vast and largely unbridgeable difference between the photographic and the visual appearance of many types of deep-sky objects, thereby leading dedicated observers to have unrealistic – and all too often disappointed – expectations.

There are in fact numerous characteristics of deep-sky objects that our eyes capture far better than photographs can, including the sheer dazzling brilliance of bright stars, the great diversity among globular clusters, and the remarkable "blinking effect" of many planetary nebulae. In the final analysis, although it naturally pleases us to be able to see photographic features visually, our eyes are not CCD cameras and they will never operate as these machines do. But they are amazing sense organs nonetheless, and what they show us in the night sky with the help of our trusty telescopes is equally wonderful, beautiful, and delightful.

***

I am an unabashed binocular enthusiast. Indeed, the idea of a powerful binocular telescope was the motive force behind this entire project. Owners of monocular telescopes (in other words almost all amateurs astronomers) might therefore wonder if this book is relevant to them. The answer is an unqualified *yes*. Whatever telescopes we use, we visual observers are all kindred spirits. Our common goal is to see our beloved astronomical objects as well as we can with our own eyes. I myself started out with an ordinary 80 mm telescope and soon graduated to a standard 10-inch Dobsonian. The 10-inch binocular telescope that I built and subsequently used for this survey was the "upgrade" I finally chose for myself.

A telescope is a telescope – an instrument useful for its magnifying and light-gathering functions. A binocular telescope is simply a particular kind of telescope. For every binocular telescope there is a monocular telescope of somewhat larger aperture that technically will show the same things, although the view will be less satisfying in some ways. Telescopes come in many forms, and the question of binocular versus monocular, like that of reflector versus refractor or apochromat versus achromat, is just one among many design variables, albeit an especially significant one.

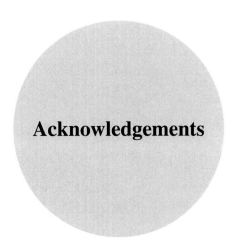

Acknowledgements

Although I observe alone, I have many intellectual and spiritual companions to thank for their contributions to my astronomical accomplishments, such as they may be. In keeping with the amateur astronomer's habit of perpetually making lists, I would like to express my special gratitude to the following persons:

Emil Bonanno, for creating and perfecting MegaStar, without which this project would not have been feasible.

The authors of telescope design and construction books, for sharing their expertise, particularly Jean Texereau, *How to Make a Telescope*; Harrie Rutten and Martin van Venrooij, *Telescope Optics*; Richard Berry, *Build Your Own Telescope*; Sam Brown, *All About Telescopes*; Peter Francis, *Newtonian Notes*; and Kenneth Novak, *Cassegrain Notes*.

The editors of, and contributors to, *Sky & Telescope*, *Astronomy*, *Deep Sky*, and *Telescope Making* magazines, for their invaluable observing and telescope making guidance.

The authors of the leading deep-sky observing guidebooks, for expanding our knowledge of the sky – critical though I may be of some aspects of their work.

The authors and editors of the major deep-sky catalogues, for putting the necessary data at our fingertips.

Wil Tirion, for pioneering the modern, elegant, and beautiful sky atlas.

John Dobson, for pioneering the sensible and efficient amateur telescope.

James Braginton of Coulter Optical, for leading the way in commercial Dobsonian telescopes.

Lee Cain, for pioneering the "mammoth" Dobsonian binocular.

Ed Beck of Enterprise Optics, for beautifully refiguring my Cassegrain mirrors.

The distributors of astronomical equipment (AstroSystems, Apogee, University, Adorama, Orion), for making our hobby possible.

Carlos Santana, for his musical inspiration.

Marcia Zumbahlen, for kindly assisting me with the reproduction of the charts and photographs in Appendices A and D.

My family, for their constant love and affection, specifically my sister Doris and my mother Mary, for providing me with an astronomical home base in the Southwest; my father Dr. Theodore Aranda, for having been such a key figure in my intellectual development; and my sister Lerna and her daughters (Afton, Haley, Ava, and Thandie), for their warmth and friendship.

My buddy, for starting out with me on this venture, being such a wonderful being, and sharing his life with me.

And last but not least, Springer and its fine staff, for taking on this book project and seeing it through to publication.

Contents

Preface vii

Acknowledgements ix

Part I   Introduction

1   **The Observing Project: An Overview** 3
    The Author at the Telescope 5

2   **Objects Chosen and Mode of Observing** 7

3   **Verbal Descriptions vs. Illustrations** 11

4   **Instrumentation** 13

5   **The GC/NGC Descriptions and the Herschels** 15

6   **Observational Parameters in the Descriptions
    of This Catalogue** 19
    Star Colors 20
    Double Star Separation 21
    Relative Brightnesses of Double Star Components 21
    Brightness of Deep-Sky Object 21
    Concentration of Galaxy or Globular Cluster 22
    Shape of Galaxy or Other Deep-Sky Object 22
    Size of Deep-Sky Object 23

	Magnitude Equivalence	24
	Distance Equivalence	24
	Resolution of Globular Cluster	24
	Number of Stars in Open Cluster	24
	Density of Open Cluster	25
7	**The Visual Appearance of Deep-Sky Objects**	27
	Stars	28
	Galaxies	29
	Globular Clusters	31
	Nebulae	32
	Open Clusters	32
	Planetary Nebulae	33
8	**Binocular Vision**	35
9	**How to Use This Catalogue**	39
	Basic Data	40
	Data Specific to Each Class of Object	40
	Abbreviations Used	44

Part II The Catalogue

10	0 to 6 Hours: FALL	49
11	6 to 12 Hours: WINTER	149
12	12 to 18 Hours: SPRING	271
13	18 to 24 Hours: SUMMER	411

Appendix A: Making a Sky Atlas	487
Appendix B: Notes on Object Descriptions	497
Appendix C: The Visibility of Galactic Detail	499
Appendix D: Building a Binocular Telescope	501
Addendum: Further Observations and Notes	529
Index	531

Part I

Introduction

Chapter 1

The Observing Project: An Overview

[B]y *applying ourselves with all our powers to the improvement of telescopes ... and turning them with assiduity to the study of the heavens, we shall ... obtain some faint knowledge of, and perhaps be able partly to delineate,* the Interior Construction of the Universe.

William Herschel, 1784

In 1783 William Herschel built the world's first truly powerful telescope and with it embarked upon a thorough exploration of the sky. The job was completed five and a half decades later by his son John. Together the Herschels discovered thousands of new "nebulae" and star clusters where only dozens had been known before. John Herschel finally published his monumental "General Catalogue of Nebulae and Clusters of Stars" in the *Philosophical Transactions* of 1864. An expanded version of this work, J. L. E. Dreyer's 1888 *New General Catalogue*, remains to this day the principal deep-sky catalogue for astronomers.

Both the GC and the NGC include concise descriptions of their listed objects, recorded by the Herschels and other renowned astronomers at the eyepieces of their telescopes before the advent of astrophotography. These old descriptions thus convey the *visual* appearance of deep-sky objects in some ways more faithfully than do those in modern observing guides, the authors of which tend to be unduly influenced by photographic imagery.

Naturally, however, the pioneer observers of the nineteenth century had a limited understanding of the true nature of deep-sky objects. Moreover, they did not use the same terminology as we do now. The aim of this project, therefore, has been to record accurately, systematically, and in modern terms the major visual characteristics of a comprehensive set of deep-sky objects as seen through medium-sized (10–14") amateur telescopes, regardless of what photographs show or views through very large telescopes reveal, and no matter how famous or iconic the objects observed.

My use of binocular instruments has helped me immensely in this endeavor, for the binocular telescopic view is far superior to the monocular view, in clarity, realism, and comfort, among other things. Thus an all-binocular approach makes for exceptionally satisfying and confident astronomical observing. I found, paradoxically, that although deep-sky objects are seen more clearly when both eyes are used, it also becomes apparent that many deep-sky objects are intrinsically ill-defined.

The observations were carried out from four southwestern U. S. locations: a site near Reserve, New Mexico (elevation 7,000 ft, latitude 34°), two neighboring sites in the Coronado National Forest of southwestern New Mexico and southeastern Arizona (5,000 ft, 31.5°), and a spot 30 miles east of Tonopah, Nevada (6,000 ft, 38°). All of these remote mountain sites are free of significant light pollution. Observations were undertaken only on clear nights when no Moon was present and no significant amount of haze impaired the view. The survey was limited to the sky above −35°. Therefore references to "the sky," as in "the brightest planetary in the sky," mean only the sky above −35°.

The Author at the Telescope

Chapter 2

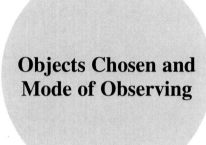

Objects Chosen and Mode of Observing

For this project I made my own specialized sky atlas using MegaStar5, a superb astronomical program (see Appendix A). The objects that I selected for observation and plotted on the atlas' charts are described here. It can be seen that my goal was more to cover the better – i.e., relatively bright and easy – objects, of which there are plenty, than to seek out observational challenges, which are practically infinite in number.

- BRIGHT STARS (200) – Stars brighter than magnitude 3.5. Bright stars that are also double stars are listed as bright stars but described as double stars, with the double star data appended.
- DOUBLE STARS (155) – Double stars with the primary brighter than magnitude 7.0, the companion brighter than magnitude 8.0, the separation between 2.0″ and 20″, and the magnitude difference smaller than the separation; plus a few wider bright doubles. There are a few closer pairs, due to the fact that although the double star list was originally culled from *Sky Catalogue 2000.0* (1985), the final separations were taken from the more recent *Washington Catalogue of Visual Double Stars* (WDS); some separations have decreased significantly over the past few years. For fast-moving binaries the separation is estimated for the year 2005.
- VARIABLE STARS (56) – Carbon stars with a maximum of 7th magnitude or brighter as shown in *Sky Atlas 2000.0,* 1st ed.; plus long-period variables with a maximum of 6th magnitude or brighter and a very wide amplitude. These classes of variables are interesting mainly for their possible deep color.
- GALAXIES (1,860) – Galaxies magnitude 13.2 or brighter.
- GLOBULAR CLUSTERS (88) – NGC, IC, Pal, and Terzan globular clusters magnitude 13.2 or brighter, plus additional globular clusters plotted in *Sky Atlas 2000.0,* 1st ed.

NEBULAE (47) – Prominent NGC nebulae.

OPEN CLUSTERS (292) – NGC, IC, Trumpler, King, Mellotte, Harvard, Stock, and Tombaugh open clusters magnitude 13.2 or brighter, plus additional open clusters plotted in *Sky Atlas 2000.0*, 1st ed.

PLANETARY NEBULAE (106) – NGC and IC planetary nebulae magnitude 13.2 or brighter, plus additional planetary nebulae plotted in *Sky Atlas 2000.0*, 1st ed.

Sources used for compiling the objects and the object data included Alan Hirshfeld, Roger Sinnott, and Francois Ochsenbein, *Sky Catalogue 2000.0, Volume 1: Stars to Magnitude 8.0*, 2nd ed. (Cambridge, MA: Sky Publishing, 1991); Alan Hirshfeld and Roger Sinnott, eds., *Sky Catalogue 2000.0, Volume 2: Double Stars, Variable Stars and Nonstellar Objects* (Cambridge, MA: Sky Publishing, 1985); Robert Burnham Jr., *Burnham's Celestial Handbook* (New York: Dover, 1978); Roger Sinnott, ed., *NGC 2000.0* (Cambridge, MA: Sky Publishing, 1988); *Deep Sky Journal #5: Summer 1993* (Aberdeen, WA: Wayfarer Publishing); MegaStar5 (Richmond, VA: Willmann-Bell, 1992–2002).

In the course of looking for and examining the 2,804 listed objects, I serendipitously encountered and described 285 additional galaxies, making a total of 3,089 examined objects altogether (excluding the handful of extra non-galactic objects I discovered). Each of these additional galaxies is flagged in the observations with an asterisk. At no point did I use either charts with additional objects plotted on them, or photographs, at the eyepiece. Although faint field objects as well as inconspicuous features of large objects can sometimes be seen if one knows they are there and *exactly* where to look for them, I preferred not to know about any such things beforehand. This made the spotting of extra "faint fuzzies" or unexpected object detail all the more gratifying. More importantly, I was thereby able to keep the relative visual prominence of the various objects and object features in proper perspective.

The objects were observed specifically for this project during the years 2004 through 2008. Numerous objects were observed a second time in the latter stages of the project due to incomplete or questionable observations the first time around. The author, an amateur astronomer with 20 years' previous experience, was the sole observer. The observations were tape-recorded in the field, then transcribed verbatim, and finally edited for clarity and consistency. They are in straightforward, non-technical language employing broad relative terms rather than non-intuitive, unrealistically precise numerical terms. I for one am not able to translate judgments of dimensions and distances into arc-minutes, or estimates of brightness into magnitudes, while looking in the eyepiece. (I did, however, provide approximate quantitative equivalents to my visual impressions; these, derived from a systematic comparison of my observations with photographs, are listed below.) Like the early observers I chose to retain the succinct staccato phrases used in the field rather than convert everything into needlessly wordy sentences.

For purposes of confirmation the descriptions of the extended objects were afterward checked against photographic images in MegaStar. (Throughout this introductory essay I will be stressing the important differences between visual observation and photography; this should in no way be construed as a lack of recognition on my part of the obvious and incalculable value of the latter to astronomy.) In those cases where observations were still incomplete or apparently inaccurate at the end of the project, and where more or better data were available from my past observations (I did a good deal of observing with the 10-inch binocular before commencing on this survey), I amended the latest observations in accordance with previous ones. Significant corrections *not* derived from my own past observations, as well as other post-observation notes and additions, appear in square brackets. The only exception to this rule is minor inaccuracies in sky directions, which were very common; these were rectified without note throughout. In the observation of galaxies, the visibility of non-NGC field galaxies brighter than magnitude 15.0 and NGC field galaxies brighter than magnitude 16.5, and within 6' of the target galaxy, was especially noted.

Chapter 3

Verbal Descriptions vs. Illustrations

It might be wondered why I did not accompany at least some of my verbal descriptions with sketches, as many guidebook authors have done. Briefly, the reasons are threefold: (1) To capture adequately in drawings the subtleties of deep-sky objects seen in eyepiece views, for example the huge range of brightness in the light of galaxies, or globular clusters' fields of outliers, requires more skill than is usually appreciated, even by many who publish their drawings. Given my own limited artistic ability, my descriptions are far more useful than any attempted sketches by me would ever be. (2) There seems to be a tendency among sketchers to draw, in the case of specific objects, the things they remember from photographs of those objects, and, in general, features that they expect to see due to their ubiquity in photographs and in the astronomical literature. This easily leads to the production of caricatures with exaggerated features, reflecting more the observer's preconceptions than what he or she clearly sees in the eyepiece. (3) Photographs are now readily available for all deep-sky objects (e.g., in MegaStar), and these, being highly detailed and objective, are generally more helpful than sketches for the purpose of comparison with verbal descriptions.

In general I would say that a well conceived and well executed set of verbal descriptions is as good a means of recording what deep-sky objects look like as there is, though of course for some observers the making of sketches is a rewarding and enjoyable exercise.

Chapter 4

Instrumentation

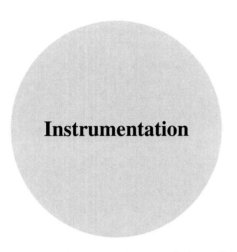

The mounted setup consisted of a 35 mm binocular finder, a 60 mm binocular telescope, and a 10-inch Cassegrain binocular telescope. (The final finder arrangement is shown in the first several photographs of Appendix D.) A handheld 6×30 binocular was additionally used for the few objects too large even for the 35 mm finder. For comparison, the 12×35 finder with its rock solid stability – being mounted atop the primary telescope – is easily the equal of a handheld 10×50 binocular. The 60 mm binocular telescope is roughly the equivalent of a 3-inch or 80 mm monocular telescope. And the large telescope is equivalent to a 12-inch monocular telescope. Technically speaking a binocular and a binocular telescope are identical things, but normally the former term is used for the familiar small, hinged instruments, while the latter term is reserved for larger and more unusual instruments.

Instrument	Field	Exit pupil	Components: objectives, eyepieces
6×30	8.3°	5.0 mm	n/a
12×35	3.8°	2.8 mm	50 mm f/3.7 achromat stopped down to 35 mm f/5.3, 15 mm Plössl (12.4×)
42×60	1.1°	1.4 mm	60 mm f/7 achromat, 10 mm Plössl
170×250	18′	1.5 mm	254 mm (10″) f/16 Classical Cassegrain, 24 mm Ultima (169×)

   Given the depth of the survey, most of the objects were observable only through the primary telescope. In any event, each object was fully described only through the instrument(s) that showed it best. The smaller telescopes were most useful in the observation of open clusters, many of which were too large to fit in the primary telescope's field.
   One summer night after having completed the survey, I reobserved most of the Messier globulars in the primary telescope at a slightly higher magnification (225×). In a few instances, further information derived from these observations was added to the globular descriptions.

Chapter 5

The GC/NGC Descriptions and the Herschels

The vast majority of the NGC's brighter objects were discovered by the Herschels, who developed for their object descriptions an admirably systematic and functional nomenclature that came to be widely used by nineteenth-century astronomers. Dreyer incorporated into his compendious NGC the abbreviated object descriptions from John Herschel's General Catalogue, which included all of John's and William's observations. In the supposed interest of accuracy, however, Dreyer "corrected" these descriptions as he saw fit, as if they consisted of objective data independent of specific observers. But insofar as today's amateurs are interested in the old descriptions at all, it is not because they are seeking abstract DSO data – they can get this from modern catalogues – but because they want to know exactly what particular early astronomers reported seeing with their particular telescopes, especially what the Herschels saw with their 18.7-inch telescopes. Since that information is in its pristine state only in the GC, I decoded the Herschels' descriptions directly from that catalogue rather than from the NGC, except in cases of obvious misprints or other glaring errors such as incorrect directions on the sky. The remainder of the descriptions (that is, those by other observers) were decoded from the NGC. In any case, for convenience I refer to all the GC/NGC objects and their descriptions, regardless of origin, as "NGC."

Below is a list of the original observers of the NGC objects listed in the current catalogue, along with the sizes of their telescopes and the number of their descriptions.

Au – Auwers	6.2" rr[a]	2
Ba – Barnard	Various	15
Bd – Bond	Various	1
Bi – Bigourdan	12.2" rr	3
Br – Bruhns	–	1
Bn – Brorsen	–	1
Bu – Burnham	Various	1
By – Borelly	7.2" rr	2
Cm – Common	36" rl	2
Cp – Copeland	Various	3
dA – d'Arrest	11" rr	42
Du – Dunér	9.6" rr	1
Ha – Harding	4" rr	1
Hi – Hind	7" rr	4
Ho – Holden	15.5" rr	6
HH – both Herschels[b]	18.7" rl	1,088
JH – J. Herschel	18.7" rl	357
Le – Leavenworth	26" rr	12
Lo – Lohse	15.5" rr	1
LR – Lord Rosse	72" rl	16
Ls – Lassell	48" rl	1
Ma – Marth	48" rl	31
Mu – Muller	26" rr	4
Pa – Palisa	Various	1
Pe – Peters	13.5" rr	3
Pi – Pickering	Various	14
Se – Secchi	9.5" rr	1
Sf – Schonfeld	6" rr	3
Sn – Stone	26" rr	9
St – Stephan	31.5" rl	37
Sv – O. Struve	15" rr	1
Sw – Swift	16" rr	32
Te – Tempel	11" rr	28
Tu – Tuttle	4" rr	1
WH – W. Herschel	18.7" rl	233
Wi – Winnecke	Various	8
All Herschels		1,678 (85.4%)
NGC total		1,966

Additional observers of non-listed objects: Dreyer (Dr), Hartwig (Hw), Schultz (Sz)

Sources: J. L. E. Dreyer, *New General Catalogue of Nebulae and Clusters of Stars* (London: Royal Astronomical Society, 1953); the online NGC/IC Project (ngcic.com). For a full discussion of NGC terms see the introduction to either the GC or *NGC 2000.0*

[a] rr = refractor, rl = reflector

[b] HH objects are those observed by both William and John Herschel, the latter having the final say

The primary mirrors of the Herschels' identical telescopes, though of excellent figure, were made from 60%-reflective, quickly tarnishing speculum metal. The frightful light loss at the surface of such mirrors prompted William to do without secondary mirrors altogether – hence the single-mirror "Herschelian" telescope design employed by him and John. Interestingly, the observational results I obtained with my 10-inch binocular at mountain sites in the American Southwest are comparable to those the Herschels obtained with their telescopes in low-lying southeastern England. This is clearly demonstrated by our very similar detection or non-detection of very faint galaxies in the fields of brighter galaxies. Usually we saw the same galaxies. In some instances I saw galaxies that they did not see and vice versa (e.g., NGC 5078, NGC 5638). Another indication of the closeness in performance of our telescopes is our respective descriptions of small, slightly elongated galaxies, the non-circular outlines of which are often difficult to discern. It turns out that in roughly equal frequency I correctly saw such galaxies as elongated while the Herschels described them as round and vice versa (e.g., NGC 4143, NGC 4138). In short, the Herschels' telescopes, though gigantic for their time (the larger telescopes listed above were built later in the century), do not greatly if at all outclass those of today's well-equipped amateurs. It can be calculated that the modern equivalent of the Herschel instruments, in light grasp, is a 15-inch monocular reflector with enhanced mirror coatings.

For their "sweeps" the Herschels used eyepieces yielding a magnification of 150× and a field of 15'. It is truly mind-boggling to consider that they examined at this relatively high power not merely specific objects conveniently plotted on sky charts, as amateurs do today when they say they are observing "the sky," but literally the entire surface of the sky, square sixteenth-degree by square sixteenth-degree.

Among the various peculiarities of the NGC, one in particular requires special explanation. This is the Herschels' frequent description of objects that we now know to be galaxies as "resolvable." Believing as they did that all non-gaseous "nebulae" were merely unresolved cousins of obvious star clusters, requiring only a little more telescopic power than they had available to them to resolve, they felt (for whatever reason; oftentimes none is apparent) that in some cases such resolution was imminent – hence "resolvable." But while some galaxies are indeed ultimately resolvable into stars photographically with modern, behemoth observatory telescopes, the Herschels did not realize that these objects are unimaginably distant systems that could not possibly be resolved into individual stars visually with the telescopes of their era. In other words they did not recognize, at least not fully, that there exist such things as galaxies – enormous objects very different from resolvable star clusters within our own Milky Way. It would be another century before the true nature of the "nebulae" would be ascertained. (One thing the Herschels did *not* mean by "resolvable" was that they were seeing small-scale galactic detail of the kind captured in photographs.)

At least with regard to galaxies, the NGC astronomers cast their nets wide and deep, so it is rather surprising to find that they missed quite a number of reasonably bright galaxies even as they captured so many extremely faint ones. The observing list in this book includes numerous non-NGC galaxies within the magnitude limit of 13.2 that turned out to be readily visible.

Chapter 6

Observational Parameters in the Descriptions of This Catalogue

While describing the various objects (speaking into a tape recorder) as I looked at them in the eyepieces, I specifically looked for certain features and mentally ticked these off as I noted them. For example, as I observed a galaxy I would cover its *brightness, concentration, shape, size, stars*. The features in the descriptions are given below for each type of object.

Bright star: COLOR.
Double star: Separation DISTANCE. Relative BRIGHTNESSES. COLORS.
Variable star: BRIGHTNESS in 35 mm finder, estimated to nearest half-magnitude. COLOR.
Galaxy: General BRIGHTNESS; surface brightness if notable. CONCENTRATION: type and degree. SHAPE. SIZE against field. Associated foreground STARS.
Globular cluster: RESOLUTION. CONCENTRATION: type. SIZE against field. General APPEARANCE if notable: nature of outliers, massive or loose, regular or irregular, bright or faint.
Nebula: BRIGHTNESS. SHAPE. SIZE against field. Associated STARS.

Open cluster: NUMBER AND BRIGHTNESS OF STARS. FIGURE of prominent stars. Overall SHAPE (if different from figure or if other than roughly round). SIZE against field. DENSITY.

Planetary nebula: BRIGHTNESS of nebula. CENTRAL STAR. DISK: sharpness of edge, distribution of light, strength of blinking effect. SHAPE. SIZE relative to other planetaries.

(Additional notes on the descriptions are contained in Appendix B.)

The ranges of the object features are given below. It need hardly be stated that descriptions rendered in such loose terms are hardly paragons of exactitude. This should not be surprising. Our eyes and our mental scales are not, and cannot be expected to be, as precise as micrometers or photoelectric meters. Nonetheless, a skilled and experienced observer utilizing fixed methods and the same optical instruments each time, at similar sites under similar conditions, can expect to obtain pretty similar results from one observation of a given object to another. In other words, although it is inevitably prone to a degree of inexactness and inconstancy, when done carefully visual observation is not entirely arbitrary. For example, a galaxy described on one occasion as "extremely faint" might be described on a subsequent occasion as "very faint," but never as "moderately bright." In general, any discrepancy in a given characteristic will be only a slight difference, not a wide divergence. In fact, the observer is struck by how he or she sees the same things over and over again.

Several of the characteristics are specific to one type of object. Only stars, for instance, display a wide range of colors. Other characteristics such as brightness pertain to several types of objects. For each characteristic listed, the type or types of objects to which it is applicable are specified. The method used here is broadly similar to that of the NGC, which is best explained in *NGC 2000.0*, p. xii–xv.

Star Colors

Blue white
Electric white
White
Soft white
Yellow white
Yellow
Yellow gold
Gold
Deep gold
Orangish (not ordinary terrestrial orange)
Orange red
Light copper red
Copper red
Deep copper red

Double Star Separation

(with approximate numerical equivalent and using primary telescope)

Not split/attached	<1.5″
Extremely close	1.5–2.5″
Very, very close	2.5–3.5″
Very close	3.5–5.0″
(Pretty/fairly) close	5.0–7.0″
Moderately close	7.0–10.0″
Moderate distance	10.0–15.0″
Moderately wide	15.0–20.0″
Wide	20.0–30.0″
Very wide	>30.0″

Relative Brightnesses of Double Star Components

- Equal pair
- Slightly unequal pair
- Unequal pair
- Primary has a fainter companion
- Primary has a much fainter companion
- Primary has a relatively very faint companion

Brightness of Deep-Sky Object

(with GC equivalent)

Not visible	n/a
Exceedingly faint/occasionally visible	Excessively faint
Extremely faint	Extremely faint
Very very faint	n/a
Very faint	Very faint
Faint	Faint
Moderately faint	Considerably faint
Moderate brightness	Pretty faint/bright
Moderately bright	Considerably bright
Bright	Bright
Very bright	Very bright
Extremely bright	Extremely bright

Concentration of Galaxy or Globular Cluster

(type and degree)

A full description of the concentration of a galaxy or a globular cluster (especially the former) has two components. The first is the type of concentration. This can readily be visualized: an object with a broad concentration has a rounded appearance; one with an even concentration brightens steadily toward the center; and one with a sharp concentration has a sudden central peak in brightness. There are, of course, intermediate types of concentration. For example, a galaxy can be evenly to sharply concentrated. The second component of the concentration is the degree. This is the extent of the difference between the intensity of the light at the edge and at the center of the galaxy, independently of the type of concentration. So the galaxy in the above example can be evenly to sharply *slightly* concentrated, or it can be evenly to sharply *very* concentrated.

Broadly
Broadly to evenly
Evenly
Evenly to sharply
Sharply

Diffuse/unconcentrated
Hardly concentrated
Slightly concentrated
Moderately concentrated
(Pretty/fairly) concentrated
Very concentrated
Extremely concentrated

Shape of Galaxy or Other Deep-Sky Object

Round
Slightly elongated
(Moderately) elongated
Pretty elongated
Very elongated
(Very) elongated to edge-on
Nearly edge-on/fat edge-on
Edge-on
Fairly thin edge-on
Thin edge-on

Size of Deep-Sky Object

Tiny
Very very small
Very small
Small
Moderately small
Moderate size
Moderately large
Large
Very large

The described size is relative to the type of object, since the various types have widely differing size ranges. Planetary nebulae are in general so small that their sizes cannot easily be estimated as fractions of the width of the field, so for them size is given in relation to other planetaries – for example: "moderately large for a planetary." To this observer galaxies appear "small" at a sixteenth of the diameter of the primary telescope field, "moderate size" at an eighth, and "large" at a quarter, as shown below. Globular clusters, nebulae, and open clusters are expected to be larger. Globulars and nebulae appear "small" at an eighth of the field, while open clusters appear "small" at a sixth of the field; beyond this the sizes of these objects are given directly as field fractions.

	1/16	1/12	1/8	1/6	1/4	1/3	1/2	2/3	3/4
ga	Small	Moderately small	Moderate size	Moderately large	Large		Very large		
gc, ne			Small						
oc				Small					

Actual size of telescope field fractions:

	1/16	1/12	1/8	1/6	1/4	1/3	1/2	2/3	3/4
12×35, 3.8°	–	–	–	–	57'	1.3°	1.9°	2.5°	2.9°
42×60, 1.1°	–	–	–	–	16.5'	22.0'	33.0'	44.0'	49.5'
170×250, 18'	1.1'	1.5'	2.3'	3.0'	4.5'	6.0'	9.0'	12.0'	13.5'
225×250, 13'	0.8'	1.1'	1.6'	2.2'	3.3'	4.3'	6.5'	8.7'	9.8'

Magnitude Equivalence

(of observed brightness of star associated with galaxy and using primary telescope; approximate)

Very/extremely faint = 15th mag.
Faint = 14th mag.
Moderately faint = 13th mag.
Moderate brightness = 12th mag.
Moderately bright = 11th mag.
Bright = 10th mag. or less

Distance Equivalence

(of star near galaxy and using primary telescope; approximate)

Off its edge/very nearby = <1'
Nearby = 1–2'
A little ways = 1.5–3'
(No qualifier) = >3'

Resolution of Globular Cluster

Not resolved
Starting to be granular
Granular
Starting to be resolved
Partly resolved
Resolved
Well resolved
Well resolved with distinct stars

Number of Stars in Open Cluster

X (count)
A handful
A few
A (modest) number
A fair number
A good number
A lot
A whole lot
Countless

Density of Open Cluster

Very sparse
Sparse
Somewhat sparse
Moderate density
Moderately dense
(Pretty/fairly) dense
Very dense
Extremely dense

Chapter 7

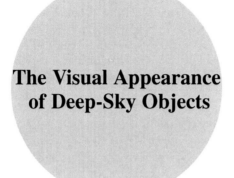

The Visual Appearance of Deep-Sky Objects

Beginning observers all have trouble merely finding deep-sky objects, much less describing them accurately and exhaustively. This difficulty is to be expected, since by terrestrial standards most deep-sky objects are faint, enigmatic things requiring serious optical power to be well seen. Yet even experienced observers with fine telescopes find that many objects stubbornly resist precise description. In fact, deep-sky objects as a whole are anything but nice and neat in appearance. For this reason the sustained air of clinical precision and the plethora of minute detail encountered in some advanced observing guides, although they might seem "scientific," are actually inappropriate and misleading in the realm of visual astronomical observation. A more credible and useful approach is to allow a degree of coarseness, vagueness, or indeterminacy to be conveyed in one's observations, since this better reflects what one really sees – oftentimes barely – in the eyepiece. Even my circumspect observations in their edited form exude more certitude than is warranted, for in order not to be tiresome I deleted countless qualifiers ("about," "more or less," "apparently," "might be," "probably," "possibly," etc.) from the original transcripts.

It might be argued that more object detail can be seen than I saw in my observations if powers higher than 170× are used and if nebula filters are utilized. This is no doubt true in some instances with some types of objects, but my goal in this survey was not to try to see everything possible at the very limits of vision, using every trick in the book and employing Herculean effort. Nor was my aim – as it apparently is for so many guidebook authors – specifically to look for and describe features found in photographs of deep-sky objects, especially famous ones, regardless of how inconspicuous these features are visually. Rather, it was to get a sense of what deep-sky objects look like in something akin to a "standard" view, and in the process record their more salient features as well as their overall appearance in unbiased relation to other objects. This required that I observe all objects in exactly the same frame of reference and in the same relaxed, unenhanced manner. Moreover, I wanted to avoid wasting valuable time juggling several twin sets of eyepieces, examining objects repeatedly, and refocusing both optical assemblies after every eyepiece switch. (Of course, some perceived characteristics of objects, such as brightness and resolution, vary with magnification and are thus harder to compare among different objects if the magnification is constantly altered.)

In any event, I generally find medium-high to be the most profitable and comfortable magnification, and I dislike the brightness and color distortions produced by filters. So for all these reasons I adopted the simplest method possible: look carefully and unhurriedly at a single, healthy magnification, describe, and move on. I am satisfied that any photographic features not evident when objects are observed in this way are simply not prominent visually. This should not be cause for lamentation; rather it is information as valuable as any other.

It goes without saying that the gaudy colors depicted in modern photographs of extended (i.e., nonstellar) deep-sky objects are not seen in the eyepiece. Only a couple of emission nebulae and a few planetary nebulae, being slightly greenish, show any color at all. Extended deep-sky objects are, overwhelmingly, shades of gray.

The following synopsis of the appearance of deep-sky objects inevitably reflects the author's own binocular experience. Binocularity per se, however, is not discussed until the following section.

Stars

Bright stars are of course the brightest objects in the sky beyond the planets, and these luminous gems are wonderful sights through the eyepiece. In fact, it is only visually that the intensity and simple beauty of bright stars can be appreciated: photography fails to convey these qualities adequately. (Due to diffraction effects, the light of the very brightest stars hitting the secondary mirror spider vanes creates impressive spikes in a cross pattern across the field of view.)

Many bright stars are quite noticeably colored, but in most cases this means only that they have a mild tint of blue or yellow overlaying their predominantly white light. So-called blue stars do not stray far from white: they are bluish at best, not truly blue. Stars toward the other end of the spectrum can be much more colorful, ranging all the way to deep gold. But no bright star, not even a "red giant" such as Betelgeuse or Antares, is actually red. Veritable red stars do exist, but only among the carbon stars, a rare class of pulsating variables none of which is close enough to Earth to be among the brightest stars in our sky. The most deeply colored carbon stars are a glowing-coal copper red (not stoplight red).

The colors of stars are notoriously difficult to pin down, due to (1) their weakness and subtlety in most cases, (2) the unusual and visually challenging conditions under which stars are observed – as point sources against a black sky, and (3) the fact that, unlike paint samples in a hardware store, bright stars obviously cannot be placed side by side in the same telescope field in order to compare their colors directly. Instead their colors must be gauged one at a time against a *mental* color scale. Unfortunately human memory regarding color is not very reliable. Hence a high degree of precision and certainty in the visual determination of the colors of individual stars is pretty much out of the question. Yet after observing a large number of bright stars through a binocular telescope one sees clearly that star colors themselves are not particularly exotic, as is sometimes reported. They are in fact soft shades in a simple continuum (blue-white-yellow) that can readily be subdivided into a few straightforward colors. The distribution, however, is greatly skewed to the yellow side of white.

In practice I have found the following rough correlation between observed color on the one hand and published B-V values and spectral types on the other, with the caveat that at the blue end of the spectrum the difference in color between plain white stars and bluer stars is too subtle, for me at least, to make out with confidence and consistency; so in the end I recorded all of these stars as simply white. (B-V is a color index indicating the difference in the intensity of the light of an object when measured photometrically through (a) blue and (b) yellow/green filters.)

Observed color	B-V	Spectral type
Blue white/electric white	<−0.10	(O,B)
White	−0.10–0.20	(B,A)
Soft white	0.20–0.50	(A,F)
Yellow white	0.50–0.80	(F,G)
Yellow	0.80–1.10	(G,K)
Yellow gold	1.10–1.40	(K)
Gold	1.40–1.70	(K,M)
Deep gold	1.70–2.00	(M)

There is not much additional to say about double stars in particular, except perhaps that they are peculiarly attractive, and color-contrast pairs are especially appealing. One notable and unexpected double star phenomenon is that the components of unequal pairs appear more and more equal as the pairs are viewed through larger and larger apertures.

Galaxies

It is often hard to make out the shape and size of galaxies even roughly, let alone with precision, since the faint light of their outer extremities typically fades out almost imperceptibly into the background sky. Not a few galaxies are of such low surface brightness throughout their extent that they are barely detectable at all, while their shape and size are practically indeterminate. In fact, galaxy light as seen through moderate-sized telescopes is remarkably diaphanous or milky, generally smooth, and always obstinately vague. The only internal variation usually seen is a continuous brightening toward the center. A *stepped* brightening with well delineated concentric zones – "halo," "core," "nucleus" – is exceptional. Nor are truly stellar nuclei very common: sharply concentrated galaxies by definition have central spikes in brightness, but these do not generally culminate in definitive starlike points. Rarer still are bars, spiral arms, dust lanes, or small irregularities such as bright patches, knots, or "mottling." Very few of the thousands of galaxies within reach of amateur instruments exhibit such features.

One might expect or hope that, with the superior vision it provides, a binocular telescope would finally enable one to see with one's own eyes the splendid intricacy displayed in photographs of galaxies. On the contrary, binocularity emphatically confirms the sober assessment given above. Yet many of today's advanced observers, using quite modest telescopes – monocular ones, no less – routinely report the dimensions of galaxies in confidently precise figures, constantly distinguish cores from halos, and frequently claim the visibility of small-scale detail. Astronomy guidebooks and magazines, for their part, blithely encourage amateurs to look for spiral arms and other internal features, as if we are likely to see these if only we look carefully enough.

Such concern to see galactic "structure" is no doubt due to the pervasive influence of astrophotography. In long-exposure observatory photographs and CCD images galaxies appear well delineated in contour, wonderfully detailed internally, and in general so bold that they almost jump out of the picture books at us. Naturally we wish to see – we feel that we *should be able* to see – similarly clear-cut spectacles, the grandest in all the heavens, with our own eyes. But the truth is that galaxies are by several orders of magnitude less sharply defined visually than photographically, for astrophotography vastly exaggerates the contrast (definition) and drastically truncates the dynamic range of the deep-sky images formed by telescopes, thereby severely distorting their brightness profiles. And the distortion is most extreme in the case of face-on spiral galaxies, photographs of which are lovely but resemble very little the corresponding eyepiece views. (It is because of photography's dynamic range limitations that the middle parts of galaxies – overexposed and rendered as uniform white blobs – so often look like distinct cores in photographs, whereas in reality most galaxies have no such feature, instead concentrating seamlessly toward the very center.)

In short, the photographic and visual appearances of galaxies are two very different things. Regardless of what photography shows, human eyes perceive galaxies as ill-defined things with little discernible detail, which is why the early observers called them "nebulae" (Latin plural for "mist"). Yet we do notice a striking and fascinating diversity among them: a huge variation in general brightness and concentration. It is in these broad features alone (disregarding size and shape, which are not straightforwardly intrinsic characteristics) that galaxies differ radically, conspicuously, and as a matter of course through telescope eyepieces. Cases of contrasting pairs in the same eyepiece field (e.g., NGC 5982 and NGC

5985) are especially instructive in this regard. Yet this doubtlessly important phenomenon – the enormous and visually unmistakable variation in brightness and concentration among galaxies – does not seem to receive much attention (certainly not emphasis) from today's professional astronomers, apparently because of their total reliance on photography, a means of detection that tends to obscure large differences in brightness even as it highlights small detail (hence the familiar morphological classification scheme).

As for those much sought-after spiral arms: in only one galaxy, M51, are they obviously visible as individual arms (actually a ring) through medium-sized telescopes. The arms of a few other galaxies – M66, M83, M99, M106, NGC 1365, NGC 5247, and NGC 5248 – can sometimes be suspected or just detected under ideal conditions, but to say the least they do not stand out. It might even be argued that for all practical purposes galaxies' spiral arms are not truly *seen* at all, if by this word we mean the obvious, unequivocal, and consistent way one sees, say, craters on the Moon. (When has anyone ever looked at the Moon through a decent telescope and not seen craters as plain as day?)

Dust lanes are more easily observed than spiral arms. Galaxies with visible ones are M31, M82, M104, NGC 891, NGC 3628, NGC 4565, and NGC 5128. It should be noted that, altogether, we are talking here about less than a dozen and a half galaxies in the entire sky, all relatively near to us and with arms and dust lanes as pronounced as they can be, the arms being of the "grand design" type and the dust lanes, with the exception of M31, being gargantuan and bisecting. Furthermore, arms or dust lanes are not necessarily the most notable features of even these galaxies. More impressive, for instance, is the concentration of M83 and the brightness of M82 and M106.

In sum, although I love galaxies and find them endlessly intriguing, I have long since come to the conclusion that in modest telescopes they will never look much like the photographs. Yet they are fine sights when we appreciate them for what they are. (See Appendix C.)

Globular Clusters

Globular clusters steal the show (!!!) in binocular telescopes. These objects are never satisfactorily viewed monocularly. The reason is that at telescopic powers the major globulars take up much of the field and simply cannot be seen as a whole with one eye (see the discussion below on binocular vision). Moreover they look flat, which we know intuitively as well as factually to be wrong. In a powerful binocular telescope, by contrast, globular clusters are a revelation. They are indeed awesome spheres of countless stars. With their combination of large overall size and extreme "detail" – all those stars! – the big globular clusters spectacularly show off both the scenic quality and the sharpness of the binocular telescopic view.

Some globulars – those with well defined, dense central balls – look intrinsically massive and are especially impressive. These tend to be clusters with high absolute magnitudes. Other globulars look comparatively loose and lightweight.

The dichotomy is so pronounced that I feel there are two main types of globulars: "massive" and "loose" (there is, of course, a gradation between the two extremes). Another major difference among globulars is their degree of regularity. Many loose globular clusters are notably irregular. Even some major ones have a disheveled appearance, M13 being a prime example: it looks downright crablike. Other globulars, for instance M2, are very regular and symmetrical in appearance. Globular clusters also vary noticeably in the character of their outlying members – the set of stars surrounding the main body of the cluster. Some have large halos of distinct outliers, others have fewer or fainter outliers. Finally, while most globulars are bright objects by deep-sky standards, some are notably faint, anomalously so in a few cases. To say the least, then, globular clusters are not as blandly similar visually as photographs make them seem.

Nebulae

Unless observed through nebula filters, most catalogued nebulae are unsuitable targets for visual observation; they are essentially photographic objects only. Very faint nebulae are in fact the most insubstantial and ill-defined of all deep-sky objects. Thus, despite their undeniable beauty in the pictures, from the visual observer's perspective they are vastly overrated. The very brightest nebulae, however – M42, M8, and M17 – are immensely impressive objects. Each of these completely fills the primary telescope field and is a tremendous binocular sight, since the two eyes working together take in the whole panorama. M42 in particular is such a tantalizingly strange and improbable beast that it starts to "make sense" only when seen through a binocular telescope.

It is worthy of note that the major emission nebulae show incomparably more internal structure than any galaxy. Their light is full-bodied, luminescent, and intricate, while that of galaxies is sheer, dull, and pretty much formless. Reflection nebulae, on the other hand, are much less consequential than their emission cousins: many are small and faint and tend merely to make their central stars look fog-enshrouded.

Open Clusters

Open clusters come in a bewildering variety, differing greatly in size, number of stars, brightness of stars, and density. Though often brilliant and beautiful, as a class they are the most difficult deep-sky objects to describe elegantly or consistently – i.e., they're sloppy! It is not surprising that hardly any two observers describe a given cluster the same way. Not only are the constituent stars of the typical open cluster distributed haphazardly, but the entire cluster is usually surrounded and pervaded by Milky Way field stars, often making it next to impossible to decide where the cluster ends and the Milky Way medium begins, or which stars within the apparent bounds of the cluster actually belong to it. (This is one reason why, except for trivial cases, I do not attempt star counts.) Indeed, all open clusters are in a sense merely concentrations in one humongous, inhomogeneous cluster of stars – the teeming Galaxy itself.

In my opinion, the designation of groups of stars as open clusters is among the more questionable areas in the astronomical catalogues. Most objects classified as open clusters are obvious and distinct agglomerations of closely related stars and as such are without question properly termed clusters. But not a few so-called open clusters are so sparse and unimpressive that they barely, if at all, distinguish themselves from the random mass of Milky Way stars. As I often note in my observations, these "open clusters" hardly deserve the name, at least based on their telescopic appearance.

As with other objects that take up a large portion of the field at high power, the great open clusters benefit hugely from binocular vision. Seeing M11 or M37 at telescopic power with both eyes is like seeing them for the first time; in each case the scene is truly breathtaking.

Planetary Nebulae

Planetaries come alive in a binocular telescope. In basic appearance they are the most varied class of objects, running the gamut from starlike to bright balls of gas to ghostly glows. Some have obvious central stars, others have none. Some have classic planetary disks, others don't look at all "planetary-like." One of the more prominent and unusual characteristics of many brighter planetaries is their "blinking effect" – the instantaneous brightening and dimming of the nebula that occurs when the observer switches from direct to averted vision and back. A number of planetary nebulae exhibit broad peculiarities in structure such as annularity or an hourglass shape. The small and wildly intricate features seen in observatory photographs, however, are for the most part absent visually.

Besides the easily recognized planetaries, there are also stellar planetaries that are indistinguishable from field stars without a prism or a nebula filter. (Many of these were discovered by Edward Pickering using a spectroscope.) Therefore the phrase "not visible" when used in a negative observation of a planetary nebula should be understood to mean "either not visible or not recognizable," since the planetary might well be bright enough to be seen if only one knew which "star" in the field it was.

When planetaries are bright they can be shockingly bright for deep-sky objects. They can also be surprisingly well defined and symmetrical. After an initial reaction of astonishment, the observer will find himself or herself staring at numerous planetaries for a long time in rapt fascination.

Chapter 8

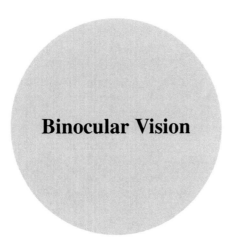

Binocular Vision

In 1995 I began to plan the building of a binocular telescope. Over the years I had become increasingly dissatisfied with the view through my 10-inch Newtonian. I could never quite get a handle on such basic things as the colors of stars or the major features of galaxies, and long observing sessions were unaccountably fatiguing. I became more and more convinced that the reason for the telescope's inadequacy was simply its monocularity; it was otherwise a fine instrument and its light grasp per se was satisfactory.

When I completed the construction of my binocular telescope and finally looked through it – *WOW!!!* – I found that both my suspicions about monocularity and my expectations for binocularity were amply confirmed. The binocular telescopic view was incomparably better than the standard view, in all observing scenarios. (The Moon was positively transformed.) In contrast to my prior disappointment stemming from an unshakable feeling of visual deficiency, I now felt that I was finally seeing things exactly as they should be seen – as they really are. And the act of observing was now perfectly comfortable. I could not have been happier with the new view. It was – and is – a dream come true.

The improvement in the performance of a binocular over that of a same-aperture monocular is a matter more of quality than of quantity. Nevertheless the difference can in some respects be measured. In careful tests with small binoculars, I found that the gain in stellar limiting magnitude from looking through both barrels versus only one is 0.5 magnitude. This corresponds to an increase in aperture by a factor of 1.25. Hence a binocular is theoretically equivalent to a monocular 1.25 times larger. But this does not take into account the light loss from the extra elements required in the binocular to redirect the light beams. Although this loss is negligible in small binoculars with their multicoated prisms, it is not insignificant in large binocular telescopes, which use aluminum-coated mirrors. For a binocular telescope with 96% reflective diagonal mirrors, the end result is an effective gain in aperture by a factor of 1.2. Thus, my 10-inch binocular should be the equal of a 12-inch monocular (in reality it seems to do a little better).

How is it that the use of both eyes can have such a salutary effect, not just on light-gathering power but on overall quality of vision?

Binocular vision is plainly superior to monocular vision. This should not be surprising: we humans are fundamentally binocular animals; binocularity is our normal and natural mode of vision. In normal binocular vision the brain takes the two eyes' imperfect raw images and integrates them to form a single clearer, sharper, brighter, and *more coherent* final image. What is going on is that (1) with two images to work with the brain is better able to recognize and suppress the omnipresent noise and spurious effects produced by our neurovisual circuitry, (2) the brain "stacks" the positive information it receives from the two eyes via the two sets of optics, thereby greatly reinforcing it while overcoming faults in both the individual eyes and the individual optics, and (3) the teamwork engaged in by the two eyes endows binocular viewing with a unique spatial coherence. This last characteristic of binocular vision, examined more closely below, makes binocularity the ideal mode for viewing the large objects and expansive vistas encountered in deep-sky observing, as well as for scanning multiple telescopic fields and sweeping about the night sky in general. With one eye out of the loop the human visual system – a remarkably well-engineered *binocular* mechanism – operates much less smoothly and effectively, and faint things in particular are seen with far less ease and certainty.

The difference between monocularity and binocularity may be characterized as that between static and dynamic modes of vision. A single eye is incapable of seeing a large angular field all at once, regardless of the eyepiece's nominal apparent field of view: one eye working alone is inherently restricted to a sort of tunnel vision. The observer can of course move his or her eye around to see more of the field, but this action is always sluggish, reluctant, and choppy; in any case the visual input remains hopelessly disjointed. In fact, with only one eye at its disposal the brain tends to keep this eye fairly still, preferring instead to rely on peripheral or averted vision to try to see a wider scene. When viewing with both eyes, on the other hand, one instantly and automatically sees more of the field away from the very center. Moving the eyes around is a quick, reflexive action, and scanning the entire field from edge to edge is a fluid, effortless process. One sees a large whole all at once rather than small fragments consecutively. And one senses that the visual system is operating at maximum efficiency – as indeed it is, since our two eyes were clearly designed by nature to work in tandem, not singly.

It is this spatial aspect of binocular vision – its intrinsically wide "field of view" – that, along with its unparalleled clarity, accounts for the sheer delight and the profound mental tranquility one experiences when looking into the eyepieces of a binocular instrument. It is this scenic quality also that is responsible for the unmistakable realism binoculars are noted for. (True three-dimensionality, which requires the presence of nearby foreground objects in the line of sight, plays no part in astronomical observation, even with binoculars.)

In sum, although it is true that binocularity is not absolutely essential to sight per se, this does not mean that the use of both eyes is a mere luxury, especially when we are trying to make sense of alien sights in the dark. The binocular advantage consists not so much in *what* we see, since technically we can see the same things through larger monocular instruments, as in *the way* we see. Compared to one-eyed observing, binocularity makes vision whole, the view starkly real, and astronomical observation much more assured, enjoyable, and physically comfortable. Quite simply, seeing with both eyes is better and more natural than seeing with only one eye. (See Appendix D for information on the building of a binocular telescope.)

Chapter 9

How to Use This Catalogue

This catalogue contains a large amount of data, expressed in a very succinct way. After you have used it for a while, and become accustomed to the way the information is set out, you will find it fast and remarkably simple to use. It is, however, essential to understand how the catalogue works.

The first point to note is that the objects are listed in order of *position* (right ascension). This is invaluable for practical observers, but means that the various types of object are not grouped together, since they occur in the catalogue according to where they are in the night sky.

To avoid wasting space with pointless repetition of headings, the data set for each class of object is laid out in a table of 9 boxes. The crucial thing to understand is that much of the data in the table differs for each class of object. (As an example, you need data for angular separation of binary stars but not for variable stars!) This means you will have to remember what kind of data is in each of the boxes – since the data vary according to the class of object – although a lot of it is self-explanatory in context.

Basic Data

The first five boxes in every entry contain the same data for *all* classes of object.

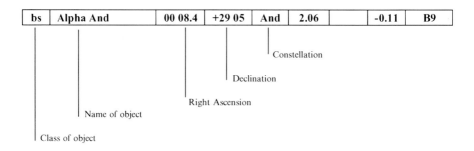

Data Specific to Each Class of Object

The remaining four boxes, 6–9, contain data that are specific to the class of object, as indicated by the abbreviation in the first box in the table (e.g., bright star = **bs**).

The actual types of data shown in boxes 6–9 of the catalogue tables (i.e., the data that is different according to the class of object) are illustrated below for each class:

Bright Star

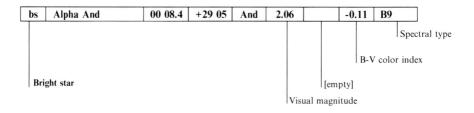

Data Specific to Each Class of Object

Double Star

| ds | STF 3053 | 00 02.6 | +66 06 | Cas | 15.1" | 5.9 | 7.3 | G8 A2 |

- Double star
- Separation
- Magnitude of primary
- Magnitude of companion
- Spectral types

Variable Star

| vs | WZ Cas | 00 01.3 | +60 21 | Cas | 7.4 | 10 | 2.5 | C9 |

- Variable star
- Magnitude at maximum brightness
- Magnitude at minimum brightness
- B-V color index
- Spectral type

Galaxy

| ga | UGCA 444 | 00 01.9 | -15 27 | Cet | 11.0b | 11.6' | 4.0' | Im |

- Galaxy
- Magnitude
- Major diameter
- Minor diameter
- Morphological type

Globular Cluster

| gc | NGC 288 | 00 52.8 | -26 35 | Scl | 8.1 | 13.0' | -6.60 | 10 |

- gc — Globular cluster
- 8.1 — Magnitude
- 13.0' — Size
- -6.60 — MV
- 10 — Concentration class

Nebula

| ne | NGC 281 | 00 52.8 | +56 37 | Cas | | 28.0' | | E |

- ne — Nebula
- [empty] — Magnitude
- 28.0' — Size
- [empty]
- E — Type

Open Cluster

| oc | NGC 103 | 00 25.3 | +61 20 | Cas | 9.8 | 5.0' | | |

- oc — Open cluster
- 9.8 — Magnitude
- 5.0' — Size
- [empty]
- [empty]

Planetary Nebula

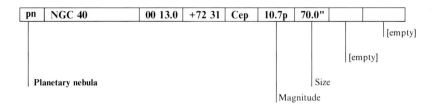

To summarize, the following object data are given:

bs: V magnitude,[1] B-V color index, spectral type
ds: separation, magnitude of primary, magnitude of companion, spectral type of each component.
vs: magnitude at maximum brightness, magnitude at minimum brightness, B-V color index, spectral type
ga: magnitude, major diameter, minor diameter, simplified morphological type.
gc: magnitude, size, MV,[2] concentration class.
ne: size, type (emission/reflection)
oc: magnitude, size
pn: magnitude, size

The data table for each object is followed by the object's NGC observer and description, if the object is NGC, and the author's observation through the indicated instrument(s). As previously mentioned, unlisted field galaxies seen near the listed target galaxies are flagged with an asterisk. A few observations are marked with a (†) note at the end, indicating that they are expanded or amended in the Addendum.

---

[1] Regarding magnitudes: V,v = visual; B,b = blue; p = photographic.
[2] MV is the total absolute visual magnitude of the cluster (the average for all globulars is about −7.50).

Abbreviations Used

All astronomical objects have catalogue designations (official names and numbers) from either astronomical catalogues of particular types of objects or well-known historical catalogues. The abbreviations of these catalogues are given below.

Double Stars

BSO	Brisbane Observatory
BU	Burnham
CPO	Cape Observatory
DUN	Dunlop
H	Herschel
HJ	John Herschel
HWE	Howe
LAL	Lalande
LCL	Lacaille
PZ	Piazzi
R	Russell
RMK	Rumker
SHJ	James South and John Herschel
STF	Wilhelm Struve
STT	Otto Struve
WNC	Winnecke

Other Deep-Sky Objects

AGC	Abell Galaxy Clusters
ANON	Anonymous
CGCG, C	Catalogue of Galaxies and Clusters of Galaxies
CR	Collinder
ESO, E	European Southern Observatory
FOR	Fornax
HAR	Harvard
IC, I	Index Catalogue
J	Jonckheere
K	King
M	Messier
MAC	Mitchell Anonymous Catalogue
MCG, M+/−	Morphological Catalogue of Galaxies
ME	Merrill
MEL	Mellotte

MKN	Markarian
NGC	New General Catalogue
PAL	Palomar
PGC, P	Principal Galaxies Catalogue
PK	Perek and Kohoutek
ST	Stock
TER	Terzan
TOM	Tombaugh
TR	Trumpler
UGC, U	Uppsala General Catalogue of Galaxies
UGCA, UA	UGC Addendum

Within the observations, objects with just a number and no catalogue designation are NGC. Where applicable, alternative names and/or catalogue designations are given in square brackets at the end of the description.

Part II

The Catalogue

Chapter 10

0 to 6 Hours: FALL

| vs | WZ Cas | 00 01.3 | +60 21 | Cas | 7.4 | 10 | 2.5 | C9 |

12x35: Mag. 7.5. *170x250:* Copper red. In the field with it are several moderately bright white stars that make good comparison stars.

| ga | UGC 12914 | 00 01.6 | +23 29 | Peg | 13.1p | 2.3' | 1.2' | Scd |

170x250: Extremely faint. Slightly concentrated. Slightly elongated NNW-SSE. Small. No involved stars. Nearby SSE is a faint star that makes a sharp triangle with two more faint stars a little ways NE. [U12915 was not seen.]

| ga | UGCA 444 | 00 01.9 | −15 27 | Cet | 11.0b | 11.6' | 4.0' | Im |

170x250: Exceedingly faint. A slight diffuse glow. Elongated N-S. 2/3 of the field in size. No involved stars. W of its N end is a moderately faint star. [WLM]

| ds | STF 3053 | 00 02.6 | +66 06 | Cas | 15.1" | 5.9 | 7.3 | G8 A2 |

42x60: A close unequal pair. *170x250:* A moderate distance slightly unequal pair. Color contrast: the primary is yellow; the companion is bluish by comparison.

| ga | NGC 7814 | 00 03.3 | +16 08 | Peg | 11.6b | 6.3' | 2.2' | Sab |

HH: considerably bright, considerably large, irregularly round, very gradually brighter middle.
170x250: Moderate brightness. Broadly to evenly moderately concentrated. Elongated NW-SE. Moderately large. The ends are very very faint. No involved stars.

| ga | MCG-2-1-14 | 00 04.0 | −11 10 | Cet | 12.2 | 1.1' | 0.8' | Sc |

170x250: Not visible.

| ga | NGC 7817 | 00 04.0 | +20 45 | Peg | 12.6p | 4.0' | 1.2' | Sbc |

HH: pretty faint, considerably large, much extended NE-SW, little brighter middle.
170x250: Moderately faint. Broadly slightly concentrated. Edge-on NE-SW. Moderate size. No involved stars. Just S of its SW end are a couple of very faint stars; the galaxy is between a sharp triangular figure of stars NNW and a star SE.

| ga | NGC 1 | 00 07.3 | +27 42 | Peg | 12.8v | 1.8' | 1.2' | Sb |

dA: faint, small, round, between an 11th mag. star and a 14th mag. star.
170x250: Faint. Sharply pretty concentrated. Round. Very very small. No involved stars. Nearby NNW is a moderately faint star; the galaxy is between a moderate brightness star NNE and a moderately faint star SSE. [2 (LR) was not seen.]

| bs | Alpha And | 00 08.4 | +29 05 | And | 2.06 | | −0.11 | B9 |

170x250: White.

| ga | NGC 14 | 00 08.8 | +15 48 | Peg | 12.7b | 2.8' | 2.0' | Im |

HH: very faint, pretty small, round, gradually little brighter middle.
170x250: Very faint. Broadly slightly concentrated. Roughly round. Small. No involved stars.

| ga | NGC 16 | 00 09.1 | +27 44 | Peg | 13.0b | 2.2' | 1.3' | S0− |

JH: pretty bright, small, round, brighter middle.
170x250: Moderate brightness. Pretty concentrated. Round. Very small. No involved stars.

| bs | Beta Cas | 00 09.2 | +59 08 | Cas | 2.27 | | 0.34 | F2 |

170x250: Soft white.

| ga | NGC 23 | 00 09.9 | +25 55 | Peg | 12.9b | 2.1' | 1.3' | Sa |

WH: 3 faint stars + nebula.
170x250: Moderate brightness. Sharply very concentrated. Round. Very small. SE of center is an involved star.

| ga | NGC 24 | 00 09.9 | –24 57 | Scl | 12.2b | 6.7' | 1.6' | Sc |

HH: very faint, considerably large, much extended, gradually brighter middle.
170x250: Faint. Broadly very slightly concentrated. A fairly thin edge-on, NE-SW. Large: a little less than a quarter of the field in size. Just E of its NE end is a moderately faint star.

| pn | NGC 40 | 00 13.0 | +72 31 | Cep | 10.7p | 70.0" | | |

HH: very faint, very small, round, very suddenly much brighter middle to a 10th mag. star, 12th mag. star WSW.
42x60: Visible. *170x250:* Bright. It has a bright central star; the nebula itself is moderately bright. A pretty good disk. The fairly sharp edge is like a thin ring: the nebula is a little darker just inside the perimeter. Weak blinking effect. Round. Moderate size for a planetary. It's in a short line with a faint star SW, an extremely faint star further SW, and a very very faint star NE; the planetary makes a large arc with two bright stars, one SSW and one ENE. A nice planetary with a conspicuous central star.

| bs | Gamma Peg | 00 13.2 | +15 11 | Peg | 2.83 | | –0.23 | B2 |

170x250: White.

| ga | MCG-1-1-52 | 00 13.6 | –05 05 | Psc | 12.4 | 1.2' | 0.6' | Sb |

170x250: Very faint. Broadly slightly concentrated. Roughly round. Very small. No involved stars. It makes a shallow arc with two stars E.

| ga | NGC 45 | 00 14.1 | –23 10 | Cet | 11.3b | 8.5' | 5.8' | Sdm |

JH: extremely faint, large, very gradually very little brighter middle, bright star in contact E.
170x250: Very very faint; extremely low surface brightness. The slightest unconcentrated glow. Roughly round. Large: a little less than a quarter of the field in size. At its edge S is a moderately bright star; WSW is the bright plotted star.

| ga | NGC 50 | 00 14.7 | −07 20 | Cet | 12.4 | 2.3' | 1.7' | S0− |

Se: very faint.
170x250: Moderately faint. Evenly moderately concentrated. Round. Small. No involved stars. It makes a long parallelogram with a faint star NW and two stars E; a little ways S is another star. [M-1-1-57 was not seen.]

| ds | 35 Psc | 00 15.0 | +08 49 | Psc | 11.5" | 6.0 | 7.6 | F0 A7 |

42x60: An unequal pair. ***170x250:*** A moderately close slightly unequal pair. Both white. [STF 12]

| ga | NGC 59 | 00 15.4 | −21 26 | Cet | 13.1b | 2.6' | 1.3' | S0− |

Sn: very faint, pretty small, irregularly round, gradually brighter middle.
170x250: Faint. Broadly to evenly moderately concentrated. Round. Small. No involved stars.

| vs | S Scl | 00 15.4 | −32 03 | Scl | 6.2 | 13.5 | 1.08 | M3 |

12x35: Mag. 7.5. ***170x250:*** Slightly orangish.

| ga | NGC 57 | 00 15.5 | +17 19 | Psc | 12.7b | 2.2' | 1.8' | E |

HH: faint, small, round, suddenly brighter middle.
170x250: Moderately faint. Evenly moderately concentrated. Round. Small. No involved stars. A little ways SW is a faint star.

| ga | MCG-1-1-64 | 00 16.8 | −05 16 | Psc | 12.6 | 1.3' | 0.4' | Sa |

170x250: Faint. Broadly slightly concentrated. Slightly elongated NE-SW. Small. No involved stars. WNW is a little triangle of faint stars.

| ga | NGC 62 | 00 17.1 | −13 29 | Cet | 12.3 | 1.0' | 0.7' | Sa |

St: faint, very small, round, gradually little brighter middle.
170x250: Faint. Evenly moderately concentrated. Round. Very small. No involved stars.

| ga | NGC 63 | 00 17.8 | +11 27 | Psc | 12.6p | 1.7' | 1.1' | S |

dA: pretty faint, small, round, suddenly brighter middle.
170x250: Moderate brightness. Broadly to evenly moderately concentrated. Slightly elongated E-W. Small. No involved stars. It's on the southern edge of a large square group of moderately faint stars.

| ga | IC 10 | 00 20.3 | +59 18 | Cas | 11.8b | 6.3' | 5.1' | Im |

170x250: Very faint. Unconcentrated. Roughly round. Moderate size. Near center is a faint involved star; E of center is a very faint involved star; SSE is a triangle of moderate brightness stars.

| ga | NGC 80 | 00 21.2 | +22 21 | And | 13.1b | 1.6' | 1.6' | S0– |

JH: faint, small, round, pretty suddenly brighter middle.
170x250: Faint. Evenly moderately concentrated. Round. Very small. No involved stars. NE is a sharp triangle of moderate brightness stars. Next to this triangle W is a smaller and fainter galaxy [*83 (JH)].

| ga | NGC 95 | 00 22.2 | +10 29 | Psc | 13.2b | 1.9' | 1.0' | Sc |

HH: faint, pretty large, round, gradually brighter middle.
170x250: Moderately faint. Evenly slightly concentrated. Round. Small. No involved stars. A little ways NE is a moderate brightness star.

| vs | R And | 00 24.0 | +38 35 | And | 5.3 | 15.1 | 1.97 | S3 |

12x35: Very faint.

| oc | NGC 103 | 00 25.3 | +61 20 | Cas | 9.8 | 5.0' | | |

JH: cluster, pretty small, pretty compressed, 11th to 18th mag. stars.
42x60: Visible. ***170x250:*** A fair number of moderately faint stars. Wedge-shaped, NNE-SSW. Small. Moderately dense.

| ga | NGC 108 | 00 26.0 | +29 12 | And | 13.1p | 1.7' | 1.5' | S0+ |

HH: pretty faint, pretty large, round, pretty suddenly little brighter middle.
170x250: Faint. Evenly moderately concentrated. Roughly round. Small. No involved stars.

| vs | AQ And | 00 27.5 | +35 35 | And | 6.9 | 8.2 | 3.79 | C5 |

12x35: Mag. 8.5. ***170x250:*** Light copper red.

| pn | PK 119-6.1 | 00 28.3 | +55 57 | Cas | 13.3p | 5.0" | | |

170x250: Not visible.[†]

| ga | NGC 125 | 00 28.8 | +02 50 | Psc | 13.1b | 1.6' | 1.4' | S0+ |

HH: very faint, small, brighter middle, double star very near, W of 2.
[See 128.]

| ga | NGC 128 | 00 29.2 | +02 51 | Psc | 12.8b | 2.7' | 0.8' | S0 |

HH: pretty bright, very small, little extended N-S, brighter middle, E of 2.
170x250: Moderately faint to moderate brightness. Evenly moderately concentrated. Very elongated N-S. Small. No involved stars. A third of the field WSW is 125. Moderately faint. Sharply fairly concentrated. Roughly round. Very small. No involved stars. Off its edge S is a pair of moderate brightness stars. [126 (LR), 127 (LR), and 130 (LR) were not seen.]

| ga | NGC 131 | 00 29.6 | −33 15 | Scl | 13.1v | 1.9' | 0.6' | Sb |

JH: faint, pretty large, pretty much extended, very gradually brighter middle, W of 2.
[See 134.]

| oc | NGC 129 | 00 29.9 | +60 13 | Cas | 6.5 | 21.0' | | |

HH: cluster, very large, pretty rich, little compressed, 9th to 13th mag. stars.
12x35: A small triangle of stars. ***42x60:*** Three bright stars and three fainter stars. ***170x250:*** A triangle of bright stars with a dozen fainter stars among and around them, plus more stars N. Half the field in size. Somewhat sparse.

| ga | NGC 134 | 00 30.4 | −33 14 | Scl | 10.3v | 7.6' | 1.7' | Sbc |

JH: very bright, large, very much extended NE-SW, pretty suddenly brighter middle, E of 2, 10th mag. star 45" NNW.
42x60: Very faintly visible. ***170x250:*** Moderately bright to bright. Sharply pretty concentrated to a tiny nucleus. Edge-on NE-SW. Large: a third to half the field in size. The ends fade out very indistinctly. At its edge NNW of center is a star; nearby SE is another star. Half the field W is 131. Very faint. Broadly slightly concentrated. Elongated ENE-WSW. Small. The entire galaxy looks like just the core of 134 but even smaller. A little ways NE is a faint star. S of both galaxies is a large square of moderately bright stars with a fainter star right in the middle.

| oc | NGC 133 | 00 31.2 | +63 21 | Cas | 9.4 | 7.0' | | |

dA: cluster, pretty large, 10th mag. and fainter stars, double star involved.
42x60: 133, 146, and K14 are a little triangle of modest clusters, each consisting of a handful of moderately faint stars. ***170x250:*** K14 is the largest. A number of moderately bright stars plus some faint stars. Rectangular N-S. Half the field in size. Somewhat sparse. 146 is similar to K14 but smaller and even less distinct. A number of mixed brightness stars: a few moderate brightness and many fainter. The brightest stars are in a NE-SW quadrilateral figure, one corner of which is a double star. A quarter of the field in size. Sparse. 133 is a "Y" of a half-dozen moderately bright stars opening up SSE, with hardly any other stars. The middle star of the stock of the "Y" is a close double. A quarter of the field in size. Very sparse.

| oc | NGC 136 | 00 31.5 | +61 31 | Cas | – | 1.2' | | |

WH: globular cluster, very faint, small, extremely compressed.
170x250: A fair number of very faint stars. Very small. Pretty dense.

| ga | NGC 145 | 00 31.8 | –05 09 | Cet | 13.2b | 1.9' | 1.5' | Sdm |

JH: faint, pretty large, very little extended, very gradually brighter middle, 8th–9th mag. star 5' ENE.
170x250: Faint. Broadly very slightly concentrated. Roughly round. Small. No involved stars.

| oc | K 14 | 00 31.9 | +63 10 | Cas | 8.5 | 7.0' | | |

[See 133.]

| oc | NGC 146 | 00 33.1 | +63 17 | Cas | 9.1 | 6.0' | | |

JH: cluster, pretty large, little compressed, 11th to 12th mag. stars, double star.
[See 133.]

| ga | NGC 147 | 00 33.2 | +48 30 | Cas | 10.5b | 13.2' | 7.7' | E5 |

JH: very faint, very large, irregularly round, gradually then suddenly much brighter middle to an 11th mag. star.
170x250: Very very faint. Diffuse. Slightly elongated NNE-SSW. Very large: a third to half the field in size. Very near center is an involved star; around its periphery are a number of faint stars.

| ga | NGC 151 | 00 34.0 | –09 42 | Cet | 12.3b | 3.7' | 1.6' | Sbc |

HH: pretty faint, pretty large, little extended E-W, very gradually little brighter middle.
170x250: Moderately faint. Evenly moderately concentrated. Slightly elongated ENE-WSW. Small. No involved stars. At its E end is a moderately faint star.

| ga | NGC 148 | 00 34.3 | –31 47 | Scl | 13.1b | 2.0' | 0.7' | S0 |

JH: very bright, small, little extended E-W, suddenly much brighter middle to an 11th mag. star.
170x250: Moderate brightness; high surface brightness. Sharply pretty concentrated. Round. Very small. No involved stars. It makes a large triangular figure with five moderate brightness stars ENE to SE.

| ga | NGC 150 | 00 34.3 | –27 48 | Scl | 12.0b | 3.9' | 1.8' | Sb |

Sw: pretty faint, pretty small, round.
170x250: Moderately faint. Broadly slightly concentrated. Slightly elongated WNW-ESE. Moderate size. No involved stars.

| ga | MCG-1-2-34 | 00 34.7 | –07 54 | Cet | 12.8 | 0.9' | 0.7' | Sc |

170x250: Very faint. Broadly very slightly concentrated. Roughly round. Very small. No involved stars.

| ga | NGC 157 | 00 34.8 | –08 24 | Cet | 11.0b | 4.2' | 2.7' | Sbc |

WH: faint, large, much extended, between 2 considerably bright stars.
42x60: Visible. ***170x250:*** Moderately bright. Broadly slightly concentrated. Slightly elongated NE-SW. Moderately large. Toward its NE end is a faint involved star; the galaxy is roughly between two bright stars, one NNW and one S.

| ga | MCG-1-2-38 | 00 35.6 | –04 44 | Cet | 13.0 | 1.1' | 0.4' | Sbc |

170x250: Very faint. Hardly concentrated. Roughly round. Very small. No involved stars. It's inside a large trapezoid of moderately bright stars.

| ga | IC 1558 | 00 35.8 | –25 22 | Scl | 12.6b | 3.4' | 2.4' | Sm |

170x250: Exceedingly faint. The slightest diffuse glow. Moderately large. No involved stars. WSW is the bright plotted star.

0 to 6 Hours: FALL

| ga | NGC 160 | 00 36.1 | +23 57 | And | 12.6v | 3.0' | 1.7' | S0+ |

HH: very faint, very small, stellar, 7th mag. star 5' NNE.
170x250: Faint. Evenly to sharply fairly concentrated. Roughly round. Small. No involved stars. A little ways SW, making a shallow arc with it, are a couple of faint stars; NNE is the bright plotted star.

| ga | NGC 169 | 00 36.9 | +23 59 | And | 13.2 | 2.6' | 0.8' | Sab |

dA: faint, pretty large, double or binuclear, 6th mag. star 4' NE.
170x250: Very faint. Broadly very slightly concentrated. Round. Small. No involved stars. A little ways SE is a faint star; SW is a moderate brightness star; NE is the bright plotted star. [I1559, attached to 169, was not seen as a separate object.]

| ga | NGC 175 | 00 37.4 | −19 56 | Cet | 12.9b | 2.1' | 1.8' | Sab |

HH: pretty bright, pretty large, extended, gradually brighter middle, resolvable.
170x250: Very faint. Broadly very slightly concentrated. Roughly round. Small. No involved stars. It makes a large triangle with two moderate brightness stars, one SSE and one ENE.

| ga | NGC 182 | 00 38.2 | +02 43 | Psc | 12.3v | 2.0' | 1.4' | Sa |

WH: very faint, small, irregularly round, very gradually brighter middle.
170x250: Very faint. Evenly slightly concentrated. Round. Small. No involved stars. It's in a line with the bright plotted star NW, a moderately faint star SE, and a brighter star further SE.

| ga | NGC 185 | 00 39.0 | +48 20 | Cas | 10.1b | 11.9' | 10.1' | E3 |

HH: pretty bright, very large, irregularly round, very gradually much brighter middle, resolvable.
42x60: Very faintly visible. ***170x250:*** Faint. Broadly slightly concentrated. Roughly round. Large: a quarter to a third of the field in size. The halo fades out indistinctly. Bounding the galaxy's edge W are three very faint stars; a little further W is a fourth star. 185 is better defined than 147.

| ga | NGC 178 | 00 39.1 | −14 10 | Cet | 13.1b | 2.3' | 0.9' | Sm |

Sn: faint, small, much extended N-S, brighter middle.
170x250: Moderately faint. Broadly moderately concentrated. Slightly elongated N-S. Small. No involved stars.

| bs | Delta And | 00 39.3 | +30 51 | And | 3.28 | | 1.28 | K3 |

170x250: Yellow gold.

| ga | NGC 194 | 00 39.3 | +03 02 | Psc | 13.2b | 1.5' | 1.3' | E |

HH: pretty bright, small, round, very gradually brighter middle.
170x250: Moderately faint. Evenly moderately concentrated. Round. Small. No involved stars. A little less than half the field NNE is a very faint galaxy [*199 (dA)]. Broadly concentrated. Round. Very small. No involved stars. 2/3 of a field NNW is another galaxy [*193 (HH)]. Moderately faint. Evenly moderately concentrated. Round. Small. Just off its edge SW is a moderately faint star that makes a triangle with two moderate brightness stars, one ESE and one S. A little more than a third of the field ESE, beyond the E star of the triangle, is a faint galaxy [204]. Evenly moderately concentrated. Roughly round. Small. No involved stars. Half the field N is a very faint galaxy [*203 (LR)]. Half the field SSE of 194 is a slightly larger galaxy [*200 (WH)]. Faint. Broadly slightly concentrated. Slightly elongated NW-SE. Moderately small. No involved stars. A third of the field SSW is another faint galaxy [*198 (WH)]. Broadly slightly concentrated. Slightly elongated ENE-WSW [round]. Small. No involved stars. An exceptionally good string of galaxies across several fields N-S.

| oc | NGC 189 | 00 39.6 | +61 05 | Cas | 8.8 | 3.7' | | |

JH: cluster, pretty large, round, 11th to 15th mag. stars.
42x60: A faint cluster. *170x250:* A number of moderate brightness to moderately faint stars. A quarter of the field in size. Sparse.

| ga | NGC 204 | 00 39.7 | +03 17 | Psc | 12.8v | 1.2' | 1.1' | S0 |

HH: pretty bright, pretty small, round, very gradually brighter middle, 2nd of 3. [See 194.]

| oc | ST 24 | 00 39.8 | +61 57 | Cas | 8.8 | 4.0' | | |

170x250: A small concentration of stars at the NE corner of a long rectangular group of bright stars that fills the field. Very small. It doesn't look like a true cluster.

| ga | NGC 205/M110 | 00 40.4 | +41 41 | And | 8.9b | 21.9' | 10.9' | E5 |

HH: very bright, very large, much extended NNW-SSE, very gradually very much brighter middle.
170x250: Bright, but much lower surface brightness than either M32 or the center of M31. Evenly pretty concentrated. Elongated N-S. Very large: 2/3 of the field in size. The halo fades out very indistinctly. SSE of center is a faint involved star; at the galaxy's edge ENE of center is another faint involved star. M110 and M32 make a sharp contrast.

| bs | Alpha Cas | 00 40.5 | +56 32 | Cas | 2.23 | | 1.17 | K0 |

170x250: Yellow.

| ga | NGC 210 | 00 40.6 | −13 52 | Cet | 11.6b | 5.0' | 3.3' | Sb |

HH: bright, pretty small, round, pretty suddenly brighter middle, resolvable, star 90" distant.
170x250: Moderately bright. Sharply pretty concentrated. Slightly elongated N-S. Moderately small. No involved stars. In line with it are two stars, one nearby WSW and another further WSW.

| ga | NGC 214 | 00 41.5 | +25 29 | And | 13.0b | 1.8' | 1.3' | Sc |

HH: pretty faint, pretty small, gradually very little brighter middle, resolvable.
170x250: Moderately faint. Evenly moderately concentrated. Round. Small. No involved stars. N is a very faint star.

| ga | NGC 227 | 00 42.6 | −01 31 | Cet | 13.1b | 1.6' | 1.2' | S0− |

WH: faint, pretty large, little brighter middle.
170x250: Moderately faint to moderate brightness. Sharply pretty concentrated. Round. Very small. No involved stars.

| ga | NGC 221/M32 | 00 42.7 | +40 51 | And | 9.0b | 8.7' | 6.4' | cE2 |

JH: ! very very bright, large, round, pretty suddenly much brighter middle to a nucleus.
170x250: Extremely bright. Sharply extremely concentrated to a small extremely bright core. Round. Moderate size. No involved stars. A little ways E is a moderate brightness star that makes an evenly spaced line with two more stars NNE of it. One of the brightest galaxies in the sky.

| ga | NGC 224/M31 | 00 42.7 | +41 16 | And | 4.4b | 192' | 62' | Sb |

JH: !!! excessively bright, extremely large, very much extended, bifid (Bd) (Andromeda Great Nebula).
12x35: Very bright. Sharply very concentrated. Edge-on NE-SW. 2/3 of the field in size. The ends fade out indistinctly. The NW edge is a little sharper than the SE edge. A third of the distance between the center and the NE end is an involved star. M110 is faint. M32 is practically stellar. *42x60:* M31, M32, and M110 are all in the same field. M31 contains several faint involved stars. M110 is somewhat faint. M32 is bright and sharply very concentrated, but clearly not stellar. *170x250:* M31 is sharply extremely concentrated to a very small nonstellar peak; the concentration is continuous, with no well demarcated core or nucleus. The light is generally smooth and vague. A first dust lane, a field width in length, starts from N of center and runs SW. Farther out on the same side of the galaxy is a second, vaguer dust lane. The SW half of the galaxy is better defined than the NE half. A little W of center is a moderately faint star. About a dozen of M31's globular clusters, all extremely faint and stellar or substellar, are visible when specifically searched for with the aid of a detailed map; otherwise they are not noticed. [Andromeda Galaxy]

| oc | NGC 225 | 00 43.4 | +61 47 | Cas | 7.0 | 12.0' | | |

HH: cluster, large, little compressed, 9th to 10th mag. stars.
12x35, 42x60: A half-disk of stars. *170x250:* Some two dozen bright stars, very uniform in brightness, in the shape of half a disk. There are hardly any other stars. 3/4 of the field in size. Sparse.

| ga | NGC 234 | 00 43.5 | +14 20 | Psc | 13.2p | 1.6' | 1.4' | Sc |

WH: faint, pretty small, irregularly little extended, brighter middle.
170x250: Faint. Broadly slightly concentrated. Roughly round. Moderately small. No involved stars.

| bs | Beta Cet | 00 43.6 | −17 59 | Cet | 2.04 | | 1.02 | G9 |

170x250: Yellow.

| oc | K 16 | 00 43.7 | +64 11 | Cas | 10.3 | 3.0' | | |

42x60: A couple of faint stars. *170x250:* A little "Y" of a few moderately faint stars opening up NW, plus a handful more very faint stars. The star at the end of the N arm of the "Y" is a double. Very small. Very sparse.

| oc | NGC 188 | 00 44.0 | +85 20 | Cep | 8.1 | 13.0' | | |

JH: cluster, very large, round, 150–200 10th to 18th mag. stars.
12x35: Very faintly visible. ***42x60:*** A few very faint stars. ***170x250:*** A number of moderately faint stars, uniform in brightness, plus a lot of threshold stars. There are empty spaces between clumps of stars. It fills the field. Moderate density.

| ga | NGC 245 | 00 46.1 | −01 43 | Cet | 13.0p | 1.3' | 1.1' | Sb |

WH: faint, pretty small, irregular figure, easily resolvable.
170x250: Moderately faint. Broadly slightly concentrated. Round. Small. No involved stars. A little ways SSW is a pair of faint stars.

| pn | NGC 246 | 00 47.0 | −11 52 | Cet | 8.0p | 4.1' | | |

HH: very faint, large, 4 stars in diffuse nebula.
170x250: Moderate brightness; fairly low surface brightness. It has a relatively bright central star that makes a triangle with two other bright stars, one well within the nebula SW and another just off its edge NW; ESE, just within the nebula, is a faint star; SW and S are two more fairly bright stars. The nebula is uneven in light: it's a little darker toward the middle but it's not conventionally annular. The edge is better defined on the W side than on the E side. Round. Very large for a planetary: a sixth of the field in size. Ghostly.

| ga | NGC 247 | 00 47.1 | −20 45 | Cet | 9.1v | 21.4' | 6.0' | Sd |

HH: faint, very large, very much extended N-S.
12x35: Extremely faintly visible. ***42x60:*** Faintly visible. ***170x250:*** Moderately bright; fairly low surface brightness. Broadly very slightly concentrated. Elongated NNW-SSE. Very large: the field width in size. The N end is broader and more diffuse than the S end, and the brightest part of the galaxy is displaced a little S of center. Near the S end are a couple of faint involved stars aligned perpendicular to the axis of the galaxy; the W of these has a companion SSW; just off the galaxy's S tip is a bright star.

| ga | NGC 254 | 00 47.5 | −31 25 | Scl | 12.6b | 2.4' | 1.4' | S0+ |

JH: very bright, pretty small, little extended, suddenly much brighter middle, 8th mag. star 5' NE.
170x250: Moderate brightness. Sharply very concentrated. Slightly elongated NW-SE. Small. No involved stars.

| ga | NGC 253 | 00 47.6 | –25 17 | Scl | 8.0b | 27.7' | 6.7' | Sc |

HH: !! very very bright, very very large, very much extended NE-SW, gradually brighter middle, 4 stars.
12x35: Easily visible. ***42x60:*** Bright. Broadly slightly concentrated. Off its edge SSW of center are two bright stars and a fainter star. ***170x250:*** Very bright. Broadly slightly concentrated. Edge-on NE-SW. Very large: one and a half field widths in size. There's conspicuous unevenness in the light in the middle part of the galaxy. In the center is a small core-like brightening but not a distinct core or nucleus. The NE end is a little brighter and better defined than the SW end, which fades out a little more indistinctly. The NW edge is a little sharper than the SE edge; this would be the closer edge with the dust lane. There are several involved stars. One of the very best galaxies in the sky, if not the best.

| ga | NGC 255 | 00 47.8 | –11 28 | Cet | 11.8v | 3.1' | 2.5' | Sbc |

HH: faint, pretty small, round, gradually brighter middle.
170x250: Moderately faint. Broadly slightly concentrated. Round. Small. No involved stars.

| bs | Eta Cas | 00 49.1 | +57 49 | Cas | 3.45 | | 0.58 | G0 |

42x60: The primary has a relatively very faint companion; nicely split. ***170x250:*** The primary has a moderate distance much fainter companion. Colorful: the primary is yellow white; the companion is orangish. [STF 60: 13.1", 3.4, 7.5, G0 K7]

| ga | NGC 266 | 00 49.8 | +32 16 | Psc | 12.5b | 3.0' | 2.8' | Sab |

HH: pretty bright, pretty small, little extended, pretty suddenly brighter middle, resolvable, 8th mag. star 4' SE.
170x250: Moderate brightness. Evenly to sharply pretty concentrated. Round. Very small. No involved stars. A little ways NNW are three faint stars in a little line; the galaxy makes a triangle with the bright plotted star SSE and a wide pair of moderate brightness stars ESE.

| ds | 65 Psc | 00 49.9 | +27 43 | Psc | 4.5" | 6.3 | 6.3 | F4 F5 |

42x60: Extremely close. ***170x250:*** A very close perfectly equal pair. Both white. [STF 61]

| ga | MCG-1-3-18 | 00 50.5 | –05 51 | Cet | 13.2 | 1.4' | 1.4' | S0– |

170x250: Faint. Evenly moderately concentrated. Round. Very small. No involved stars. Nearby SW is a pair of stars, moderate brightness and very faint; S is a brighter star.

| ga | NGC 270 | 00 50.5 | −08 39 | Cet | 12.9 | 1.9' | 1.7' | S0+ |

HH: pretty faint, very small, irregularly round, pretty gradually brighter middle.
170x250: Faint. Evenly moderately concentrated. Round. Very small. No involved stars. W is a moderately bright star.

| ga | NGC 271 | 00 50.7 | −01 54 | Cet | 12.9p | 2.1' | 1.6' | Sab |

HH: pretty faint, small, little extended, pretty suddenly brighter middle, 8th mag. star E.
170x250: Moderately faint. Evenly moderately concentrated. Roughly round. Very small. No involved stars. Nearby ESE is the bright plotted star.

| oc | K 2 | 00 51.0 | +58 11 | Cas | – | 5.0' | | |

12x35, 42x60, 170x250: There's no cluster in the position. [It's a small dense cluster of extremely faint stars.]

| ga | NGC 274 | 00 51.0 | −07 03 | Cet | 12.8p | 1.5' | 1.4' | S0− |

HH: pretty bright, pretty small, suddenly much brighter middle, [NW] of double nebula.
170x250: 274 and 275 are a double galaxy aligned NW-SE. 274 is a little brighter and more concentrated. Moderate brightness. Sharply very concentrated. Round. Very very small. No involved stars. 275 is moderately faint. Broadly slightly concentrated. Slightly elongated NW-SE, toward 274. Very small. No involved stars. A somewhat contrasting pair: a brighter concentrated galaxy and a slightly fainter, less concentrated galaxy. A little more than half the field NNW is another galaxy [*273 (WH)]. Faint to moderately faint. Evenly moderately concentrated. Slightly elongated E-W. Very small. Off its edge NW is a very faint star.

| ga | NGC 275 | 00 51.1 | −07 03 | Cet | 13.2p | 1.5' | 1.0' | Scd |

JH: very faint, small, round, [SE] of double nebula.
[See 274.]

| oc | NGC 272 | 00 51.4 | +35 49 | And | 8.5 | 0.5' | | |

dA: cluster, large, little compressed.
12x35, 42x60: In the position is a moderately faint star. ***170x250:*** A moderately bright star with several moderate brightness to moderately faint stars surrounding it: one right next to it NW, two in line with it NE, and four in a small deep arc SE. [A quarter of the field in size. Very sparse.]

| ga | NGC 278 | 00 52.1 | +47 33 | Cas | 11.5b | 2.2' | 2.2' | Sb |

HH: considerably bright, pretty large, round, 2 10th mag. stars near.
170x250: Bright. Sharply very concentrated to a substellar nucleus. Round. Small. No involved stars. It makes a trapezoid with the bright plotted star N and two stars NE.

| ga | NGC 289 | 00 52.7 | −31 12 | Scl | 11.7b | 5.1' | 3.6' | Sbc |

JH: very bright, large, pretty much extended, gradually little brighter middle, 11th mag. star NW.
170x250: Moderately bright. Broadly to evenly moderately concentrated. Slightly elongated NW-SE. Moderate size. At its NW end is a faint star.

| ne | NGC 281 | 00 52.8 | +56 37 | Cas | | 28.0' | | E |

Ba: faint, very large, diffuse, faint triple star on NW edge.
170x250: At first sight: a very small very faint nebula around a very small sharp triangle of stars. Then: a lot more extremely diffuse nebulosity is visible, gray compared to the surrounding black sky. It takes up half the field, mostly NE of the triple star. [No separate open cluster (I1590) was noticed.]

| gc | NGC 288 | 00 52.8 | −26 35 | Scl | 8.1 | 13.0' | −6.60 | 10 |

HH: globular cluster, bright, large, little extended, 12th to 16th mag. stars.
12x35, 42x60: Visible. *170x250:* Well resolved with distinct stars, especially away from the center. Broadly very slightly concentrated. A little less than half the field in size. The highest concentration of stars is displaced a little W of center. Somewhat triangular. Irregular. Fairly loose; skimpy. It doesn't have a central ball.

| bs | Gamma Cas | 00 56.7 | +60 43 | Cas | 2.39 | | −0.10 | B0 |

170x250: White.

| ga | NGC 309 | 00 56.7 | −09 54 | Cet | 12.5b | 3.0' | 2.4' | Sc |

Te: pretty bright, pretty large, 12th–13th mag. star N.
170x250: Very faint. Broadly very slightly concentrated. Round. Moderately small. No involved stars. Nearby W is a very faint star; a little ways NNE is a moderate brightness star.

| ga | NGC 315 | 00 57.8 | +30 21 | Psc | 12.2b | 3.2' | 2.2' | E+ |

HH: pretty bright, pretty large, round, gradually brighter middle, 9th mag. star 3' SE.
170x250: Moderate brightness to moderately bright. Evenly fairly concentrated. Roughly round. Small. No involved stars. It makes a shallow arc with two faint stars WSW. A third of the field SW is a smaller and fainter galaxy [*311 (JH)]. Concentrated. Round. Very small. No involved stars. It makes a curved "Y" figure with nearby stars, opening up ESE. [318 (LR) was not seen.]

| ga | NGC 336 | 00 58.8 | -18 45 | Cet | 12.0p | 0.1' | 0.1' | - |

Le: very faint, very small, round, suddenly brighter middle.
170x250: [The observed galaxy is *E541-4; there is no galaxy in the exact position of 336.] Very faint. Hardly concentrated. Round. Moderately small. No involved stars. It's inside a large triangle of stars, the S star of which makes a very small isosceles triangle with two fainter stars.

| ga | NGC 337 | 00 59.8 | -07 34 | Cet | 12.1b | 2.9' | 1.8' | Sd |

HH: pretty faint, large, extended N-S, gradually little brighter middle, 10th mag. star E.
170x250: Moderate brightness. Broadly slightly concentrated. Slightly elongated NW-SE. Moderately small. No involved stars.

| ds | STF 79 | 01 00.1 | +44 43 | And | 7.5" | 6.0 | 6.8 | B9 A2 |

42x60: A very close unequal pair. *170x250:* A moderately close slightly unequal pair. Both white.

| ga | ESO 351-30 | 01 00.2 | -33 43 | Scl | 10.5b | 39.8' | 30.8' | E |

12x35, 42x60: Not visible. [Sculptor dwarf]

| ga | NGC 337A | 01 01.6 | -07 35 | Cet | 12.7b | 6.9' | 5.4' | Sdm |

170x250: Not visible.

| ga | NGC 357 | 01 03.4 | -06 20 | Cet | 12.0v | 2.4' | 1.7' | S0/a |

HH: faint, small, irregularly round, suddenly brighter middle, 14th mag. star 20" NE.
170x250: Moderately faint. Evenly moderately concentrated. Roughly round. Small. At its edge E is a very faint star. [355 (Ma) was not seen.]

| ga | IC 1613 | 01 04.8 | +02 07 | Cet | 9.9b | 16.3' | 14.5' | Im |

170x250: Exceedingly faint. The slightest unconcentrated brightening in the sky. 2/3 of the field in size. No clearly involved stars. It's surrounded by moderate brightness and moderately faint stars.

| ga | MCG-1-3-85 | 01 05.1 | –06 12 | Cet | 11.6v | 4.2' | 3.6' | Sd |

170x250: Extremely faint. The slightest glow. Roughly round. Moderate size. No involved stars. E is a moderately bright star that makes an evenly spaced line with two fainter stars SSW of it. [M-1-3-88 was not seen.]

| oc | NGC 358 | 01 05.2 | +62 02 | Cas | – | 3.0' | | |

dA: cluster, very little rich.
42x60: Two or three faint stars. *170x250:* A small parallelogram of four moderately faint stars. Not really a cluster.

| ds | Psi 1 Psc | 01 05.6 | +21 28 | Psc | 29.8" | 5.6 | 5.8 | A1 A0 |

12x35: Easily split. A very slightly unequal pair. *42x60:* A wide equal pair. *170x250:* A very wide equal pair. Both white. [STF 88]

| ga | NGC 379 | 01 07.3 | +32 31 | Psc | 12.8v | 1.4' | 0.9' | S0 |

HH: pretty faint, small, round, brighter middle, 1st of 3.
170x250: In the position is a N-S string of galaxies the field width in length. All are sharply pretty concentrated and have no involved stars. The brightest is in the middle [*383 (HH)]. Moderately faint. Round. Small. Right next to it SSW is a companion galaxy [*382 (LR)]. Very faint. Very very small. Nearby SSW is a faint star. A third of the field S are two galaxies [384, 385]. Both are faint. Round. Very small. A third of the field NNW of the middle galaxy are two more galaxies [379, *380 (HH)] almost identical to the two southern galaxies. The S of these [380] is a little brighter. Both are faint. Round. Very small. [375 (LR), 386 (LR), and 388 (LR) were not seen.]

| ga | NGC 384 | 01 07.4 | +32 17 | Psc | 13.1v | 1.1' | 0.9' | E3 |

LR: pretty faint, pretty small, SW of 2.
[See 379.]

| ga | NGC 385 | 01 07.5 | +32 19 | Psc | 12.9v | 1.2' | 1.0' | S0– |

LR: pretty faint, pretty small, round, NE of 2.
[See 379.]

| oc | NGC 381 | 01 08.3 | +61 35 | Cas | 9.3 | 6.0' | | |

WH: cluster, pretty compressed.
42x60: Faintly visible. ***170x250:*** A fair number of faint stars. A quarter of the field in size. Moderate density.

| bs | Eta Cet | 01 08.6 | −10 10 | Cet | 3.45 | | 1.16 | K2 |

170x250: Yellow gold.

| ga | NGC 404 | 01 09.4 | +35 43 | And | 11.2b | 3.4' | 3.4' | S0− |

HH: pretty bright, considerably large, round, gradually brighter middle, Beta And near.
170x250: Bright. Sharply pretty concentrated. Round. Moderately small. No involved stars. Tangent to its edge N is a narrow ENE-WSW houselike figure of moderately faint stars; a little more than a third of the field SE is Beta And.

| bs | Beta And | 01 09.7 | +35 37 | And | 2.06 | | 1.58 | M0 |

170x250: Gold. A little more than a third of the field NW is NGC 404.

| ga | NGC 418 | 01 10.6 | −30 13 | Scl | 13.1b | 2.1' | 1.9' | Sc |

JH: faint, pretty large, round, very gradually little brighter middle, W of 2.
170x250: Very faint. Hardly concentrated. Round. Small. No involved stars. 3/4 of a field SSE is a second galaxy [*I1637] pretty much identical to 418 in every respect.

| ga | NGC 410 | 01 11.0 | +33 09 | Psc | 12.5b | 2.4' | 1.7' | E+ |

WH: pretty large, E of double nebula.
170x250: Moderate brightness. Evenly fairly concentrated. Round. Very small. No involved stars. [407 (WH) and 414 (Sz) were not seen.]

| ga | NGC 420 | 01 12.2 | +32 07 | Psc | 13.1p | 1.9' | 1.6' | S0 |

HH: faint, pretty small, round, brighter middle.
170x250: Faint. Evenly moderately concentrated. Round. Very very small. No involved stars.

ga	NGC 428	01 12.9	+00 58	Cet	11.9b	4.1'	3.1'	Sm

WH: faint, large, round, brighter middle, easily resolvable.
170x250: Faint. Broadly slightly concentrated. Slightly elongated NW-SE. Moderately small. No involved stars. It makes a flat triangle with two moderately faint stars, one NW and one SSW.

ds	Zeta Psc	01 13.7	+07 35	Psc	23.2"	5.6	6.5	A7 –

12x35, 42x60: Easily split. ***170x250:*** A wide slightly unequal pair. Both white. [STF 100]

ga	NGC 439	01 13.8	–31 44	Scl	12.5p	2.4'	1.5'	S0–

JH: pretty bright, small, round, gradually brighter middle.
170x250: Moderate brightness. Sharply fairly concentrated. Round. Small. No involved stars. A little ways SSE is a slightly smaller and slightly fainter galaxy [*441 (JH)]. Faint to moderately faint. Sharply concentrated. Round. Very small. No involved stars. The two galaxies make an irregular kite figure with two faint stars, one E and one W.

ga	IC 1657	01 14.1	–32 39	Scl	13.1p	2.3'	0.5'	Sbc

170x250: Faint to moderately faint. Broadly to evenly moderately concentrated. A thin edge-on, N-S. Moderately small. No involved stars.

ga	NGC 448	01 15.3	–01 37	Cet	13.1p	1.6'	0.7'	S0–

Sw: pretty bright, very small, little extended.
170x250: Moderate brightness to moderately bright; fairly high surface brightness. Sharply pretty concentrated. Slightly elongated WNW-ESE. Very small. No involved stars.

ga	NGC 450	01 15.5	–00 51	Cet	12.2p	3.1'	2.3'	Scd

WH: very faint, very large.
170x250: Very faint. Hardly concentrated. Slightly elongated E-W. Moderate size. No involved stars. S is a nearly equilateral triangle of stars.

oc	NGC 436	01 15.6	+58 49	Cas	8.8	5.0'		

WH: cluster, small, irregular figure, pretty compressed.
12x35: Visible. ***42x60:*** A faint cluster of a few faint stars NW of a line of three brighter stars. ***170x250:*** A half-dozen moderately faint stars plus fainter stars. The three moderately bright stars SE are undoubtedly not cluster members. Small. Moderate density.

| vs | Z Psc | 01 16.1 | +25 46 | Psc | 7.0 | 7.9 | 2.62 | C7 |

12x35: Mag. 7.5. *170x250:* Orange red.

| ga | IC 93 | 01 19.0 | –17 03 | Cet | 12.5 | 1.3' | 0.7' | Sb |

170x250: Faint. Broadly slightly concentrated. Elongated N-S. Very small. No involved stars. [M-3-4-39 was not seen.]

| oc | NGC 457 | 01 19.1 | +58 20 | Cas | 6.4 | 13.0' | | |

HH: cluster, bright, large, pretty rich, 7th, 8th, and 10th mag. stars.
12x35: A few faint stars (excluding Phi Cas and its very wide companion, which undoubtedly are not cluster members). *42x60:* An elongated cluster of faint stars with two bright stars at its NW end. *170x250:* A good number of moderately bright and moderate brightness stars. Most of the stars are in a big "X": a NW-SE linear concentration crossed by a longer NE-SW arc of very widely spaced stars. 3/4 of the field in size. Moderately dense. A nice cluster.

| ga | NGC 467 | 01 19.2 | +03 18 | Psc | 12.9b | 1.6' | 1.5' | S0 |

HH: pretty bright, pretty large, round, gradually much brighter middle, W of 2.
170x250: Moderately faint. Evenly moderately concentrated. Round. Small. No involved stars. It makes a line with the bright plotted star E and another star further E.

| ga | NGC 470 | 01 19.7 | +03 24 | Psc | 12.5b | 2.8' | 1.7' | Sb |

WH: considerably bright, very large, irregularly round, pretty bright star E. [See 474.]

| ga | NGC 474 | 01 20.1 | +03 25 | Psc | 12.4b | 7.0' | 6.2' | S0 |

WH: pretty bright, small, suddenly much brighter middle, E of 2.
170x250: Bright. Sharply very concentrated to a bright substellar nucleus. Round. Very small. No involved stars. NE is a shallow arc of three moderately faint stars. A third of the field W is 470. Moderately faint. Broadly slightly concentrated. Slightly elongated NNW-SSE. Moderately small. No involved stars. It's on the side of a triangle of faint stars.

| ga | NGC 491 | 01 21.4 | –34 03 | Scl | 13.2b | 1.3' | 0.9' | Sb |

JH: bright, small, very little extended, brighter middle, very faint star near.
170x250: Moderately faint. Broadly slightly concentrated. Round. Small. No involved stars. Off its edge WSW is a faint star.

| ga | UGC 863 | 01 21.7 | +78 37 | Cep | 13.2 | 1.8' | 1.8' | Scd |

170x250: Not visible.

| ga | NGC 488 | 01 21.8 | +05 15 | Psc | 10.2v | 6.6' | 5.3' | Sb |

HH: pretty bright, large, round, suddenly very much brighter middle, 7th mag. star E.
42x60: Visible. *170x250:* Bright. Evenly fairly concentrated. Round. Moderately small. It has a very faint outer halo. No involved stars. Nearby SSE is a moderate brightness star that makes an ENE-WSW line with three other moderate brightness stars.

| ga | NGC 493 | 01 22.2 | +00 56 | Cet | 12.9b | 3.3' | 1.0' | Scd |

HH: very faint, large, much extended ENE-WSW, little brighter middle.
170x250: Very very faint. Hardly concentrated. Edge-on ENE-WSW. Moderate size to moderately large. No involved stars. Tenuous; very tantalizing.

| ga | NGC 499 | 01 23.2 | +33 27 | Psc | 12.1v | 1.8' | 1.2' | S0– |

HH: pretty bright, pretty large, round, 3rd of 3.
170x250: Moderate brightness. Sharply fairly concentrated. Slightly elongated ENE-WSW. Small. No involved stars. It's among a number of stars in a long WNW-ESE linear group, including a moderately bright star E and a sharp triangle of faint stars W. Just inside the triangle of stars is a very small very faint galaxy [*495 (WH)]. [496 (WH) and 501 (LR) were not seen.]

| ga | NGC 507 | 01 23.7 | +33 15 | Psc | 11.2v | 2.5' | 2.5' | S0 |

HH: very faint, pretty large, round, brighter middle, W of 2.
170x250: Moderate brightness. Evenly to sharply fairly concentrated. Round. Very small. No involved stars. Nearby N is a very small very faint galaxy [*508 (HH)]. A quarter of the field SW is another galaxy [*504 (JH)]. Sharply very concentrated. Very small. [503 (dA) and I1687 were not seen.]

| ga | NGC 514 | 01 24.1 | +12 54 | Psc | 12.2b | 4.2' | 2.7' | Sc |

HH: faint, large, little extended, very gradually little brighter middle, double star E.
170x250: Faint. Broadly very slightly concentrated. Slightly elongated E-W. Moderate size. No involved stars. A little ways E is a small quadrilateral of stars: a bright star and three very faint stars.

| ga | NGC 520 | 01 24.6 | +03 47 | Psc | 12.2b | 4.5' | 1.8' | Pec |

HH: faint, considerably large, extended NW-SE.
170x250: Moderate brightness. Broadly slightly concentrated. A fat edge-on, NW-SE. Moderately small. No involved stars.[†]

| ga | NGC 521 | 01 24.6 | +01 43 | Cet | 12.6b | 3.1' | 2.8' | Sbc |

HH: faint, pretty large, round, gradually brighter middle.
170x250: Faint to moderately faint. Evenly to sharply moderately concentrated. Round. Moderately small. No involved stars. Alongside it W is a N-S line of moderately faint stars.

| ga | NGC 524 | 01 24.8 | +09 32 | Psc | 11.3b | 2.7' | 2.7' | S0+ |

HH: very bright, pretty large, much brighter middle, 4 faint stars near.
170x250: Moderately bright. Evenly moderately concentrated. Round. Small. No involved stars. It's surrounded by moderate brightness to moderately faint stars.

| ga | NGC 533 | 01 25.5 | +01 45 | Cet | 12.4b | 3.8' | 2.3' | E3 |

HH: pretty bright, pretty large, round, gradually much brighter middle.
170x250: Moderate brightness. Evenly moderately concentrated. Roughly round. Small. No involved stars. WNW is a moderately faint star.

| ga | NGC 529 | 01 25.7 | +34 42 | And | 13.1b | 2.4' | 2.0' | S0– |

JH: pretty bright, very small, suddenly brighter middle, W of 2.
[See 536.]

| ga | NGC 541 | 01 25.7 | –01 22 | Cet | 13.0b | 1.8' | 1.8' | S0– |

dA: faint, small, round, brighter middle.
[See 547.]

| bs | Delta Cas | 01 25.8 | +60 14 | Cas | 2.68 | | 0.13 | A5 |

170x250: White.

| ga | NGC 545 | 01 26.0 | –01 20 | Cet | 13.2p | 2.4' | 1.5' | S0– |

WH: stellar, W of double nebula.
[See 547.]

| ga | NGC 547 | 01 26.0 | −01 20 | Cet | 13.2p | 1.0' | 0.9' | E1 |

WH: stellar, E of double nebula.
170x250: 547, 545, 541 and a fourth galaxy [*535 (dA)] are in an ENE-WSW linear group a quarter of the field in length. 545 and 547 are a very close identical pair – a double galaxy. Both are moderate brightness. Sharply concentrated. Round. Very small. No involved stars. SW is 541. Moderately faint. Evenly moderately concentrated. Roughly round. Very small. No involved stars. Further SW is the fourth galaxy. Very faint to faint. Slightly concentrated. Slightly elongated ENE-WSW. Very small. No involved stars. [The observed galaxies are the brightest members of a large cluster of galaxies (AGC 194) extending farther SW; 543 (dA) and C385-129 were not seen.]

| ga | NGC 536 | 01 26.4 | +34 42 | And | 12.3v | 3.6' | 1.3' | Sb |

JH: pretty bright, pretty large, gradually brighter middle, E of 2.
170x250: Moderately faint. Sharply very concentrated to a faint substellar nucleus. Round. Very small. Very near center is a very very faint involved star; nearby S are a couple of extremely faint stars in line with the galaxy. Half the field W is 529, which is pretty similar. Faint. Evenly to sharply moderately concentrated. Round. Very small. No involved stars. It makes a flat triangle with two relatively bright stars S; E of the closer star is a very faint star. [531 (LR) and 542 (LR) were not seen.]

| vs | R Scl | 01 27.0 | −32 33 | Scl | 6.1 | 8.8 | 3.83 | C6 |

12x35: Mag. 6.5. ***170x250:*** Orangish.

| oc | NGC 559 | 01 29.5 | +63 19 | Cas | 9.5 | 4.4' | | |

HH: cluster, bright, pretty large, pretty rich, stars of mixed magnitudes.
12x35: Faintly visible. ***42x60:*** A couple of stars with a string of stars leading away from them NW. ***170x250:*** A couple of brighter stars and a lot of faint stars. The long string of bright stars seen in the 60 mm is unrelated. Slightly elongated NE-SW. A third of the field in size. Pretty dense.

| ga | NGC 578 | 01 30.5 | −22 39 | Cet | 11.4b | 4.9' | 3.0' | Sc |

JH: bright, large, pretty much extended, gradually pretty much brighter middle.
170x250: Moderately faint. Broadly slightly concentrated. Slightly elongated E-W. Moderate size. No involved stars. Nearby E is an isosceles triangle of faint stars.

| ga | NGC 584 | 01 31.3 | −06 52 | Cet | 11.4b | 4.1' | 2.2' | E4 |

HH: very bright, pretty large, round, much brighter middle, W of 2.
170x250: Very bright. Sharply very concentrated. Round. Small. No involved stars. A quarter of the field ESE is a small very faint galaxy [*586 (HH)].

0 to 6 Hours: FALL 73

| ga | NGC 596 | 01 32.9 | −07 01 | Cet | 11.8b | 3.2' | 2.0' | E+ |

HH: pretty bright, round, brighter middle, resolvable, 6th mag. star E.
170x250: Bright. Sharply very concentrated. Round. Very small to small. No involved stars.

| ga | NGC 600 | 01 33.1 | −07 18 | Cet | 12.9b | 3.3' | 2.8' | Sd |

WH: excessively faint.
170x250: Very faint. Diffuse. Roughly round. Moderate size. No involved stars.

| oc | NGC 581/M103 | 01 33.2 | +60 43 | Cas | 7.4 | 6.0' | | |

JH: cluster, bright, round, rich, pretty large, 10th to 11th mag. stars.
12x35: A little NW-SE line of stars with a hint of the glow of more stars. ***42x60:*** A bright star, three moderate brightness stars, and more threshold stars. ***170x250:*** A fair number of mixed brightness stars, including four brighter stars. Somewhat triangular. A third of the field in size. Moderate density.

| ga | NGC 598/M33 | 01 33.9 | +30 39 | Tri | 6.3b | 65.6' | 38.0' | Scd |

HH: ! extremely bright, extremely large, round, very gradually brighter middle to a nucleus, partially resolved.
12x35: Easily visible. ***42x60:*** Moderately bright. Broadly slightly concentrated. Very slightly elongated N-S. A third of the field in size. At its edge N is a very faint involved star. ***170x250:*** Bright; fairly low surface brightness. Evenly slightly concentrated to a small brighter core and a tiny nucleus. Elongated N-S. The faint halo overflows the field, but the brightest part is the field width in size. There's vague unevenness in the light. 2/3 of a field NE of center is a very conspicuous HII region [604 (HH)] that resembles a small bright galaxy. Nearby SE of this is a moderately bright star. A number of faint stars are involved, especially near center.

| ga | NGC 613 | 01 34.3 | −29 24 | Scl | 10.7b | 5.5' | 4.1' | Sbc |

HH: very bright, very large, very much extended WNW-ESE, suddenly brighter middle, star NE.
42x60: Visible next to a star. ***170x250:*** Bright. Sharply pretty concentrated. Edge-on WNW-ESE. Moderately large. No involved stars. A little ways NE is a bright star.

| ga | NGC 615 | 01 35.1 | −07 20 | Cet | 12.5b | 3.6' | 1.4' | Sb |

HH: pretty bright, pretty large, irregularly little extended, gradually much brighter middle, resolvable, 8th mag. star 10' NW.
170x250: Moderate brightness to moderately bright. Sharply pretty concentrated. Slightly elongated NNW-SSE. Small. No involved stars.

| oc | TR 1 | 01 35.7 | +61 17 | Cas | 8.1 | 4.5' | | |

12x35: In the position is a star. *42x60:* Two or three stars. *170x250:* A number of mixed brightness stars: a very short NE-SW line of four evenly spaced moderate brightness stars, a second line of three stars S of the first line and at a slight angle to it, and more faint to threshold stars, especially around the first line of stars. Very small. Moderate density. [CR15]

| ds | Tau Scl | 01 36.1 | −29 54 | Scl | 0.8" | 6.0 | 7.1 | F2 − |

170x250: Not split. [HJ 3447]

| ga | NGC 628/M74 | 01 36.7 | +15 47 | Psc | 10.0b | 10.5' | 9.5' | Sc |

JH: globular cluster, faint, very large, round, very gradually then pretty suddenly much brighter middle, partially resolved.
42x60: Visible. *170x250:* Moderate brightness. Evenly pretty concentrated. Roughly round. Large: a quarter of the field in size. It has a very faint outer halo and a brighter core. Halfway to the galaxy's edge ENE is an extremely faint involved star; off its edge E and W are several stars.

| oc | NGC 609 | 01 37.2 | +64 34 | Cas | 11.0 | 3.0' | | |

dA: cluster, small, pretty rich, 14th mag. and fainter stars.
170x250: A very faint cloud of a lot of threshold stars; best with averted vision. Small. Dense to very dense. Near its edge SW is a wide unequal pair of stars; SE is a bright star; farther SSE is another bright star.

| ga | NGC 636 | 01 39.1 | −07 30 | Cet | 12.4b | 2.8' | 2.1' | E3 |

HH: pretty bright, very small, round, much brighter middle, resolvable.
170x250: Moderately bright. Sharply very concentrated. Round. Very small. No involved stars.

| ga | NGC 642 | 01 39.1 | −29 54 | Scl | 12.9v | 2.0' | 1.1' | Sc |

JH: very faint, pretty small, round, gradually brighter middle, star E, near.
170x250: Very very faint. Broadly very slightly concentrated. Roughly round. Small. At its edge SE is a moderately faint star that makes a long shallow arc with two moderate brightness stars, one NE and one S. Nearby WSW is a smaller, extremely faint galaxy [*639 (JH)]. Elongated to edge-on NNE-SSW.

0 to 6 Hours: FALL

ga	UGC 1183	01 40.5	+02 40	Psc	12.7	1.2'	0.6'	S

170x250: Not visible.

ds	STF 147	01 41.7	−11 19	Cet	1.0"	6.1	7.4	F5 −

170x250: Not split.

ga	NGC 658	01 42.2	+12 36	Psc	13.1b	3.0'	1.5'	Sb

St: pretty faint, pretty small, much extended, much brighter middle.
170x250: Faint. Broadly to evenly moderately concentrated. Round. Small. No involved stars. It's just off a line between a moderately bright star NNE and a moderate brightness star SW.

pn	NGC 650/M76	01 42.3	+51 34	Per	12.2p	2.8'		

HH: very bright, W of double nebula.
12x35: Very very faintly visible. *42x60:* Faintly visible. *170x250:* Moderately bright. No central star. It's two-lobed NE-SW [651, 650], with very faint nebulosity perpendicular to the axis of the lobes. On either side of middle is a dark indentation; otherwise the object is pretty much rectangular. No blinking effect. Pretty large for a planetary. Somewhat like a miniature M27, except that it's more rectangular. An unconventional planetary. [Little Dumbbell Nebula]

oc	NGC 637	01 42.9	+64 00	Cas	8.2	3.5'		

WH: cluster, pretty small, bright and very faint stars.
12x35: A handful of stars. *42x60:* A little triangular group of stars. *170x250:* Mixed brightness stars: a half-dozen moderately bright stars in a triangular figure with a concentration of fainter stars at the SW corner. A little less than a quarter of the field in size. Moderate density.

ga	NGC 660	01 43.0	+13 38	Psc	12.0b	8.3'	3.1'	Sa

WH: pretty bright, pretty large, extended, brighter middle, resolvable.
170x250: Faint. Broadly slightly concentrated. Very elongated NE-SW. Moderate size. No involved stars. Nearby ESE is a very faint star with an extremely faint companion N.

| oc | NGC 654 | 01 44.1 | +61 53 | Cas | 6.5 | 5.0' | | |

HH: cluster, irregular figure, rich, one 6th–7th mag. star, 11th to 14th mag. stars.
12x35: Faintly visible. *42x60:* A little cloud of stars. *170x250:* A fair number of moderately faint and faint stars. Square. The bright plotted star next to it SSE doesn't look like a member. A quarter of the field in size. Moderately dense.

| bs | Tau Cet | 01 44.1 | –15 56 | Cet | 3.50 | | 0.72 | G8 |

170x250: Yellow white.

| oc | NGC 659 | 01 44.2 | +60 42 | Cas | 7.9 | 5.0' | | |

WH: cluster, small, little rich, bright stars.
12x35: Very faintly visible. *42x60:* A faint little group of stars. *170x250:* A small pentagon-shaped group of moderately faint stars within a slightly larger group of fainter stars. Somewhat triangular overall. A quarter of the field in size. Somewhat sparse.

| ga | NGC 661 | 01 44.2 | +28 42 | Tri | 13.2b | 1.7' | 1.4' | E+ |

HH: faint, small, round, brighter middle, resolvable.
170x250: Moderately bright. Sharply very concentrated. Round. Very small. No involved stars. Nearby S are two very faint stars; nearby NE are two more very faint stars.

| ga | NGC 665 | 01 44.9 | +10 25 | Psc | 13.2b | 2.4' | 1.6' | S0 |

WH: faint, small, little extended, brighter middle, resolvable.
170x250: Faint. Evenly moderately concentrated. Round. Small. No involved stars. It makes a right triangle with two faint stars, one S and one E.

| oc | NGC 663 | 01 46.0 | +61 15 | Cas | 7.1 | 16.0' | | |

WH: cluster, bright, large, extremely rich, pretty bright stars.
12x35: A number of faint stars. *42x60:* Four moderately faint stars in two wide pairs, plus fainter stars. *170x250:* A number of moderately bright and moderate brightness stars including the brightest four, plus numerous moderately faint stars. It concentrates toward the middle; the outer parts are sparse and scattered. Elongated NW-SE. It takes up the entire field, but the main concentration is a little less than half the field in size. Moderate density overall.

| ga | IC 1727 | 01 47.5 | +27 19 | Tri | 12.1b | 6.9' | 3.0' | Sm |

[See 672.]

0 to 6 Hours: FALL

| ga | NGC 672 | 01 47.9 | +27 25 | Tri | 11.5b | 7.3' | 2.5' | Scd |

HH: faint, extended, between 2 stars.
170x250: Moderately faint. Broadly very slightly concentrated. Edge-on ENE-WSW. Large: a quarter of the field in size. No involved stars. The galaxy fits comfortably inside a sharp E-W triangle of moderately faint stars. Half the field SW is I1727. Very very faint. Diffuse. Very elongated NW-SE, at an angle to 672. Large. No involved stars. A little ways E are a couple of moderately faint stars; S is a moderately bright star.

| ga | NGC 673 | 01 48.4 | +11 31 | Ari | 13.2b | 2.1' | 1.6' | Sc |

WH: faint, pretty large, extended, little brighter middle, star 2' NE.
170x250: Very faint to faint. Broadly slightly concentrated. Round. Moderately small. No involved stars. It makes a triangle with a moderate brightness star ENE and a faint star SSE.

| ga | NGC 676 | 01 49.0 | +05 54 | Psc | 10.4 | 4.0' | 1.2' | S0/a |

HH: very faint, very much extended NNW-SSE, suddenly brighter middle to a 9th mag. star.
12x35, 42x60: Visible as a faint star. *170x250:* Moderate brightness. Slightly concentrated. Edge-on N-S. Moderate size. The ends are very faint. A bright involved star in the center interferes; a little ways W is a faint star; N is a moderate brightness star; ESE is another moderate brightness star.

| ga | NGC 677 | 01 49.2 | +13 03 | Ari | 13.2b | 2.0' | 2.0' | E |

Sw: most extremely faint, small, round, NE of 2.
170x250: Faint. Sharply pretty concentrated. Round. Very small. No involved stars. Nearby WNW is a faint star. [675 (Sw) was not seen.]

| ga | NGC 681 | 01 49.2 | –10 25 | Cet | 12.8b | 2.5' | 1.5' | Sab |

HH: pretty faint, considerably large, round, gradually little brighter middle, faint star 90" W.
170x250: Faint. Broadly to evenly moderately concentrated. Slightly elongated ENE-WSW. Small. No involved stars. At its edge NW is a faint star; this is the closest of several moderately faint to faint stars in a pentagonal figure next to the galaxy SW.

| ga | UGCA 21 | 01 49.2 | –10 03 | Cet | 12.7 | 2.2' | 1.2' | Sd |

170x250: Very faint. An unconcentrated glow. Roughly round. Moderately small. No involved stars. A little ways ESE is a moderately faint star.

| ds | STF 162 | 01 49.3 | +47 54 | Per | 2.0" | 6.5 | 7.0 | A3 – |

42x60: The primary has a very faint companion. *170x250:* The primary has a somewhat wide much fainter companion and is itself an extremely close very slightly unequal pair. Both stars of the close pair are white. [AB-C: 19.2", 6.5, 8.4; AB: 2.0", 6.5, 7.0]

| ga | UGC 1281 | 01 49.5 | +32 35 | Tri | 12.9p | 4.5' | 0.5' | Sdm |

170x250: Exceedingly faint. The very slightest diffuse glow. Moderate size. No involved stars. Along its E side is a slightly crooked NNE-SSW line of four evenly spaced moderately faint stars. [C503-27, right next to U1281, was not seen separately; the two galaxies may have been seen together as one amorphous glow.]

| ga | NGC 680 | 01 49.8 | +21 58 | Ari | 11.9v | 2.0' | 1.6' | E+ |

WH: pretty bright, small, irregularly round, much brighter middle, 2nd of 2.
170x250: Moderate brightness. Sharply very concentrated. Round. Very small. No involved stars. A third of the field WNW is an almost identical galaxy [*678 (WH)]. Evenly to sharply pretty concentrated.

| ds | 1 Ari | 01 50.1 | +22 17 | Ari | 2.9" | 6.2 | 7.4 | K1 A6 |

170x250: The primary has a very very close fainter companion. The primary is yellowish; the companion is white. [STF 174]

| ga | NGC 693 | 01 50.5 | +06 08 | Psc | 13.2b | 2.3' | 1.0' | S0/a |

HH: pretty faint, small, extended E-W, very gradually little brighter middle, 10th mag. star NE.
170x250: Moderately faint. Broadly slightly concentrated. A fat edge-on, WNW-ESE. Small. No involved stars. Nearby ENE is a moderately bright star.

| ga | NGC 687 | 01 50.6 | +36 22 | And | 12.3v | 1.4' | 1.4' | S0 |

WH: very faint, stellar.
170x250: Moderately faint. Sharply pretty concentrated. Round. Very small. No involved stars. It's inside a triangle of stars, the NW of which has a wide companion.

| ga | NGC 691 | 01 50.7 | +21 45 | Ari | 11.4v | 3.4' | 2.2' | Sbc |

WH: faint, considerably large, very gradually little brighter middle.
170x250: Faint. Broadly slightly concentrated. Round. Small. No involved stars. Nearby NE is a moderately bright star–a close double.

| ga | IC 167 | 01 51.1 | +21 54 | Ari | 13.1v | 2.6' | 1.4' | Sc |

170x250: Exceedingly faint. The slightest diffuse glow. Small. No involved stars. A third of the field NNW is a brighter galaxy [*694 (dA)]. Faint. Sharply fairly concentrated. Round. Very small. No involved stars. S to ESE are three moderate brightness to moderately faint stars.

| ga | NGC 701 | 01 51.1 | −09 42 | Cet | 12.8b | 2.4' | 1.1' | Sc |

HH: faint, pretty large, extended, very gradually very little brighter middle, resolvable.
170x250: Faint. Broadly slightly concentrated. Elongated NE-SW. Moderate size. No involved stars

| ga | NGC 697 | 01 51.3 | +22 21 | Ari | 12.8b | 4.4' | 1.4' | Sc |

WH: faint, considerably large, extended, much brighter middle.
170x250: Moderate brightness. Broadly moderately concentrated. A fat edge-on, WNW-ESE. Moderately small to moderate size. No involved stars. Nearby E are two moderate brightness stars; these are part of a large houselike N-S figure of stars E of the galaxy.

| ga | NGC 706 | 01 51.8 | +06 17 | Psc | 13.2b | 1.8' | 1.3' | Sbc |

HH: faint, small, brighter middle, 13th mag. star 1' N.
170x250: Moderately faint. Broadly slightly concentrated. Roughly round. Small. No involved stars. Off its edge N is a moderate brightness star.

| oc | IC 166 | 01 52.4 | +61 51 | Cas | 11.7 | 4.5' | | |

42x60, 170x250: There's no cluster in the position. [It's a cluster of very faint stars.]

| ga | NGC 703 | 01 52.7 | +36 10 | And | 13.2v | 1.2' | 0.9' | S0− |

HH: very faint, stellar, 1st of 4.
170x250: In the position are four galaxies in a small triangular group [703, *704 (WH), *705 (WH), *708 (HH)]. All are very faint to faint. Sharply concentrated. Round. Very small. No involved stars. The brightest and most concentrated is the easternmost [708]. Between the two southernmost galaxies [704, 705] is a very faint star; N of the group is a moderately bright star; NE is a moderate brightness star. A third of the field SSE is an extremely faint galaxy [*710 (LR)]. Diffuse. Small. [709 (LR) was not seen.]

ga	UGC 1347	01 52.8	+36 37	And	12.8v	1.3'	1.3'	Sc

170x250: Not visible.

ga	NGC 720	01 53.0	−13 44	Cet	11.2b	4.6'	2.3'	E5

HH: considerably bright, pretty large, little extended, pretty suddenly much brighter middle.
42x60: Visible. *170x250:* Bright. Sharply pretty concentrated. Slightly elongated NW-SE. Moderately small. No involved stars.

bs	Alpha Tri	01 53.1	+29 34	Tri	3.41		0.49	F6

170x250: Soft white.

ga	NGC 718	01 53.2	+04 11	Psc	12.6b	2.3'	2.0'	Sa

HH: pretty bright, small, irregularly round, pretty suddenly much brighter middle.
170x250: Moderately bright. Sharply very concentrated to a substellar nucleus. Round. Small. No involved stars.

ds	Gamma Ari	01 53.5	+19 18	Ari	7.8"	4.8	4.8	A1 B9

42x60: A very close perfectly equal pair; nicely split. *170x250:* A close perfectly equal pair. Both white. A very nice double in both scopes. [STF 180]

ga	NGC 714	01 53.5	+36 13	And	13.1v	1.6'	0.4'	S0/a

LR: faint, very small, round, 2 13th mag. stars W and NW.
170x250: Faint to moderately faint. Sharply pretty concentrated. Slightly elongated WNW-ESE. Very small. No involved stars. Making a very small flat triangle with it are two faint stars W. [717 (LR) was not seen.]

ga	NGC 723	01 53.8	−23 45	Cet	13.2p	1.4'	1.2'	Sbc

HH: pretty faint, very small, round, very gradually brighter middle.
170x250: Moderately faint. Broadly slightly concentrated. Round. Very small. No involved stars. A little ways S is a moderate brightness star.

bs	Epsilon Cas	01 54.4	+63 40	Cas	3.37		−0.14	B2

170x250: White.

| bs | Beta Ari | 01 54.6 | +20 48 | Ari | 2.64 | | 0.13 | A5 |

170x250: White.

| ga | NGC 731 | 01 54.9 | −09 00 | Cet | 13.0b | 1.7' | 1.7' | E+ |

WH: extremely faint, stellar.
170x250: Moderate brightness to moderately bright. Sharply pretty concentrated. Round. Very small. No involved stars.

| ga | IC 171 | 01 55.2 | +35 16 | Tri | 13.2p | 2.5' | 2.1' | E |

170x250: Moderately faint. Evenly moderately concentrated. Round. Small. Just off its edge ENE is a bright star that makes a triangle with a moderately faint star NNW and a double star W.

| ga | NGC 741 | 01 56.4 | +05 37 | Psc | 12.2b | 2.9' | 2.8' | E0 |

HH: pretty faint, small, round, W of 2.
170x250: Moderate brightness. Evenly moderately concentrated. Roughly round. Small. E of center is a faint stellar companion galaxy [*742 (HH)] that at first sight looks like an involved star. The larger galaxy makes a shallow arc with a moderate brightness star NW and another star farther NNW.

| ga | NGC 755 | 01 56.4 | −09 03 | Cet | 13.1b | 3.4' | 1.1' | Sb |

HH: very faint, pretty small, very little extended.
170x250: Faint. Broadly slightly concentrated. Elongated NE-SW. Moderate size. No involved stars.

| ga | NGC 736 | 01 56.7 | +33 02 | Tri | 12.1v | 1.8' | 1.5' | E+ |

JH: pretty bright, round, brighter middle, 13th mag. star NW.
170x250: Moderate brightness. Sharply fairly concentrated. Round. Very small. N of center is a faint involved star. [738 (LR) and 740 (LR) were not seen.]

| ga | NGC 750 | 01 57.5 | +33 12 | Tri | 12.9p | 1.5' | 1.2' | E |

HH: faint, pretty large, much extended, resolvable, E of 2.
170x250: Moderate brightness. Evenly moderately concentrated. Slightly elongated N-S. Small. No involved stars. [751 (LR), completely merged with 750, was not seen as a separate galaxy.]

| pn | IC 1747 | 01 57.6 | +63 19 | Cas | 13.6p | 13.0" | | |

170x250: Moderately faint to moderate brightness. No central star. A very small disk. Moderate blinking effect. Round. Nearby NNW is a faint star; nearby SW is another faint star.

| ga | NGC 753 | 01 57.7 | +35 54 | And | 13.0b | 3.0' | 1.9' | Sbc |

dA: pretty bright, pretty large, round, gradually much brighter middle.
170x250: Faint. Evenly moderately concentrated. Roughly round. Small. No involved stars.

| oc | NGC 752 | 01 57.8 | +37 51 | And | 5.7 | 49.0' | | |

HH: cluster, very very large, rich, scattered bright stars.
12x35: A good number of moderately faint and faint stars. SW to W of it are the bright plotted stars in a shallow arc, including 56 And; this is a foreground group. *42x60:* A fair number of moderate brightness and moderately faint stars. 2/3 of the field in size. Moderate density.

| oc | NGC 744 | 01 58.4 | +55 29 | Per | 7.9 | 11.0' | | |

JH: cluster, pretty large, pretty rich, irregular figure, 11th to 13th mag. stars.
12x35: Visible. *42x60:* A few faint stars. *170x250:* A little quadrilateral figure of some dozen moderate brightness stars. Small. Somewhat sparse.

| ga | NGC 772 | 01 59.3 | +19 00 | Ari | 11.1b | 7.2' | 4.2' | Sb |

HH: bright, considerably large, round, gradually brighter middle, resolvable.
42x60: Very faintly visible. *170x250:* Moderately faint to moderate brightness. Evenly fairly concentrated. Elongated WNW-ESE. Large: a quarter of the field in size. It has a faint outer halo and a brighter core. No involved stars. [770 (LR) was not seen.]†

| ga | NGC 779 | 01 59.7 | −05 57 | Cet | 12.0b | 4.0' | 1.1' | Sb |

HH: considerably bright, large, much extended NNW-SSE, much brighter middle.
170x250: Very bright. Sharply pretty concentrated. Edge-on NNW-SSE. Moderate size. No involved stars.

| ga | NGC 776 | 01 59.9 | +23 38 | Ari | 13.2p | 1.7' | 1.6' | Sb |

dA: faint, pretty large.
170x250: Faint. Broadly slightly concentrated. Roughly round. Small. No involved stars. A sixth of the field SSE is a smaller and fainter galaxy [*I180].

| ga | NGC 777 | 02 00.2 | +31 25 | Tri | 12.5b | 2.4' | 1.9' | E1 |

HH: pretty bright, pretty large, round, gradually little brighter middle.
170x250: Moderate brightness. Evenly to sharply pretty concentrated. Round. Small. No involved stars. A little less than half the field S is a fainter galaxy [*778 (St)]. Slightly concentrated. Round. Small. No involved stars. It's inside a large trapezoidal figure of bright stars.

| ga | NGC 783 | 02 01.1 | +31 52 | Tri | 12.8p | 1.7' | 1.3' | Sc |

St: extremely faint, small, irregularly round, very faint stars attached.
170x250: Moderately faint. Broadly slightly concentrated. Elongated E-W. Small. At its edge W is a moderately faint star; nearby ESE is another moderately faint star. Half the field ESE is a small faint galaxy [*785 (St)]. Concentrated. Round. No involved stars. It's between two moderately bright stars, one SSE and one NNW.

| ga | NGC 788 | 02 01.1 | –06 48 | Cet | 13.0b | 1.6' | 1.4' | S0/a |

HH: pretty faint, pretty small, round, brighter middle.
170x250: Moderately faint to moderate brightness. Evenly fairly concentrated. Round. Small. No involved stars.

| ga | NGC 784 | 02 01.3 | +28 50 | Tri | 12.2b | 7.3' | 1.5' | Sdm |

dA: very faint, large, extended.
170x250: Very faint. Hardly concentrated. Elongated N-S. Moderate size. No involved stars. Nearby are two moderately faint stars, one on either side of center E and NNW; just NE of the galaxy's N end is a very faint star.

| ga | NGC 799 | 02 02.2 | –00 06 | Cet | 13.0v | 2.0' | 1.7' | Sa |

Sw: most extremely faint, pretty small, round, N of 2.
170x250: In the position are two galaxies next to each other N-S [799, *800 (Sw)]. Both are roughly round. Small. No involved stars. The N galaxy is extremely faint. Slightly concentrated. The S galaxy is exceedingly faint. Hardly concentrated.

| ga | NGC 803 | 02 03.8 | +16 01 | Ari | 13.2b | 4.3' | 2.0' | Sc |

HH: very faint, small, irregularly round, gradually little brighter middle, 10th mag. star W.
170x250: Very faint. Broadly very slightly concentrated. Roughly round. Small. No involved stars. Off its edge WSW is a moderate brightness star.

bs	Gamma And	02 03.9	+42 20	And	2.26		1.37	K3

42x60: The primary has a fainter companion; nicely split. ***170x250:*** A moderately close unequal pair. Color contrast: the primary is yellow gold; the companion is white. A beautiful double. [STF 205: 9.8", 2.3, 4.8, K3 B9]

ga	NGC 812	02 06.9	+44 34	And	12.2p	2.7'	1.0'	S

St: extremely faint, pretty large, extended NE-SW, brighter middle.
170x250: Faint. Broadly very slightly concentrated. Edge-on NNW-SSE. Moderately small. No involved stars. Nearby SW is a relatively bright star that interferes a little bit.

bs	Alpha Ari	02 07.2	+23 27	Ari	2.00		1.15	K2

170x250: Yellow.

ga	NGC 821	02 08.4	+10 59	Ari	11.7b	2.5'	1.5'	E6

HH: bright, very small, very little extended, suddenly very much brighter middle, 10th mag. star 55" NW.
170x250: Moderate brightness. Sharply pretty concentrated. Round. Very small. No involved stars. Nearby NW is a moderately bright star; a little farther S is a moderate brightness star.

ga	NGC 818	02 08.7	+38 46	And	13.2b	3.0'	1.2'	Sc

HH: pretty bright, considerably large, little extended, much brighter middle.
170x250: Faint. Broadly slightly concentrated. Slightly elongated E-W. Moderately small. No involved stars. Nearby SE is a very faint star.

ga	ESO 478-6	02 09.3	−23 24	Cet	13.2p	1.8'	0.9'	Sbc

170x250: Moderately faint. Broadly moderately concentrated. Round. Small. No involved stars. In line with it is a pair of faint stars a little ways E; nearby SW is a very faint star; WNW is a brighter star.

ga	NGC 833	02 09.3	−10 07	Cet	12.7v	1.4'	0.8'	Sa

HH: faint, small, round, 1st of 4.
[See 835.]

| ga | NGC 835 | 02 09.4 | −10 08 | Cet | 12.1v | 1.4' | 1.1' | Sab |

HH: faint, small, round, 2nd of 4.
170x250: In the position are four very similar galaxies [833, 835, 838, 839] making an arc around a moderately bright star NNW to E. All are moderately faint. Sharply pretty concentrated. Round. Very small. No involved stars. 3/4 of a field SE is another galaxy [848]. Faint. Broadly slightly concentrated. Slightly elongated NW-SE. [Very small.] No involved stars. Nearby NNE is a moderate brightness star.

| bs | Beta Tri | 02 09.5 | +34 59 | Tri | 3.00 | | 0.14 | A5 |

170x250: White.

| ga | IC 206 | 02 09.5 | −06 58 | Cet | 13.1 | 1.0' | 0.4' | S |

[See I207.]

| ga | NGC 838 | 02 09.6 | −10 08 | Cet | 13.1v | 1.2' | 0.9' | S0 |

HH: very faint, very small, round, 3rd of 4.
[See 835.]

| ga | IC 207 | 02 09.7 | −06 55 | Cet | 12.8 | 1.8' | 0.5' | S0+ |

170x250: I207 and I206 are a little more than a sixth of the field apart NE-SW. Both are exceedingly faint. Hardly concentrated. Very small. No involved stars. I207 is elongated E-W; I206 is elongated NW-SE. NNW of both galaxies is a large nearly isosceles triangle of moderate brightness stars.

| ga | NGC 839 | 02 09.7 | −10 11 | Cet | 13.1v | 1.6' | 0.7' | S0 |

HH: very faint, pretty small, round, 4th of 4.
[See 835.]

| ga | IC 1783 | 02 10.1 | −32 56 | For | 13.2b | 1.9' | 0.7' | Sb |

170x250: Moderately faint. Evenly moderately concentrated. Edge-on N-S. Small. No involved stars. A little ways WSW is a faint star.

| ga | NGC 828 | 02 10.2 | +39 11 | And | 13.2b | 2.8' | 2.1' | Sa |

HH: pretty bright, small, irregularly round, double star E.
170x250: Faint. Broadly slightly concentrated. Round. Very small. No involved stars. A little ways ESE is a small sharp triangle of stars.

| ga | NGC 848 | 02 10.3 | −10 19 | Cet | 13.0v | 1.5' | 1.0' | Sab |

Sw: most extremely faint, pretty large, very difficult, star NE.
[See 835.]

| ds | 59 And | 02 10.9 | +39 02 | And | 16.8" | 6.1 | 6.8 | B9 A1 |

42x60: A slightly unequal pair. *170x250:* A moderate distance very slightly unequal pair. Both white. [STF 222]

| ga | NGC 846 | 02 12.2 | +44 34 | And | 13.0p | 1.9' | 1.4' | Sab |

St: extremely faint, very small, round, gradually brighter middle.
170x250: Faint. Broadly very slightly concentrated. Roughly round. Small. No involved stars. Off its edge NE is a faint star; off its edge NW is another faint star.

| ds | 6 Tri | 02 12.4 | +30 18 | Tri | 3.9" | 5.3 | 6.9 | G5 F6 |

170x250: A very close unequal pair. Color contrast: the primary is yellow; the companion is white. [STF 227]

| ds | 66 Cet | 02 12.8 | −02 24 | Cet | 16.6" | 5.7 | 7.5 | F8 G4 |

42x60: Well separated. *170x250:* A wide unequal pair. Both white. [STF 231]

| ga | NGC 864 | 02 15.5 | +06 00 | Cet | 11.4b | 4.7' | 3.5' | Sc |

HH: extremely faint, considerably large, round, gradually brighter middle, 12th mag. star attached SE.
170x250: Faint. Broadly very slightly concentrated. Roughly round. Moderately small. At its edge E is a relatively bright star; further E is a NNE-SSW curved line of three stars.

| oc | ST 2 | 02 15.6 | +59 32 | Cas | 4.4 | 60.0' | | |

12x35: A lot of moderately faint stars. *42x60:* A lot of moderate brightness stars of uniform brightness. Somewhat triangular, with a large empty space on the N side. It fills the field. Moderate density.

| ga | IC 1788 | 02 15.8 | −31 12 | For | 12.3v | 2.6' | 1.1' | Sbc |

170x250: Faint. Broadly slightly concentrated. Edge-on NNE-SSW. Moderately small. No involved stars. Nearby SW is a very faint star.

| ga | NGC 873 | 02 16.5 | −11 20 | Cet | 13.2 | 1.5' | 1.1' | Sc |

HH: faint, pretty large, round, very gradually little brighter middle.
170x250: Faint. Broadly slightly concentrated. Roughly round. Small. No involved stars.

| ga | NGC 877 | 02 18.0 | +14 32 | Ari | 12.6b | 2.4' | 1.8' | Sbc |

HH: pretty faint, pretty large, little extended, pretty gradually brighter middle, 9th mag. star 5' S, faint star 1' SE.
170x250: Faint. Broadly very slightly concentrated. Slightly elongated NW-SE. Moderately small. No involved stars. Just off its SE end is a moderately faint star; SSE is the bright plotted star. [876 (LR) was not seen.]

| ga | NGC 881 | 02 18.8 | −06 38 | Cet | 13.2b | 2.2' | 1.4' | Sc |

HH: faint, pretty small, extended, brighter middle, 2 or 3 stars near.
170x250: Faint. Broadly slightly concentrated. Round. Very small. No involved stars. Nearby NNE is a moderately faint star; a little ways NW is a faint star; WNW is the bright plotted star, which makes an arc with two more bright stars NE of it.

| oc | NGC 869 | 02 19.0 | +57 09 | Per | 5.3 | 29.0' | | |

HH: ! cluster, very very large, very rich, 7th to 14th mag. stars.
12x35: 869 and 884 both have a lot of bright stars. *42x60:* Both clusters have very mixed brightness stars, including a lot of bright stars. 869 is more concentrated. With the outer members each cluster is half the field in size; together they fit nicely in the same field. *170x250:* 869 has a very distinctively shaped concentration in the center: a bright yellowish star with a little arc of stars ESE of it and a larger, more irregular arc WNW of it; both arcs are concave WNW and the figure is like a broad, rounded arrow pointing ESE. A little ways NNE is another bright yellowish star. These two bright stars, along with the group attending the first, are the heart of the cluster. NW is a second, looser concentration. The brightest stars of 884 are in a ring; the cluster is somewhat empty in the middle. The highest concentration of stars is a rich grouping in two clumps at the SW edge of the main body of the cluster. Both clusters have very indistinct boundaries, but with each of them the main part, including most of the bright stars, fits in one field. 869 is very dense. 884 is pretty dense. [Double Cluster]

| ga | NGC 883 | 02 19.1 | –06 47 | Cet | 13.2 | 1.6' | 1.2' | S0– |

HH: pretty faint, pretty small, very little extended, brighter middle, double star near.
170x250: Moderately faint. Sharply pretty concentrated. Round. Very small. No involved stars. Nearby WNW is a faint star. Half the field SW is another galaxy [*I219]. Faint. Sharply pretty concentrated. Round. Very very small. No involved stars.

| vs | Omicron Cet | 02 19.3 | –02 59 | Cet | 3 | 9.5 | 1.42 | M5 |

naked eye: 4th mag. ***170x250:*** Yellow.

| ga | NGC 887 | 02 19.5 | –16 04 | Cet | 12.8 | 1.9' | 1.4' | Sc |

HH: faint, small, irregularly round, pretty gradually brighter middle.
170x250: Faint. Evenly moderately concentrated. Round. Very small. E of center is a faint involved star; the galaxy is on the side of a large triangle of stars.

| ga | NGC 897 | 02 21.1 | –33 43 | For | 12.8b | 2.1' | 1.2' | Sa |

JH: pretty bright, small, round, pretty suddenly brighter middle, 10th mag. star 35" E.
170x250: Moderately faint to moderate brightness. Sharply pretty concentrated. Round. Small. No involved stars. Just off its edge E is a relatively bright star.

| ga | NGC 895 | 02 21.6 | –05 31 | Cet | 12.3b | 3.6' | 2.5' | Scd |

HH: faint, very large, irregularly round, gradually brighter middle.
170x250: Very faint to faint. Broadly very slightly concentrated. Slightly elongated NW-SE. Moderate size. No involved stars. Nearby E is a very faint star.

| ga | NGC 899 | 02 21.9 | –20 49 | Cet | 13.1p | 1.8' | 1.2' | Im |

JH: pretty bright, small, gradually brighter middle, resolvable, double star W.
170x250: Moderately faint. Broadly to evenly moderately concentrated. Round. Small. No involved stars. Nearby WSW is a pair of stars. [I223 was not seen.]

| ga | NGC 890 | 02 22.0 | +33 15 | Tri | 12.2b | 2.7' | 1.8' | S0– |

HH: bright, small, round, brighter middle, 3 faint stars SW.
170x250: Moderately bright. Sharply pretty concentrated. Slightly elongated NE-SW. Small. No involved stars. A little ways W is a triangle of moderate brightness to moderately faint stars.

| oc | NGC 884 | 02 22.4 | +57 07 | Per | 6.1 | 29.0' | | |

HH: ! cluster, very large, very rich, ruby star in middle.
[See 869.]

| ga | NGC 891 | 02 22.6 | +42 21 | And | 10.8b | 14.3' | 2.4' | Sb |

HH: ! bright, very large, very much extended NNE-SSW.
42x60: Very faintly visible with an involved star. *170x250:* Moderately faint. Broadly very slightly concentrated. Edge-on NNE-SSW. Very large: a little more than half the field in size. The bisecting dust lane is visible, especially in the middle part of the galaxy; clearly seen at times with averted vision but not conspicuous; not completely steady. N of center is a moderate brightness involved star; farther NNE is a faint involved star; just before the S tip is a very faint involved star; at the S tip is a moderately faint involved star; nearby NW of center is a moderately faint star; nearby SSE of center is a pair of stars.

| ga | NGC 907 | 02 23.0 | –20 42 | Cet | 13.2b | 1.8' | 0.5' | Sdm |

HH: faint, small, extended E-W, gradually brighter middle.
170x250: Moderately faint. Broadly slightly concentrated. A fat edge-on, E-W. Small. No involved stars.

| ga | NGC 908 | 02 23.1 | –21 13 | Cet | 10.8b | 6.0' | 2.6' | Sc |

WH: considerably bright, very large, extended.
42x60: Visible. *170x250:* Moderately bright; moderate surface brightness. Broadly slightly concentrated. Elongated ENE-WSW; the galaxy is irregular in shape; somewhat rectangular. Large: a quarter of the field in size. There's some vague unevenness in the light. Just S of its E end is a very faint star; N is an E-W elongated group of stars.

| ga | NGC 922 | 02 25.1 | –24 47 | For | 12.5b | 2.0' | 1.7' | Scd |

HH: considerably faint, pretty large, round, gradually pretty much brighter middle.
170x250: Moderately faint to moderate brightness. Sharply fairly concentrated. Slightly elongated N-S. Small. No involved stars. It makes a square with three faint stars N to W.

| ga | NGC 910 | 02 25.4 | +41 49 | And | 12.1v | 1.4' | 1.4' | E+ |

WH: extremely faint, stellar.
170x250: Faint. Evenly moderately concentrated. Round. Small. No involved stars. It makes a small triangle with two faint stars, one WNW and one SSW. [912 (St) and U1866 were not seen.]

| ga | NGC 918 | 02 25.8 | +18 29 | Ari | 13.1b | 3.5' | 2.4' | Sc |

JH: pretty faint, large, round, 10th mag. star 3' SE.
170x250: Extremely faint. Hardly concentrated. Slightly elongated NNW-SSE. Moderate size. No involved stars. A little ways SSE is a moderate brightness star.

| ga | UGC 1886 | 02 26.0 | +39 28 | And | 12.7p | 3.7' | 2.0' | Sbc |

170x250: Very very faint. Broadly slightly concentrated. Round. Small. No involved stars. It's between two nearby moderate brightness stars, one WNW and one a little farther ESE.

| ga | UGCA 32 | 02 26.4 | −24 17 | For | 12.9p | 2.7' | 1.4' | Sdm |

170x250: Faint. Broadly slightly concentrated. Edge-on NE-SW. Moderate size. No involved stars. Next to it E is a sharp triangle of moderate brightness stars.

| ga | NGC 925 | 02 27.3 | +33 34 | Tri | 10.7b | 10.5' | 5.9' | Sd |

HH: considerably faint, considerably large, extended, very gradually brighter middle, 2 13th mag. stars NW.
42x60: Very faintly visible. ***170x250:*** Moderately faint; the outer parts are low surface brightness. Broadly slightly concentrated. Very elongated WNW-ESE. Large: a quarter of the field in size. Along its N and S edges are several stars.

| ga | NGC 936 | 02 27.6 | −01 09 | Cet | 11.1b | 4.7' | 4.0' | S0+ |

HH: very bright, very large, round, much brighter middle to a nucleus.
42x60: Faintly visible. ***170x250:*** Bright. Sharply pretty concentrated. Slightly elongated E-W. Moderate size. It has a faint extended halo. No involved stars.

| ga | UGC 1933 | 02 28.3 | +38 25 | And | 12.5 | 0.9' | 0.7' | S |

170x250: Not visible.

| ga | NGC 941 | 02 28.5 | −01 09 | Cet | 12.9b | 2.6' | 1.9' | Sc |

HH: very faint, considerably large, round, E of 2.
170x250: Very very faint. Hardly concentrated. Round. Moderately small. No involved stars.

| ga | NGC 947 | 02 28.5 | −19 02 | Cet | 13.2b | 2.0' | 1.0' | Sc |

JH: pretty bright, extended, gradually brighter middle.
170x250: Faint. Broadly very slightly concentrated. Round. Small. No involved stars.

| ga | NGC 945 | 02 28.6 | −10 32 | Cet | 12.8b | 2.4' | 1.9' | Sc |

HH: very faint, large, irregularly round, gradually little brighter middle.
170x250: Very faint. Broadly very slightly concentrated. Roughly round. Moderately small. No involved stars. Nearby SE is a faint star. [948 (Sw) was not seen.]

| ds | Iota Cas | 02 29.1 | +67 24 | Cas | 2.7" | 4.6 | 6.9 | A3 F5 |

170x250: The primary has a close much fainter companion. The primary is white. [STF 262]

| ga | NGC 942 | 02 29.2 | −10 50 | Cet | 12.2 | 1.0' | 0.5' | S0+ |

Mu: very faint, round, nebulous double star?
[See 943.]

| ga | NGC 943 | 02 29.2 | −10 49 | Cet | 12.4 | 1.2' | 0.6' | I0 |

Mu: very faint, round, nebulous double star?
170x250: 943 and 942 look at first sight like a single faint galaxy, slightly elongated NNW-SSE. But the "object" is binuclear: it's actually two galaxies, so close together and so faint that they're not well distinguished. Both are evenly to sharply concentrated. Together they're still small. No involved stars.

| vs | R For | 02 29.3 | −26 06 | For | 7.5 | 13.0 | − | C4 |

12x35: Mag. 8.0. *170x250:* Hardly colored.

| ga | NGC 955 | 02 30.6 | −01 06 | Cet | 12.9b | 2.7' | 0.6' | Sab |

HH: pretty bright, small, extended, pretty suddenly brighter middle.
170x250: Moderately bright. Sharply pretty concentrated. A fairly thin edge-on, NNE-SSW. Small. No involved stars. A little ways SSE is a moderate brightness star.

| ga | NGC 958 | 02 30.7 | −02 56 | Cet | 12.9b | 2.9' | 1.0' | Sc |

HH: pretty faint, irregularly little extended N-S, brighter middle.
170x250: Moderately faint to moderate brightness. Broadly slightly concentrated. Edge-on N-S. Moderately small. No involved stars. Near its N end is a faint star.

| ga | NGC 949 | 02 30.8 | +37 08 | Tri | 12.4b | 2.4' | 1.2' | Sb |

HH: considerably bright, large, extended, very gradually brighter middle.
170x250: Moderately faint. Broadly slightly concentrated. Elongated NW-SE. Small to moderately small. No involved stars. Nearby W, almost parallel to the galaxy, is a NNW-SSE line of stars.

| bs | Alpha UMi | 02 31.8 | +89 15 | UMi | 2.02 | | 0.60 | F7 |

170x250: The primary has a slightly wide relatively very faint companion. The primary is yellow white. [Polaris; STF 93: 17.8", 2.0, 9.0]

| ga | NGC 959 | 02 32.4 | +35 29 | Tri | 13.0b | 3.1' | 2.0' | Sdm |

St: extremely faint, pretty large, little extended, little brighter middle.
170x250: Very faint. Broadly very slightly concentrated. Slightly elongated ENE-WSW. Moderately small. No involved stars.

| oc | NGC 956 | 02 32.5 | +44 39 | And | 8.9 | 7.0' | | |

JH: cluster, pretty rich, 9th to 15th mag. stars.
12x35, 42x60: A little row of three stars. ***170x250:*** A slightly curved E-W row of three fairly bright stars, and another bright star SSE of these. Between the easternmost star of the row and the star S is probably the actual cluster: some ten moderately faint stars. A quarter of the field in size. Sparse.

| oc | MEL 15 | 02 32.7 | +61 27 | Cas | 6.5 | 21.0' | | |

12x35, 42x60: A handful of widely spaced fairly bright stars. ***170x250:*** Four bright stars in a large slanted "L" figure. Surrounding the corner star is a slight concentration of fainter stars a quarter of the field in size. Overall the cluster is fairly sparse. [CR26]

| oc | NGC 957 | 02 33.6 | +57 32 | Per | 7.6 | 11.0' | | |

JH: cluster, pretty large, pretty rich, 13th to 15th mag. stars.
42x60: An elongated cluster of a few faint stars and a couple of bright stars.
170x250: A long ENE-WSW cluster of a number of moderate brightness stars, with a couple of bright stars along its S edge. Half the field in size. Somewhat sparse.

| ga | NGC 989 | 02 33.8 | −16 30 | Cet | 13.2 | 0.6' | 0.6' | S0 |

Le: faint, very small, round, brighter middle to a nucleus.
170x250: Not visible.

| ds | Omega For | 02 33.8 | −28 14 | For | 10.8" | 5.0 | 7.7 | B9 A3 |

42x60: The primary has a relatively very faint companion. *170x250:* The primary has a moderate distance relatively faint companion. The primary is white. [HJ 3506]

| ga | NGC 968 | 02 34.1 | +34 28 | Tri | 13.2p | 2.7' | 1.5' | E |

St: pretty faint, pretty small, round, brighter middle.
170x250: Faint. Evenly moderately concentrated. Round. Small. No involved stars. It makes a flat triangle with a moderate brightness star a little ways ESE and a faint star N.

| ga | NGC 972 | 02 34.2 | +29 18 | Ari | 12.3b | 3.6' | 1.7' | Sab |

HH: pretty bright, considerably large, little extended, gradually much brighter middle, 3 stars S.
170x250: Moderate brightness. Broadly slightly concentrated. Elongated NNW-SSE. Small. No involved stars. It makes a small triangle with two bright stars SW; making a long arc with these bright stars are several faint stars; the arc curves around the S end of the galaxy and goes ENE.

| ga | NGC 978 | 02 34.8 | +32 50 | Tri | 13.2 | 1.4' | 0.9' | S0− |

JH: pretty bright, round, 3rd of 3.
170x250: Faint to moderately faint. Evenly moderately concentrated. Roughly round. Small. No involved stars. Half the field NW are two more galaxies. The W galaxy [*969 (JH)] is faint. Concentrated. Round. Small. No involved stars. Very nearby SSE is a faint star. The E galaxy [*974 (JH)] is very faint. Slightly concentrated. Roughly round. Small. At its edge SSW is a faint star; at its edge NNE is another faint star.

| ga | NGC 988 | 02 35.5 | −09 21 | Cet | 10.9v | 5.3' | 1.8' | Scd |

St: nebulous 7.5 mag. star.
170x250: Moderately faint to moderate brightness. Broadly slightly concentrated. Elongated WNW-ESE. Moderate size. Toward its W end, well within the galaxy, is the very bright plotted star; the galaxy is easily seen despite this star.

| ga | NGC 991 | 02 35.5 | −07 09 | Cet | 12.4p | 2.7' | 2.4' | Sc |

HH: very faint, considerably large, irregular figure, very little brighter middle.
170x250: Very faint. Hardly concentrated. Round. Moderately small. No involved stars. Off its edge S is a moderately faint star.

| ga | UGC 2069 | 02 35.6 | +37 38 | And | 13.0p | 2.3' | 1.4' | Sd |

170x250: Very faint. Roughly round. Small. At its center is a moderately faint involved star that interferes; E to SE are several moderately bright to moderate brightness stars.

| oc | K 4 | 02 36.0 | +59 01 | Cas | 10.5 | 3.0' | | |

170x250: A few faint stars. Small. Somewhat sparse to moderate density.

| ga | IC 239 | 02 36.5 | +38 58 | And | 11.8b | 4.5' | 4.1' | Scd |

170x250: Exceedingly faint. A diffuse glow. Large: a quarter of the field in size. A number of very faint stars are involved with the galaxy; at its edge S is the bright plotted star; this is the easternmost of a N-S kite-like figure of stars along the galaxy's W side: a diamond with a tail.

| ga | UGCA 34 | 02 36.6 | +59 39 | Cas | 11.4v | 1.8' | 1.3' | S0– |

170x250: Not visible. In the position is a little group of faint stars that may mask the galaxy. [The galaxy is behind these stars.]

| oc | TR 2 | 02 36.9 | +55 54 | Per | 5.9 | 20.0' | | |

12x35: An elongated cluster of a few moderately faint stars. *42x60:* A curved line of some half-dozen moderate brightness stars. *170x250:* A number of widespread moderately bright stars in an E-W linear group. The field width in size. Somewhat sparse. [CR29]

| vs | R Tri | 02 37.0 | +34 16 | Tri | 5.7 | 12.5 | 1.55 | M4 |

12x35: Mag. 7.5. *170x250:* Slightly yellowish.

| ga | NGC 1015 | 02 38.2 | –01 19 | Cet | 13.0p | 2.0' | 1.5' | Sa |

Te: very faint, small.
170x250: Moderately faint. Sharply pretty concentrated. Round. Very small. No involved stars. Covering most of the field N is a very large and narrow E-W triangular group of stars.

| ga | NGC 1034 | 02 38.2 | –15 48 | Cet | 12.5 | 1.0' | 0.7' | Irr |

Le: very faint, very small, little extended, little brighter middle, 2 bright stars W.
170x250: Faint. Evenly moderately concentrated. Round. Very very small. No involved stars.

| ga | NGC 1016 | 02 38.3 | +02 07 | Cet | 12.6b | 2.4' | 2.4' | E |

Ma: faint, small, round, pretty suddenly brighter middle.
170x250: Moderately faint. Sharply fairly concentrated. Round. Small. No involved stars. It makes a trapezoidal figure with four faint stars SE to WSW.

| ga | NGC 1022 | 02 38.5 | −06 40 | Cet | 12.1b | 2.4' | 1.9' | Sa |

HH: considerably bright, pretty large, round, much brighter middle.
170x250: Moderate brightness to moderately bright. Broadly moderately concentrated. Roughly round. Moderately small. No involved stars. A little ways NE is a faint star.

| ga | IC 1830 | 02 39.1 | −27 26 | For | 13.2b | 1.6' | 1.3' | S0+ |

170x250: Faint to moderately faint. Broadly slightly concentrated. Round. Small. No involved stars. Nearby W is a moderate brightness star; this is the nearest of five stars in a very large NNE-SSW wedge-shaped figure on the W side of the galaxy.

| ga | NGC 1024 | 02 39.2 | +10 50 | Ari | 13.1b | 4.1' | 1.4' | Sab |

HH: pretty faint, small, little extended, brighter middle, 11th mag. star 25" NE.
170x250: Faint to moderately faint. Evenly moderately concentrated. Elongated NNW-SSE. Small. Very nearby NE is a moderate brightness star that makes a large triangular figure with more stars E of the galaxy. A little ways SE of the easternmost star of this figure is a second, fainter galaxy [*1029 (Ma)]. Evenly moderately concentrated. Elongated ENE-WSW. Small. No involved stars.

| ga | NGC 1003 | 02 39.3 | +40 52 | Per | 12.0b | 5.5' | 1.8' | Scd |

HH: pretty faint, large, extended E-W, much brighter middle, resolvable.
170x250: Faint to moderately faint. Broadly very slightly concentrated. Elongated E-W. Moderate size. At its edge NE of center is a star; a little ways WSW is a bright star; farther SSW is another star.

| ga | NGC 1012 | 02 39.3 | +30 09 | Ari | 13.0p | 3.1' | 1.5' | S0/a |

HH: faint, pretty small, irregularly round, brighter middle, stars involved.
170x250: Moderately faint. Broadly slightly concentrated. Elongated NNE-SSW. Small. SE of center is a faint involved star.

ga	NGC 1032	02 39.4	+01 05	Cet	12.6b	3.3'	1.1'	S0/a

HH: pretty bright, small, very little extended, brighter middle, forms a trapezium with 3 stars.
170x250: Moderate brightness. Evenly moderately concentrated. Slightly elongated ENE-WSW. Small. No involved stars. It makes a nearly square quadrilateral with three moderately faint stars NNW to ENE.

ga	NGC 1035	02 39.5	−08 07	Cet	12.9b	2.6'	0.9'	Sc

HH: pretty faint, large, much extended, resolvable, 17th mag. star attached SE.
170x250: Moderate brightness. Broadly slightly concentrated. Edge-on NNW-SSE. Moderate size. At its SE end is a faint involved star.

gc	NGC 1049	02 39.7	−34 17	For	12.6	0.8'	−	−

JH: pretty bright, small, round, stellar.
170x250: Concentrated. Almost stellar. A little ways SSE is a faint star. Fairly bright. An interesting globular: there are probably none within the Milky Way that look like this: it's so small and evidently distant. [This globular lies outside of the Milky Way, in the Fornax dwarf galaxy.]

ga	ESO 356-4	02 39.9	−34 28	For	9.0b	20.0'	13.8'	dE3

12x35, 42x60: Not visible. Of this galaxy's globular clusters apart from 1049: FOR2, FOR4, and FOR5 are visible in the 10-inch when specifically looked for with a special chart. All are very faint at best, with no detail made out. FOR1 and FOR6 are not visible (there is no FOR3). [Fornax dwarf]

ga	NGC 1023	02 40.4	+39 03	Per	10.4b	8.7'	2.3'	S0−

HH: very bright, very large, very much extended, very very much brighter middle.
42x60: Visible. *170x250:* Bright. Sharply pretty concentrated. Very elongated E-W. Moderately large. The ends fade out very indistinctly. No involved stars. It's in an arc with two bright stars SSW and a couple of fainter stars NNW. [1023A was not seen as a separate object.]

ga	NGC 1041	02 40.4	−05 26	Cet	13.1	1.6'	1.1'	S0−

St: pretty faint, pretty small, irregularly round, brighter middle.
170x250: Very faint. Broadly to evenly slightly concentrated. Round. Very small. No involved stars. It's on the side of a large triangular figure of moderate brightness stars.

| ga | NGC 1042 | 02 40.4 | –08 26 | Cet | 11.6b | 5.3' | 3.6' | Scd |

Sw: most extremely faint, large, round, NW of 2.
170x250: Faint. Hardly concentrated. Roughly round. Moderately large. At its edge ESE is a faint star; a little ways N is another faint star. 1042 contrasts greatly with 1052.

| ga | NGC 1045 | 02 40.5 | –11 16 | Cet | 12.9 | 2.3' | 1.2' | S0– |

HH: faint, small, round, brighter middle.
170x250: Moderate brightness. Sharply pretty concentrated. Round. Very small. No involved stars.

| ga | NGC 1052 | 02 41.1 | –08 15 | Cet | 10.4v | 3.0' | 2.4' | E4 |

HH: bright, pretty large, round, much brighter middle to a 12th mag. star.
42x60: Visible. ***170x250:*** Bright. Sharply pretty concentrated. Round. Small. No involved stars. Nearby SW is a very small triangle of very faint stars.

| ga | NGC 1055 | 02 41.8 | +00 26 | Cet | 11.4b | 7.6' | 2.6' | Sb |

HH: pretty faint, considerably large, irregularly extended E-W, brighter middle, pretty bright star near.
170x250: Moderately faint to moderate brightness. Broadly slightly concentrated. Very elongated WNW-ESE. Moderate size to moderately large. The ends fade out very indistinctly. No involved stars. Very nearby NW is a moderately bright star that makes a triangle with two fainter stars further NW.

| oc | NGC 1039/M34 | 02 42.0 | +42 47 | Per | 5.2 | 35.0' | | |

JH: cluster, bright, very large, little compressed, scattered 9th mag. stars.
12x35: A fair number of stars in a square, with a concentration in the middle. ***42x60:*** A good number of bright and moderate brightness stars; some of the bright stars make a large outer square. The inner concentration is a N-S upside-down stickman-like figure with widespread eyes at the S end. Half the field in size. Moderate density. ***170x250:*** A good number of bright stars. A field and a half in size.

| oc | NGC 1027 | 02 42.7 | +61 33 | Cas | 6.7 | 20.0' | | |

WH: cluster, large, scattered stars, one 10th mag. star.
12x35, 42x60: A handful of faint stars around a brighter star. ***170x250:*** A fair number of mixed brightness stars: the bright star in the middle plus moderate brightness and fainter stars. 2/3 of the field in size. Moderate density.

| ga | NGC 1068/M77 | 02 42.7 | −00 00 | Cet | 9.6b | 7.1' | 6.0' | Sb |

JH: very bright, pretty large, irregularly round, suddenly brighter middle to a partially resolved nucleus, star 2' SE.
12x35: Visible. ***42x60:*** Visible with a faint star next to it. ***170x250:*** Very bright. Sharply pretty concentrated. Roughly round. Moderately small. At times there's some indefinite unevenness in the light; the galaxy is not quite regular. No involved stars. Just off its edge ESE is a relatively bright star.

| bs | Gamma Cet | 02 43.3 | +03 14 | Cet | 3.47 | | 0.09 | A3 |

170x250: The primary has an extremely close greatly fainter companion. The primary is white. [STF 299: 2.6", 3.5, 7.3]

| ga | NGC 1060 | 02 43.3 | +32 25 | Tri | 11.8v | 2.3' | 1.7' | S0− |

HH: faint, pretty large, round, little brighter middle, 7th–8th mag. star [ESE].
170x250: Moderately faint. Evenly moderately concentrated. Round. Small. No involved stars. A quarter of the field NNW is an extremely faint galaxy [*1057 (LR)]. Half the field ENE is a very very faint galaxy [*1066 (HH)]. Broadly very slightly concentrated. Roughly round. Small. No involved stars. [1061 (LR) was not seen.]

| ga | NGC 1070 | 02 43.4 | +04 58 | Cet | 12.7p | 2.3' | 1.9' | Sb |

HH: pretty faint, small, irregularly round, gradually brighter middle.
170x250: Faint. Broadly to evenly moderately concentrated. Round. Small. No involved stars. It makes a shallow arc with two moderate brightness stars SSW.

| ga | NGC 1058 | 02 43.5 | +37 20 | Per | 11.8b | 3.0' | 2.7' | Sc |

HH: pretty faint, considerably large, round, gradually little brighter middle.
170x250: Moderately faint. Broadly very slightly concentrated. Round. Moderately small. At its edge WNW is a very very faint involved star.

| ga | NGC 1076 | 02 43.5 | −14 45 | Cet | 13.1 | 1.9' | 1.0' | S0/a |

Sw: very faint, pretty small, round, bright star E.
170x250: Moderately faint. Evenly moderately concentrated. Slightly elongated WNW-ESE. Very small. No involved stars.

| ga | NGC 1073 | 02 43.7 | +01 22 | Cet | 11.5b | 4.9' | 4.4' | Sc |

WH: very faint, large, little brighter middle, easily resolvable.
170x250: Very faint to faint; very low surface brightness. Broadly very slightly concentrated. Roughly round. Large: a quarter of the field in size. No involved stars. WSW is a large triangle of bright stars.

| ga | NGC 1079 | 02 43.7 | −29 00 | For | 12.4b | 5.0' | 3.4' | S0/a |

JH: bright, pretty large, pretty much extended, suddenly brighter middle.
170x250: Moderately faint to moderate brightness. Sharply fairly concentrated. Slightly elongated NW-SE. Small. No involved stars.

| ga | NGC 1098 | 02 44.9 | −17 39 | Eri | 12.5v | 1.6' | 1.4' | S0− |

Le: faint, very small, round, brighter middle to a nucleus, 1st of 3.
170x250: Moderately faint. Evenly to sharply fairly concentrated. [Round.] Small. No involved stars. NNE is the bright plotted star. Half the field ESE are 1099 and another galaxy [*1100 (Le)]. Both are faint. Broadly concentrated. Roughly round. Small. No involved stars. A little more than half the field NE of 1098 are a couple of very small galaxies close together. The E galaxy [*1092 (Le)] is a little brighter. Very faint. The W galaxy [*1091 (Le)] is very very faint.

| ga | NGC 1099 | 02 45.3 | −17 42 | Eri | 13.1v | 2.5' | 0.8' | Sb |

Le: faint, pretty small, little extended, brighter middle to a nucleus, 2nd of 3.
[See 1098.]

| ga | NGC 1084 | 02 46.0 | −07 34 | Eri | 11.3b | 3.2' | 1.7' | Sc |

HH: very bright, pretty large, extended, gradually pretty much brighter middle.
170x250: Bright to very bright. Broadly slightly concentrated. Slightly elongated NNE-SSW. Moderately small. No involved stars. A somewhat exceptional galaxy: broadly concentrated yet quite bright.

| ga | NGC 1097 | 02 46.3 | −30 16 | For | 10.2b | 12.7' | 9.4' | Sb |

HH: very bright, large, very much extended NNW-SSE, very much brighter middle to a nucleus.
42x60: Visible. ***170x250:*** Bright. Sharply very concentrated to a small bright core. Very elongated to edge-on NNW-SSE. Large: a quarter of the field in size. No involved stars. Off its NW end is a small faint galaxy [*1097A].

| ga | NGC 1085 | 02 46.4 | +03 36 | Cet | 13.1p | 2.9' | 2.0' | Sbc |

dA: faint, small, round, little brighter middle, between 2 stars.
170x250: Faint. Evenly slightly concentrated. Round. Small. No involved stars.

| ga | NGC 1087 | 02 46.4 | −00 30 | Cet | 11.5b | 3.7' | 2.2' | Sc |

HH: pretty bright, considerably large, little extended, much brighter middle.
170x250: Moderate brightness to moderately bright. Broadly very slightly concentrated. Slightly elongated N-S. Moderately small to moderate size. No involved stars. It makes a triangle with two moderate brightness stars, one NE and one ESE.

| ga | NGC 1090 | 02 46.6 | −00 14 | Cet | 12.5b | 4.0' | 1.7' | Sbc |

HH: very faint, pretty large, irregularly round, brighter middle.
170x250: Very faint. Hardly concentrated. Slightly elongated WNW-ESE. Moderately small to moderate size. No involved stars.

| ga | UGC 2296 | 02 49.2 | +18 20 | Ari | 13.1 | 0.6' | 0.6' | S |

170x250: Not visible.

| oc | IC 1848 | 02 51.2 | +60 26 | Cas | 6.5 | 12.0' | | |

12x35, 42x60: A couple of moderately bright stars. ***170x250:*** A couple of fairly bright stars, each with a very small moderately dense group of faint stars surrounding it.

| ga | NGC 1122 | 02 52.9 | +42 12 | Per | 12.9p | 2.1' | 1.5' | Sb |

Sw: very faint, pretty small, round, star near N.
170x250: Faint. Broadly slightly concentrated. Roughly round. Small. ESE of center is a very faint involved star; nearby W are a few faint stars; a little ways N is a moderate brightness star.

| ga | IC 1864 | 02 53.7 | −34 11 | For | 12.6v | 1.1' | 0.6' | E |

170x250: Moderately faint; high surface brightness. Sharply very concentrated. Round. Tiny. No involved stars.

| ga | NGC 1134 | 02 53.7 | +13 00 | Ari | 13.1p | 2.5' | 0.9' | S |

WH: faint, small, irregularly round, resolvable.

170x250: Moderately faint. Broadly slightly concentrated. Slightly elongated NW-SE. Small. Very nearby ENE is a moderately faint star. A little more than half the field S is I267. Very very faint. Unconcentrated. Elongated NNW-SSE. Moderately small. No involved stars.

| ga | IC 267 | 02 53.8 | +12 50 | Ari | 12.9v | 2.1' | 1.6' | Sb |

[See 1134.]

| ga | NGC 1137 | 02 54.0 | +02 57 | Cet | 13.2p | 2.1' | 1.2' | Sb |

Sw: very faint, pretty small, round, little brighter middle.
170x250: Faint. Evenly slightly concentrated. Round. Small. No involved stars. A little ways SSW is a faint star.

| ga | NGC 1140 | 02 54.6 | −10 01 | Eri | 12.8b | 1.6' | 0.8' | Im |

HH: [pretty bright], small, round, stellar.
170x250: Moderately bright; fairly high surface brightness. Sharply very concentrated. Slightly elongated N-S. Very small. No involved stars.

| ga | MCG-2-8-33 | 02 57.8 | −10 10 | Eri | 13.2 | 2.7' | 0.7' | S0/a |

170x250: Very faint. Evenly moderately concentrated. Slightly elongated NW-SE. Very very small. No involved stars.

| ga | IC 1870 | 02 57.9 | −02 20 | Eri | 13.1 | 2.7' | 1.5' | Sm |

170x250: Very faint. Hardly concentrated. Roughly round. Very small. No involved stars. It makes a small triangle with a moderate brightness star NNW and the bright plotted star ENE, which interferes.

| ga | NGC 1156 | 02 59.7 | +25 14 | Ari | 12.3b | 3.3' | 2.4' | Im |

WH: pretty bright, considerably large, pretty much extended N-S, resolvable, star 1' N.
170x250: Moderately faint. Broadly slightly concentrated. Elongated NNE-SSW. Small. At its edge N is a moderate brightness star.

| ds | STF 331 | 03 00.9 | +52 21 | Per | 12.0" | 5.3 | 6.7 | B7 B9 |

42x60: A fairly close unequal pair. A nice double. *170x250:* A somewhat wide unequal pair. Both white.

| ga | NGC 1161 | 03 01.2 | +44 53 | Per | 12.1b | 2.8' | 2.0' | S0 |

HH: faint, pretty small, little extended, suddenly brighter middle, double star W.
170x250: Moderate brightness to moderately bright. Evenly moderately concentrated. Round. Very small. No involved stars. It makes a little arc with two moderately bright stars W; a little ways E are two moderate brightness stars in line with the galaxy. A little less than a quarter of the field N is a slightly larger galaxy [*1160 (WH)]. Faint. Hardly concentrated. Slightly elongated NE-SW. No involved stars. A little ways N is a very small arc of three stars.

| ga | NGC 1172 | 03 01.6 | −14 50 | Eri | 12.7b | 2.3' | 1.7' | E+ |

HH: pretty faint, pretty large, round, pretty suddenly brighter middle.
170x250: Moderately faint. Evenly moderately concentrated. Round. Very small. No involved stars. It's just inside a right triangle of bright to moderate brightness stars.

| bs | Alpha Cet | 03 02.3 | +04 05 | Cet | 2.53 | | 1.64 | M2 |

170x250: Gold.

| ga | NGC 1179 | 03 02.6 | −18 53 | Eri | 12.6b | 4.9' | 3.8' | Scd |

Sn: extremely faint, pretty small, gradually brighter middle, 12th mag. star 1' E.
170x250: Exceedingly faint. The slightest diffuse glow. Moderately large. ESE of center is a faint involved star.

| ga | NGC 1187 | 03 02.6 | −22 52 | Eri | 11.3b | 5.5' | 4.0' | Sc |

HH: pretty faint, considerably large, pretty much extended, gradually brighter middle to a 16th mag. star, resolvable.
170x250: Moderately faint. Broadly to evenly slightly concentrated. Slightly elongated NW-SE. Moderately large. In or near the middle are a couple of extremely faint involved stars, one of which may actually be a stellar nucleus. [E480-20 was not seen.]

| ga | NGC 1169 | 03 03.6 | +46 23 | Per | 12.2b | 4.2' | 2.8' | Sb |

HH: pretty faint, pretty small, irregular figure, suddenly brighter middle.
170x250: Moderate brightness. Sharply pretty concentrated. Roughly round. Small. No involved stars. It makes a small flat triangle with a faint star E and a very faint star SSW.

| ga | NGC 1199 | 03 03.6 | −15 36 | Eri | 11.3v | 2.4' | 1.9' | E3 |

HH: considerably bright, pretty small, irregularly round, suddenly much brighter middle.
170x250: Moderately bright. Evenly pretty concentrated. Roughly round. Small. No involved stars. [1189 (Le), 1190 (Le), 1191 (Le), and 1192 (Le), all forming a little group with 1199, were not seen.]

| ga | NGC 1171 | 03 04.0 | +43 23 | Per | 13.0p | 2.6' | 1.1' | Scd |

St: very faint, pretty large, irregular figure.
170x250: Very faint. Broadly very slightly concentrated. Elongated NNW-SSE. Small. No involved stars. A little ways NNW is a faint star.

| ga | NGC 1201 | 03 04.1 | −26 04 | For | 11.7b | 3.6' | 2.1' | S0 |

HH: considerably bright, pretty small, very little extended N-S, resolvable, faint star near.
170x250: Bright to very bright. Sharply very concentrated. Slightly elongated NNE-SSW. Small. No involved stars. It makes a large "Y" figure with fairly bright stars NE to NW.

| bs | Gamma Per | 03 04.8 | +53 30 | Per | 2.90 | | 0.73 | G5 |

170x250: Yellow white.

| bs | Rho Per | 03 05.2 | +38 50 | Per | 3.39 | | 1.65 | M3 |

170x250: Gold. A highly colored bright star.

| ga | NGC 1186 | 03 05.5 | +42 50 | Per | 12.2p | 3.1' | 1.1' | Sbc |

HH: faint, much extended, suddenly much brighter middle to a faint star.
170x250: Very faint. Broadly very slightly concentrated. Elongated NW-SE. Moderately small. Near center is a relatively bright involved star; several faint stars surround the galaxy's SE end.

| oc | NGC 1193 | 03 05.9 | +44 23 | Per | 12.6 | 1.5' | | |

WH: faint, considerably large, easily resolvable.
42x60: Very faintly visible. A little glow. ***170x250:*** A faint cloud of a good number of very faint and threshold stars between two moderate brightness stars, one E and one WNW. Very small. Dense.

| ga | NGC 1209 | 03 06.0 | −15 36 | Eri | 12.4b | 2.3' | 1.1' | E6 |

HH: bright, small, considerably extended, pretty suddenly brighter middle.
170x250: Bright; pretty high surface brightness. Sharply pretty concentrated. Slightly elongated E-W. Small. No involved stars.

| ga | IC 284 | 03 06.2 | +42 22 | Per | 12.5b | 4.4' | 2.0' | Sdm |

170x250: Very very faint. Hardly concentrated. Elongated NNE-SSW. Moderately small. NE of center is an extremely faint involved star; nearby NE is a faint star; a little ways SW are a couple of stars; the galaxy is in a large E-W arc of moderately bright and moderate brightness stars crossing the field.

| bs | Beta Per | 03 08.2 | +40 57 | Per | 2.12 | | −0.05 | B8 |

170x250: White.

| ga | NGC 1222 | 03 08.9 | −02 57 | Eri | 13.1b | 1.1' | 0.8' | S0− |

St: very faint star in pretty faint, small, round nebula.
170x250: Moderately faint. Sharply fairly concentrated. Round. Very small. No involved stars. Making a crooked line with it are a wide pair of stars WSW and another star further WSW.

| ga | NGC 1232 | 03 09.8 | −20 34 | Eri | 10.5b | 7.4' | 6.4' | Sc |

HH: pretty bright, considerably large, round, gradually brighter middle, resolvable.
42x60: Visible. *170x250:* Moderately faint to moderate brightness; somewhat low surface brightness. Evenly concentrated to a small, brighter core. Roughly round. Large: a quarter of the field in size. At its edge NNE is a faint star; further NNE is another faint star.

| pn | IC 289 | 03 10.3 | +61 19 | Cas | 12.3p | 48.0" | | |

170x250: Faint. No central star. No clear disk. Round. Moderate size for a planetary. It looks like a faint galaxy. It makes a very small triangle with two faint stars, one NE and one farther E; a little ways S is a moderately bright star.

| ga | NGC 1241 | 03 11.3 | −08 55 | Eri | 12.0v | 3.6' | 2.2' | Sb |

HH: faint, pretty large, round, very gradually little brighter middle, 9th mag. star [N].
170x250: Faint. Broadly slightly concentrated. Slightly elongated NW-SE. Moderately small. No involved stars. It makes a sharp triangle with a moderately bright star N and a moderate brightness star NNE. Between 1241 and the NNE star is another galaxy [*1242 (WH)]. Extremely faint. Small.

| vs | V623 Cas | 03 11.4 | +57 54 | Cas | 7.6 | – | 2.07 | C4 |

12x35: Mag. 7.5. ***170x250:*** Light copper red.

| oc | NGC 1220 | 03 11.7 | +53 20 | Per | 11.8 | 1.6' | | |

JH: cluster, very small, faint stars.
42x60: Extremely faintly visible. ***170x250:*** A compact group of faint stars. Elongated N-S. Very small. Moderately dense.

| oc | HAR 1 | 03 12.0 | +63 11 | Cas | 7.0 | 23.0' | | |

12x35: A number of stars. ***42x60:*** A dozen moderate brightness stars. ***170x250:*** A fair number of widespread moderately bright stars of uniform brightness plus some fainter stars. The main part of the cluster is somewhat rectangular N-S; NE is a little extension. There are large empty gaps within it. It takes up the entire field. Somewhat sparse. [CR36]

| ds | Alpha For | 03 12.1 | −28 59 | For | 5.3" | 4.0 | 7.0 | F8 G7 |

42x60: Extremely close. ***170x250:*** The primary has a very close relatively faint companion. The primary is yellow white. [HJ 3555]

| ga | NGC 1255 | 03 13.5 | −25 43 | For | 11.4b | 4.1' | 2.5' | Sbc |

Ba: faint, pretty large, faint star close W.
170x250: Moderately faint. Broadly very slightly concentrated. Slightly elongated NW-SE. Moderately large. No involved stars. Nearby SSW is a moderate brightness star.

| ga | NGC 1253 | 03 14.2 | −02 49 | Eri | 12.3b | 5.2' | 2.3' | Scd |

WH: star with 90" long nebula attached.
170x250: Moderately faint. Broadly slightly concentrated. Very elongated E-W. Moderate size. At its W end is a moderately bright involved star. [1253A was not seen.]

| oc | K 5 | 03 14.7 | +52 43 | Per | − | 7.0' | | |

170x250: A faint cluster of a good number of very faint stars. A quarter of the field in size. Moderately dense.

| oc | NGC 1245 | 03 14.7 | +47 15 | Per | 8.4 | 10.0' | | |

HH: cluster, pretty large, rich, compressed, irregularly round, 12th to 15th mag. stars.
12x35: Very faintly visible. ***42x60:*** A faint cloud with a handful of stars made out.
170x250: A lot of stars, mostly faint. Along its edge N is a shallow arc of five stars. SSE is the bright plotted star. A third of the field in size. Dense.

| ga | NGC 1250 | 03 15.4 | +41 21 | Per | 12.8v | 2.2' | 0.8' | S0 |

Sw: very faint, very small, round.
170x250: Very faint. Evenly moderately concentrated. Round. Small. No involved stars. Nearby N is a very faint star.

| oc | ST 23 | 03 16.3 | +60 02 | Cam | − | 14.0' | | |

12x35: Four moderately bright stars in the shape of a quarter of an oval. ***42x60:*** Four fairly bright stars plus a handful more stars. ***170x250:*** Four bright stars and a couple dozen moderate brightness stars. The middle star of the curved part of the figure is a double. 3/4 of the field in size. Sparse. Fairly well defined; a fairly nice cluster.

| ga | IC 310 | 03 16.7 | +41 19 | Per | 12.7v | 1.4' | 1.4' | S0 |

170x250: Very faint. Evenly moderately concentrated. Round. Small. No involved stars. Nearby SW is a very faint star.

| ga | ESO 481-17 | 03 17.1 | −22 51 | Eri | 13.2p | 2.2' | 2.0' | Sa |

170x250: Moderately faint. Evenly moderately concentrated. Round. Very small. No involved stars. ESE, in line with the galaxy, are three progressively brighter stars.

| ga | NGC 1288 | 03 17.2 | −32 34 | For | 12.8b | 2.2' | 1.8' | Sc |

JH: very faint, large, round, very gradually little brighter middle.
170x250: Very faint. Broadly very slightly concentrated. Roughly round. Moderately small. At its center is either an extremely faint involved star or a substellar nucleus.

| ga | NGC 1284 | 03 17.8 | −10 17 | Eri | 12.9 | 1.7' | 1.5' | S0 |

HH: extremely faint, very small, 2 stars 2'–3' S.
170x250: Very faint. Broadly very slightly concentrated. Roughly round. Small. No involved stars. It makes a right triangle with a pair of faint stars a little ways SSE and a moderately bright star ESE.

| ga | NGC 1292 | 03 18.2 | −27 36 | For | 12.8b | 2.9' | 1.2' | Sc |

Ba: faint, pretty small, little extended, very gradually brighter middle, faint double star near.
170x250: Faint to moderately faint. Broadly slightly concentrated. Elongated NNE-SSW. Moderate size. No involved stars. NNE is a little triangular group of moderate brightness stars.

| ga | NGC 1265 | 03 18.3 | +41 51 | Per | 12.1v | 1.7' | 1.4' | E+ |

Bi: very faint, very small, much brighter middle.
170x250: Very very faint. Hardly concentrated. Roughly round. Very small. At its center is a moderately faint involved star (it doesn't look like a nucleus); making a flat triangle with the galaxy are two moderate brightness stars, one S and one a little farther NE.

| ga | NGC 1270 | 03 19.0 | +41 28 | Per | 13.1v | 1.0' | 0.8' | E |

dA: very faint, small, round.
[See 1275.]

| ga | NGC 1297 | 03 19.2 | −19 06 | Eri | 12.8p | 2.2' | 1.9' | S0 |

Ba: faint, pretty small.
170x250: Moderately faint. Evenly moderately concentrated. Slightly elongated N-S. Small to moderately small. At its N end is an involved star.

| ga | NGC 1272 | 03 19.4 | +41 29 | Per | 11.7v | 1.8' | 1.8' | E+ |

dA: faint, small, round.
[See 1275.]

| ga | NGC 1273 | 03 19.4 | +41 32 | Per | 13.2v | 1.0' | 0.8' | S0 |

dA: very faint, very small.
[See 1275.]

| ga | NGC 1300 | 03 19.7 | −19 24 | Eri | 10.3v | 5.9' | 4.9' | Sbc |

JH: considerably bright, very large, very much extended, pretty suddenly very much brighter middle.
170x250: Moderately faint. Sharply moderately concentrated to a small core. Slightly elongated WNW-ESE. Moderately large to large. The outer halo is very low surface brightness; it fades out indistinctly. No involved stars.

| ga | NGC 1275 | 03 19.8 | +41 30 | Per | 11.9v | 2.2' | 1.8' | Pec |

dA: faint, small.
170x250: Several members of this cluster of galaxies [AGC 426, the Perseus Galaxy Cluster] are visible within 2/3 of a field. All are round and have no involved stars. Starting at the W end: 1270 is faint. Sharply pretty concentrated. Very small. It forms a semicircle with several moderately faint stars. 1272 is faint. Evenly slightly concentrated. Small. Very nearby ESE is a very faint star. 1273 is very faint to faint. Sharply pretty concentrated. Very small. A sixth of the field NNE, in a little group with four or five faint stars, is a very faint galaxy [*I1907]. 1275 is moderately faint. Evenly moderately concentrated. Small. Very nearby are two very faint stars, one WNW and one NE. 1278 is very faint. Evenly slightly concentrated. Small. Nearby NW is another galaxy [*1277 (LR)]. Of all the large and famous Abell clusters of galaxies, this is by far the best for visual observation. [1267 (dA), 1268 (dA), 1274 (LR), and 1281 (Dr) were not seen; these additional members can be found with the aid of a very detailed chart.]

| ga | NGC 1278 | 03 19.9 | +41 33 | Per | 12.4v | 1.4' | 1.0' | E |

HH: pretty bright, pretty small, round, brighter middle, 6th of 7.
[See 1275.]

| ga | NGC 1302 | 03 19.9 | −26 03 | For | 11.6b | 3.7' | 3.7' | S0/a |

Ba: small, round, pretty suddenly very much brighter middle, 9th mag. star 1' NW.
170x250: Bright. Sharply very concentrated. Round. Small. No involved stars.

| ga | NGC 1309 | 03 22.1 | −15 24 | Eri | 12.0b | 2.1' | 1.9' | Sbc |

HH: pretty bright, considerably large, irregularly round, gradually brighter middle, 7th mag. star SSW.
170x250: Moderately bright. Broadly moderately concentrated. Roughly round. Moderately small. No involved stars. SSW is the bright plotted star.

0 to 6 Hours: FALL

| ga | UGCA 67 | 03 22.9 | −11 12 | Eri | 13.1 | 2.5' | 1.7' | Sd |

170x250: Not visible.

| bs | Alpha Per | 03 24.3 | +49 51 | Per | 1.79 | | 0.48 | F5 |

170x250: Soft white.

| oc | MEL 20 | 03 24.3 | +49 51 | Per | 2.3 | 184' | | |

12x35: A fair number of bright stars with fainter stars scattered about. Alpha Per is at the N end of the cluster. Winding SSE from Alpha to Sigma is a very distinctive "S" or serpent-like figure of stars; most of these are plotted but they look more serpentine in the sky. 2/3 of the field in size. Somewhat sparse to moderate density. A neat cluster. [CR39]

| ga | NGC 1325 | 03 24.4 | −21 32 | Eri | 12.2b | 4.7' | 1.5' | Sbc |

HH: faint, much extended ENE-WSW, cometic, 9th–10th mag. star attached.
170x250: Moderate brightness. Broadly slightly concentrated. A fairly thin edge-on, NE-SW. Moderately large. ENE of center is a relatively bright involved star; nearby ESE is a faint star. A little less than half the field W is a very small faint galaxy [*1319 (JH)].

| ga | NGC 1332 | 03 26.3 | −21 20 | Eri | 11.3b | 4.6' | 1.4' | S0− |

WH: very bright, small, extended, suddenly much brighter middle to a nucleus, NW of 2.
170x250: Very bright. Sharply very concentrated to a small bright nucleus. Elongated WNW-ESE. Moderate size. At its edge SW of center is a very faint star. Off its E end is a small faint galaxy [*1331 (WH)].

| ga | UGC 2729 | 03 27.0 | +68 35 | Cas | 12.2 | 3.4' | 2.4' | − |

170x250: Not visible.

| ga | NGC 1337 | 03 28.1 | −08 23 | Eri | 12.5b | 5.7' | 1.4' | Scd |

Sw: extremely faint, very large, much extended N-S.
170x250: Very faint. Hardly concentrated. Edge-on NNW-SSE. Moderately large. No involved stars. A little ways ESE is a very faint star.

| ga | NGC 1339 | 03 28.1 | −32 17 | For | 12.5b | 1.9' | 1.3' | E+ |

JH: considerably bright, pretty small, round, pretty suddenly brighter middle, double star W.
170x250: Moderately bright; high surface brightness. Sharply very concentrated. Slightly elongated N-S. Very small. No involved stars.

| ga | NGC 1344 | 03 28.3 | −31 04 | For | 11.3b | 6.0' | 3.4' | E5 |

HH: considerably bright, pretty large, irregularly round, very gradually brighter middle.
170x250: Very bright. Sharply very concentrated. Slightly elongated NNW-SSE. Small. No involved stars.

| ne | NGC 1333 | 03 29.3 | +31 25 | Per | | 4.8' | | R |

Sf: faint, large, 10th mag. star NE.
170x250: A faint vague nebulosity surrounding a moderate brightness star and arcing SW, ending in a small concentration. The nebula is most prominent right around the star but overall it's faint; the SW extension is very faint. A quarter of the field in size.

| ds | STF 389 | 03 30.2 | +59 22 | Cam | 2.6" | 6.5 | 7.5 | A2 − |

170x250: The primary has a very very close fainter companion. The primary is white.

| ga | NGC 1351 | 03 30.5 | −34 51 | For | 11.5v | 2.8' | 1.7' | S0 |

JH: pretty bright, pretty small, round, pretty suddenly brighter middle.
170x250: Moderate brightness to moderately bright. Sharply pretty concentrated. Round. Small. No involved stars.

| ga | NGC 1350 | 03 31.1 | −33 37 | For | 11.2b | 5.8' | 2.7' | Sab |

JH: bright, large, much extended, very much brighter middle to a round nucleus.
42x60: Faintly visible. ***170x250:*** Moderately bright. Sharply pretty concentrated. Elongated NNE-SSW. Moderate size. The outer halo fades out indistinctly. SW of center is a faint involved star.

| ds | STF 401 | 03 31.3 | +27 34 | Tau | 11.4" | 6.4 | 6.9 | A2 A3 |

42x60: A fairly close slightly unequal pair. NNW is a wider, fainter pair [STF 7: 44.1", 7.4, 8.4]. ***170x250:*** A moderate distance pretty much equal pair. The other double is much wider and a little fainter; the two doubles are not quite parallel. All the stars are white. A nice scene.

| oc | NGC 1342 | 03 31.6 | +37 20 | Per | 6.7 | 14.0' | | |

HH: cluster, very large, about 60 stars.
12x35: A number of faint stars. *42x60:* Some dozen moderate brightness stars. Elongated NE-SW. *170x250:* Some dozen fairly bright stars and about the same number of moderate brightness stars. Elongated NE-SW. It's roughly in the shape of a narrow "U" opening up NE. It takes up the whole field. Somewhat sparse. A fairly nice cluster.

| ga | NGC 1353 | 03 32.1 | −20 49 | Eri | 12.4b | 3.3' | 1.3' | Sb |

HH: pretty bright, considerably large, irregularly extended, much brighter middle.
170x250: Moderate brightness. Evenly moderately concentrated. Elongated NW-SE. Moderately small. No involved stars. A little ways SE, in line with the axis of the galaxy, is a moderate brightness star.

| ga | UGC 2765 | 03 32.1 | +68 22 | Cam | 12.5 | 2.4' | 1.5' | − |

170x250: Not visible.

| ga | NGC 1357 | 03 33.2 | −13 39 | Eri | 12.4b | 3.2' | 2.5' | Sab |

HH: pretty faint, pretty large, round, little brighter middle, pretty bright star 5' NE.
170x250: Moderately faint. Evenly moderately concentrated. Round. Moderately small. No involved stars. It makes a triangle with two bright stars, one WNW and one NNE.

| pn | NGC 1360 | 03 33.3 | −25 51 | For | 9.6p | 6.4' | | |

Sw: 8th mag. star in bright, large nebula, extended N-S.
12x35: Visible. *42x60:* It has a faint central star. *170x250:* Fairly bright. It has a relatively very bright central star. No disk. The nebula is faint and diffuse; it has a milky, soft-light appearance. Oval NNE-SSW. Extremely large for a planetary: half the field in size. It looks more like a regular nebula around a star than a typical planetary. An interesting planetary.

| ga | IC 1953 | 03 33.7 | −21 28 | Eri | 12.2b | 2.7' | 2.0' | Sd |

170x250: Faint. Broadly very slightly concentrated. Roughly round. Moderately small to moderate size. No involved stars. [E548-40 was not seen.]

| ga | NGC 1358 | 03 33.7 | −05 05 | Eri | 13.0b | 2.5' | 1.9' | S0/a |

HH: very faint, small, between 2 stars.

170x250: Moderately faint. Sharply moderately concentrated. Round. Very small. No involved stars. It's between two pairs of moderate brightness stars: a close pair a little ways ENE and a very wide pair a little farther W; these stars are part of a long E-W line of evenly spaced stars going right across the field. A little more than a third of the field NW is another galaxy [*1355 (dA)], fairly similar. Moderately faint. Sharply moderately concentrated. Slightly elongated ENE-WSW. Very small. No involved stars.

| ga | NGC 1359 | 03 33.8 | –19 29 | Eri | 12.6b | 2.4' | 1.7' | Sm |

JH: faint, large, round, very gradually little brighter middle.
170x250: Very faint. Hardly concentrated. Roughly round. Moderate size. At its edge E is a faint star. Half the field ENE is a very small very faint galaxy [*E548-44].

| ga | NGC 1366 | 03 33.9 | –31 11 | For | 12.0v | 2.1' | 0.9' | S0 |

HH: very faint, small, irregular figure, little brighter middle.
170x250: Moderately bright to bright; high surface brightness. Sharply very concentrated to a substellar nucleus. Elongated N-S. Very small. No involved stars. It makes a triangle with two faint stars E.

| ga | NGC 1361 | 03 34.3 | –06 15 | Eri | 13.1 | 1.5' | 1.3' | E+ |

Sn: extremely faint, extremely small, gradually brighter middle to a nucleus.
170x250: Very faint. Moderately concentrated. Round. Very very small. No involved stars.

| ga | NGC 1371 | 03 35.0 | –24 56 | For | 11.6b | 5.8' | 4.6' | Sa |

HH: pretty bright, pretty large, very little extended, pretty suddenly brighter middle.
170x250: Bright. Sharply pretty concentrated. Slightly elongated NW-SE. Moderately small. It has a very faint outer halo. No involved stars.

| ga | IC 335 | 03 35.5 | –34 26 | For | 12.9p | 2.5' | 0.6' | S0 |

170x250: Moderate brightness. Evenly moderately concentrated. A fairly thin edge-on, E-W. Small. No involved stars.

| ga | NGC 1380 | 03 36.5 | –34 58 | For | 10.9b | 4.8' | 2.7' | S0 |

JH: very bright, large, round, pretty suddenly brighter middle.
170x250: Moderately bright. Evenly pretty concentrated. Elongated N-S. Moderate size. Just off its edge SW is a very faint star. [1380 and 1380A are the northernmost members of the Fornax Galaxy Cluster, the second largest cluster of bright galaxies in the sky after Virgo.]

0 to 6 Hours: FALL

| ga | NGC 1380A | 03 36.8 | −34 44 | For | 12.4v | 2.6' | 0.8' | S0 |

170x250: Very faint. Broadly slightly concentrated. Very elongated to edge-on N-S. Moderately small. No involved stars.

| ga | NGC 1376 | 03 37.1 | −05 02 | Eri | 12.8p | 1.6' | 1.6' | Scd |

HH: extremely faint, pretty large, irregularly round, brighter middle, resolvable.
170x250: Moderately faint. Broadly slightly concentrated. Round. Moderately small. No involved stars.

| ga | NGC 1385 | 03 37.5 | −24 30 | For | 11.5b | 3.4' | 2.0' | Scd |

HH: pretty bright, pretty small, round, gradually pretty much brighter middle.
170x250: Bright. Broadly moderately concentrated. Roughly round. Moderately small. No involved stars. N to SW is a linear group of moderate brightness stars.

| ga | NGC 1383 | 03 37.7 | −18 20 | Eri | 12.5v | 2.7' | 0.9' | S0 |

JH: pretty faint, small, round, pretty suddenly much brighter middle.
170x250: Moderately faint. Evenly to sharply pretty concentrated. Round. Very small. No involved stars. It's right in between two nearby stars: a moderately faint star WSW and a very faint star ENE.

| ga | NGC 1395 | 03 38.5 | −23 01 | Eri | 10.6b | 5.9' | 4.4' | E2 |

HH: bright, pretty small, extended, pretty suddenly much brighter middle.
42x60: Visible. *170x250:* Bright. Sharply pretty concentrated. Round. Small. WSW of center is a very faint involved star.

| ga | NGC 1393 | 03 38.6 | −18 25 | Eri | 12.0v | 1.5' | 1.1' | S0 |

HH: faint, small, round, gradually little brighter middle.
170x250: Moderate brightness. Sharply pretty concentrated. Round. Small. No involved stars. 2/3 of a field NE is 1394. Moderately faint. Sharply fairly concentrated. Slightly elongated N-S. Small. No involved stars. It makes a triangle with two faint stars, one NNW and one WSW. Right in between 1393 and 1394 is a small faint galaxy [*1391 (Le)].

| ga | NGC 1398 | 03 38.9 | −26 20 | For | 10.6b | 7.1' | 5.3' | Sab |

Wi: considerably bright, considerably large, round, very much brighter middle.
42x60: Visible. *170x250:* Very bright. Sharply very concentrated. Round. Moderately small. No involved stars.

| ga | NGC 1394 | 03 39.1 | –18 17 | Eri | 12.8v | 1.5' | 0.5' | S0 |

Le: very faint, very small, extended N-S, suddenly brighter middle to a nucleus.
[See 1393.]

| ga | NGC 1401 | 03 39.4 | –22 43 | Eri | 13.1b | 2.4' | 0.6' | S0 |

HH: very faint, very small, round.
170x250: Moderately faint to moderate brightness. Sharply fairly concentrated. Roughly round. Very small. Off its edge N is a faint star.

| ga | NGC 1406 | 03 39.4 | –31 19 | For | 12.4b | 3.8' | 0.7' | Sbc |

JH: faint, considerably large, very much extended, very gradually little brighter middle, 7th mag. star NW.
170x250: Moderate brightness. Broadly slightly concentrated. Edge-on NNE-SSW. Moderate size. No involved stars.

| ga | NGC 1400 | 03 39.5 | –18 41 | Eri | 10.9v | 2.5' | 2.4' | S0– |

HH: considerably bright, pretty small, round, pretty suddenly much brighter middle.
[See 1407.]

| ga | UGC 2800 | 03 40.1 | +71 24 | Cam | 13.0 | 2.2' | 1.0' | Im |

170x250: Not visible.

| ga | NGC 1407 | 03 40.2 | –18 34 | Eri | 10.7b | 4.5' | 4.1' | E0 |

HH: very bright, large, round, suddenly very much brighter middle to a nucleus.
42x60: 1407 and 1400 are both visible. ***170x250:*** 1407 is very bright. Evenly to sharply very concentrated. Round. Moderately small. It has a very faint extremely low surface brightness outer halo. No involved stars. 2/3 of a field WSW is 1400. Bright. Sharply very concentrated. Round. Small. No involved stars. A little ways WSW is a faint star. Half the field N of 1400, making an isosceles triangle with it and 1407, is a tiny faint galaxy [*1402 (Le)].

| ga | NGC 1415 | 03 41.0 | –22 33 | Eri | 12.8b | 3.8' | 1.7' | S0/a |

HH: pretty bright, small, little extended, pretty gradually little brighter middle, star 2' SE.
170x250: Moderate brightness. Sharply fairly concentrated. Slightly elongated NW-SE. Very small. No involved stars. It's on the long leg of a sharp right triangle of moderate brightness stars.

| ga | NGC 1417 | 03 42.0 | –04 42 | Eri | 12.8b | 2.7' | 1.6' | Sb |

HH: pretty faint, pretty large, little extended, little brighter middle, star SE, 2nd of 3.
170x250: Moderately faint. Broadly slightly concentrated. Slightly elongated N-S. Small. No involved stars. Nearby SE is a moderate brightness star. A quarter of the field ESE is a very faint galaxy [*1418 (HH)]. Hardly concentrated. Very small. It makes a triangle with two faint stars, one SSW and one E. A little less than half the field WNW of 1417 is an extremely faint galaxy [*I344].

| ga | ESO 548-81 | 03 42.1 | –21 14 | Eri | 12.8p | 1.1' | 1.1' | Sa |

170x250: Moderately faint. Sharply moderately concentrated. Roughly round. Very small. Tagged right next to it NW is the bright plotted star, which interferes.

| ga | MCG-1-10-23 | 03 42.2 | –06 45 | Eri | 13.1 | 1.7' | 1.3' | S0/a |

170x250: Extremely faint. Broadly concentrated. Round. Tiny. No involved stars. SE is a large group of moderately bright and moderate brightness stars.

| ga | NGC 1425 | 03 42.2 | –29 53 | For | 11.3b | 6.6' | 2.8' | Sb |

WH: faint, pretty large, irregularly round, gradually brighter middle.
170x250: Moderately bright. Evenly to sharply fairly concentrated. Slightly elongated NW-SE. Moderately small. No involved stars. ENE to NW, almost parallel to the galaxy, is a line of three widely spaced moderately bright to moderate brightness stars.

| ga | NGC 1421 | 03 42.5 | –13 29 | Eri | 12.0b | 3.5' | 0.8' | Sbc |

HH: faint, considerably large, much extended N-S, resolvable.
170x250: Moderate brightness to moderately bright. Hardly concentrated. Edge-on N-S. Moderately large. No involved stars. A little ways SE is a faint star; WNW is a moderate brightness star.

| ga | NGC 1426 | 03 42.8 | –22 06 | Eri | 12.3b | 2.8' | 1.8' | E4 |

HH: pretty faint, small, little extended, brighter middle.
170x250: Moderately bright. Sharply pretty concentrated. Round. Very small. No involved stars.

| bs | Delta Per | 03 42.9 | +47 47 | Per | 2.99 | | –0.11 | B5 |

170x250: White.

| oc | IC 348 | 03 44.6 | +32 10 | Per | 7.3 | 8.0' | | |

170x250: A faint diffuse nebulosity around a moderately bright star and its companion. Slightly elongated WNW-ESE. Small. The two stars are in an oval figure with a half-dozen moderate brightness stars W to S.

| ga | NGC 1439 | 03 44.8 | −21 55 | Eri | 12.3b | 2.4' | 2.2' | E1 |

HH: faint, pretty small, gradually pretty much brighter middle.
170x250: Moderately faint to moderate brightness. Sharply pretty concentrated. Round. Very small. No involved stars.

| ga | NGC 1440 | 03 45.0 | −18 15 | Eri | 12.6b | 2.1' | 1.5' | S0 |

HH: pretty bright, pretty small, round, suddenly much brighter middle to a 13th mag. star.
170x250: Moderately bright. Sharply very concentrated. Round. Small. No involved stars. It makes a triangle with two faint stars, one WNW and another a little farther ENE.

| ga | IC 334 | 03 45.3 | +76 38 | Cam | 12.5b | 3.2' | 2.5' | S |

170x250: Faint to moderately faint. Evenly moderately concentrated. Round. Small. S of center is an extremely faint involved star; nearby N is a very faint star; a little ways NW are three moderate brightness to moderately faint stars.

| ga | NGC 1438 | 03 45.3 | −23 00 | Eri | 13.2b | 2.0' | 0.8' | S0/a |

Sn: extremely faint, much extended, nucleus, 10th mag. star 1' E.
170x250: Faint. Broadly slightly concentrated. Slightly elongated ENE-WSW. Small. No involved stars. Nearby E is a moderately bright star; a little ways S is a faint star.

| ga | NGC 1452 | 03 45.4 | −18 38 | Eri | 12.8b | 2.6' | 1.7' | S0/a |

WH: faint, round, little brighter middle.
170x250: Moderate brightness. Evenly pretty concentrated. Round. Small. No involved stars. NNW are two faint stars.

| ga | ESO 358-63 | 03 46.3 | −34 56 | For | 12.6b | 5.1' | 1.2' | I0 |

170x250: Moderately faint. Broadly slightly concentrated. Edge-on NW-SE. Moderately small. No involved stars. A little ways N are two faint stars; the galaxy is inside a very large E-W trapezoidal figure of moderate brightness stars.

| ga | NGC 1453 | 03 46.4 | –03 58 | Eri | 12.6b | 2.4' | 1.9' | E2-3 |

HH: pretty bright, small, round, 17th mag. star in middle.
170x250: Moderate brightness. Sharply pretty concentrated. Round. Very small. No involved stars.

| ga | MCG-3-10-45 | 03 46.6 | –16 33 | Eri | 12.5 | 1.6' | 0.9' | Im |

170x250: Moderately faint. Evenly moderately concentrated. Elongated NE-SW. Very small. No involved stars. It's right in between a pair of moderately bright stars NW and a moderately bright star SE.

| ga | IC 342 | 03 46.8 | +68 05 | Cam | 9.1b | 21.6' | 21.1' | Scd |

170x250: A relatively bright nucleus with an exceedingly faint diffuse glow immediately around and NE of it. A third of the field in size. In a triangle around the nucleus are several involved stars: a moderately faint star nearby NNW of center, a moderate brightness star a little ways ENE, and a NW-SE line of six stars W.

| oc | M45 | 03 47.0 | +24 07 | Tau | 1.2 | 110' | | |

12x35: A fair number of mixed brightness stars, including seven or eight bright stars and numerous moderate brightness stars. There aren't many faint stars. The bright stars make a small dipper figure slightly elongated E-W. In the middle of the bowl of the dipper is a very sharp NE-SW triangle of stars; going S from Eta Tau is a long curved string of stars. A little less than half the field in size. [Moderate density.] ***42x60:*** A very nice field of widely spaced very bright and moderately bright stars. There aren't many faint stars. The brightest stars take up the field exactly. ***170x250:*** A little ways WNW of Eta is a small triangle of moderately bright stars (also visible in the smaller scopes). [Pleiades/Seven Sisters]

| ga | IC 1993 | 03 47.1 | –33 42 | For | 12.4b | 2.4' | 2.2' | Sb |

170x250: Very faint. Hardly concentrated. Roughly round. Small. No involved stars. Nearby N is a faint star; a little farther WNW is a bright star.

| bs | Eta Tau | 03 47.5 | +24 06 | Tau | 2.90 | | –0.12 | B7 |

170x250: White.

| pn | IC 351 | 03 47.5 | +35 03 | Per | 12.4p | 7.0" | | |

170x250: Moderately bright. Almost stellar: a slightly fat star. Next to it WNW is a very very faint star; making an isosceles triangle with the planetary is a star S and a little trio of stars SE.

| oc | TOM 5 | 03 47.7 | +59 05 | Cam | 8.4 | 17.0' | | |

12x35: Faintly visible. *42x60:* A faint cluster of faint stars. *170x250:* A good number of widely scattered moderately faint and faint stars. Somewhat square. 3/4 of the field in size. Moderate density.

| ga | NGC 1461 | 03 48.5 | −16 23 | Eri | 12.8b | 3.0' | 0.9' | S0 |

HH: pretty bright, small, little extended, much brighter middle to a nucleus.
170x250: Moderately bright. Evenly pretty concentrated. Slightly elongated NNW-SSE. Moderately small. No involved stars. It makes a triangle with a moderately bright star NW and a moderate brightness star farther N.

| oc | NGC 1444 | 03 49.4 | +52 40 | Per | 6.6 | 4.0' | | |

HH: cluster of about 30 12th to 14th mag. stars.
170x250: About ten faint to very faint stars surrounding the bright plotted star, mostly W of it; the bright star has a fairly close faint companion. Small. Sparse. A concentration of stars, but not much of a cluster.

| pn | PK 171-25.1 | 03 53.5 | +19 27 | Tau | 13.9p | 48.0" | | |

170x250: Not visible.

| bs | Zeta Per | 03 54.1 | +31 53 | Per | 2.93 | | 0.10 | B1 |

170x250: The primary has a moderate distance relatively very faint companion. The primary is white. [STF 464: 11.9", 2.9, 9.5]

| ds | 32 Eri | 03 54.3 | −02 57 | Eri | 6.9" | 4.8 | 6.1 | G8 A2 |

42x60: The primary has an extremely close relatively faint companion. *170x250:* The primary has a close fainter companion. Color contrast: the primary is yellow; the companion is white. A very nice double. [STF 470]

| ga | NGC 1482 | 03 54.7 | −20 30 | Eri | 13.1b | 2.4' | 1.3' | S0+ |

HH: faint, small, very little extended, 2 10th mag. stars near, E of 2.
170x250: Very faint. Broadly slightly concentrated. Round. Very small. No involved stars. It makes a triangle with two bright stars, one ENE and one NNW. [1481 (JH) was not seen.]

| pn | IC 2003 | 03 56.4 | +33 51 | Per | 12.6p | 9.0" | | |

42x60: Visible. *170x250:* Moderately bright. Almost stellar: a slightly fat star. Pretty strong blinking effect. Next to it SW is a very faint star. It's among some moderately bright and moderate brightness stars that make good comparison stars: the planetary is approximately the same brightness but it has a blinking effect and a softer light, and it's clearly not as sharp.

| bs | Epsilon Per | 03 57.9 | +40 00 | Per | 2.88 | | −0.17 | B0 |

170x250: The primary has a moderately close much fainter companion. The primary is white. [STF 471: 9.7", 2.9, 8.1]

| bs | Gamma Eri | 03 58.0 | −13 30 | Eri | 2.97 | | 1.60 | M0 |

170x250: Gold. A well colored bright star; pretty.

| bs | Lambda Tau | 04 00.7 | +12 29 | Tau | 3.47 | | −0.12 | B3 |

170x250: White.

| ne | NGC 1499 | 04 00.7 | +36 37 | Per | | 160' | | E |

Ba: very faint, very large, extended N-S, diffuse.
12x35, 42x60: Not visible. [California Nebula]

| ne | NGC 1491 | 04 03.4 | +51 19 | Per | | 21.0' | | E |

WH: very bright, small, irregular figure, brighter middle, resolvable, star involved.
42x60: Very very faintly visible with an involved star. *170x250:* Faint. It has a relatively bright central star. The nebulosity is a little brighter on the W side. A quarter of the field in size. Surrounding the nebula are a number of faint stars.

| oc | NGC 1496 | 04 04.5 | +52 39 | Per | 9.6 | 6.0' | | |

JH: cluster, segment of a ring.
42x60: A few faint stars. *170x250:* A half-dozen moderately faint stars in a semi-circle plus a handful more fainter stars. Small. Sparse.

| ga | NGC 1507 | 04 04.5 | −02 11 | Eri | 12.9b | 3.6' | 0.8' | Sm |

WH: very faint, pretty large, much extended, very little brighter middle, easily resolvable.
170x250: Faint. Broadly slightly concentrated. Edge-on NNE-SSW. Moderate size. No involved stars. Nearby W of center is a faint star.

| vs | UV Cam | 04 05.9 | +61 48 | Cam | 8.2 | 8.9 | 2.26 | C5 |

12x35: Mag. 7.5. ***170x250:*** Orangish.

| ga | NGC 1518 | 04 06.8 | −21 10 | Eri | 12.3b | 3.0' | 1.3' | Sdm |

JH: bright, large, pretty much extended, gradually brighter middle, 8th mag. star SW.
170x250: Moderately faint. Broadly slightly concentrated. A fat edge-on, NE-SW. Moderate size. No involved stars. Nearby WSW is a moderately bright star.

| pn | NGC 1501 | 04 07.0 | +60 55 | Cam | 13.3p | 52.0" | | |

WH: planetary nebula, pretty bright, pretty small, very little extended, 1' in diameter.
42x60: Visible. ***170x250:*** Moderately faint to moderate brightness. It has a very very faint central star. A well defined disk and edge. At times there's very small-scale rippling in the light: a rough texture. Practically no blinking effect. Perfectly round. Moderately large for a planetary. A very nice and intriguing planetary.

| oc | NGC 1502 | 04 07.7 | +62 20 | Cam | 6.9 | 7.0' | | |

WH: cluster, pretty rich, considerably compressed, irregular figure.
12x35: The line of stars [Kemble's Cascade] leading to 1502 is very nice. It's straighter and more continuous than it appears on the chart. 1502 is a tiny cloud of stars around the bright plotted star. ***42x60:*** The bright star is a double. A few more faint stars are made out. ***170x250:*** The cluster is in the shape of an arrowhead pointing NW. The double star – a wide equal pair – is in the bottom half. Besides this double and the point star of the arrowhead, which are bright, the cluster is a fair number of moderately bright stars. Half the field in size. Moderate density. An attractive cluster.

| ga | IC 356 | 04 07.8 | +69 48 | Cam | 11.4b | 5.3' | 3.9' | Sab |

170x250: Faint. Evenly slightly concentrated. Roughly round. Moderate size. No involved stars. It's between a wide pair of moderate brightness stars SSW and the bright plotted star N.

| ga | NGC 1516 | 04 08.1 | −08 49 | Eri | 12.6 | 1.4' | 0.9' | Sbc |

HH: excessively faint, small, extended, pretty suddenly much brighter middle, easily resolvable.
170x250: Very faint. Broadly very slightly concentrated. Slightly elongated NNW-SSE. Very small. No involved stars. [1516 is a very close double galaxy consisting of 1516A and 1516B; the two components were not distinguished.]

| ga | NGC 1521 | 04 08.3 | −21 03 | Eri | 12.4b | 2.7' | 1.6' | E3 |

JH: pretty bright, round, brighter middle.
170x250: Moderate brightness. Evenly fairly concentrated. Round. Very small. No involved stars. Nearby WSW is a moderately faint star.

| pn | NGC 1514 | 04 09.2 | +30 46 | Tau | 10.0p | 1.9' | | |

HH: 8th mag. star in nebula 3' in diameter.
12x35: Visible as a faint star. ***42x60:*** The nebula is faintly visible. ***170x250:*** A moderate brightness star with a faint nebula around it. Round. Large for a planetary. It looks more like a reflection nebula than a planetary.

| oc | NGC 1513 | 04 10.0 | +49 31 | Per | 8.4 | 9.0' | | |

WH: cluster, large, very rich, pretty compressed, very bright stars.
12x35: Very faintly visible. ***42x60:*** A few faint stars. ***170x250:*** A fair number of stars: a big "S" figure of moderate brightness stars that just fits within a N-S rectangle of more numerous fainter stars. The brightest star is at the cluster's edge N. Half the field in size. Moderate density.

| ga | NGC 1531 | 04 12.0 | −32 51 | Eri | 13.2p | 1.4' | 0.9' | S0− |

JH: pretty bright, pretty large, round, brighter middle, NW of 2.
[See 1532.]

| ga | NGC 1532 | 04 12.1 | −32 53 | Eri | 9.8v | 15.2' | 2.4' | Sb |

JH: bright, very large, very much extended NNE-SSW, pretty suddenly much brighter middle.
42x60: Faintly visible. ***170x250:*** Moderate brightness. Evenly to sharply fairly concentrated. A thin edge-on, NE-SW. Large: a third of the field in size. The ends fade out very indistinctly. Toward the S tip is a small slight brightening. No involved stars. Next to it NW of center is 1531. Moderately faint. Evenly moderately concentrated. Slightly elongated NW-SE, perpendicular to 1532. Very small. No involved stars.

| ga | NGC 1537 | 04 13.7 | −31 38 | Eri | 11.5b | 3.9' | 2.5' | S0− |

JH: very bright, pretty small, little extended, pretty suddenly very much brighter middle.
170x250: Bright. Sharply very concentrated. Round. Very small. No involved stars.

| pn | NGC 1535 | 04 14.2 | −12 44 | Eri | 9.6p | 60.0" | | |

HH: globular cluster, very bright, small, round, pretty suddenly then very suddenly brighter middle, resolvable.
12x35, 42x60: Visible as a star. ***170x250:*** Very bright. No central star. It has a soft perimeter, a little fainter than the interior; not a clean disk. Moderate blinking effect. Round. Moderately small for a planetary. A little ways W is a group of faint stars; ESE is a kite-shaped group of brighter stars.

| oc | NGC 1528 | 04 15.4 | +51 14 | Per | 6.4 | 23.0' | | |

WH: cluster, bright, very rich, considerably compressed.
12x35: A cluster of faint stars. ***42x60:*** A good number of moderately faint stars. ***170x250:*** A good number of moderate brightness stars, uniform in brightness. Roughly triangular. Exactly the field width in size. Moderate density.

| oc | IC 361 | 04 19.0 | +58 18 | Cam | 11.7 | 6.0' | | |

170x250: A very faint cluster of a good number of very very faint stars plus additional threshold stars and the light of unresolved stars. A quarter of the field in size. Dense. Tangent to its edge NNE is a broken WNW-ESE line of stars.

| ga | NGC 1550 | 04 19.6 | +02 24 | Tau | 13.1b | 2.2' | 1.9' | S0− |

dA: very faint, small, round, 13th mag. star near.
170x250: Faint. Evenly moderately concentrated. Round. Very small. No involved stars. A little ways SW is a quadrilateral of moderately faint stars.

| oc | NGC 1545 | 04 20.9 | +50 15 | Per | 6.2 | 18.0' | | |

WH: cluster, pretty rich, little compressed, bright stars.
12x35: A wide double star. ***42x60:*** A wide bright double and a few widely scattered faint stars. ***170x250:*** Widely scattered mixed brightness stars around the two bright stars. It fills the field. Sparse.

| ga | UGCA 90 | 04 21.2 | −21 50 | Eri | 12.7b | 5.6' | 2.1' | Sd |

170x250: Very very faint. Diffuse. Slightly elongated E-W [NW-SE]. Moderate size. W of center is a moderately faint involved star that interferes; SW is a moderate brightness star; farther W is another moderate brightness star.

| ne | NGC 1555 | 04 21.8 | +19 32 | Tau | | 1.5' | | R |

Hi: !!! very faint, small, variable.
170x250: A small very very faint nebula around the plotted star. Inconspicuous; it looks like fog on the optics. It must be a reflection nebula. [1555=1554]

| ds | Chi Tau | 04 22.6 | +25 38 | Tau | 19.5" | 5.5 | 7.6 | B9 − |

42x60: The primary has a close faint companion. *170x250:* The primary has a moderate distance fainter companion. The primary is white. [STF 528]

| ga | NGC 1530 | 04 23.4 | +75 17 | Cam | 12.3b | 4.6' | 2.4' | Sb |

Te: pretty bright, large.
170x250: Very faint. Broadly very slightly concentrated. Roughly round. Moderate size. No involved stars. NNE is a moderately faint star; S is a faint star.

| ga | NGC 1575 | 04 26.4 | −10 06 | Eri | 13.0 | 0.4' | 0.3' | S |

Mu: very faint, pretty small, round, 9.5 mag. star 2' S.
170x250: Very faint. Broadly very slightly concentrated. Round. Small. No involved stars. It makes a flat triangle with a moderately bright star a little ways S and a faint star a little ways NW. [1575=1577]

| oc | Hyades | 04 27.0 | +16 00 | Tau | 0.5 | 330' | | |

naked eye: The sprawling Hyades makes quite a contrast with the relatively well defined Pleiades, small and compact and fairly nearby in the sky. *6x30:* A good number of bright stars, roughly uniform in brightness, in a "V" shape, plus a second tier of moderately faint stars. Aldebaran, which is much brighter than the other stars, doesn't look like a member. N is a smaller group, including Upsilon, Kappa, and Omega Tau; this looks like a detached part of the cluster. Farther SSW is a little group of three or more stars, including Mu; these look like they might also be part of the cluster, but this is more questionable. The main part of the cluster including the "V" is 3/4 of the field in size. Moderate density. [MEL25]

| bs | Theta 2 Tau | 04 28.7 | +15 52 | Tau | 3.40 | | 0.18 | A7 |

170x250: Theta 1 and Theta 2 are aligned N-S and easily fit in the same main scope field. Theta 2 is a little brighter. Theta 2 is white; Theta 1 is yellow (B-V = 0.95; G7) – a very clear though not drastic color difference. The best pair of bright stars in a single main scope field.

| ne | NGC 1579 | 04 30.2 | +35 16 | Per | | 12.0' | | R |

HH: pretty bright, very large, irregularly round, much brighter middle, 8th mag. star 2' N.
170x250: A moderately faint diffuse nebula. Roughly round. Small. It looks a little like a slightly concentrated galaxy. No involved stars. It makes a triangle with two stars NE and a moderate brightness star a little farther NNW.

| ga | NGC 1587 | 04 30.7 | +00 39 | Tau | 11.7v | 1.8' | 1.6' | E |

HH: faint, pretty small, round, resolvable, W of double nebula.
170x250: In the position is a double galaxy [*1588 (HH), 1587] with the components right next to each other E-W. The W galaxy is moderate brightness. Evenly to sharply fairly concentrated. Round. Very small. No involved stars. The E galaxy is faint. Concentrated. Round. Very very small. No involved stars.

| ga | NGC 1569 | 04 30.8 | +64 50 | Cam | 11.9b | 3.6' | 1.7' | Im |

WH: pretty bright, small, little extended, brighter middle to a nucleus, pretty bright star N.
170x250: Moderate brightness. Broadly moderately concentrated. Very elongated WNW-ESE. Moderately small. No involved stars. Nearby N is a bright star.

| ga | NGC 1589 | 04 30.8 | +00 51 | Tau | 12.8b | 3.1' | 1.0' | Sab |

HH: faint, pretty large, little extended NW-SE, star 80" NE.
170x250: Moderately faint. Evenly moderately concentrated. Roughly round. Small. No involved stars. It makes a little arc with a moderately faint star ENE and a faint star SSW.

| oc | NGC 1582 | 04 31.7 | +43 45 | Per | 7.0 | 37.0' | | |

WH: cluster, very large, pretty rich, little compressed, bright stars.
12x35: A half-dozen stars in two short arcs. *42x60:* The W arc consists of a half-dozen stars; the E arc consists of three stars; and there are more stars connecting the two arcs. *170x250:* A number of bright stars – those comprising the arcs – plus more moderate brightness and fainter stars. Elongated E-W. Two field widths in size. Somewhat sparse to moderate density.

| ga | NGC 1600 | 04 31.7 | −05 05 | Eri | 10.9v | 3.0' | 2.5' | E3 |

HH: pretty bright, pretty large, round, gradually much brighter middle.
170x250: Moderately faint. Broadly moderately concentrated. Round. Very small. No involved stars. It makes a flat triangular figure with a slightly crooked NNW-SSE line of moderate brightness stars off its W side. Nearby NNE is a tiny extremely faint galaxy [*1601 (LR)]. A little ways E is another similar galaxy [*1603 (LR)].

| ds | 1 Cam | 04 32.0 | +53 55 | Cam | 10.3" | 5.7 | 6.8 | B0 B1 |

42x60: A pretty close unequal pair; nicely split. A nice double. ***170x250:*** A moderately close slightly unequal pair. Both white. [STF 550]

| ga | NGC 1560 | 04 32.8 | +71 53 | Cam | 12.2b | 9.8' | 1.7' | Sd |

Te: very faint, large, extended, 9.3 mag. star SW.
170x250: Very faint. Hardly concentrated. A thin edge-on, NNE-SSW. Moderate size to moderately large. N of center is a faint involved star; nearby S is a moderately bright star; NNW is another moderately bright star.

| ds | STF 559 | 04 33.5 | +18 01 | Tau | 3.0" | 6.9 | 7.0 | B9 − |

170x250: A very very close perfectly equal pair. Both white. A faint double.

| ga | MCG-2-12-39 | 04 33.6 | −13 15 | Eri | 12.6 | 2.1' | 1.6' | E+ |

170x250: Very faint. Broadly slightly concentrated. Roughly round. Very small. No involved stars. ESE is a moderate brightness star.

| ga | MCG-2-12-41 | 04 33.9 | −11 42 | Eri | 12.8 | 3.2' | 1.0' | Sab |

170x250: Very very faint. Roughly round. Very very small. No involved stars. WNW is a moderately bright star with a faint star just E of it.

| oc | NGC 1605 | 04 34.9 | +45 16 | Per | 10.7 | 5.0' | | |

WH: cluster, very faint, pretty small, compressed, extremely faint stars.
170x250: A very faint cluster of a fair number of very faint to threshold stars. Somewhat empty in the middle: the brighter stars are on the periphery, mainly E. A quarter of the field in size. Somewhat sparse. An interesting cluster.

| ga | NGC 1573 | 04 35.1 | +73 15 | Cam | 12.8b | 1.9' | 1.3' | E |

Te: very faint, small, 9.5 mag. star E.
170x250: Faint. Evenly moderately concentrated. Round. Small. No involved stars. E is a moderately bright star along with a WNW-ESE line of three faint stars. [U3069 was not seen.]

| ds | STF 570 | 04 35.2 | −09 44 | Eri | 12.8" | 6.7 | 7.7 | A1 A7 |

42x60: A close slightly unequal pair. ***170x250:*** A moderately close very slightly unequal pair. Both white.

| bs | Alpha Tau | 04 35.9 | +16 30 | Tau | 0.85 | | 1.54 | K5 |

170x250: Gold. [Aldebaran]

| ga | NGC 1618 | 04 36.1 | −03 09 | Eri | 12.7v | 2.8' | 0.9' | Sb |

HH: faint, small, irregular figure, little brighter middle, 2 stars SE.
[See 1622.]

| ga | NGC 1620 | 04 36.6 | −00 08 | Eri | 13.1b | 3.4' | 1.0' | Sbc |

HH: very faint, pretty large, much extended, bright double star NE.
170x250: Very faint. Broadly very slightly concentrated. A fat edge-on, NNE-SSW. Moderate size. Toward its N end is an extremely faint involved star.

| ga | NGC 1622 | 04 36.6 | −03 11 | Eri | 12.5v | 4.0' | 0.8' | Sab |

LR: very faint, small, 20th mag. star W.
170x250: Faint. Evenly moderately concentrated. Elongated NE-SW. Very small. No involved stars. Half the field WNW is 1618. Very faint. Broadly slightly concentrated. Slightly elongated NNE-SSW. Small. No involved stars. A little ways E are four faint stars in a little diamond figure.

| ga | NGC 1625 | 04 37.1 | −03 18 | Eri | 12.2v | 2.7' | 0.7' | Sb |

JH: very faint, extended, suddenly brighter middle, bright star W.
170x250: Faint. Broadly very slightly concentrated. Edge-on NW-SE. Moderately small. At its NW end is a faint involved star.

| ga | IC 382 | 04 37.9 | −09 31 | Eri | 13.0 | 2.3' | 1.3' | Sc |

170x250: Very faint. Broadly slightly concentrated. Round. Very small. No involved stars. It's almost in line with four moderately bright to moderate brightness stars WNW.

| ga | UGC 3108 | 04 38.5 | +44 02 | Per | 13.2 | 1.4' | 1.0' | S |

170x250: Not visible.

| ne | NGC 1624 | 04 40.4 | +50 27 | Per | | 1.9' | | E |

WH: faint, considerably large, irregular figure, 6 or 7 stars + nebula.
42x60: Faintly visible. *170x250:* A very small very faint nebula surrounding a little equilateral triangle of moderately faint stars. The triangle is at the apex of a large wide "V" of 6–8 stars pointing S.

| ga | NGC 1636 | 04 40.7 | −08 36 | Eri | 12.8 | 1.2' | 0.8' | Sab |

HH: very faint, pretty small, round, very gradually brighter middle, resolvable, star 1' NE.
170x250: Very faint. Broadly slightly concentrated. Roughly round. Very small. No involved stars. It makes a shallow arc with a faint star nearby NNE and a moderate brightness star N.

| ga | NGC 1637 | 04 41.5 | −02 51 | Eri | 11.5b | 4.0' | 3.2' | Sc |

HH: considerably bright, large, round, very gradually brighter middle, easily resolvable.
170x250: Moderately faint to moderate brightness. Broadly to evenly slightly concentrated. Round. Moderately small. At its edge NW is an extremely faint star; nearby NNE is a moderately faint star.

| ga | NGC 1638 | 04 41.6 | −01 48 | Eri | 12.9b | 2.0' | 1.4' | S0 |

WH: faint, pretty large, little extended.
170x250: Moderately faint. Evenly to sharply fairly concentrated. Slightly elongated ENE-WSW. Small. No involved stars.

| ga | NGC 1640 | 04 42.2 | −20 26 | Eri | 12.4b | 2.6' | 2.0' | Sb |

Sn: very faint, pretty small, extended NE-SW, gradually brighter middle.
170x250: Faint to moderately faint. Broadly slightly concentrated. Slightly elongated NE-SW. Small. No involved stars. It makes a small triangle with two moderate brightness stars, one WSW and one SSW.

| ds | 55 Eri | 04 43.6 | −08 48 | Eri | 9.3" | 6.7 | 6.8 | G8 F4 |

42x60: A close equal pair. *170x250:* A moderately close perfectly equal pair. Both white. [STF 590]

| ga | NGC 1645 | 04 44.1 | −05 27 | Eri | 13.0 | 2.2' | 0.9' | S0+ |

dA: very faint, pretty small, round.
170x250: Faint. Evenly to sharply moderately concentrated. Round. Very small. No involved stars.

| ga | IC 381 | 04 44.5 | +75 38 | Cam | 13.1b | 2.4' | 1.3' | Sbc |

170x250: Very faint. Broadly slightly concentrated. Roughly round. Small. Just off its edge NNE is a faint star.

| ga | NGC 1650 | 04 45.2 | −15 52 | Eri | 12.9 | 2.2' | 1.2' | E+ |

Le: very faint, pretty small, extended N-S, brighter middle to a nucleus.
170x250: Very faint. Broadly slightly concentrated. Roughly round. Very small. No involved stars. It's in a slightly curved ENE-WSW line with three widely spaced moderately faint stars; NNW is a fourth star. Nearby SSE is a very very faint galaxy [*uncatalogued].

| ga | NGC 1653 | 04 45.8 | −02 23 | Eri | 12.9b | 1.4' | 1.4' | E+ |

WH: faint, considerably small, round, little brighter middle.
170x250: Moderate brightness. Sharply fairly concentrated. Round. Small. No involved stars.

| ga | MCG-3-13-4 | 04 45.9 | −17 16 | Eri | 13.1 | 1.8' | 1.2' | Sbc |

170x250: Very faint. Broadly concentrated. Roughly round. Very small. No involved stars.

| oc | NGC 1647 | 04 46.0 | +19 05 | Tau | 6.4 | 45.0' | | |

WH: cluster, very large, scattered bright stars.
12x35: A large cluster of faint stars. ***42x60:*** A number of widely scattered moderately faint stars, mostly uniform in brightness, plus some threshold stars. S of the main part of the cluster are two bright stars that don't look related. Somewhat triangular. A little more than half the field in size. Somewhat sparse.

| gc | PAL 2 | 04 46.1 | +31 22 | Aur | 13.0 | 2.2' | − | 9 |

170x250: Not visible.

| ga | NGC 1659 | 04 46.5 | −04 47 | Eri | 13.1b | 1.6' | 1.1' | Sbc |

HH: pretty faint, pretty small, irregularly extended E-W, brighter middle.
170x250: Moderately faint. Broadly to evenly moderately concentrated. Slightly elongated NE-SW. Small. No involved stars. Nearby SSE is a very faint star.

| ga | MCG-2-13-9 | 04 48.2 | −13 40 | Eri | 13.2 | 1.4' | 1.2' | E+ |

170x250: Very faint. Concentrated. Round. Tiny. No involved stars. Nearby NE is a faint double star.

| oc | NGC 1662 | 04 48.5 | +10 55 | Ori | 6.4 | 20.0' | | |

HH: cluster of scattered bright and faint stars.
12x35: A compact little cluster. ***42x60:*** A group of stars in a narrow wedge. In the middle is a small concentration in the shape of another wedge perpendicular to the rest of the group. ***170x250:*** A NNW-SSE linear group of stars with a small concentration in the middle: a little diamond figure of stars perpendicular to the other stars. W are more stars in a large semicircle. With the semicircle the cluster takes up the whole field and is empty in the middle; the linear group on the E side is 2/3 of the field in size. Very sparse. Nice-looking; the stars are bright.

| ga | NGC 1667 | 04 48.6 | −06 19 | Eri | 12.8b | 1.7' | 1.3' | Sc |

St: pretty faint, pretty small, round, resolvable?
170x250: Moderately faint. Broadly slightly concentrated. Round. Very small. No involved stars. A little ways WSW is a faint star; this is the first of several stars in that direction.

| ga | UGCA 95 | 04 49.2 | −29 12 | Cae | 13.1b | 3.4' | 2.8' | Sm |

170x250: Exceedingly faint. The very slightest diffuse glow. [Moderate size.] No involved stars. E is a little isosceles triangle of moderate brightness stars; SSW is a large arc of moderate brightness stars.

| bs | Pi 3 Ori | 04 49.8 | +06 57 | Ori | 3.19 | | 0.45 | F6 |

170x250: Soft white.

| ga | NGC 1679 | 04 49.9 | −31 58 | Cae | 12.0p | 2.7' | 2.0' | Sm |

JH: very bright, large, irregularly round, 4 stars involved.
170x250: Faint to moderately faint. Broadly slightly concentrated. Slightly elongated NNW-SSE. Moderately small. E of center is a faint involved star; at the galaxy's NW end is a moderately bright star; very nearby WSW is another moderately bright star; the two brighter stars interfere.

| oc | NGC 1664 | 04 51.1 | +43 42 | Aur | 7.6 | 18.0' | | |

WH: cluster, little rich, little compressed, pretty large.
12x35: Visible. ***42x60:*** A number of faint stars. ***170x250:*** Some two dozen moderate brightness stars in a NNW-SSE kite-like figure: a diamond head and a long curved tail passing W of the bright plotted star. The figure takes up 2/3 of the field; with additional scattered stars the cluster is the entire field width in size. Moderate density.

| vs | ST Cam | 04 51.2 | +68 10 | Cam | 7.0 | 8.4 | 2.2 | C5 |

12x35: Mag. 7.5. ***170x250:*** Light copper red. A pretty nice carbon star.

| ga | NGC 1682 | 04 52.3 | −03 06 | Ori | 12.6 | 0.8' | 0.8' | E/S0 |

dA: very faint, very small.
[See 1684.]

| ga | MCG-1-13-30 | 04 52.4 | −03 33 | Ori | 13.1 | 1.6' | 1.4' | S0 |

170x250: Faint. Evenly moderately concentrated. Round. Very very small. No involved stars. Nearby N is a moderate brightness star; this is the southeasternmost of an E-W trapezoidal figure of stars NW of the galaxy.

| ga | NGC 1684 | 04 52.5 | −03 06 | Ori | 12.5 | 2.4' | 1.6' | E+ |

HH: pretty faint, small, round, brighter middle, 7th mag. star SW.
170x250: Moderately faint to moderate brightness. Evenly moderately concentrated. Round. Small. No involved stars. A sixth of the field W is 1682. Moderately faint. Sharply pretty concentrated. Round. Very small. No involved stars.

| ga | NGC 1691 | 04 54.6 | +03 16 | Ori | 13.0p | 2.3' | 1.8' | S0/a |

St: faint, small, 11th mag. star involved.
170x250: Faint. Sharply pretty concentrated. Roughly round. Very small. No involved stars. It makes a small flat triangle with two faint stars, one WSW and one NNE; these are among a number of faint stars in the field that together look almost like a very loose cluster. The galaxy looks much like the individual stars except that it has a little halo.

| ga | IC 2104 | 04 56.3 | −15 47 | Lep | 13.1 | 2.3' | 1.4' | Sbc |

170x250: Extremely faint. Broadly slightly concentrated. Slightly elongated E-W. Small. No involved stars. It makes a long slightly curved and evenly spaced line with three stars SSW.

| ga | NGC 1700 | 04 56.9 | −04 51 | Eri | 12.2b | 3.3' | 2.0' | E4 |

HH: considerably bright, small, much brighter middle to a star.
170x250: Moderately bright. Sharply pretty concentrated. Round. Small. No involved stars.

| bs | Iota Aur | 04 57.0 | +33 09 | Aur | 2.70 | | 1.52 | K3 |

170x250: Gold.

| ga | IC 391 | 04 57.4 | +78 11 | Cam | 13.0b | 1.1' | 1.0' | Sc |

170x250: Faint. Broadly very slightly concentrated. Round. Small. No involved stars. Nearby N are a couple of faint stars; a little ways SSW are four moderately faint to faint stars.

| ga | IC 396 | 04 58.0 | +68 19 | Cam | 13.0p | 2.1' | 1.4' | S |

170x250: Faint. Sharply slightly concentrated. Roughly round. Small. No involved stars. Nearby SE are a couple of faint stars; nearby WNW are a couple of very faint stars.

| ga | NGC 1721 | 04 59.3 | −11 07 | Eri | 13.1 | 2.5' | 1.6' | S0 |

Ba: very faint, very small, round.
170x250: 1721, 1725, and a third galaxy [*1728 (Ba)] are in a little arc W to E. All are evenly to sharply fairly concentrated. Very small. No involved stars. 1721 is faint. Roughly round. Very nearby ENE is a faint star. 1725 is very faint. Roughly round. The easternmost galaxy is very faint. Slightly elongated N-S.

| ga | NGC 1723 | 04 59.4 | −10 58 | Eri | 12.5 | 3.2' | 1.9' | Sa |

Te: faint, between 2 9th and 10th mag. stars N and S, 3rd star E.
170x250: Moderately faint. Broadly slightly concentrated. Round. Small. No involved stars. It's on the side of an isosceles triangle of moderately bright stars.

| ga | NGC 1725 | 04 59.4 | −11 07 | Eri | 13.1 | 1.9' | 1.2' | E/S0 |

Ba: extremely faint, very small, round.
[See 1721.]

| ga | NGC 1730 | 04 59.5 | −15 49 | Lep | 13.1 | 2.4' | 1.2' | Sa |

Sw: faint, pretty small, little extended, between 2 faint stars.
170x250: Faint. Evenly moderately concentrated. Round. Very small. No involved stars.

| vs | R Lep | 04 59.6 | −14 48 | Lep | 5.9 | 11 | 5.74 | C6 |

12x35: Mag. 7.5. ***170x250:*** Copper red. A pretty good carbon star.

| ga | NGC 1726 | 04 59.7 | −07 45 | Eri | 12.7b | 1.4' | 1.0' | S0 |

JH: faint, round, 13th mag. star S.
170x250: Moderately faint to moderate brightness. Evenly moderately concentrated. Roughly round. Very small. No involved stars. Very nearby S is a moderate brightness star. Half the field SW is another galaxy [*1720 (dA)]. Faint. Broadly slightly concentrated. Elongated E-W. Small; a little larger than 1726. No involved stars.

| ga | NGC 1744 | 05 00.0 | −26 01 | Lep | 11.6b | 8.2' | 4.4' | Sd |

JH: faint, very large, very much extended, very gradually very little brighter middle.
170x250: Faint. Broadly very slightly concentrated. Elongated N-S. Moderately large. SW of center is an extremely faint involved star; off the galaxy's N end is a pair of very faint stars; these are the middle stars of a WNW-ESE line of stars curving around the N end of the galaxy.

| ga | NGC 1729 | 05 00.3 | −03 21 | Ori | 13.1 | 1.6' | 1.3' | Sc |

WH: very faint, pretty large, 2 bright stars very near.
170x250: Faint. Broadly slightly concentrated. Roughly round. Small. No involved stars. It makes a very small triangle with a moderately bright star E and a moderate brightness star N.

| bs | Epsilon Aur | 05 02.0 | +43 49 | Aur | 2.99 | | 0.54 | A9 |

170x250: Soft white.

| ds | STF 630 | 05 02.0 | +01 37 | Ori | 14.4" | 6.5 | 7.7 | B8 A1 |

42x60: The primary has a fainter companion. *170x250:* A moderate distance unequal pair. Both white.

| ga | ESO 486-19 | 05 03.2 | −22 49 | Lep | 12.5v | 1.4' | 0.4' | S0− |

170x250: Moderate brightness; high surface brightness. Extremely concentrated to a stellar nucleus. The galaxy is almost completely stellar; it looks like a star with a tiny halo. Round. No involved stars. It makes a sharp triangle with two faint stars a little ways NNE. Very nearby NNW is a faint stellar galaxy [*E486-17].

| oc | NGC 1746 | 05 03.6 | +23 49 | Tau | 6.1 | 41.0' | | |

dA: cluster, poor.
12x35: A large scattered group of stars. *42x60:* A good number of scattered mixed brightness stars: moderately bright and fainter. The cluster is more concentrated toward the E side. Rectangular E-W. It fills the field. Moderate density.

| ga | NGC 1779 | 05 05.3 | −09 08 | Eri | 13.0b | 2.3' | 1.2' | S0/a |

HH: pretty bright, small, round, gradually pretty much brighter middle.
170x250: Faint. Evenly moderately concentrated. Round. Small. No involved stars.

| ga | NGC 1784 | 05 05.4 | −11 52 | Lep | 12.4b | 4.6' | 2.7' | Sc |

JH: pretty bright, pretty large, very little extended, very gradually brighter middle, among stars.
170x250: Faint. Broadly slightly concentrated. Slightly elongated E-W. Moderately small. No involved stars. It's in the middle of a curved ENE-WSW group of five moderate brightness stars.

| vs | W Ori | 05 05.4 | +01 11 | Ori | 6.5 | 10 | 3.44 | C5 |

12x35: Mag. 5.5. *170x250:* Orange red. A very good carbon star; nice and bright.

| bs | Epsilon Lep | 05 05.5 | −22 22 | Lep | 3.19 | | 1.46 | K5 |

170x250: Gold.

| pn | J 320 | 05 05.6 | +10 42 | Ori | 12.9p | 26.0" | | |

170x250: Moderately bright. Almost stellar. Moderate blinking effect. It makes an arc with two moderate brightness stars a little ways WSW. [PK190-17.1]

| ga | NGC 1800 | 05 06.4 | −31 57 | Col | 13.1b | 2.0' | 1.0' | Im |

JH: pretty bright, pretty much extended, gradually pretty much brighter middle, 13th mag. star E.
170x250: Faint. Broadly slightly concentrated. Slightly elongated WNW-ESE. Small. No involved stars. Nearby ENE is a faint star.

| bs | Eta Aur | 05 06.5 | +41 14 | Aur | 3.20 | | −0.21 | B3 |

170x250: White.

| ne | NGC 1788 | 05 06.9 | −03 21 | Ori | | 5.5' | | R |

WH: bright, considerably large, round, brighter middle to a 15th mag. triple star, 10th mag. star NW.
170x250: Moderately faint. It has a moderately faint central star. The nebula surrounds this star and a moderately bright star a little ways NW in a NW-SE oval, and doesn't go much beyond; it's more obvious around the fainter star. At first sight it looks like a faint galaxy with a nuclear concentration.

| bs | Beta Eri | 05 07.9 | −05 05 | Eri | 2.79 | | 0.13 | A3 |

170x250: White.

| oc | NGC 1778 | 05 08.1 | +37 03 | Aur | 7.7 | 6.0' | | |

HH: cluster, pretty compressed, little rich, irregular figure, bright stars.
12x35, 42x60: A handful of faint stars. *170x250:* A number of mixed brightness stars: moderate brightness and moderately faint. The main figure is some dozen moderate brightness stars in a long NW-SE oval half the field in size. NE are more stars in a few small clumps, making the cluster roughly triangular overall and the field width in size. Somewhat sparse.

| oc | NGC 1807 | 05 10.7 | +16 31 | Tau | 7.0 | 17.0' | | |

JH: cluster, pretty rich, bright and faint stars.
12x35: 1807 and 1817 are a few faint stars each. *42x60:* 1807 is a large irregular "X" of some half-dozen moderately bright stars. 1817 is a cluster of very faint stars, with four brighter stars in a NNW-SSE linear figure on the W side. *170x250:* 1807 is some dozen widespread fairly bright stars in two axes, a NE-SW one exactly the field width in length and a shorter one N-S; plus a handful of moderately faint stars. Toward the middle is a bit of a concentration including a triangle of stars, one corner of which is a double. Very sparse. 1817 is a good number of moderately faint stars plus the four moderately bright stars on the W side, the NW of which is a wide

double. Like 1807 it completely takes up the field but it's a denser, more regular cluster. Moderate density. The two clusters make a good contrast.

| ga | UGCA 103 | 05 10.8 | −31 35 | Col | 13.1b | 2.9' | 2.4' | Sdm |

170x250: Not visible.

| oc | NGC 1798 | 05 11.7 | +47 41 | Aur | 10.0 | 3.0' | | |

Ba: small, cluster or cluster + nebula.
170x250: A faint cluster of a fair number of mostly very faint stars. Slightly elongated E-W. At its E end is a moderate brightness star, probably an unrelated field star. A quarter of the field in size. Moderately dense.

| ga | UGCA 104 | 05 11.7 | −14 47 | Lep | 13.0b | 2.7' | 2.0' | Scd |

170x250: Exceedingly faint. A slight glow. Moderate size. No involved stars. It's between a moderate brightness star WNW and a faint star E.

| ga | UGCA 106 | 05 12.0 | −32 58 | Col | 12.9p | 3.5' | 2.9' | Sm |

170x250: Not visible.

| oc | NGC 1817 | 05 12.1 | +16 41 | Tau | 7.7 | 15.0' | | |

HH: cluster, large, rich, little compressed, 11th to 14th mag. stars.
[See 1807.]

| ga | NGC 1832 | 05 12.1 | −15 41 | Lep | 12.0b | 2.5' | 1.6' | Sbc |

WH: pretty bright, irregularly round, much brighter middle, considerably faint star 1' NE.
170x250: Moderate brightness. Broadly to evenly moderately concentrated. Round. Small. No involved stars. Very nearby E is a moderate brightness star.

| bs | Mu Lep | 05 12.9 | −16 12 | Lep | 3.31 | | −0.11 | B9 |

170x250: White.

| bs | Beta Ori | 05 14.5 | −08 12 | Ori | 0.12 | | −0.03 | B8 |

170x250: The primary has a close relatively very faint companion. The primary is white. A brilliant star. [Rigel; STF 668: 9.5", 0.1, 6.8, B8 --]

| ga | UGCA 108 | 05 15.3 | −30 31 | Col | 13.1p | 2.8' | 0.7' | Sd |

170x250: Very very faint. Hardly concentrated. A fat edge-on, NNE-SSW. Moderate size. No involved stars. Nearby SE is a very faint star; NNE are two moderately bright stars aligned perpendicular to the axis of the galaxy; WNW are two more moderately bright stars in line with the galaxy.

| ds | 14 Aur | 05 15.4 | +32 41 | Aur | 14.3" | 5.1 | 7.4 | A9 A2 |

42x60: The primary has a relatively very faint companion. ***170x250:*** A moderate distance unequal pair. The primary is white. Closer to the primary than the main companion is a very very faint star. [STF 653]

| bs | Alpha Aur | 05 16.7 | +45 59 | Aur | 0.08 | | 0.80 | G5 |

170x250: Yellow white. A brilliant star, with diffraction spikes going all the way across the field. [Capella]

| pn | IC 2120 | 05 18.2 | +37 36 | Aur | – | 47.0" | | |

170x250: Not visible.

| ga | UGC 3253 | 05 19.7 | +84 03 | Cep | 13.2p | 1.6' | 1.0' | Sb |

170x250: Very faint. Diffuse. Small. It's among a group of about ten moderately faint and faint stars in a N-S quadrilateral figure, one of which is at the galaxy's edge NNW; these stars interfere.

| ga | NGC 1879 | 05 19.8 | −32 08 | Col | 13.2b | 2.4' | 1.6' | Sm |

JH: very faint, large, round, very gradually very little brighter middle, 12th mag. star W.
170x250: Exceedingly faint. The very slightest unconcentrated glow. Roughly round. Moderate size. No involved stars. NW is a bright star.

| oc | NGC 1857 | 05 20.2 | +39 21 | Aur | 7.0 | 5.0' | | |

HH: cluster, pretty rich, pretty compressed, 7th mag. and fainter stars.
42x60: Faintly visible around a bright star. ***170x250:*** A fair number of faint stars around the bright plotted star plus three moderately bright stars: two ESE and one NW. Elongated N-S. Half the field in size. Moderate density.

| ds | HJ 3752 | 05 21.8 | −24 46 | Lep | 3.4" | 5.4 | 6.6 | G7 A7 |

170x250: The primary has an extremely close relatively faint companion. The primary is yellow.

| ga | NGC 1888 | 05 22.6 | −11 30 | Lep | 12.8b | 3.5' | 1.0' | Sc |

HH: pretty bright, pretty large, round, resolvable.
170x250: Moderately faint to moderate brightness. Evenly moderately concentrated. Elongated NW-SE. Small. NE of center is a very faint involved star or some other object [this is *1889 (LR), a stellar galaxy].

| oc | NGC 1893 | 05 22.7 | +33 24 | Aur | 7.5 | 11.0' | | |

JH: cluster, large, rich, little compressed.
12x35: A few stars. *42x60:* An elongated cluster of 8–10 stars. *170x250:* A fair number of mixed brightness stars: moderately bright, moderate brightness, and fainter. Very elongated N-S; it widens and is rounded toward the N end, like a lightbulb. The S half of the cluster is a N-S linear group of brighter stars; the N half contains more numerous, fainter stars. The cluster is the field width in size. Moderate density.

| ds | STF 701 | 05 23.3 | −08 25 | Ori | 6.1" | 6.0 | 7.8 | B8 A5 |

42x60: The primary has a very very close relatively faint companion. *170x250:* A moderately close unequal pair. Both white.

| ds | WNC 2 | 05 23.9 | −00 52 | Ori | 3.0" | 6.1 | 7.1 | F7 − |

42x60: Elongated. *170x250:* A very close perfectly equal pair. Both white.

| bs | Eta Ori | 05 24.5 | −02 23 | Ori | 3.35 | | −0.17 | B1 |

170x250: An extremely close unequal pair; the companion is tagged right next to the primary. Both white. [DA 5: 1.9", 3.8, 4.8]

| gc | NGC 1904/M79 | 05 24.5 | −24 33 | Lep | 8.0 | 9.6' | −7.65 | 5 |

HH: globular cluster, pretty large, extremely rich, extremely compressed, well resolved.
12x35, 42x60: Visible. *170x250:* Resolved throughout into faint stars, except for the very middle. Evenly concentrated. A quarter of the field in size. It has a small central ball. The outliers form a partial ring around the main body of the cluster; the ring is broken on the SW side. Pretty regular, especially the central part.

| bs | Gamma Ori | 05 25.1 | +06 20 | Ori | 1.64 | | −0.22 | B2 |

170x250: White.

| ga | MCG+1-14-37 | 05 25.7 | +06 35 | Ori | 12.0 | 0.8' | 0.8' | – |

170x250: Not visible.

| oc | NGC 1883 | 05 25.9 | +46 32 | Aur | 12.0 | 2.5' | | |

WH: cluster, very faint, pretty rich, pretty compressed, irregular figure.
170x250: A faint cluster of very faint stars. Slightly elongated NNE-SSW. Small. Moderately dense.

| bs | Beta Tau | 05 26.3 | +28 36 | Tau | 1.65 | | −0.13 | B7 |

170x250: White. A nice bright star.

| pn | IC 418 | 05 27.5 | −12 41 | Lep | 10.7p | 12.0" | | |

12x35, 42x60: It looks like an ordinary star, moderate brightness in the 60 mm.
170x250: Very bright; very high surface brightness. It has a moderately faint central star. A pretty well defined disk. Fairly strong blinking effect. Round. Small.

| oc | NGC 1907 | 05 28.0 | +35 19 | Aur | 8.2 | 6.0' | | |

HH: cluster, pretty rich, pretty compressed, round, 9th to 12th mag. stars.
12x35: Very faintly visible. *42x60:* A faint little cloud. *170x250:* A fair number of moderately faint stars. Very small. Moderately dense.

| bs | Beta Lep | 05 28.2 | −20 45 | Lep | 2.84 | | 0.82 | G5 |

170x250: Yellow.

| oc | NGC 1912/M38 | 05 28.7 | +35 50 | Aur | 6.4 | 21.0' | | |

HH: cluster, bright, very large, very rich, irregular figure, bright and faint stars.
12x35: A lot of closely packed faint stars. *42x60:* A good number of moderately faint stars, fairly uniform in brightness, in a kite shape. Detached from the main body of the cluster is an outer ring of stars, most substantial S and SE. The cluster is half the field in size with these outer members. *170x250:* A lot of moderate brightness to moderately faint stars, pretty uniform in brightness. At its edge ENE is one slightly brighter star. There aren't a lot of faint stars. The main body of the cluster takes up the entire field. Moderate density to moderately dense.

0 to 6 Hours: FALL

| ds | 118 Tau | 05 29.3 | +25 09 | Tau | 4.9" | 5.8 | 6.6 | B8 A1 |

42x60: Extremely close: as close as can be in the 60 mm with the stars still distinctly separate. Slightly unequal. *170x250:* A close slightly unequal pair. Both white. [STF 716]

| oc | NGC 1931 | 05 31.4 | +34 15 | Aur | 10.1 | 6.0' | | |

HH: very bright, large, round, bright triple star in middle.
12x35: Faintly visible. *42x60:* A half-dozen widely spaced faint stars. *170x250:* A number of moderate brightness stars, uniform in brightness, and some much fainter stars, all in an E-W crescent shape. The field width in size. Sparse. At the cluster's W end, around a tight little trio of stars, is a small nebula. Moderate brightness. Round. An interesting combination of cluster and nebula.

| bs | Delta Ori | 05 32.0 | –00 17 | Ori | 2.24 | | –0.22 | O9 |

170x250: The primary has a very wide relatively faint companion. The primary is white. [STF 14: 53.2", 2.2, 6.3]

| ga | IC 421 | 05 32.1 | –07 55 | Ori | 12.3 | 3.2' | 2.7' | Sbc |

170x250: Not visible.

| ds | STF 730 | 05 32.2 | +17 03 | Tau | 9.7" | 6.0 | 6.5 | B7 B7 |

12x35: Filling the field around STF 730 is a large group of stars in a crescent shape; it looks like a loose cluster [this is CR65]. *42x60:* STF 730 is a nearly equal pair. *170x250:* A moderate distance equal pair. Both white.

| bs | Alpha Lep | 05 32.7 | –17 49 | Lep | 2.58 | | 0.21 | F0 |

170x250: Soft white.

| ga | NGC 1954 | 05 32.8 | –14 03 | Lep | 12.4b | 4.2' | 2.2' | Sbc |

HH: very faint, small, round, suddenly much brighter middle.
170x250: Very faint to faint. Sharply concentrated. Round. Very small. It makes a little arc with two faint stars NNW. A quarter of the field SSE is a very faint galaxy [*1957 (Le)]. Sharply concentrated. Very small. Nearby SSW is a moderately faint star. Half the field NNW of 1954 is a very very faint galaxy [*I2132]. Broadly slightly concentrated. Roughly round. Small. No involved stars. It's on the long side of a trapezoid of faint stars.

| ga | NGC 1964 | 05 33.4 | −21 56 | Lep | 11.6b | 5.6' | 2.1' | Sb |

HH: faint, very small, round, very suddenly very much brighter middle to a 12th mag. star, 3 stars involved.
170x250: Bright. Sharply extremely concentrated. Elongated NNE-SSW. Small. It has an extremely faint halo and a very small bright core. At its edge WSW is a faint double star; at its S end is a faint star; NW is a sharp triangle of bright stars.

| ga | NGC 1979 | 05 34.0 | −23 18 | Lep | 12.8p | 2.1' | 1.7' | S0 |

WH: very faint, very small, stellar.
170x250: Moderately faint. Sharply pretty concentrated. Round. Very very small. No involved stars. A little ways ESE is a moderate brightness star.

| ne | NGC 1952/M1 | 05 34.5 | +22 01 | Tau | | 6.0' | | E |

JH: very bright, very large, extended NW-SE, very gradually little brighter middle, resolvable.
12x35: Visible. *42x60:* Fairly bright. *170x250:* Pretty bright. Elongated NW-SE. Pretty even in light, except that the SE end is a little more diffuse, wispy. Most of the nebula is roughly oval, but a very faint extension at the SE end makes it more pointed: football-shaped. The extension has a dark indentation on its N side. The nebula is a third of the field in size. At its edge NE is a very faint involved star; at its edge W is another very faint involved star. [Crab Nebula]

| ga | ESO 423-24 | 05 34.7 | −29 13 | Col | 13.1p | 1.7' | 1.7' | S0 |

170x250: Moderately faint to moderate brightness; fairly high surface brightness. Sharply very concentrated. Round. Very small. No involved stars. Off its edge NW is an extremely faint star.

| bs | Lambda Ori | 05 35.1 | +09 56 | Ori | 3.3 | | − | O8 |

170x250: The primary has a very close fainter companion. Both white. A nice double star. [STF 738: 4.3", 3.6, 5.5, O8 B0]

| oc | NGC 1981 | 05 35.2 | −04 25 | Ori | 4.2 | 24.0' | | |

JH: cluster, very bright, little rich, scattered bright stars.
12x35: A half-dozen stars in two roughly parallel N-S shallow arcs, each consisting of three stars of uniform brightness. The stars of the E arc are brighter. *42x60:* A very loose cluster of a dozen stars. Very sparse. *170x250:* It just fills the field.

| bs | Iota Ori | 05 35.4 | –05 54 | Ori | 2.77 | | –0.24 | O9 |

170x250: The primary has a moderately close much fainter companion. The primary is white. [STF 752: 10.9", 2.8, 6.9]

| ne | NGC 1976/M42 | 05 35.4 | –05 27 | Ori | | 60.0' | | E+R |

JH: !!! Theta 1 Ori and the great nebula.
12x35: The broad outline and the major features of M42 are visible in miniature: the umbrella-like nebula, the dark cloud, including the little snout pointing in toward the Trapezium, and the Trapezium itself, which is clearly multiple. *42x60:* The arc-shaped nebula is very bright. There's a big difference between the SE side, which is sharper, and the NW side, which is much more diffuse. The Trapezium is resolved into four stars. Nearby SE are two more bright stars [Theta 2 A and B] and a faint star in a little E-W line. The Trapezium and these three additional stars are the major involved stars within the bright part of the nebula. The nebula is half the field in size. *170x250:* The NW wing of the nebula takes up an entire field with variegated nebulosity. The SE wing also takes up an entire field but is sharper: a remarkably well defined, relatively thin arclike cloud, the inside edge of which is especially sharp. The relatively small, brightest part of the nebula around the Trapezium is notably square; two sides of it, SE and SW, are quite straight. There's obvious curdling in this part of the nebula and broader unevenness throughout the rest. The nebula is slightly greenish. The dark nebula N, between M42 and 1977, is very wide. Elongated NW-SE, in the same direction as the bright nebula, but larger. It has large wings like the bright nebula, as well as appendages, one of which goes N and bisects M43 into a western bright part with a bright star in the middle and fainter nebulosity E. The two additional components of the Trapezium, E and F, are well seen; all six stars are clear and sharp. Truly an exceptional deep-sky object in the amount and sharpness of its detail and in the brightness of its features. [Orion Nebula]

| oc | NGC 1980 | 05 35.4 | –05 54 | Ori | 2.5 | 13.0' | | |

HH: very faint, very very large, Iota Ori involved.
42x60: A cluster consisting of Iota Ori, a wide pair SW, and a half-dozen fainter stars, all in a long NNE-SSW quadrilateral figure. A quarter of the field in size. Very sparse. *170x250:* A very loose cluster of some 20 very widespread bright and moderately bright stars in a triangular figure. There are hardly any fainter stars. It's somewhat empty in the middle; Iota itself is a little E of center. It fills the field exactly.

| oc | NGC 1977 | 05 35.5 | –04 49 | Ori | – | 20.0' | | |

WH: !! 42 Ori and nebula.

12x35: A couple of bright stars and a couple more faint stars. *42x60:* Two bright stars and three relatively faint stars, with nebulosity surrounding all of these stars and going NW toward two more stars. *170x250:* There are more stars around the bright stars, making a wedge or triangle shape. A lot of surprisingly bright nebulosity surrounds these stars and extends W. N is a wide dark lane with an island of bright nebulosity within it. NW, on the other side of the dark lane, is more bright nebulosity and three stars in a NE-SW line, the northernmost of which is a double. Altogether the cluster is 3/4 of the field in size; the nebula fills the field. The cluster is just a handful of very widespread stars but the nebula is pretty neat.

ne	NGC 1982/M43	05 35.5	−05 16	Ori		7.0'		E

WH: ! very bright, very large, round with tail, much brighter middle to an 8th–9th mag. star.
[See M42; M43 is not really a separate object.]

oc	NGC 1960/M36	05 36.1	+34 08	Aur	6.0	12.0'		

JH: cluster, bright, very large, very rich, little compressed, scattered 9th to 11th mag. stars.
12x35: A compact cluster of stars. *42x60:* A compact cluster of moderately bright stars. *170x250:* A good number of moderately bright stars, very uniform in brightness, plus some fainter stars. The outer parts are very sparse; the cluster concentrates toward the middle. The field width in size. Moderately dense. A nice cluster.

bs	Epsilon Ori	05 36.2	−01 12	Ori	1.70		−0.19	B0

6x30: Surrounding the belt stars of Orion and going N toward Gamma is a substantial open cluster of moderately bright and fainter stars [CR70]. It's oval NW-SE around the belt stars – matching the alignment of the stars – but since it turns N it's bent overall. The brightest stars are densest right around Epsilon, forming a ring around it. *12x35:* The cluster surrounds the belt stars and overflows the field. It's mostly around Epsilon – not so much around Zeta – and extends N from Delta. *170x250:* Epsilon is white.

ne	NGC 1999	05 36.5	−06 42	Ori		21.5'		E+R

WH: bright star involved in nebula.
42x60: A very faint very vague nebulosity. *170x250:* The faint nebulosity surrounds a star and fills the field, but immediately around the star is a smaller, relatively bright nebula that looks a little like a planetary. Within this small round nebula, right next to the star W, is a small dark space.

0 to 6 Hours: FALL 143

| bs | Zeta Tau | 05 37.6 | +21 08 | Tau | 3.03 | | −0.19 | B4 |

170x250: White.

| ne | NGC 1985 | 05 37.7 | +32 00 | Aur | | 1.0' | | R |

HH: considerably faint, small, round, pretty suddenly brighter middle.
170x250: A very faint nebula. Round. Very small. It looks like a galaxy except that it has a faint central star like a nebula or a planetary. SSW is a flat triangle of stars; ENE are a couple of fairly bright stars.

| ds | Sigma Ori | 05 38.7 | −02 36 | Ori | 13.0" | 3.7 | 7.5 | O9 B2 |

12x35: The primary has a fainter companion. Two relatively faint stars NW and one SE form a very sharp wedge with the primary. *42x60:* The primary has a second fainter companion. The southern star of the two NW is a double. *170x250:* The primary has a still closer third companion. In sum: WSW of the primary is a moderately close relatively very faint companion [11.4", 10.3], E is a moderate distance fainter companion [12.9", 7.5], and ENE is a very wide fainter companion [42.6", 6.5]. The primary is white. A very nice quadruple star. [STF 762] The double star NW is a moderate distance equal pair [STF 761: 9.0", 8.4, 8.6].

| oc | NGC 2017 | 05 39.4 | −17 51 | Lep | − | 10.0' | | |

JH: cluster of bright stars.
12x35, 42x60: A little NW-SE "Y" figure of four stars; the brightest is in the middle and the faintest is SE. *170x250:* The stars of the "Y" are bright. On the E side of the figure, making a trapezoid with the three easternmost stars of the "Y," are two faint stars. A little less than a quarter of the field in size. A very attractive group in all scopes.

| bs | Alpha Col | 05 39.6 | −34 04 | Col | 2.60 | | −0.08 | B7 |

170x250: White.

| ds | LAL 194 | 05 39.7 | −20 26 | Lep | 10.9" | 6.9 | 7.9 | B8 − |

42x60: An unequal pair. *170x250:* A moderate distance slightly unequal pair. Both white.

| bs | Zeta Ori | 05 40.8 | −01 57 | Ori | 1.70 | | −0.10 | O9 |

170x250: The primary has an extremely close fainter companion. The primary is white. [STF 774: 2.3", 1.9, 4.0, O9 B0]

ne	NGC 2023	05 41.6	–02 14	Ori		10.0'		E+R

WH: bright star in middle of large, little extended nebula.
42x60: A small very faint nebula surrounding the plotted star. *170x250:* A pretty faint reflection nebula surrounding the star. Round. Small.

ne	NGC 2024	05 41.9	–01 51	Ori		30.0'		E

WH: ! irregular, bright, very very large, black [lane].
42x60: Very faintly visible. *170x250:* A large dark cloud with a moderately faint emission nebula surrounding it W, N, and E. There's obvious unevenness in the E part of the nebula. At the S end of this eastern part is a dark bar jutting out ENE from the main dark lane, with a parallel bright counterpart right below it. The nebula W of the dark lane also has some unevenness but it's not as distinct as that on the E side. Partly straddling the dark lane is a large isosceles triangle of moderate brightness stars pointing S. The nebula as a whole takes up the entire field. A very interesting nebula despite its relative faintness; it would be much more impressive if Zeta were not so nearby casting glare.

ga	NGC 1961	05 42.1	+69 22	Cam	11.7b	4.5'	2.9'	Sc

WH: considerably faint, pretty large, irregular figure, much brighter middle, easily resolvable, star involved.
170x250: Faint. Broadly very slightly concentrated. Roughly round. Small. At its edge SSE is a faint star that is the crux of a long "Y" of faint stars lying across the galaxy and opening up NW.

pn	NGC 2022	05 42.1	+09 05	Ori	12.4p	35.0"		

HH: planetary nebula, pretty bright, very small, very little extended, resolvable?
170x250: Moderately bright. No central star. A small disk. Not a very clean edge. Even in brightness. Moderate blinking effect. Roughly round.

vs	Y Tau	05 45.7	+20 41	Tau	7.1	9.5	3.03	C5

12x35: Mag. 7.5. *170x250:* Orangish.

ne	NGC 2068/M78	05 46.7	+00 03	Ori		8.4'		R

JH: bright, large, wisp-shaped, very gradually much brighter middle to a nucleus, 3 stars involved, resolvable.
12x35: Easily visible. *42x60:* A moderate brightness nebula around two stars.
170x250: Moderate brightness. Irregularly oval N-S. There's some broad unevenness in the nebula. The N edge is sharper than the S edge; dark nebulosity probably

borders the nebula N. A quarter of the field in size. On the N side are two moderately bright stars, one of which is close to the edge; just inside the S edge is a faint star.

| ga | NGC 2090 | 05 47.0 | –34 15 | Col | 12.0b | 6.3' | 2.9' | Sc |

JH: globular cluster, bright, pretty large, irregularly round, gradually brighter middle.
170x250: Moderate brightness. Broadly moderately concentrated. Very elongated NNE-SSW. Moderately small. It's closely surrounded by faint stars: curving E from its N end is a little arc of stars and on either side of its S end is a star.

| ne | NGC 2071 | 05 47.1 | +00 18 | Ori | | 8.0' | | R |

WH: star with very faint, large chevelure.
170x250: A faint nebulosity around a moderately bright star with a distant faint companion. Roughly round. Small. NW is another moderately bright star.

| bs | Kappa Ori | 05 47.8 | –09 40 | Ori | 2.06 | | –0.17 | B0 |

170x250: White.

| ga | NGC 2089 | 05 47.9 | –17 36 | Lep | 12.9p | 1.8' | 1.1' | S0– |

WH: very faint, extremely small, stellar.
170x250: Moderately faint. Sharply pretty concentrated. Round. Small. No involved stars. Closely surrounding it on the E side are three faint stars; SSE are two moderate brightness stars in a short evenly spaced line with the galaxy.

| oc | K 8 | 05 49.3 | +33 37 | Aur | 11.2 | 7.0' | | |

42x60: A couple of stars. ***170x250:*** A few faint stars between a star NW and a wide unequal double star SSE, plus threshold stars and the light of unresolved stars. Very small. Moderate density.

| ga | PGC 17965 | 05 50.3 | –10 17 | Ori | 12.9 | 2.2' | 1.6' | Sb |

170x250: Not visible.

| ga | MCG-3-15-21 | 05 50.8 | –18 10 | Lep | 13.0 | 1.7' | 1.4' | S0 |

170x250: Very very faint. Slightly concentrated. Round. Very small. No involved stars. Next to it SSE is a bright star.

| ga | NGC 2106 | 05 50.8 | –21 34 | Lep | 13.1p | 2.7' | 1.3' | S0 |

JH: very faint, small, very little extended, gradually brighter middle.
170x250: Moderately faint. Evenly moderately concentrated. Round. Very small. No involved stars. It's on the side of a nearly isosceles triangle of moderately faint stars.

| ga | UGCA 114 | 05 50.9 | –14 46 | Lep | 12.6 | 2.9' | 2.1' | Sd |

170x250: Very very faint. Hardly concentrated. Roughly round. Moderately small. It takes up much of the gap between a moderately faint star nearby ESE and a moderate brightness star a little farther WNW; a little E of the western star is a faint star. [M-2-15-12 was not seen.]

| ga | NGC 2110 | 05 52.2 | –07 27 | Ori | 13.2 | 1.7' | 1.2' | S0– |

HH: extremely faint, considerably small, little extended, pretty suddenly little brighter middle, easily resolvable.
170x250: Moderately faint. Sharply fairly concentrated. Round. Small. No involved stars.

| oc | NGC 2099/M37 | 05 52.4 | +32 33 | Aur | 5.6 | 23.0' | | |

JH: cluster, rich, pretty compressed middle, bright and faint stars.
12x35: A cluster of very faint stars. Broadly concentrated. It looks a lot like a globular cluster. ***42x60:*** The stars are pretty uniform in brightness and relatively faint for a major open cluster. Broadly concentrated. Extremely dense for an open cluster. For all these reasons it looks a little like a globular cluster. ***170x250:*** A whole lot of moderate brightness stars, very uniform in brightness, with very very dense little clumps. In the middle is a slightly brighter yellow star. With outer members it's a little more than the field width in size, but the main body of the cluster fits comfortably in one field. Very dense. Because of its density and the uniformity of its stars it doesn't look like a typical open cluster; it looks halfway to a globular cluster – a loose and very well resolved one. A very nice and interesting cluster.

| ga | IC 438 | 05 53.0 | –17 52 | Lep | 12.7p | 2.8' | 2.1' | Sc |

170x250: Faint. Broadly slightly concentrated. Slightly elongated NE-SW. Moderately small. Just off its NE end is a faint star; nearby NW is an extremely faint star; a little ways N is a moderately faint star.

| oc | NGC 2112 | 05 53.9 | +00 24 | Ori | 9.1 | 11.0' | | |

HH: cluster, pretty large, little rich, pretty compressed, faint stars.
12x35: Very faintly visible. *42x60:* A handful of very faint stars in a shallow arc. *170x250:* A brighter star and a half-dozen moderately faint stars in a "Y" figure opening up SE, plus a good number of very faint and threshold stars. Half the field in size. Moderate density.

| bs | Alpha Ori | 05 55.2 | +07 24 | Ori | 0.50 | | 1.85 | M2 |

170x250: Deep gold. A brilliant star, with diffraction spikes all the way across the field. A little more deeply colored than Aldebaran. Beautiful. [Betelgeuse]

| vs | U Ori | 05 55.8 | +20 10 | Ori | 5.4 | 12.5 | – | M6 |

12x35: Mag. 9.0. *170x250:* Copper red.

| pn | IC 2149 | 05 56.3 | +46 07 | Aur | 11.2p | 34.0" | | |

42x60: Easily visible. *170x250:* Fairly bright. No central star. A very small disk. Pretty strong blinking effect. Round. In the field with it are a number of good comparison stars: bright, obvious stars with no blinking effect.

| bs | Beta Aur | 05 59.5 | +44 56 | Aur | 1.85 | | 0.08 | A2 |

170x250: White.

| bs | Theta Aur | 05 59.7 | +37 12 | Aur | 2.62 | | −0.08 | A0 |

170x250: White.

Chapter 11

6 to 12 Hours: WINTER

| oc | NGC 2129 | 06 01.0 | +23 18 | Gem | 6.7 | 6.0' | | |

HH: cluster, pretty large, 40 or 50 8th to 15th mag. stars.
12x35: A couple of stars. *42x60:* Several more faint stars are made out. *170x250:* A modest number of mixed brightness stars: two bright stars, about a dozen moderate brightness stars, and a few fainter stars. A quarter of the field in size. Somewhat sparse.

| ga | NGC 2139 | 06 01.1 | −23 40 | Lep | 12.0b | 2.6' | 1.9' | Scd |

WH: faint, small.
170x250: Moderately faint. Broadly slightly concentrated. Round. Moderately small. No involved stars. NNW is a sharp triangular figure of stars that points toward the galaxy.

| oc | NGC 2126 | 06 03.0 | +49 54 | Aur | 10.2 | 6.0' | | |

WH: cluster, not rich, one 7th mag. star.
42x60: A few stars around the bright plotted star are resolved with averted vision. *170x250:* A number of moderately faint and faint stars NW to SSW of the bright star. Somewhat square. A third of the field in size. Somewhat sparse.

| oc | NGC 2141 | 06 03.1 | +10 26 | Ori | 9.4 | 10.0' | | |

Ba: faint, pretty small, diffuse.
170x250: A cluster of a number of very faint and threshold stars plus the light of more unresolved stars. It's within an irregular ENE-WSW diamond figure of brighter stars; three of the four corners of the figure are wide pairs. The brightest stars of the cluster, which are still very faint, are in a small NNW-SSE linear concentration. A quarter of the field in size. Pretty dense. Not well made out; somewhat tantalizing.

| ne | NGC 2149 | 06 03.5 | –09 44 | Mon | | 3.0' | | R |

St: faint, 12th mag. star involved.
170x250: A faint reflection nebula around a moderately faint star. Round. Small. The star makes a triangle with two stars WNW, one moderately bright and one faint.

| oc | IC 2157 | 06 05.0 | +24 00 | Gem | 8.4 | 6.0' | | |

12x35: Faintly visible. *42x60:* A few faint stars. *170x250:* Six moderately faint stars in a small trapezoidal figure plus some faint stars. Small. Sparse.

| ga | IC 2158 | 06 05.3 | –27 51 | Col | 12.9p | 1.7' | 1.3' | Sab |

170x250: Faint to moderately faint. Broadly slightly concentrated. Roughly round. Small. At its edge SSE is an involved star; W is a long slightly curved N-S line of three stars.

| oc | NGC 2158 | 06 07.5 | +24 05 | Gem | 8.6 | 5.0' | | |

HH: cluster, pretty small, much compressed, very rich, triangle of stars near, extremely faint stars.
12x35: Faintly visible. *42x60:* A faint cloud. *170x250:* A whole lot of closely packed faint stars. Small. Very dense.

| ne | NGC 2170 | 06 07.5 | –06 24 | Mon | | 4.7' | | R |

WH: 9th mag. star in very faint, pretty large nebula, extended N-S.
42x60: Very faintly visible around a star. *170x250:* A moderate brightness reflection nebula around a star. Round. Small. It makes a short curved line with two faint stars, one N and one S; it doesn't reach either star.

ga	NGC 2179	06 08.0	−21 44	Lep	13.2b	1.6'	1.1'	S0/a

JH: faint, pretty small, very much extended, gradually little brighter middle.
170x250: Faint. Evenly moderately concentrated. Round. Very small. No involved stars. Making a very short line with it are two faint stars, one NNW and one SSE.

oc	NGC 2169	06 08.4	+13 57	Ori	5.9	6.0'		

HH: cluster, small, little rich, pretty much compressed, double star (STF 848).
12x35: Four stars in a little trapezoid, a little like Corvus, plus one or two more fainter stars. ***42x60:*** About nine stars, including the trapezoid and an additional bright star S that makes the figure triangular overall. ***170x250:*** An attractive cluster of about 16 bright and moderately bright stars in two groups aligned NW-SE, plus the single star S. The SE group is larger than the NW group. Altogether the cluster is a third of the field in size. Sparse, but a nice cluster of bright stars.

oc	NGC 2168/M35	06 08.9	+24 20	Gem	5.1	28.0'		

JH: cluster, very large, considerably rich, pretty compressed, 9th to 16th mag. stars.
12x35: A lot of stars. Somewhat elongated. ***42x60:*** About half a dozen moderate brightness stars and a lot more moderately faint stars. The main body of the cluster, which is elongated NW-SE, is a third of the field in size; overall the cluster is half the field in size. Pretty dense. ***170x250:*** A very nice field of a lot of bright stars. The main body, containing the brightest stars, slightly overflows the field.

ne	NGC 2182	06 09.5	−06 19	Mon		2.0'		R

HH: pretty bright double star and bright star in nebula, extended E-W.
170x250: A small faint nebula surrounding a bright star. Round. Very nearby E is a faint star that the nebula just about reaches; a little ways WNW is a star; ENE is a brighter star. All of the associated stars form an E-W arc.

ne	NGC 2174	06 09.7	+20 29	Ori		29.0'		E

St: extremely faint, between 3 very faint stars.
12x35: There's nebulosity next to the bright plotted star. ***42x60:*** NE of the bright star is a N-S string of five stars along with very faint nebulosity. ***170x250:*** Faint nebulosity, very vague and uneven, fills the field and gives it a grayish cast. The line of five stars turns sharply E at the N end; surrounding the S end are 10–12 fainter stars.

| oc | NGC 2175 | 06 09.8 | +20 19 | Ori | 6.8 | 18.0' | | |

Br: 8th mag. star in nebula.
[See 2174.]

| ga | NGC 2188 | 06 10.1 | –34 06 | Col | 12.1b | 4.3' | 1.1' | Sm |

JH: pretty faint, pretty large, very much extended, gradually very little brighter middle.
170x250: Faint to moderately faint. Hardly concentrated. A very thin edge-on, N-S. Moderately large. Right at its N end is a faint star.

| ne | NGC 2183 | 06 10.8 | –06 13 | Mon | | 2.0' | | R |

dA: extremely faint, small, little extended, 11th–12th mag. star SW.
[See 2185.]

| vs | TU Gem | 06 10.9 | +26 01 | Gem | 7.5 | 8.4 | 2.77 | C4 |

12x35: Mag. 7.5. *170x250:* Copper red. A very nice carbon star.

| ga | MCG-3-16-18 | 06 11.1 | –15 29 | Lep | 13.1 | 0.9' | 0.5' | Sc |

170x250: Not visible.

| ne | NGC 2185 | 06 11.1 | –06 13 | Mon | | 2.0' | | R |

HH: 11th mag. star and 4 faint stars in very faint, large nebula.
170x250: A faint indistinct nebula surrounding five moderate brightness stars in a NE-SW linear figure. A quarter of the field in size. A quarter of the field W is 2183. Very faint. Round. Very small; much smaller than 2185. It looks like a small faint galaxy. No involved stars.

| ds | 41 Aur | 06 11.6 | +48 43 | Aur | 7.6" | 6.3 | 7.0 | A3 A8 |

42x60: A very close unequal pair. *170x250:* A moderately close slightly unequal pair. Both soft white. [STF 845]

| oc | NGC 2186 | 06 12.2 | +05 27 | Ori | 8.7 | 4.0' | | |

HH: cluster, pretty large, pretty rich, pretty compressed, bright and faint stars.
12x35: A couple of very faint stars. *42x60:* A shallow arc of three stars: two NW and one SE. *170x250:* Surrounding the two stars NW is a very small cluster of a number of very faint stars; the star SE is a close unequal double. The cluster is elongated E-W. Moderate density. The bright stars of the arc are probably foreground stars.

| ga | NGC 2196 | 06 12.2 | −21 48 | Lep | 11.8b | 2.8' | 2.1' | Sa |

HH: pretty faint, pretty small, very little extended, pretty much brighter middle, stars near.
170x250: Moderate brightness. Evenly fairly concentrated. Round. Small. No involved stars. WNW is a crooked WNW-ESE line of faint stars.

| oc | NGC 2194 | 06 13.8 | +12 48 | Ori | 8.5 | 10.0' | | |

WH: cluster, large, rich, gradually very much compressed middle.
12x35: Very faintly visible. ***42x60:*** A very faint cluster of threshold stars. ***170x250:*** A lot of faint, very faint, and threshold stars within a trapezoid of brighter members. A third of the field in size. Pretty dense.

| bs | Eta Gem | 06 14.9 | +22 30 | Gem | 3.28 | | 1.61 | M3 |

170x250: Gold.

| oc | NGC 2192 | 06 15.2 | +39 51 | Aur | 10.9 | 5.0' | | |

WH: cluster, considerably large, compressed, irregular figure, very faint stars.
42x60: Extremely faintly visible. ***170x250:*** A fair number of faint and very faint stars. A quarter of the field in size. Moderate density.

| ds | STF 872 | 06 15.6 | +36 09 | Aur | 11.4" | 6.9 | 7.9 | F4 G5 |

42x60: A close slightly unequal pair. ***170x250:*** A moderately close to moderate distance slightly unequal pair. The primary is yellow white; the companion is white.

| oc | NGC 2204 | 06 15.7 | −18 38 | CMa | 8.6 | 12.0' | | |

WH: cluster, large, pretty rich, little compressed.
42x60: A very faint cloud. ***170x250:*** A fair number of moderately faint and faint stars in a NNW-SSE elongated cluster. At the N end is a bright field star; at the S end is a wide pair of moderate brightness stars; SW is another star. A little more than a third of the field in size. Moderate density.

| ga | NGC 2206 | 06 16.0 | −26 45 | CMa | 12.9b | 2.4' | 1.2' | Sbc |

JH: faint, pretty small, very little extended, pretty suddenly little brighter middle.
170x250: Moderately faint. Sharply concentrated. Round. Small. Near center is a faint involved star; the galaxy makes a small triangle with a moderately faint star WNW and a faint star SW; E is a brighter star. [E489-28 was not seen.]

| ga | NGC 2207 | 06 16.4 | –21 22 | CMa | 11.6p | 3.2' | 2.2' | Sbc |

JH: pretty bright, pretty large, much extended E-W, pretty suddenly little brighter middle to a round nucleus.
170x250: Moderate brightness. Sharply pretty concentrated. Roughly round. Small. SW of center is a faint involved star. Right next to 2207 E, and apparently interacting with it, is I2163. Faint. Broadly very slightly concentrated. Slightly elongated E-W. Moderately small. No involved stars. A bit of a contrast in galaxies.

| ga | IC 2163 | 06 16.5 | –21 22 | CMa | 12.6p | 2.2' | 1.0' | Sc |

[See 2207.]

| ga | NGC 2146 | 06 18.7 | +78 21 | Cam | 11.4b | 5.3' | 3.3' | Sab |

Wi: pretty bright, 2' long, little extended.
170x250: Moderate brightness to moderately bright. Broadly moderately concentrated. Elongated NW-SE. Moderately large. No involved stars. A little ways E is a pair of stars; further E are several bright stars.

| bs | Zeta CMa | 06 20.3 | –30 03 | CMa | 3.00 | | –0.17 | B2 |

170x250: White.

| ga | UGCA 127 | 06 20.9 | –08 29 | Mon | 12.9 | 3.9' | 1.1' | Scd |

170x250: Not visible.

| oc | NGC 2215 | 06 21.0 | –07 17 | Mon | 8.4 | 11.0' | | |

HH: cluster, considerably large, pretty rich, pretty compressed, 11th to 15th mag. stars.
12x35: Faintly visible. ***42x60:*** A number of faint stars. ***170x250:*** About a dozen moderately faint stars of uniform brightness and some fainter stars. Somewhat square. Half the field in size. Somewhat sparse.

| ga | UGCA 128 | 06 21.1 | –20 02 | CMa | 12.6p | 2.6' | 1.8' | Sa |

170x250: Faint. Broadly slightly concentrated. Round. Moderately small. At its edge SSE is a moderately faint star; nearby WNW is another moderately faint star.

| pn | IC 2165 | 06 21.7 | −12 59 | CMa | 12.9p | 9.0" | | |

42x60: Visible. *170x250:* Bright; high surface brightness. No central star. A tiny disk. Strong blinking effect. Nearby NNE is a star; the planetary makes a large triangular figure with three moderately faint stars: one NNE, one ENE, and one ESE.

| ga | NGC 2217 | 06 21.7 | −27 14 | CMa | 11.7b | 4.5' | 4.5' | S0+ |

JH: very bright, small, round, pretty suddenly much brighter middle, resolvable.
170x250: Bright. Sharply pretty concentrated. Round. Small. No involved stars. Nearby W is a pair of stars.

| bs | Beta CMa | 06 22.7 | −17 57 | CMa | 1.98 | | −0.23 | B1 |

170x250: White.

| bs | Mu Gem | 06 23.0 | +22 30 | Gem | 2.88 | | 1.64 | M3 |

170x250: Gold.

| ds | Epsilon Mon | 06 23.8 | +04 36 | Mon | 12.4" | 4.5 | 6.5 | A5 F5 |

42x60: An unequal pair. *170x250:* A moderate distance unequal pair. Both white. A nice double. [STF 900]

| ga | NGC 2223 | 06 24.6 | −22 50 | CMa | 12.4p | 3.2' | 2.7' | Sb |

JH: faint, pretty large, round, very gradually little brighter middle, 2 stars involved.
170x250: Faint. Sharply concentrated. Roughly round. Moderately small. N of center is an involved star that interferes; the galaxy is just off the hypotenuse of a large isosceles right triangle of relatively bright stars.

| vs | BL Ori | 06 25.5 | +14 43 | Ori | 6.3 | 7.0 | 2.34 | C6 |

12x35: Mag. 6.0. *170x250:* Orange red to light copper red. A nice carbon star.

| pn | J 900 | 06 25.9 | +17 47 | Gem | 12.4p | 9.0" | | |

170x250: Moderate brightness. No central star. A tiny disk. Strong blinking effect. It makes a "double star" with a moderately faint star SSW. [PK194+2.1]

ga	NGC 2227	06 26.0	−22 00	CMa	13.2b	2.1'	1.1'	Sc

JH: extremely faint, round, double star 90" W.
170x250: Very faint. Hardly concentrated. Round. Moderately small. No involved stars. Passing just off its edge S is a WNW-ESE curved line of stars.

ga	IC 2166	06 26.9	+59 04	Lyn	13.2p	3.0'	2.1'	Sbc

170x250: Very faint. Hardly concentrated. Slightly elongated WNW-ESE. Moderately small to moderate size. At its edge WSW is a moderately faint involved star; E is a bright star with a very wide companion; WNW is another bright star.

oc	NGC 2232	06 28.0	−04 50	Mon	4.2	29.0'		

WH: bright star (10 Mon) + cluster.
12x35, 42x60: About eight stars in a wedge-shaped figure pointing N; the brightest star is at the point. Half the field in size in the 60 mm. Very sparse. ***170x250:*** The figure stars are bright. There are also a few very faint stars, especially around the point star.

ds	Beta Mon	06 28.8	−07 02	Mon	7.2"	4.7	5.2	B3 B3

42x60: A very close equal double. One of the components looks unfocused. ***170x250:*** The component that looks fuzzy in the 60 mm is a very very close equal double (BC). The double split in the 60 mm (A-BC) is a moderately close slightly unequal pair, with the single star (A) being the brighter component. All the stars are white. A unique triple star: the three stars are all relatively close together, unlike the more common situation in which one star orbits two or two stars orbit one at a considerable distance. [STF 919 AB: 7.2", 4.7, 5.2; AC: 9.9", 4.7, 6.1; BC: 2.8", 5.2, 6.1]

ga	MCG-4-16-6	06 29.6	−26 37	CMa	11.9	1.1'	1.0'	S0

170x250: Not visible.

oc	NGC 2236	06 29.7	+06 50	Mon	8.5	6.0'		

HH: cluster, pretty rich, pretty compressed, 10th mag. star and 12th to 15th mag. stars.
42x60: A star with a very slight glow of more stars surrounding it. ***170x250:*** A fairly bright star with a good number of faint and very faint stars surrounding it. The cluster is roughly in the shape of a backward comma pointing S. A quarter of the field in size. The N section right around the brighter star is pretty dense; the cluster overall is moderately dense.

| oc | NGC 2243 | 06 29.8 | −31 16 | CMa | 9.4 | 5.0' | | |

JH: pretty bright, considerably large, round, very gradually little brighter middle, 4' in diameter.

42x60: A faint little cloud of stars. ***170x250:*** A lot of closely packed very faint stars plus threshold stars. Off its edge W is a little arc of three brighter stars that probably aren't members; the cluster is inside a triangle of stars pointing NW. Small. Very dense.

| oc | NGC 2244 | 06 32.4 | +04 52 | Mon | 4.8 | 23.0' | | |

WH: cluster, beautiful, scattered stars (12 Mon).

12x35: A very long NNW-SSE trapezoid of five stars: four corner stars and one star in the middle. ***42x60:*** There are a handful more stars in the middle. ***170x250:*** The figure stars are very bright. Right around and E of the middle star is a concentration of a dozen moderately faint stars; scattered about the field are a number of moderate brightness stars. The whole cluster is the field width in size. Sparse. The cluster itself occupies black space, but very vague gray nebulosity surrounds it in a ring half a field width in thickness and two field widths in diameter. The nebulosity is most conspicuous off the NW end of the cluster. [Rosette Nebula]

| ne | NGC 2245 | 06 32.7 | +10 10 | Mon | | 2.0' | | R |

HH: pretty large, cometic, much brighter middle to an almost stellar nucleus SE of center, 7th–8th mag. star NE.

170x250: Moderate brightness. The object consists of a brighter nucleus, with a central star, and a fainter nebula fanning out SW. Cometic, a little like 2261 but not as symmetrical; the fan shape is not as pronounced or as well defined. Very small. Nearby ENE is a moderately bright star.

| ne | NGC 2247 | 06 33.2 | +10 20 | Mon | | 2.0' | | R |

LR: nebulous star in extremely faint, extremely large nebulosity.

170x250: A very small nebula surrounding a moderately bright star. 2247 is smaller, fainter, and vaguer than 2245, but its central star is brighter.

| oc | NGC 2250 | 06 33.8 | −05 04 | Mon | 8.9 | 7.0' | | |

JH: cluster, pretty rich, little compressed, irregular figure, 8th mag. star and 12th to 14th mag. stars.

42x60: A couple of faint stars. ***170x250:*** Some dozen very widespread moderately faint stars W of the plotted star, with hardly any concentration. Elongated E-W. Half the field in size. Very sparse. Hardly a cluster.

| oc | NGC 2252 | 06 34.4 | +05 22 | Mon | 7.7 | 15.0' | | |

HH: cluster, very large, pretty rich, little compressed, faint stars.
12x35: Very faintly visible. ***42x60:*** A number of faint stars with one brighter star at the N end. ***170x250:*** Some dozen and a half moderate brightness and moderately faint stars in a linear group snaking first SE and then S from the bright star. Going E from near the N end is an extension of more widespread stars: some six brighter stars and a few more faint stars. A very unusually shaped cluster: a very large "Y" opening up SE, taking up the entire field exactly. The linear concentration is somewhat sparse; the cluster overall is sparse. An interesting and somewhat attractive cluster.

| oc | NGC 2251 | 06 34.7 | +08 22 | Mon | 7.3 | 10.0' | | |

HH: cluster, very large, extended, rich, little compressed.
12x35: A couple of stars with a hint of more stars. ***42x60:*** Two moderately faint and several faint stars. Elongated NW-SE. ***170x250:*** A very elongated cluster of a number of moderate brightness stars in the shape of half a raindrop, split down the middle: one side is flat and the other side is a curve and a tail; a fairly geometric figure. The stars are mainly on the perimeter but in the very middle is a little group with empty space around it. 3/4 of the field in size. Somewhat sparse. A somewhat attractive cluster, mainly because of its shape.

| oc | NGC 2254 | 06 35.8 | +07 40 | Mon | 9.1 | 4.0' | | |

HH: cluster, small, pretty compressed, irregular figure, 11th to 15th mag. stars.
42x60: Faintly visible. ***170x250:*** A concentration of a few faint stars in a small semicircle, with more very faint and threshold stars in and around it, especially NE. The cluster is a quarter of the field in size overall. Moderately dense.

| oc | TR 5 | 06 36.5 | +09 28 | Mon | 10.9 | 7.0' | | |

170x250: In the position is something of a concentration of faint stars along with some bright stars, but no recognizable cluster. [It's an extremely dense cluster of very faint stars.] [CR105]

| vs | UU Aur | 06 36.5 | +38 27 | Aur | 5.1 | 7 | 2.60 | C5 |

12x35: Mag. 5.5. ***170x250:*** Copper red. A fantastic carbon star, possibly the very best in the sky, due to the combination of its brightness and its color: very bright for a carbon star and definitely copper red, unlike some others that are paler. Startling.

| ga | MCG+1-17-3 | 06 36.7 | +05 35 | Mon | 13.0 | 1.6' | 1.6' | – |

170x250: Not visible.

bs	Gamma Gem	06 37.7	+16 23	Gem	1.90		0.03	A0

170x250: White.

oc	NGC 2262	06 38.4	+01 12	Mon	11.3	3.5'		

HH: cluster, very compressed, irregularly round, brighter middle, extremely faint stars.
170x250: A faint cluster of very very faint to threshold stars. A quarter of the field in size. Pretty dense.

ga	NGC 2263	06 38.5	−24 50	CMa	12.9p	2.8'	2.0'	Sab

JH: pretty faint, little extended, between 2 very faint stars, pretty suddenly little brighter middle.
170x250: Faint. Broadly slightly concentrated. Round. Small. No involved stars. It's among a NE-SW linear group of faint stars; the galaxy is between two of these stars, one N and one S.

oc	NGC 2259	06 38.6	+10 54	Mon	10.8	4.5'		

WH: cluster, considerably rich, extremely compressed, irregular figure, extremely faint stars.
170x250: A faint cluster of very faint stars. Small. Moderate density. W is the bright plotted star.

ne	NGC 2261	06 39.2	+08 44	Mon		2.0'		E+R

HH: bright, very much extended NNW-SSE, cometic nucleus = 11th mag. star.
42x60: Very faintly visible. *170x250:* Moderately bright. It has a stellar nucleus. The object looks like a comet head with a fairly symmetrical fan sweeping N and fading. Very small. Nearby NE is a moderate brightness star. [Hubble's Variable Nebula]

ga	UGC 3504	06 40.1	+60 04	Lyn	13.0p	2.6'	2.1'	Scd

170x250: Exceedingly faint. The slightest glow; hardly concentrated. Roughly round. Moderately small to moderate size. Off its edge N are four or five very faint stars in a diamond figure; SSW is a brighter star.

| ga | NGC 2267 | 06 40.9 | –32 28 | CMa | 13.2p | 1.6' | 1.2' | S0 |

JH: pretty bright, small, round, 2 or 3 stars very near.
170x250: Moderately faint. Sharply pretty concentrated. Round. Tiny; the whole galaxy is somewhat stellar. No involved stars. It makes a tiny triangle with two faint stars W.

| oc | NGC 2264 | 06 41.1 | +09 53 | Mon | 4.1 | 20.0' | | |

HH: 15 Mon, cluster, double star, nebulous?
12x35: Some dozen moderately faint stars in an upside-down Christmas-tree shape (pointing S), with the brightest star at the base N. ***42x60:*** Some dozen and a half moderate brightness stars, mainly on the outline of the figure; there's a big hole in the middle. A little less than half the field in size. Sparse. ***170x250:*** The N star is very bright; most of the other stars are pretty bright. Two field widths in size. There's a lot of empty space, but it's a nice cluster with a nice shape. [Christmas Tree Cluster/Cone Nebula]

| ga | NGC 2272 | 06 42.7 | –27 27 | CMa | 12.7p | 2.4' | 1.5' | S0– |

JH: pretty faint, pretty small, very little extended, brighter middle, resolvable.
170x250: Moderately faint. Evenly moderately concentrated. Round. Very small. No involved stars. It makes a flat equilateral triangle with two faint stars, one NW and one E.

| ga | NGC 2271 | 06 42.9 | –23 28 | CMa | 13.2p | 2.1' | 1.4' | S0– |

JH: pretty faint, small, round, gradually brighter middle, among stars.
170x250: Moderately faint. Evenly moderately concentrated. Round. Very small. No involved stars. It's just inside a triangle of faint stars.

| oc | NGC 2266 | 06 43.2 | +26 58 | Gem | 9.5 | 6.0' | | |

HH: cluster, pretty small, extremely compressed, rich, 11th to 15th mag. stars.
12x35: A tiny cloud. ***42x60:*** A faint cloud of stars. On the SW side is the bright plotted star; making a shallow little arc with this bright star are a couple more stars NE of it. ***170x250:*** A fair number of faint stars in addition to the brighter stars, which are in a string curving ENE from the SE corner of the cluster. A quarter of the field in size. Moderately dense.

| oc | NGC 2269 | 06 43.3 | +04 37 | Mon | 10.0 | 4.0' | | |

HH: cluster, very much compressed, not rich, very faint stars.

42x60: Faintly visible. *170x250:* A good number of stars in a flat ENE-WSW triangular figure. At the corners are moderate brightness stars; toward the center are some half-dozen moderately faint stars, including a relatively close pair, plus a fair number of very faint to threshold stars. Half the field in size. Moderate density.

ga	UGC 3511	06 43.7	+65 12	Cam	13.1p	1.4'	1.0'	Scd

170x250: Moderately faint. Broadly slightly concentrated. Round. Small. No involved stars. Nearby SSE is a pair of faint stars; a little ways ESE is a pair of very faint stars; a little farther N is a faint star.

bs	Epsilon Gem	06 43.9	+25 07	Gem	2.98		1.40	G8

170x250: Yellow gold.

ga	ESO 490-37	06 44.4	−26 06	CMa	12.9p	2.0'	1.2'	S0/a

170x250: Faint. Evenly moderately concentrated. Slightly elongated N-S. Very small. Just off its S end is a very faint star.

ga	NGC 2280	06 44.8	−27 38	CMa	10.9b	6.3'	3.0'	Scd

JH: pretty faint, pretty large, little extended, gradually brighter middle.
170x250: Moderately faint. Broadly slightly concentrated. Very elongated NNW-SSE. Moderately small. No involved stars. On either side of its N end is a faint star; off its S end are two relatively bright stars.

bs	Alpha CMa	06 45.1	−16 42	CMa	−1.47		0.01	A1

12x35: Very bright. Right across the field is a neat line due to the diagonal prisms. *42x60:* It has a little violet halo due to chromatic aberration from the objective lens. Beautiful. *170x250:* White. Overwhelmingly bright; a very brilliant star. Diffraction spikes make a nice "X" right across the field. (The star is so bright that it produces "special effects" in all three scopes.) [Sirius]

bs	Xi Gem	06 45.3	+12 53	Gem	3.40		0.39	F5

170x250: Soft white.

ga	NGC 2283	06 45.9	−18 12	CMa	12.9p	3.6'	2.7'	Scd

WH: 3 or 4 faint stars + nebula.

170x250: Very faint to faint. Hardly concentrated. Roughly round. Moderately small. It's inside a very small triangle of moderately faint to faint stars that interfere.

ga	NGC 2273B	06 46.5	+60 20	Lyn	13.1b	2.7'	1.4'	Scd

170x250: Not visible.

ne	NGC 2282	06 46.9	+01 19	Mon		4.0'		R

Ba: 10th mag. star in faint, round nebulosity.
170x250: A very faint inconspicuous nebula surrounding a moderately bright star. Moderately small.

oc	NGC 2287/M41	06 47.0	−20 43	CMa	4.5	38.0'		

JH: very large, bright, little compressed, 8th mag. and fainter stars.
12x35: A cluster of bright stars, including a wide pair in the middle, plus pinpoint fainter stars. *42x60:* A lot of bright stars plus fainter stars. The main body of the cluster is rectangular NE-SW, but the cluster overall is elongated NW-SE. Half the field in size. Fairly dense. *170x250:* A field full of relatively very bright stars. In the middle is a little semicircle of stars. A very nice cluster in all scopes.

ga	NGC 2274	06 47.3	+33 34	Gem	13.1p	1.2'	1.0'	E

HH: faint, small, brighter middle.
170x250: Moderately faint. Evenly moderately concentrated. Roughly round. Small. No involved stars. A little ways N is a fainter galaxy [*2275 (III)]. Faint. Evenly slightly concentrated. Roughly round. Small. No involved stars.

oc	NGC 2286	06 47.6	−03 09	Mon	7.5	14.0'		

HH: cluster, large, compressed, about 100 9th to 15th mag. stars.
42x60: A handful of widely scattered stars. *170x250:* The cluster is a number of faint stars inside the larger group of relatively bright stars seen in the 60mm. A third of the field in size. Somewhat sparse.

ga	NGC 2292	06 47.7	−26 44	CMa	11.8p	3.5'	2.8'	S0

JH: extremely faint, round, gradually brighter middle, one of double nebula, among stars.
[See 2293.]

| ga | NGC 2293 | 06 47.7 | −26 45 | CMa | 12.3b | 4.2' | 3.3' | S0+ |

JH: pretty bright, round, gradually brighter middle, one of double nebula, among stars.
170x250: 2292 and 2293 are right next to each other NW-SE. 2293 is brighter. Moderately faint to moderate brightness. Sharply pretty concentrated. Round. Very small. No involved stars. 2292 is faint. Concentrated. Round. Very small. No involved stars. A quarter of the field WNW is another galaxy with some involved stars [*2295 (JH)].

| ga | NGC 2258 | 06 47.8 | +74 28 | Cam | 13.0b | 2.3' | 1.5' | S0 |

Te: faint, 2 10th–11th mag. stars E.
170x250: Moderate brightness. Evenly moderately concentrated. Round. Small. No involved stars. It makes a very small triangle with a moderately bright star ENE and a moderate brightness star SE.

| ds | STF 958 | 06 48.2 | +55 42 | Lyn | 4.7" | 6.3 | 6.3 | F6 F5 |

42x60: Extremely close: as close as two stars can be in the 60mm and still be distinctly separated. ***170x250:*** A very close perfectly equal pair. Both white.

| oc | NGC 2281 | 06 49.3 | +41 04 | Aur | 5.4 | 14.0' | | |

WH: cluster, pretty rich, very little compressed, pretty bright stars.
12x35: A compact little group of stars, including a couple that are relatively bright.
42x60: An arrowhead-shaped little group of about eight stars. The SE corner star is the brightest. ***170x250:*** The arrowhead, which points W, consists of some 20 moderately bright stars; this group, a third of the field in size, is the central concentration in a larger cluster the field width in size. The rest of the cluster stars are widely spaced and in a partial ring, the NE section of which is empty; there's a large space between the ring and the central concentration. Overall the cluster is sparse. An attractive cluster.

| ga | NGC 2273 | 06 50.1 | +60 50 | Lyn | 12.6b | 3.2' | 2.4' | Sa |

Du: faint, small, irregularly round, resolvable?
170x250: Moderately faint. Sharply very concentrated to a faint substellar nucleus. Round. Small. No involved stars. It makes a large triangle with the two bright plotted stars NNE and ENE.

| ga | NGC 2289 | 06 50.9 | +33 28 | Gem | 13.2v | 1.1' | 0.7' | S0 |

HH: extremely faint, very small.
170x250: [The first galaxy observed is *2290.] Very faint. Evenly slightly concentrated. Round. Very small. No involved stars. It makes a right triangle with a relatively

bright star S and a fainter unequal pair of stars SE. A little ways NNW is another galaxy [2289]. Extremely faint. Small. Very nearby N is a very faint star. [2288, 2291, and 2294 were not seen; there is confusion in the NGC about the identity and the observers of the galaxies in this field.]

oc	NGC 2301	06 51.8	+00 29	Mon	6.0	12.0'		

HH: cluster, rich, large, irregular figure, bright and faint stars.
12x35: At least half a dozen stars in a ragged N-S line. A very conspicuous, very linear cluster – one of the most linear in the sky. *42x60:* At the N end is a concentration of much fainter stars. A pretty neat cluster. *170x250:* Six bright stars in a sinewy N-S line, a good number of moderate brightness stars in a concentration at the N end of the line, an "L"-shaped extension of moderate brightness stars E, and fainter stars all around, especially in the middle of the concentration. The field width in size altogether, assuming that the three southernmost bright stars are part of the cluster, which is a little questionable. The concentration is moderately dense. A neat and interesting cluster.

oc	NGC 2302	06 51.9	–07 02	Mon	8.9	2.5'		

HH: cluster, large, poor, little compressed.
12x35: Faintly visible. *42x60:* A handful of stars, mainly in a little line. *170x250:* A small cluster consisting of a shallow little arc of three moderate brightness stars and about ten moderately faint stars in two small groups beside it: one of five stars SE of the arc and a smaller one of four or five stars a little farther E. Sparse. It lies in a field of moderately bright stars.

ga	UGC 3574	06 53.2	+57 10	Lyn	13.2p	4.1'	3.5'	Scd

170x250: Exceedingly faint. The slightest unconcentrated glow. Roughly round. Moderately small. Closely surrounding it in a third of a circle NNE to W are four moderately faint stars and a couple of fainter stars; SSW is a brighter double star.

ds	38 Gem	06 54.6	+13 11	Gem	7.2"	4.7	7.7	F0 G4

42x60: The primary has a very close very faint companion. *170x250:* The primary has a moderately close relatively faint companion. The primary is white. [STF 982]

oc	NGC 2304	06 55.0	+18 01	Gem	10.0	5.0'		

HH: cluster, pretty large, rich, much compressed, very faint stars.
42x60: Extremely faintly visible. *170x250:* A faint cluster of a good number of faint stars of uniform brightness. A little less than a quarter of the field in size. Dense.

| ga | UGC 3580 | 06 55.5 | +69 33 | Cam | 12.7p | 3.4' | 1.8' | Sa |

170x250: Moderately faint. Evenly moderately concentrated. Round. Small. Right at its edge E is a faint star; nearby NNW is a very faint star; the galaxy makes a zigzag line with three stars SW.

| oc | NGC 2309 | 06 56.2 | –07 11 | Mon | 10.5 | 3.0' | | |

HH: cluster, pretty large, pretty rich, much compressed, 13th mag. stars.
42x60: A little cloud of threshold stars. *170x250:* A good number of moderately faint to very faint stars. Small. Fairly dense.

| oc | NGC 2311 | 06 57.8 | –04 34 | Mon | 9.6 | 6.0' | | |

HH: cluster, little compressed, not rich.
42x60: A number of threshold stars. *170x250:* A dozen faint stars and a few more very faint stars in a raindrop shape pointing SE. A third of the field in size. Sparse.

| vs | RV Mon | 06 58.4 | +06 10 | Mon | 7.0 | 8.9 | 2.65 | C5 |

12x35: Mag. 7.5. *170x250:* Light copper red.

| bs | Epsilon CMa | 06 58.6 | –28 58 | CMa | 1.50 | | –0.20 | B2 |

170x250: White.

| ga | IC 456 | 07 00.3 | –30 09 | CMa | 12.9p | 2.1' | 1.7' | S0 |

170x250: Faint to moderately faint. Evenly moderately concentrated. Round. Very small. No involved stars. Nearby ENE is a moderately bright star.

| bs | Sigma CMa | 07 01.7 | –27 56 | CMa | 3.47 | | 1.73 | K4 |

170x250: Deep gold. One of the most deeply colored bright stars.

| ga | NGC 2325 | 07 02.7 | –28 41 | CMa | 12.4p | 3.3' | 1.8' | E4 |

JH: pretty bright, pretty large, little extended, gradually brighter middle.
170x250: Moderate brightness. Broadly slightly concentrated. Elongated N-S. Small to moderately small. At its edge SSE is a faint involved star; nearby SE is a moderate brightness star.

| bs | Omicron 2 CMa | 07 03.0 | –23 50 | CMa | 3.00 | | –0.06 | B3 |

170x250: White.

| oc | NGC 2323/M50 | 07 03.2 | –08 20 | Mon | 5.9 | 16.0' | | |

JH: ! cluster, very large, rich, pretty compressed, extended, 12th to 16th mag. stars.

12x35: A compact cluster of faint stars. *42x60:* A good number of moderately faint and faint stars; the five brightest are three in a triangle at the cluster's edge NNE and two at its edge S. *170x250:* A good number of moderately bright and moderate brightness stars. Aside from the five brightest, the stars are fairly uniform in brightness. E and W, in two large sparse arcs, are distant members, with mostly empty space between them and the main body of the cluster. With the outer members the cluster is a field and a half in size; the main body is a little more than half the field in size and is moderately dense.

| ne | NGC 2327 | 07 04.0 | –11 19 | CMa | | 3.0' | | E |

HH: pretty bright double star involved in small, very faint nebula.

170x250: A very very small moderate brightness nebula around a moderate brightness star with a companion E. The nebula makes an arc with four moderately faint stars E to NE. Filling the field S of the nebula is an uncatalogued fairly substantial cluster of moderately bright stars, elongated N-S. Somewhat sparse.

| oc | NGC 2324 | 07 04.2 | +01 03 | Mon | 8.4 | 7.0' | | |

HH: cluster, large, rich, considerably compressed, 12th to 16th mag. stars.

42x60: A very faint cloud with a large sparse group of stars next to it NE. *170x250:* A lot of faint to threshold stars. Somewhat triangular. Half the field in size. [Dense.] Near the middle of this faint cluster is a moderate brightness star; this is the southwesternmost star of a separate, very sparse foreground cluster of some dozen widely scattered moderately bright stars filling the field NE. Interesting: two very different kinds of clusters apparently at different distances, slightly overlapping in the line of sight.

| ds | HJ 3928 | 07 05.5 | –34 47 | Pup | 2.7" | 6.5 | 7.6 | F2 – |

170x250: The primary has an extremely close fainter companion. The primary is white.

| ga | NGC 2320 | 07 05.7 | +50 34 | Lyn | 12.9b | 1.7' | 1.2' | E |

HH: pretty bright, small, irregularly round, gradually brighter middle, 8th mag. star ESE.

170x250: Faint. Evenly moderately concentrated. Roughly round. Small. Very nearby N is a faint star; nearby ENE is a bright star. A third of the field SE is a second galaxy [*2322 (WH)]. Very faint. Evenly moderately concentrated. Elongated NW-SE. No involved stars. It makes an arc with a very faint star nearby E and a moderately faint star a little ways SE.

| ds | STF 1009 | 07 05.7 | +52 45 | Lyn | 4.3" | 6.9 | 7.0 | A3 A2 |

42x60: Extremely close. *170x250:* A very close practically equal pair. Both white.

| oc | NGC 2335 | 07 06.6 | −10 03 | Mon | 7.2 | 12.0' | | |

WH: cluster, large, little compressed.
12x35: Very faintly visible. *42x60:* Three or four very faint stars plus more threshold stars. *170x250:* A fair number of moderately faint stars plus fainter stars. Slightly elongated ENE-WSW. 2/3 of the field in size. Somewhat sparse to moderate density.

| oc | NGC 2331 | 07 07.0 | +27 17 | Gem | 8.5 | 18.0' | | |

HH: cluster, large, very little compressed, small cluster involved.
12x35: Faintly visible. A few stars are made out. *42x60:* A few somewhat widespread faint stars. *170x250:* A very loose cluster of over a dozen very widespread moderate brightness stars of pretty uniform brightness. Elongated N-S. There are hardly any fainter stars in the main part of the cluster. At its edge SE is a small concentration: a little ring of some half-dozen stars. Overall the cluster is a field width in size. Very sparse.

| ga | UGC 3691 | 07 08.0 | +15 10 | Gem | 12.6p | 2.1' | 1.0' | Scd |

170x250: Not visible.

| vs | W CMa | 07 08.1 | −11 55 | CMa | 7.0 | 8 | 2.53 | C6 |

12x35: Mag. 6.5. *170x250:* Orangish.

| ga | NGC 2339 | 07 08.3 | +18 46 | Gem | 12.5b | 2.8' | 2.0' | Sbc |

HH: pretty bright, pretty large, round, gradually little brighter middle.
170x250: Faint. Broadly very slightly concentrated. Roughly round. Moderately small. At its edge ENE is a very faint involved star; the galaxy is inside a triangle of stars: a faint star nearby W, another faint star a little ways SSE, and a moderate brightness star NNE.

oc	NGC 2343	07 08.3	−10 38	Mon	6.7	6.0'		

WH: cluster, considerably large, poor, little compressed.
12x35: A handful of faint stars. *42x60:* A half-dozen moderately faint stars in the shape of a little triangle. *170x250:* About 14 moderate brightness stars in a roughly square figure. A third of the field in size. Sparse.

oc	NGC 2345	07 08.3	−13 09	CMa	7.7	12.0'		

JH: cluster, pretty large, pretty rich, gradually brighter middle, 10th to 14th mag. stars.
12x35: An elongated cluster of a few very faint stars. *42x60:* Some eight stars in a group SSW of the plotted star. Elongated NE-SW. *170x250:* A number of moderate brightness stars plus many fainter stars in a very elongated house-shaped figure pointing NE. 3/4 of the field in size. Somewhat sparse.

bs	Delta CMa	07 08.4	−26 23	CMa	1.80		0.72	F8

170x250: Yellow white.

vs	R CMi	07 08.7	+10 01	CMi	7.3	11	2.40	C7

12x35: Mag. 8.5. *170x250:* Yellow.

ga	NGC 2329	07 09.1	+48 36	Lyn	12.4v	1.4'	1.1'	S0−

HH: very faint, very small, stellar.
170x250: Faint. Evenly slightly concentrated. Round. Small. No involved stars. It's in the middle of a very large diamond-shaped group of stars. A little ways ENE is another galaxy [*U3696]. Very small; somewhat stellar.

ga	UGC 3685	07 09.1	+61 35	Lyn	12.8p	3.3'	2.7'	Sb

170x250: Faint. Broadly slightly concentrated. Round. Small to moderately small. WSW of center is a faint involved star; the galaxy is in the middle of a long NW-SE line of relatively bright stars, including a group SE.

ga	NGC 2342	07 09.3	+20 38	Gem	13.1b	1.2'	1.0'	S

Ma: pretty faint, small, little extended, very little brighter middle.
170x250: Very faint. Broadly very slightly concentrated. Roughly round. Small. No involved stars. A little ways SW is a second galaxy [*2341 (Ma)]. Faint. Small. Next to it N is a faint star; a little ways NW is another star; the galaxy and these two stars form a little arc resembling Aries.

| pn | NGC 2346 | 07 09.4 | −00 48 | Mon | 11.8p | 2.0' | | |

WH: 9th mag. star with small, very faint nebula.
42x60: *A faint star.* ***170x250:*** *A small faint nebula around a relatively bright central star. Slight blinking effect. Round. Closely surrounding it are faint stars; it makes a large triangle with a moderately bright star W and a triangular group of moderate brightness stars NW.*

| ga | PGC 20274 | 07 09.6 | −05 25 | Mon | 13.2 | 2.1' | 1.5' | Im |

170x250: *Not visible.*

| ga | NGC 2337 | 07 10.2 | +44 27 | Lyn | 12.9p | 2.2' | 1.6' | Im |

St: extremely faint, small, extended.
170x250: *Faint. Broadly slightly concentrated. Slightly elongated WNW-ESE. Moderately small. No involved stars. It's closely surrounded by faint stars; W is a triangular group of brighter stars; SE is another group of relatively bright stars.*

| ga | NGC 2314 | 07 10.5 | +75 19 | Cam | 13.2b | 1.7' | 1.3' | E3 |

Te: very faint, small, round.
170x250: *Moderate brightness. Sharply pretty concentrated. Round. Very small. No involved stars. It's at the intersection of a large "X" figure of stars, the NW point of which is a little triangle of stars. [I2174 was not seen.]*

| ga | NGC 2340 | 07 11.2 | +50 10 | Lyn | 11.6v | 2.4' | 1.3' | E |

HH: pretty faint, small, round, gradually little brighter middle, resolvable.
170x250: *In the position is 2340 along with three more galaxies in a NE-SW line. 2340 is faint. Broadly moderately concentrated. Roughly round. Small. No involved stars. A little ways WNW are two stars. A little ways SSW is a very faint galaxy [*I464]. Somewhat concentrated. Round. Very small. No involved stars. A little more than a third of the field SW is another galaxy [*I458]. Faint; a little fainter than 2340. Very concentrated. Round. Very small. Next to it S is a star. A third of the field NE of 2340 is a very faint galaxy [*I465]. Evenly moderately concentrated. Round. Very small. No involved stars. It makes a sharp triangle with a faint star ENE and a brighter star farther N.*

| ga | ESO 492-2 | 07 11.7 | −26 42 | CMa | 13.0p | 2.1' | 1.4' | Sb |

170x250: *Very faint. Broadly slightly concentrated. Round. Small. At its edge NW is an extremely faint star; very nearby SE is a very faint star; a little ways in each direction NW, W, and SSE is a brighter star.*

| ga | NGC 2344 | 07 12.5 | +47 10 | Lyn | 12.8b | 2.2' | 2.0' | Sc |

Sw: pretty bright, pretty small, round, little brighter middle.
170x250: Faint. Broadly slightly concentrated. Round. Small. No involved stars. It makes a little arc with three moderately faint to faint stars N to NNE.

| ga | UGC 3714 | 07 12.6 | +71 45 | Cam | 12.8p | 1.8' | 1.5' | S |

170x250: Faint. Broadly to evenly moderately concentrated. Round. Very small. No involved stars. It's inside a large NW-SE rectangular figure of moderate brightness stars.

| ga | NGC 2268 | 07 14.0 | +84 22 | Cam | 12.2b | 3.2' | 1.9' | Sbc |

By: pretty faint, pretty large, little extended.
170x250: Moderately faint. Evenly moderately concentrated. Very elongated ENE-WSW. Moderate size. Toward its W end is a very very faint involved star; S are several very faint stars.

| oc | NGC 2354 | 07 14.3 | –25 43 | CMa | 6.5 | 20.0' | | |

HH: cluster, considerably rich, little compressed.
12x35: A cluster of faint stars. ***42x60:*** A fair number of moderately faint and faint stars. ***170x250:*** A good number of moderate brightness and fainter stars. It has a N-S oblong hole in the middle; otherwise it fills the field. Moderate density.

| ga | UGC 3730 | 07 14.3 | +73 28 | Cam | 13.2b | 2.8' | 1.4' | – |

170x250: Very faint. Sharply concentrated. Round. Very small. No involved stars. Nearby NW is a very faint star that makes a long trapezoid with three moderately faint stars SW.

| oc | NGC 2353 | 07 14.6 | –10 18 | Mon | 7.1 | 20.0' | | |

WH: cluster, large, little compressed, one very bright star.
12x35, 42x60: A few faint stars N of the bright plotted star. ***170x250:*** A modest number of moderate brightness stars N and E of the bright star. Elongated NW-SE. 2/3 of the field in size. Somewhat sparse.

| ga | ESO 428-11 | 07 15.5 | –29 21 | CMa | 12.9b | 1.4' | 1.1' | S0– |

170x250: Moderately faint. Sharply concentrated. Round. Very small. No involved stars. Very nearby ENE is a moderately faint star; very nearby SSW is a very faint star; a little ways WNW is a pair of stars; a little ways E is a faint star.

| ga | NGC 2347 | 07 16.1 | +64 42 | Cam | 13.2b | 1.7' | 1.2' | Sb |

WH: very faint, small, round, little brighter middle.
170x250: Faint to moderately faint. Evenly moderately concentrated. Round. Small. No involved stars. A little ways WSW is a moderately bright star; N is the bright plotted star.

| ds | HJ 3945 | 07 16.6 | –23 19 | CMa | 26.8" | 4.8 | 6.8 | K3 F0 |

42x60: An unequal pair. ***170x250:*** A wide unequal pair. Color contrast: the primary is yellow gold; the companion is basically white but slightly bluish by comparison. One of the better color contrasting pairs; a very nice double.

| oc | NGC 2355 | 07 16.9 | +13 46 | Gem | 9.7 | 9.0' | | |

HH: cluster, pretty small, pretty rich, much compressed, 15th to 16th mag. stars.
12x35: Visible. ***42x60:*** A little cloud of stars next to a moderately faint star. ***170x250:*** A fair number of moderately faint stars. At its edge SE is a bright star. Elongated N-S. Half the field in size. Moderate density.

| oc | NGC 2360 | 07 17.8 | –15 37 | CMa | 7.2 | 12.0' | | |

HH: cluster, very large, rich, pretty compressed, 9th to 12th mag. stars.
12x35: Fairly bright. ***42x60:*** A good number of faint stars. Fairly rich. ***170x250:*** A lot of moderate brightness to moderately faint stars, pretty uniform in brightness. The central part of the cluster is half the field in size; with outer members the cluster is the field width in size. Moderately dense. A nice cluster.

| ne | NGC 2359 | 07 18.6 | –13 12 | CMa | | 13.0' | | E |

HH: !! very faint, very very large, very irregular figure.
42x60: Vaguely visible. ***170x250:*** Faint. The brighter S part of the nebula is elongated ENE-WSW. At its southern edge, where there is a moderately bright star, it runs into a large dark nebula adjoining it. N is a detached piece of nebulosity with four involved stars in an E-W row and a fifth star S of them, forming a triangle. Overall the nebula is a little less than half the field in size. Irregular; a strange, convoluted nebula.

| oc | NGC 2362 | 07 18.8 | –24 56 | CMa | 3.8 | 6.0' | | |

HH: cluster, pretty large, rich.
12x35: A little cloud around the bright plotted star. ***42x60:*** A neat little cluster of moderately faint stars around the bright star. ***170x250:*** A fair number of moderate

brightness stars, uniform in brightness, around the bright star, which makes a tiny arc with two faint stars E of it. Half the field in size. Moderately dense. An attractive cluster.

| oc | NGC 2367 | 07 20.1 | −21 53 | CMa | 7.9 | 3.5' | | |

WH: cluster, small, poor, little compressed.
42x60: A little triangle of stars. ***170x250:*** A number of mixed brightness stars. The brightest stars are in a sharp triangle, the NW corner of which is a double; there are a few more stars, mostly N. Elongated N-S. A little more than a quarter of the field in size. Somewhat sparse.

| ga | ANON 0718-34 | 07 20.8 | −34 07 | Pup | 13.0 | 2.3' | 1.2' | S0− |

170x250: Not visible. [P20731]

| oc | NGC 2368 | 07 21.0 | −10 22 | Mon | 11.8 | 5.0' | | |

JH: cluster, small, pretty rich, 15th mag. stars.
42x60: Faintly visible. ***170x250:*** A number of moderately faint stars, pretty uniform in brightness, in a table-like figure: two parallel NNW-SSE concentrations – the sides – with an ENE-WSW top. In the middle is an empty space. Irregular. A quarter of the field in size. The concentrations are moderate density; the cluster overall is somewhat sparse.

| ga | UGC 3804 | 07 22.6 | +71 36 | Cam | 13.1p | 1.7' | 1.0' | Scd |

170x250: Moderately faint. Broadly to evenly slightly concentrated. Round. Small. No involved stars. N is a moderately faint star.

| ds | 19 Lyn | 07 22.9 | +55 17 | Lyn | 14.8" | 5.6 | 6.5 | B8 B9 |

12x35: Extremely close. ***42x60:*** A moderately close unequal pair. ***170x250:*** A moderate distance slightly unequal pair. Both white. [STF 1062]

| ga | NGC 2380 | 07 23.9 | −27 31 | CMa | 12.3b | 1.8' | 1.8' | S0 |

JH: pretty faint, pretty small, round, very suddenly much brighter middle, among stars.
170x250: Moderately faint. Evenly moderately concentrated. Round. Very small. No involved stars. Nearby N and S are several faint stars.

| oc | NGC 2374 | 07 24.0 | −13 15 | CMa | 8.0 | 19.0' | | |

HH: cluster, very large, pretty rich, little compressed, bright stars.
12x35: A number of very faint stars. ***42x60:*** A fair number of faint stars in two parts aligned ENE-WSW. ***170x250:*** The SW part is a concentration of a number of moderate brightness stars. A quarter of the field in size. Moderate density. The NE part consists of much more widely spaced moderately bright stars. Half the field in size. Somewhat sparse. Overall the object is 3/4 of the field in size and somewhat sparse. The two parts have different characters; they might actually be separate clusters nearly in the same line of sight, the SW cluster being the more distant one.

| bs | Eta CMa | 07 24.1 | −29 18 | CMa | 2.40 | | −0.03 | B5 |

170x250: White.

| ga | UGC 3828 | 07 24.6 | +57 58 | Lyn | 12.9p | 1.7' | 0.9' | Sb |

170x250: Faint. Evenly slightly concentrated. Round. Small. No involved stars. It makes a triangle with two faint stars, one W and one NNW.

| oc | NGC 2383 | 07 24.8 | −20 55 | CMa | 8.4 | 5.0' | | |

JH: cluster, pretty small, pretty much compressed, 12th mag. stars.
[See 2384.]

| oc | NGC 2384 | 07 25.2 | −21 01 | CMa | 7.4 | 2.5' | | |

JH: cluster, little compressed, bifid, double star.
12x35: 2384 is a wide pair of stars. 2383 is a closer and fainter pair of stars. ***42x60:*** Each cluster has a couple more faint stars. ***170x250:*** 2384 consists of two little groups aligned E-W; each is about half a dozen stars surrounding a bright star. A quarter of the field in size. Sparse. Hardly more than a concentration in the field. 2383, half the field NW, is a fair number of moderately bright and faint stars. The four brightest stars are in a very sharp E-W triangular figure; within this figure is a N-S concentration of faint stars. A quarter of the field in size. Moderate density.

| pn | NGC 2371 | 07 25.6 | +29 29 | Gem | 13.0p | 55.0" | | |

HH: bright, small, round, brighter middle to a nucleus, W of double nebula.
42x60: Very faintly visible. ***170x250:*** A NE-SW two-lobed planetary [2372, 2371], like a wide double star consisting of very fat stars. Each lobe is moderate brightness, but the SW lobe is a little brighter. Each is vaguely round but not disklike. No central star. Together they're moderate size for a planetary.

| oc | TR 6 | 07 26.4 | −24 12 | CMa | 10.0 | 6.0' | | |

42x60: A very faint group of very faint stars. *170x250:* A modest number of moderately faint and faint stars. Rectangular N-S. A little more than a quarter of the field in size. Somewhat sparse. [CR145]

| ga | NGC 2276 | 07 27.0 | +85 45 | Cep | 11.9b | 2.8' | 2.6' | Sc |

Te: faint, 60", little brighter middle.
[See 2300.]

| ga | NGC 2336 | 07 27.1 | +80 10 | Cam | 11.1b | 7.0' | 3.8' | Sbc |

Te: pretty bright, pretty large, round, 2 11th mag. stars near.
170x250: Moderately faint. Evenly to sharply moderately concentrated. Slightly elongated N-S. Moderately large. It has an extremely faint outer halo that fades out very indistinctly. No involved stars. A little ways ESE is a faint star; the galaxy makes a quadrilateral with a bright star NNW, another bright star SW, and a moderately bright star W.

| oc | NGC 2395 | 07 27.1 | +13 35 | Gem | 8.0 | 12.0' | | |

WH: cluster, pretty rich, compressed.
42x60: Three moderately faint stars plus much fainter stars. *170x250:* Three moderately bright stars, a number of moderate brightness stars, and some faint stars, all in a narrow NW-SE diamond figure stretching across the entire field. Somewhat sparse.

| bs | Beta CMi | 07 27.2 | +08 17 | CMi | 2.90 | | −0.09 | B8 |

170x250: White.

| oc | TR 7 | 07 27.4 | −23 58 | Pup | 7.9 | 5.0' | | |

12x35: A couple of stars. *42x60:* Two or three brighter stars in a little arc, with a linear extension S. *170x250:* A modest number of mixed brightness stars: moderate brightness to faint. Very elongated NNW-SSE. A third of the field in size. Somewhat sparse. [CR146]

| oc | NGC 2396 | 07 28.0 | −11 43 | Pup | 7.4 | 10.0' | | |

HH: cluster, very large, very little compressed.
42x60: A handful of stars around a brighter star. *170x250:* Some 16 moderate brightness stars, very uniform in brightness, plus some fainter stars. The brightest

star is at the cluster's edge W; it might not be a member. Half the field in size. Sparse. Irregular.

| ga | NGC 2363 | 07 28.5 | +69 11 | Cam | 13.0 | 1.8' | 1.0' | Irr |

LR: nebulous star or very faint, very small nebula.
[See 2366.]

| ds | DUN 49 | 07 28.9 | −31 51 | Pup | 8.9" | 6.5 | 7.2 | B2 B4 |

42x60: A slightly unequal pair. ***170x250:*** A moderately close very slightly unequal pair. Both white.

| ga | NGC 2366 | 07 28.9 | +69 12 | Cam | 11.5b | 8.2' | 3.3' | Irr |

WH: very bright, pretty large, round, much brighter middle, resolvable, very faint star involved.
170x250: In the position are two galaxies aligned NE-SW. The SW galaxy is relatively bright; high surface brightness. Sharply very concentrated. Round. Very small. No involved stars. It makes a very narrow diamond with three stars N. Nearby NE is a much much larger galaxy. Low surface brightness. Diffuse. Moderate size. It's unclear whether the two objects are attached or separate, but they are very disparate in appearance. [The "two galaxies" are actually one galaxy (2366) consisting of a bright knot or a greatly displaced nucleus at the SSW end of a faint very elongated halo. 2363, a detached part of 2366 a little W of the bright knot, was not seen. The whole thing is very peculiar.]

| pn | NGC 2392 | 07 29.2 | +20 54 | Gem | 9.9p | 50.0" | | |

HH: bright, small, round, 8th mag. star in middle.
12x35: It looks like a double star. ***42x60:*** The planetary is the fuzzy southern component of a wide "double star." ***170x250:*** Very bright. It has an obvious moderately faint central star. A very well defined and symmetrical disk. There's a hint of detail inside. Moderate blinking effect. Very round. Moderate size for a planetary. Nearby NW is a faint star that makes a triangle with the planetary and the bright star N. A superb, very beautiful planetary; one of the very best in the sky. It's exceptional in combining an obvious and distinct central star with a bright nebula, which usually don't go together: ordinarily there's either a bright central star with a faint nebula or a bright nebula with no discernible central star. Classic-looking: this is what one expects a planetary to look like. [Eskimo Nebula]

| oc | NGC 2401 | 07 29.4 | −13 58 | Pup | 12.6 | 2.0' | | |

HH: cluster, small, considerably rich, considerably compressed, very faint stars.

12x35, 42x60: A couple of faint stars aligned N-S. *170x250:* A cluster of a fair number of very faint and threshold stars NNE of the brighter stars seen in the smaller scopes, which are unrelated. Very small. Fairly dense. Apparently a very distant cluster.

ds	STF 1104	07 29.4	–15 00	Pup	1.9"	6.4	7.5	F7 –

170x250: An extremely close unequal pair. Both white.

ga	IC 467	07 30.3	+79 52	Cam	13.2b	3.2'	1.2'	Sc

170x250: Very faint. Hardly concentrated. Elongated E-W. Moderately small. At its edge SW of center is an extremely faint star; nearby N is a moderately faint star that makes a triangle with two moderately bright stars further N; SE is a little group of faint stars.

ga	NGC 2300	07 32.0	+85 42	Cep	12.1b	2.0'	2.0'	S0

Wi: pretty bright, pretty large, little extended, brighter middle.
170x250: Bright. Sharply very concentrated. Round. Small. No involved stars. A little more than a third of the field WNW is 2276. Faint. Broadly very slightly concentrated. Round. Moderately small. No involved stars. Nearby W is a pair of stars, moderately faint and bright; a little ways S are a couple of very faint stars. 2/3 of a field SSE of 2300 is another galaxy [*I455]. Moderately faint. Sharply very concentrated. Round. Very small. No involved stars.

oc	NGC 2414	07 33.2	–15 27	Pup	7.9	4.0'		

HH: cluster, poor, little compressed, 9th mag. and fainter stars.
12x35: In the position is the plotted star. *42x60:* The light of additional threshold stars is visible. *170x250:* A very small cluster of 8–10 faint stars in addition to the bright plotted star, which is at the cluster's edge SE. About five of the faint stars form a small half-circle NW of the bright star, concave away from it.

ds	n Pup	07 34.3	–23 28	Pup	9.8"	5.8	5.9	F4 F6

42x60: A close perfectly equal pair. *170x250:* A moderately close perfectly equal pair. Both white. [H 19]

bs	Alpha Gem	07 34.6	+31 53	Gem	1.94		0.00	A1

42x60: Elongated. *170x250:* A very close slightly unequal pair. Both white. Both stars are very bright. An exceptional double; one of the very best in the sky. [STF 1110: 4.4", 1.9, 2.9, A1 A]

| oc | NGC 2421 | 07 36.3 | −20 36 | Pup | 8.3 | 10.0' | | |

HH: cluster, large, considerably rich, 11th to 13th mag. stars.
12x35: Visible. ***42x60:*** A number of faint stars. ***170x250:*** A good number of moderately faint and faint stars. Roughly triangular. Half the field in size. Moderately dense.

| ga | NGC 2418 | 07 36.6 | +17 53 | Gem | 13.2p | 1.8' | 1.4' | E |

St: very faint, extremely small, brighter middle.
170x250: Moderate brightness. Evenly to sharply pretty concentrated. Round. Small. No involved stars. Off its edge SSE is a very faint star; off its edge WNW is another very faint star; a little ways NE are two slightly brighter stars.

| oc | NGC 2422/M47 | 07 36.6 | −14 29 | Pup | 4.4 | 29.0' | | |

HH: cluster, bright, very large, pretty rich, bright and faint stars.
12x35: M47 is a number of moderately bright stars. M46 is a cloud of very faint stars. A very nice contrasting pair of clusters; exceptional. ***42x60:*** M47 is a cluster of mixed brightness stars: a half-dozen fairly bright stars plus moderate brightness and fainter stars. The main part of the cluster is roughly in a house-shaped figure pointing NE. E are two stars, somewhat detached but probably members. Overall, with the detached stars, the cluster is half the field in size. Somewhat sparse. ***170x250:*** A handful of bright stars, more moderate brightness stars, and faint stars. The most prominent stars form a large triangle. The two stars E are connected to the main part of the cluster by other stars. The cluster is two field widths in size. A nice couple of fields of stars; a very nice cluster.

| ga | NGC 2403 | 07 36.9 | +65 36 | Cam | 8.9b | 22.1' | 12.4' | Sc |

WH: !! considerably bright, extremely large, very much extended, very gradually much brighter middle to a 7' diameter nucleus.
12x35: Easily visible. ***42x60:*** Bright. ***170x250:*** Bright. Broadly slightly concentrated. Elongated NW-SE. Very large: the field width in size. The ends are indistinct, especially the SE end, which is extremely faint. The galaxy extends farther on the NW side. There's vague, broad unevenness in the central part of the galaxy; E of center is a small knot. Near center are several involved stars: a moderate brightness star very close S, a relatively bright star WSW, and a bright star ESE; there are several other involved stars within the halo, all faint.

| ga | NGC 2415 | 07 36.9 | +35 14 | Lyn | 12.8b | 1.0' | 1.0' | Im |

HH: pretty bright, considerably small, round, very gradually very little brighter middle, resolvable, almost planetary.

170x250: Moderately faint. Broadly moderately concentrated. Round. Small. No involved stars. Very nearby NNW is a faint star; nearby SSE is another faint star; a little ways ENE is a relatively bright star.

| oc | NGC 2423 | 07 37.1 | –13 51 | Pup | 6.7 | 19.0' | | |

HH: cluster, very large, rich, pretty compressed, very faint stars.
12x35: A couple of faint stars surrounded by threshold stars. *42x60:* A handful of faint stars and many more very faint stars. *170x250:* A lot of moderate brightness stars, mostly pretty uniform in brightness, with little clumps of stars and gaps between them. It takes up the whole field. Moderate density.

| oc | MEL 71 | 07 37.5 | –12 03 | Pup | 7.1 | 9.0' | | |

12x35: Visible. *42x60:* A cloud of threshold stars. At its edge SW is a sharp triangle of moderately faint stars. *170x250:* A good number of moderately faint and faint stars. The easternmost of the brighter triangle stars is a close double. Half the field in size. Fairly dense. [CR155]

| gc | NGC 2419 | 07 38.1 | +38 53 | Lyn | 10.4 | 4.6' | –9.57 | 2 |

HH: pretty bright, pretty large, little extended E-W, very gradually brighter middle, 7th–8th mag. star [W].
42x60: Very faintly visible. *170x250:* Not resolved. Broadly concentrated. Small. It's surrounded by a half-dozen faint field stars; in line with it W are the two bright plotted stars. Faint to moderately faint for a globular. It looks like a galaxy.

| oc | MEL 72 | 07 38.5 | –10 41 | Mon | 10.1 | 9.0' | | |

42x60: Faintly visible. *170x250:* A good number of faint stars. Triangular. A third of the field in size. Pretty dense. MEL72 is like a more distant version of MEL71. [CR156]

| oc | NGC 2420 | 07 38.5 | +21 34 | Gem | 8.3 | 10.0' | | |

HH: cluster, considerably large, rich, compressed, 11th to 18th mag. stars.
12x35: Faintly visible. *42x60:* A little group of very faint stars. *170x250:* Some ten moderately faint stars plus a lot of much fainter stars. A third of the field in size. Dense to very dense.

| ds | k Pup | 07 38.8 | –26 48 | Pup | 9.9" | 4.5 | 4.7 | B6 B5 |

12x35: Elongated. *42x60:* A close perfectly equal pair. *170x250:* A moderately close perfectly equal pair. Both white. [H 27]

| bs | Alpha CMi | 07 39.3 | +05 13 | CMi | 0.34 | | 0.40 | F5 |

170x250: Soft white. A brilliant star, with diffraction spikes all the way across the field. It's fairly isolated but there are a number of widespread fainter stars in the field, making it look like it's in the foreground. A nice sight. [Procyon]

| oc | NGC 2439 | 07 40.8 | −31 38 | Pup | 6.9 | 10.0' | | |

JH: cluster, bright, pretty rich, pretty large, little compressed, 9th mag. star and 12th to 14th mag. stars.
12x35: Faintly visible. *42x60:* A little group of stars SW of a relatively bright star. *170x250:* A good number of moderate brightness to moderately faint stars, including an oval of stars SW of the bright star and a larger outer ring of stars. Half the field in size. Moderately dense. A pretty good cluster.

| oc | NGC 2432 | 07 40.9 | −19 04 | Pup | 10.2 | 7.0' | | |

HH: cluster, pretty large, pretty compressed, extended N-S, bright and faint stars.
42x60: Faintly visible. Linear N-S. *170x250:* A fair number of moderately faint stars mainly in a N-S linear concentration, with more stars on either side. Half the field in size. The concentration is fairly dense; the cluster is moderate density overall.

| oc | NGC 2437/M46 | 07 41.8 | −14 48 | Pup | 6.1 | 27.0' | | |

JH: ! cluster, very bright, very rich, very large, involved planetary nebula.
42x60: A lot of moderately faint stars. A third of the field in size. Dense. *170x250:* A lot of moderate brightness stars, pretty uniform in brightness; there are only a couple of slightly brighter stars. It just overflows the field. Dense. The planetary, 2438, is moderate brightness. A little W of center is an involved star that does not appear to be the central star. The disk is quite round but doesn't have a very sharp edge. Pretty uniform in brightness except for a small darker hole in the middle. Practically no blinking effect. Moderate size for a planetary. Just off its edge SE is a moderate brightness star. The cluster looks distant, so the planetary would have to be humongous to be a member and yet appear so large; it's probably a foreground object [correct].

| pn | NGC 2438 | 07 41.8 | −14 44 | Pup | 10.1p | 64.0" | | |

HH: planetary nebula, pretty bright, pretty small, very little extended, resolvable.
[See M46.]

| pn | NGC 2440 | 07 41.9 | −18 12 | Pup | 10.8p | 70.0" | | |

HH: planetary nebula, considerably bright, not very well defined.
12x35: Visible as a very faint star. *42x60:* A faint "star" with a blinking effect. *170x250:* Bright; pretty high surface brightness. No central star. A nice disk but not a very sharp edge. There's an outer nebulosity beyond the main nebula, elongated ENE-WSW. The disk itself is uniform in brightness. Slight blinking effect. Moderately small for a planetary. A pretty nice planetary.

| oc | NGC 2447/M93 | 07 44.6 | −23 51 | Pup | 6.2 | 22.0' | | |

JH: cluster, large, pretty rich, little compressed, 8th to 13th mag. stars.
12x35: A dense cluster. *42x60:* It's in a wedge or knife-blade shape pointing SW. Dense. *170x250:* A lot of moderately bright to moderate brightness stars plus fainter stars. The dense part of the cluster is half the field in size; with bright outer members the cluster takes up the entire field. The middle is dense; overall the cluster is moderately dense. A nice cluster.

| bs | Beta Gem | 07 45.3 | +28 01 | Gem | 1.15 | | 1.00 | K0 |

170x250: Yellow. [Pollux]

| ds | 2 Pup | 07 45.5 | −14 41 | Pup | 16.8" | 6.1 | 6.8 | A2 A8 |

12x35: An extremely close slightly unequal pair. *42x60:* A close very slightly unequal pair. *170x250:* A moderate distance very slightly unequal pair. Both white. [STF 1138]

| pn | NGC 2452 | 07 47.4 | −27 20 | Pup | 12.6p | 30.0" | | |

JH: planetary nebula, faint, small, little extended, among 60 stars.
170x250: Moderate brightness. No central star. A small disk; not a very clean edge. Weak blinking effect. Round. Small. Very nearby SW is a faint star; nearby ESE is a moderate brightness star.

| oc | NGC 2453 | 07 47.8 | −27 13 | Pup | 8.3 | 5.0' | | |

JH: cluster, small, pretty rich, pretty compressed.
12x35: Visible. *42x60:* A very small cluster of stars SE of a moderate brightness star. *170x250:* A fair number of moderately faint stars. Along its W side is a bent line of five moderately bright to moderate brightness stars. Very small. Very dense.

| ds | 5 Pup | 07 47.9 | −12 12 | Pup | 1.6" | 5.6 | 7.7 | F5 − |

170x250: Not split. [STF 1146]

| oc | NGC 2455 | 07 49.0 | −21 16 | Pup | 10.2 | 7.0' | | |

JH: cluster, considerably large, pretty rich, little compressed, 12th mag. stars.
12x35: Very faintly visible. *42x60:* A faint cluster of very faint stars. *170x250:* A modest number of moderately faint and fainter stars, most of which are in an arrowhead shape pointing NW. A third of the field in size. Moderate density.

| bs | Xi Pup | 07 49.3 | −24 51 | Pup | 3.35 | | 1.25 | G3 |

170x250: Yellow gold.

| ga | UGC 4028 | 07 50.8 | +74 21 | Cam | 13.1b | 1.1' | 0.8' | Sc |

170x250: Moderately faint to moderate brightness. Evenly moderately concentrated. Round. Small. No involved stars.

| ga | NGC 2441 | 07 51.9 | +73 00 | Cam | 13.0b | 2.0' | 1.7' | Sb |

Te: very faint, pretty small.
170x250: Very faint. Broadly very slightly concentrated. Round. Moderately small. No involved stars.

| oc | NGC 2467 | 07 52.6 | −26 22 | Pup | 7.1 | 15.0' | | |

HH: pretty bright, very large, round, easily resolvable, 8th mag. star in middle.
12x35: 2467 and HAR2 are visible in the same field. *42x60:* 2467 is visible around a moderate brightness star. *170x250:* [I describe the object as a planetary nebula because this is what it most looks like.] Bright; moderate surface brightness. There is a centrally located faint star, but it's not clear that this is the central star: there are at least three more very faint stars in the vicinity of the center; at its edge N is the relatively bright star seen in the 60 mm. Not a well defined disk. Milky; even in light. Round. Very large for a planetary: a quarter of the field in size. Half the field NE is a cluster of a number of mixed brightness stars in a triangular figure [Haffner 18]. The stars are on the perimeter; the figure is empty inside. A little more than a quarter of the field in size. Sparse.

| oc | NGC 2482 | 07 54.9 | −24 17 | Pup | 7.3 | 12.0' | | |

HH: cluster, large, considerably rich, very little compressed.

12x35: A faint cloud of stars, just resolved. *42x60:* A number of faint stars. *170x250:* A fair number of moderate brightness to moderately faint stars. The stars are bunched in separate groups, including a linear group going SE from the middle of the cluster. 2/3 of the field in size. Moderate density.

| oc | NGC 2479 | 07 55.1 | –17 42 | Pup | 9.6 | 7.0' | | |

WH: *cluster, pretty large, pretty rich, pretty compressed, faint stars.*
12x35: Extremely faintly visible. *42x60:* A faint cluster of a fair number of faint stars. *170x250:* A fair number of moderately faint stars, very uniform in brightness. The stars in the middle are in an incomplete oval ring; overall the cluster is somewhat triangular in shape. 2/3 of the field in size. Moderate density.

| ga | ESO 561-2 | 07 55.4 | –21 20 | Pup | 13.2 | 2.2' | 1.4' | Sc |

170x250: Not visible.

| oc | HAR 2 | 07 55.7 | –25 53 | Pup | 8.7 | 5.0' | | |

42x60: A group of faint stars E of a very sharp triangular figure of moderate brightness stars. *170x250:* A number of moderate brightness and fainter stars. It's empty in the middle. A quarter of the field in size. Somewhat sparse. NE is a bright star; W is the sharp triangle of bright stars. [CR168]

| oc | NGC 2483 | 07 55.7 | –27 53 | Pup | 7.6 | 10.0' | | |

JH: *cluster, large, little compressed.*
12x35: Visible. *42x60:* A few faint stars. *170x250:* A number of mixed brightness stars: moderately bright to very faint. In the middle is a bit of a concentration elongated N-S. There's lots of space between the brighter stars. Half the field in size. Somewhat sparse to moderate density.

| oc | NGC 2489 | 07 56.2 | –30 03 | Pup | 7.9 | 8.0' | | |

HH: *cluster, pretty large, considerably rich, pretty compressed, 11th to 13th mag. stars.*
12x35: Faintly visible. *42x60:* A little group of very faint stars. *170x250:* A fair number of moderately faint stars, pretty uniform in brightness. Half the field in size. Moderate density.

| ga | NGC 2485 | 07 56.8 | +07 28 | CMi | 13.1p | 1.4' | 1.4' | Sa |

Ma: *nebulous 12th mag. star.*

170x250: Moderately faint. Sharply very concentrated. Round. Very small. No involved stars. Very nearby S is a moderately faint star; nearby SE is a slightly fainter star.

| ga | NGC 2460 | 07 56.9 | +60 21 | Cam | 12.7b | 2.4' | 1.8' | Sa |

Te: faint, small, round, faint star in center.
170x250: Moderate brightness. Broadly to evenly moderately concentrated. Round. Small. No involved stars. A third of the field WSW is another galaxy [*I2209]. Very very faint. Hardly concentrated. Roughly round. Small. No involved stars.

| pn | PK 164+31.1 | 07 57.8 | +53 24 | Lyn | 14.0p | 6.3' | | |

170x250: Not visible.

| ga | NGC 2487 | 07 58.3 | +25 09 | Gem | 13.2b | 2.6' | 2.1' | Sb |

Ma: very faint, small, gradually brighter middle.
170x250: In the position are two very similar galaxies a third of the field apart E-W [2487, *2486 (Ma)]. Both are very faint. Small. No involved stars. The E galaxy is broadly very slightly concentrated. Round. Nearby S is a bright star. The W galaxy is broadly slightly concentrated. Roughly round.

| oc | NGC 2506 | 08 00.2 | −10 47 | Mon | 7.6 | 6.0' | | |

HH: cluster, pretty large, very rich, compressed, 11th to 20th mag. stars.
12x35: Visible. *42x60:* A faint cluster of threshold stars. *170x250:* A whole lot of extremely faint and threshold stars, mainly between two wide pairs of moderately faint stars, one E and one W. The main concentration is a quarter of the field in size; with averted vision outer members N and S expand the cluster to half the field in size. Very dense.

| ga | NGC 2493 | 08 00.4 | +39 49 | Lyn | 13.0b | 1.9' | 1.9' | S0 |

HH: considerably bright, small, round, suddenly brighter middle.
170x250: Moderately faint. Evenly to sharply fairly concentrated. Round. Very small. No involved stars. Nearby NNW is a very faint star; S is an arc of three stars. [2495 (LR) was not seen.]

| oc | NGC 2509 | 08 00.7 | −19 03 | Pup | 9.3 | 8.0' | | |

WH: cluster, bright, pretty rich, little compressed, faint stars.
12x35: Very faintly visible. *42x60:* A faint cluster of threshold stars. *170x250:* A good number of faint and very faint stars in an ENE-WSW rectangular shape,

with the main concentration comprising the N edge. A third of the field in size. The concentration is very dense; the cluster as a whole is pretty dense.

| ga | NGC 2507 | 08 01.6 | +15 42 | Cnc | 13.2p | 2.4' | 1.7' | S0/a |

HH: pretty bright, pretty large, irregularly round, very gradually brighter middle, easily resolvable, star 60" SW.
170x250: Faint. Broadly very slightly concentrated. Slightly elongated NE-SW. Small. No involved stars. Nearby SW is a moderate brightness star; WSW are more stars.

| ga | NGC 2500 | 08 01.9 | +50 44 | Lyn | 12.2b | 2.6' | 2.6' | Sd |

HH: faint, large, round, very gradually brighter middle, resolvable, among stars.
170x250: Very faint. Broadly very slightly concentrated. Roughly round. Moderate size. It's surrounded by stars, including three bright stars N and two bright stars S.

| ga | NGC 2513 | 08 02.4 | +09 24 | Cnc | 12.6b | 2.5' | 2.0' | E |

HH: faint, small, round, pretty suddenly much brighter middle, resolvable.
170x250: Moderately faint. Evenly moderately concentrated. Round. Small. No involved stars. It makes a triangle with two moderate brightness stars, one NW and one farther N. [2510 (LR) and 2511 (LR) were not seen.]

| ga | NGC 2517 | 08 02.8 | −12 19 | Pup | 12.7b | 1.4' | 1.0' | S0 |

JH: faint, very small, round, between 3 13th–14th mag. stars.
170x250: Very faint to faint. Evenly to sharply moderately concentrated. Round. Very small. No involved stars. Nearby NNW is a moderately faint star; nearby S is another moderately faint star with a faint companion.

| ga | UGC 4151 | 08 04.3 | +77 49 | Cam | 13.2p | 1.3' | 1.2' | Sdm |

170x250: Very faint to faint. Hardly concentrated. Round. Small. No involved stars. E, crossing the field, is a NNW-SSE string of moderately faint stars.

| oc | NGC 2527 | 08 05.3 | −28 08 | Pup | 6.5 | 15.0' | | |

HH: cluster, very large, pretty rich, little compressed, 10th to 15th mag. stars.
12x35: A number of faint stars. *42x60:* A triangular figure of about eight moderately faint stars plus some fainter stars. *170x250:* A loose cluster of about a dozen moderate brightness stars, pretty uniform in brightness, plus more fainter stars. The brighter stars form a triangular figure, with the brightest star at the NE corner. On the SE side of the cluster is a small linear concentration. The cluster is 2/3 of the field in size overall. Sparse.

| ga | NGC 2525 | 08 05.6 | −11 25 | Pup | 12.3b | 2.9' | 1.9' | Sc |

HH: considerably faint, pretty large, round, very gradually very little brighter middle, among stars.
170x250: Very faint. Hardly concentrated. Round. Moderate size. No involved stars. Starting from near its edge N and going N is a string of six faint stars with a moderately bright star at the far end; SSE is a little group of very faint stars; farther S is a slightly zigzag line of four moderately bright stars.

| ds | STF 1177 | 08 05.6 | +27 32 | Cnc | 3.8" | 6.6 | 7.5 | A0 A1 |

42x60: Attached. *170x250:* A very close unequal pair. Both white.

| ga | ESO 494-26 | 08 06.2 | −27 31 | Pup | 12.5p | 4.7' | 3.1' | Sb |

170x250: Not visible.

| oc | NGC 2533 | 08 07.0 | −29 52 | Pup | 7.6 | 3.5' | | |

JH: cluster, pretty large, rich, compressed, 9th mag. star and 13th to 14th mag. stars.
12x35: In the position is a star or two. *42x60:* A handful of moderately faint stars in a N-S elongated group. *170x250:* Some ten moderate brightness stars in a slightly curved N-S figure. Interspersed with the cluster stars are faint field stars. Half the field in size. Very sparse. Not much of a cluster.

| bs | Rho Pup | 08 07.5 | −24 18 | Pup | 2.81 | | 0.43 | F2 |

170x250: Soft white.

| ga | NGC 2532 | 08 10.3 | +33 57 | Lyn | 13.0b | 2.1' | 1.7' | Sc |

HH: pretty bright, pretty large, round, very gradually little brighter middle, resolvable, 2 stars NE.
170x250: Very very faint. Broadly very slightly concentrated. Round. Small. No involved stars. It's on the side of a triangle of moderate brightness stars; a little ways SE is a very faint star.

| oc | NGC 2539 | 08 10.7 | −12 49 | Pup | 6.5 | 21.0' | | |

HH: cluster, very large, rich, little compressed, 11th to 13th mag. stars.
12x35: A cluster of threshold stars next to the bright plotted star. *42x60:* A good number of faint stars. A third of the field in size. *170x250:* A field and a half full of moderately faint stars. Moderately dense.

| ds | Zeta Cnc | 08 12.2 | +17 39 | Cnc | 6.0" | 5.6 | 6.2 | F8 G5 |

42x60: A very close almost equal pair; nicely split. ***170x250:*** A fairly close very slightly unequal pair. Both yellow white. A nice double. [STF 1196]

| ga | NGC 2543 | 08 13.0 | +36 15 | Lyn | 12.7p | 2.3' | 1.3' | Sb |

HH: faint, pretty large, irregularly round, very gradually brighter middle, double star near.

170x250: Very faint. Broadly very slightly concentrated. Roughly round. Moderately small. No involved stars. Nearby SSE is a very faint star; WNW are two moderately bright stars.

| ga | NGC 2537 | 08 13.2 | +45 59 | Lyn | 11.7v | 1.9' | 1.7' | Sm |

HH: globular cluster, pretty bright, pretty large, round, well resolved, 20th mag. stars.

170x250: Moderately faint. Broadly very slightly concentrated. Round. Small. No involved stars. Nearby ENE is a very faint star; a little ways ESE is a moderately bright star.

| oc | NGC 2548/M48 | 08 13.8 | −05 47 | Hya | 5.8 | 54.0' | | |

HH: cluster, very large, pretty rich, pretty much compressed, 9th to 13th mag. stars.

12x35: A lot of faint stars, with a little linear concentration in the middle. ***42x60:*** A good number of moderate brightness stars. Remarkably triangular in shape, with three lines of stars making the three sides of the triangle. Except for a NNE-SSW linear concentration in the middle, the cluster is mostly empty inside. Half the field in size. Moderate density. ***170x250:*** Two fields of relatively bright stars.

| ga | IC 2233 | 08 14.0 | +45 44 | Lyn | 13.1b | 4.7' | 0.5' | Sd |

170x250: Not visible.

| ga | NGC 2545 | 08 14.2 | +21 21 | Cnc | 13.2b | 2.0' | 1.1' | Sab |

HH: faint, small, little extended NE-SW, 8th mag. star 4' NW.

170x250: Faint. Sharply moderately concentrated. Elongated N-S. Moderately small. No involved stars. Nearby NNW is a very faint star; WNW is a triangle of bright stars.

| ga | NGC 2541 | 08 14.7 | +49 03 | Lyn | 12.3b | 6.3' | 3.1' | Scd |

HH: faint, large, extended, very gradually brighter middle.
170x250: Very faint. Diffuse. Slightly elongated N-S. Moderate size. No involved stars. WSW is a moderate brightness star; farther NNE is a bright star.

| ga | NGC 2523 | 08 15.0 | +73 34 | Cam | 12.6b | 2.9' | 1.7' | Sbc |

Sw: pretty bright, pretty large, little extended, little brighter middle, star near.
170x250: Faint. Broadly slightly concentrated. Roughly round. Small. No involved stars. [Nearby SSW is a moderate brightness star.]

| ga | UGC 4289 | 08 15.7 | +58 19 | Lyn | 13.1v | 1.4' | 1.2' | E |

170x250: Very very faint. Evenly moderately concentrated. Round. Very small. No involved stars. Nearby SSW is another very very faint galaxy [*C287-61].

| ga | NGC 2559 | 08 17.1 | −27 27 | Pup | 11.7p | 3.7' | 1.6' | Sbc |

JH: faint, pretty large, gradually much brighter middle, among 60 stars.
170x250: Moderately faint. Hardly concentrated. Roughly round. Moderately small. It's between two stars right at its edge: a bright star SE and a moderately faint star NW; the bright star makes a large hook figure with more stars N and S.

| ga | UGCA 137 | 08 17.7 | −30 07 | Pup | 12.4 | 3.1' | 2.2' | Sc |

170x250: Not visible.

| ga | NGC 2554 | 08 17.9 | +23 28 | Cnc | 12.9b | 3.2' | 2.3' | S0/a |

HH: faint, small, round, much brighter middle, resolvable.
170x250: Moderately faint. Evenly to sharply fairly concentrated. Round. Very small. No involved stars. It makes a small flat triangle with two faint stars, one SSE and one NNE.

| ga | NGC 2555 | 08 17.9 | +00 44 | Hya | 13.1p | 1.9' | 1.3' | Sab |

HH: very faint, considerably small, irregular figure, 3 faint stars involved?
170x250: Faint to moderately faint. Broadly slightly concentrated. Roughly round. Small. WNW of center is a very faint involved star; the galaxy is on the side of a small triangle of stars.

| oc | NGC 2567 | 08 18.6 | −30 38 | Pup | 7.4 | 10.0' | | |

HH: cluster, pretty large, pretty rich, little compressed, irregularly round, 11th to 14th mag. stars.
12x35: *Visible.* ***42x60:*** *A few faint stars.* ***170x250:*** *Mixed brightness stars: on the SW side are about seven moderate brightness stars in a diamond figure; on the E side is a string of 6–8 moderately faint stars in a slightly wavy line; N and S are faint to very faint stars. Half the field in size. Moderately dense. An unusual and interesting cluster.*

| ga | IC 2311 | 08 18.8 | −25 22 | Pup | 12.5b | 1.9' | 1.9' | E0 |

[See 2566.]

| ga | NGC 2566 | 08 18.8 | −25 29 | Pup | 11.8b | 3.4' | 2.2' | Sab |

WH: very faint, considerably large, easily resolvable.
170x250: *Moderately faint. Sharply very concentrated. [Elongated ENE-WSW.] Small. E of center is a faint involved star. Half the field N is I2311. Also moderately faint. Evenly moderately concentrated. Round. Small. No involved stars. [E495-1 was not seen.]*

| oc | NGC 2571 | 08 18.9 | −29 44 | Pup | 7.0 | 13.0' | | |

HH: cluster, very large, considerably rich, little compressed, 9th mag. and fainter stars.
12x35: *A bright little cluster.* ***42x60:*** *A couple of moderate brightness stars and more faint stars surrounding them.* ***170x250:*** *Mixed brightness stars: a bright pair, a fair number of moderately bright stars, and more moderate brightness and faint stars. Elongated NW-SE. It fills the field. Somewhat sparse.*

| ga | NGC 2549 | 08 19.0 | +57 48 | Lyn | 12.2b | 4.3' | 1.4' | S0 |

JH: pretty bright, small, much extended N-S, pretty suddenly much brighter middle.
170x250: *Moderate brightness. Sharply very concentrated. Very elongated to edge-on N-S. Moderately small. The ends fade out very indistinctly. No involved stars.*

| ga | UGC 4305 | 08 19.2 | +70 43 | UMa | 11.1b | 7.9' | 6.2' | Dw I |

170x250: *Exceedingly faint. Involved is a relatively bright star that makes a very small triangle with two faint stars, one SSW and one SE.*

| ga | NGC 2552 | 08 19.3 | +50 00 | Lyn | 12.6b | 3.4' | 2.2' | Sm |

WH: extremely faint, considerably large, little extended NE-SW.
170x250: Extremely faint. Diffuse. Elongated ENE-WSW. Moderate size. Near its edge N and E are two or three very faint stars; NE is a relatively bright star; W is an arc of three stars.

| ga | NGC 2563 | 08 20.6 | +21 04 | Cnc | 13.2b | 2.5' | 2.2' | S0 |

HH: considerably faint, small, round, brighter middle.
170x250: In the position are two very similar galaxies a quarter of the field apart NW-SE [*2562 (HH), 2563]. Both are moderately faint. Sharply fairly concentrated. Roughly round. Very small. No involved stars. They make a triangle with a moderately faint star a little ways W of the SE galaxy.

| oc | NGC 2580 | 08 21.6 | −30 17 | Pup | 9.7 | 7.0' | | |

JH: cluster, considerably large, pretty rich, pretty compressed, round, 12th mag. stars.
12x35: Very faintly visible. ***42x60:*** A few faint to very faint stars. ***170x250:*** Some ten widely spaced moderately faint stars plus more faint stars among them. A third of the field in size. Moderate density.

| oc | NGC 2588 | 08 23.2 | −32 58 | Pup | 11.8 | 2.0' | | |

JH: cluster, faint, small, round, gradually brighter middle, 15th mag. stars.
170x250: A faint cluster of a fair number of faint to very faint stars. Very small. Moderate density.

| ga | UGC 4375 | 08 23.2 | +22 39 | Cnc | 12.8b | 2.4' | 1.5' | Sc |

170x250: Faint. Small. Superimposed right in the center [just E of center] is an interfering moderately faint involved star; nearby E is another star; SSW is a triangle of moderate brightness to moderately faint stars. Half the field SW is a better seen galaxy [*2577 (WH)]. Moderately faint. Evenly to sharply fairly concentrated. Round. Very small. No involved stars. A little ways NNE are two very faint stars almost in line with the galaxy.

| oc | NGC 2587 | 08 23.5 | −29 29 | Pup | 9.2 | 9.0' | | |

JH: cluster, pretty much compressed middle, irregular figure, 9th mag. star and 10th to 13th mag. stars.
12x35: Faintly visible. ***42x60:*** A faint cluster of very faint stars with a brighter star at its edge SE. ***170x250:*** A number of moderately faint stars between the relatively

bright star ESE and a moderate brightness star WNW. The cluster's stars are in two parallel N-S lines. A quarter of the field in size. Moderate density.

| ga | NGC 2551 | 08 24.8 | +73 24 | Cam | 13.1b | 1.6' | 1.0' | S0/a |

Te: very faint, small, faint star in center.
170x250: Moderately faint. Sharply fairly concentrated. Very elongated NE-SW. Small. No involved stars. [A little ways ENE is a moderate brightness star.]

| ds | Phi 2 Cnc | 08 26.8 | +26 56 | Cnc | 5.1" | 6.3 | 6.3 | A3 A6 |

42x60: Extremely close. *170x250:* A pretty close perfectly equal pair. Both white. A very nice double. [STF 1223]

| ga | MCG-2-22-17 | 08 27.6 | −12 45 | Pup | 12.6 | 2.3' | 1.5' | Scd |

170x250: Not visible.

| ga | NGC 2595 | 08 27.7 | +21 28 | Cnc | 12.9b | 3.1' | 2.3' | Sc |

HH: very faint, pretty large, irregular figure, resolvable, double star 2' SW.
170x250: Very faint. Hardly concentrated. Roughly round. Small. At its edge NE is a very faint involved star; a little ways WSW is a relatively bright star with a wide companion.

| bs | Omicron UMa | 08 30.3 | +60 43 | UMa | 3.40 | | 0.80 | G4 |

170x250: Yellow.

| ga | NGC 2599 | 08 32.2 | +22 33 | Cnc | 13.1b | 1.8' | 1.6' | Sa |

HH: very faint, small, stellar.
170x250: Moderately faint. Sharply pretty concentrated. Roughly round. Small. No involved stars. A little ways WSW is a faint star; SE is a little triangle of moderate brightness to moderately faint stars.

| ga | NGC 2604 | 08 33.4 | +29 32 | Cnc | 13.0p | 2.0' | 2.0' | Scd |

HH: very faint, pretty large, round, little brighter middle, resolvable, double star near.
170x250: Very very faint. Hardly concentrated. Roughly round. Small to moderately small. No involved stars. A little ways E is a faint star; a little ways S is another faint star; the galaxy is in the middle of a very large isosceles triangle of bright stars, the S star of which is a wide double.

| pn | NGC 2610 | 08 33.4 | −16 09 | Hya | 13.6p | 58.0" | | |

HH: faint, small, attached to 13th mag. star, 7th mag. star NE.
170x250: Pretty faint. No central star. Round, but not a disk. No blinking effect. Moderately small for a planetary. At its edge NE is a star that makes a small "L" figure with a line of three stars nearby ENE; further NE is the bright plotted star. An unimpressive planetary.

| ga | NGC 2613 | 08 33.4 | −22 58 | Pyx | 11.2b | 7.2' | 1.7' | Sb |

HH: considerably bright, large, very much extended WNW-ESE.
42x60: Very very faintly visible. ***170x250:*** Moderately bright. Broadly moderately concentrated. Very elongated to edge-on WNW-ESE. Moderately large. The ends fade out very indistinctly. No involved stars. Nearby N is a moderately faint star; nearby S are three faint stars in a shallow arc; off the galaxy's W end is a moderate brightness star.

| ga | NGC 2608 | 08 35.3 | +28 28 | Cnc | 13.0b | 2.2' | 1.3' | Sb |

HH: faint, very little extended, much brighter middle, resolvable.
170x250: Faint. Broadly very slightly concentrated. Elongated E-W. Small to moderately small. No involved stars. It's surrounded by very faint stars.

| ds | STF 1245 | 08 35.8 | +06 37 | Cnc | 10.2" | 6.0 | 7.2 | F8 G5 |

42x60: A faint double. ***170x250:*** A moderate distance slightly unequal pair. Both yellowish.

| ga | NGC 2618 | 08 35.9 | +00 42 | Hya | 13.0p | 2.4' | 1.9' | Sab |

HH: extremely faint, pretty large, irregular figure.
170x250: Very very faint. Broadly slightly concentrated. Round. Small. No involved stars. Nearby E is a moderately faint star that makes a sharp triangle with two more stars, one NE and one ENE.

| ga | ESO 495-21 | 08 36.3 | −26 24 | Pyx | 12.4p | 1.7' | 1.3' | I0 |

170x250: Moderate brightness; high surface brightness. Sharply very concentrated. Round. Very small. No involved stars. Passing S of it in a shallow arc are a moderate brightness star nearby W, a faint star nearby SSE, and another faint star ESE.

| oc | NGC 2627 | 08 37.3 | −29 56 | Pyx | 8.4 | 11.0' | | |

HH: cluster, considerably large, pretty rich, pretty compressed, 11th to 13th mag. stars.

12x35: Faintly visible. *42x60:* A faint cluster of very faint stars. *170x250:* A good number of moderately faint and faint stars. Slightly elongated E-W. A third of the field in size. Moderately dense.

| ga | NGC 2591 | 08 37.4 | +78 01 | Cam | 12.9p | 3.0' | 0.6' | Scd |

dA: faint, small, extended, little brighter middle.
170x250: Very very faint. Hardly concentrated. Elongated NNE-SSW. Small. No involved stars.

| ga | NGC 2619 | 08 37.5 | +28 42 | Cnc | 13.2p | 2.3' | 1.4' | Sbc |

HH: faint, pretty small, round, brighter middle, resolvable.
170x250: Faint. Broadly very slightly concentrated. Roughly round. Small. No involved stars.

| oc | NGC 2635 | 08 38.5 | -34 45 | Pyx | 11.2 | 3.0' | | |

JH: cluster, pretty much compressed, irregularly triangular, 13th mag. and fainter stars.
170x250: A very small compact cluster of a number of faint to very faint stars. Moderate density.

| oc | NGC 2632/M44 | 08 40.1 | +19 59 | Cnc | 3.1 | 95.0' | | |

JH: Praesepe.
12x35: A fair number of bright and moderately bright stars. The brightest stars are in a central "V" pointing WSW, and there are quite a few more bright stars. A third of the field in size. Moderate density. *42x60:* It slightly overflows the field with nice bright stars. [Beehive Cluster]

| ga | NGC 2648 | 08 42.7 | +14 17 | Cnc | 12.7p | 3.2' | 1.0' | Sa |

HH: faint, small, very little extended NW-SE, pretty suddenly brighter middle.
170x250: Moderately bright; fairly high surface brightness. Sharply pretty concentrated. Roughly round. Very small. No involved stars. A little ways E is a moderate brightness star.

| oc | NGC 2658 | 08 43.4 | -32 38 | Pyx | 9.2 | 12.0' | | |

JH: cluster, pretty small, little rich, little compressed, very irregular figure, 12th–13th mag. stars.
42x60: Very faintly visible. *170x250:* A number of faint stars in a narrow "V" pointing SW; off the tip of the "V" is a moderate brightness star. Very small. Moderate density.

| ga | NGC 2639 | 08 43.6 | +50 12 | UMa | 12.6b | 2.1' | 1.3' | Sa |

HH: considerably bright, small, extended NW-SE, pretty suddenly much brighter middle to a star?
170x250: Moderately faint. Evenly moderately concentrated. Elongated NW-SE. Moderately small. No involved stars. Making a triangle with it are two faint stars, one SE and another a little farther SW.

| ga | NGC 2649 | 08 44.1 | +34 43 | Lyn | 13.1p | 1.5' | 1.4' | Sbc |

HH: faint, large, round, resolvable.
170x250: Extremely faint. Hardly concentrated. Roughly round. Small. At its edge N is a very faint involved star; the galaxy makes a long shallow arc with a faint star NW and two more faint stars farther SSE.

| ga | NGC 2663 | 08 45.1 | –33 47 | Pyx | 11.9p | 3.4' | 2.3' | E |

Sw: pretty faint, pretty small, little extended.
170x250: Moderate brightness. Broadly moderately concentrated. Round. Small. No involved stars. It makes a small quadrilateral with faint to very faint stars N to W; further NW is a brighter star.

| ds | STF 1270 | 08 45.3 | –02 36 | Hya | 4.7" | 6.4 | 7.4 | F2 – |

42x60: Extremely close. ***170x250:*** A moderately close slightly unequal pair. Both white.

| ga | NGC 2665 | 08 46.0 | –19 18 | Hya | 12.9b | 2.0' | 1.4' | Sa |

Mu: faint, small, round, gradually brighter middle to a nucleus.
170x250: Moderately faint. Broadly slightly concentrated. Roughly round. Small. No involved stars. It's among a number of relatively bright stars N, S, and E.

| ds | Iota 1 Cnc | 08 46.7 | +28 46 | Cnc | 30.6" | 4.2 | 6.6 | G7 A3 |

12x35: Easily split. ***42x60:*** Somewhat wide. ***170x250:*** The primary has a very wide fainter companion. Color contrast: the primary is yellow; the companion is white. A pretty nice double. [STF 1268]

| bs | Epsilon Hya | 08 46.8 | +06 25 | Hya | 3.38 | | 0.68 | F0 |

170x250: The primary has an extremely close much fainter companion. The primary is yellow white. A fairly nice double. [STF 1273: 2.7", 3.4, 6.8]

| ga | ESO 371-16 | 08 47.1 | −33 45 | Pyx | 12.8p | 3.3' | 2.7' | S0/a |

170x250: Not visible.

| ga | NGC 2633 | 08 48.1 | +74 06 | Cam | 12.9b | 2.4' | 1.5' | Sb |

Te: faint, small, little extended.
170x250: 2633 and 2634 are half the field apart N-S. 2633 is very slightly bigger and brighter, but otherwise the two galaxies are pretty similar. Both are moderately faint. Sharply moderately concentrated. Roughly round. Small. No involved stars. [2634A was not seen.]

| ga | NGC 2634 | 08 48.4 | +73 58 | Cam | 12.9b | 1.7' | 1.6' | E1 |

Te: faint, small, little extended.
[See 2633.]

| ga | NGC 2654 | 08 49.2 | +60 13 | UMa | 12.7b | 4.2' | 1.0' | Sab |

Te: pretty faint, small, faint star in middle, faint star close SW.
170x250: Moderate brightness. Evenly moderately concentrated. Edge-on ENE-WSW. Moderately small to moderate size. No involved stars. A little ways ESE is a very faint star; a little ways S is another very faint star.

| ga | NGC 2672 | 08 49.3 | +19 04 | Cnc | 12.7b | 2.9' | 2.7' | E1-2 |

HH: bright, pretty large, little extended N-S or binuclear, much brighter middle to a star.
170x250: Moderately faint. Evenly moderately concentrated. Slightly elongated E-W [round]. Small. Just E of center is a very faint involved star [this is *2673 (LR), a stellar galaxy]; at its edge WNW is a very very faint involved star; nearby SE is a very faint star; a little ways NNE is a faint star.

| ga | NGC 2646 | 08 50.4 | +73 27 | Cam | 13.1p | 1.3' | 1.2' | S0 |

Te: very faint, small, 2 faint stars 2.5' SE.
170x250: Faint to moderately faint. Evenly moderately concentrated. Round. Very small. No involved stars. A little ways SSE are a couple of moderately bright stars.

| oc | NGC 2682/M67 | 08 50.4 | +11 49 | Cnc | 6.9 | 29.0' | | |

JH: ! cluster, very bright, very large, extremely rich, little compressed, 10th to 15th mag. stars.
12x35: Starting to be resolved. *42x60:* Mostly faint stars. The brighter central part is elongated NE-SW. Off the NE end of this concentration is one brighter star.

170x250: A good number of stars, mostly moderate brightness to moderately faint. The central concentration is elongated NE-SW, but the cluster as a whole is elongated E-W. It takes up the entire field. Moderately dense. A nice cluster.

ga	ESO 563-31	08 52.3	−17 44	Hya	13.2	1.6'	1.3'	S0

170x250: Not visible.

ga	NGC 2683	08 52.7	+33 25	Lyn	10.6b	10.5'	2.5'	Sb

HH: very bright, very large, very much extended NE-SW, gradually much brighter middle.
42x60: Visible. *170x250:* Bright. Evenly moderately concentrated. Edge-on NE-SW. Large: a third of the field in size. The ends fade out very indistinctly. Crossing its NE end is a shallow arc of three moderate brightness stars; at times the galaxy extends past the middle star of this arc, but it doesn't quite reach a very faint star further NE, even with averted vision; a little ways S is a moderately faint star with a wide very faint companion. A nice edge-on galaxy.

ga	NGC 2681	08 53.5	+51 18	UMa	11.1b	3.6'	3.2'	S0/a

HH: very bright, large, very gradually then very suddenly much brighter middle to a 10th mag. star.
42x60: Visible. *170x250:* Bright. Sharply very concentrated to a stellar nucleus. Round. Small. No involved stars. Nearby WNW are two stars; nearby E is a star; a little ways SE is another star.

ga	IC 520	08 53.7	+73 29	Cam	12.6b	1.9'	1.5'	Sab

170x250: Moderately faint. Evenly moderately concentrated. Slightly elongated N-S. Moderately small. Near its edge SSE is a very faint involved star; a little ways NNW is a faint star.

ga	NGC 2695	08 54.5	−03 03	Hya	12.8b	2.1'	1.7'	S0

HH: pretty faint, considerably small, extended E-W, between 2 stars.
170x250: In the position of 2695 and 2708 is a widespread NW-SE group of galaxies. The westernmost is 2695. Moderately faint. Evenly moderately concentrated. Elongated E-W [round]. Small. Just E of center is a very faint involved star; just off its W end is a moderate brightness star. Half the field ENE is a faint galaxy [*2697 (LR)]. Hardly concentrated. Round. Small. No involved stars. At the SE end of the group is 2708. Moderately faint. Broadly to evenly moderately concentrated. Elongated NNE-SSW. Small. At its NE end is a moderately faint star; nearby NW

is a moderate brightness star. A little less than half the field N is a small extremely faint galaxy [*2709 (LR)]. A quarter of the field ESE of the bright plotted star between 2695 and 2708 is another galaxy [*2698 (JH)]. Moderate brightness. Sharply very concentrated. Round. Very small. Just SW of center is a faint involved star. A quarter of the field NE is a galaxy [*2699 (dA)] very similar to the previous one except that it's not quite as bright. Moderately faint. Sharply pretty concentrated. Round. Very small. No involved stars. Very nearby SW is a very faint star; a little farther NW is a moderately faint star.

| vs | X Cnc | 08 55.4 | +17 14 | Cnc | 6.2 | 7.5 | 3.36 | C5 |

12x35: Mag. 6.5. *170x250:* Orangish.

| bs | Zeta Hya | 08 55.4 | +05 56 | Hya | 3.11 | | 1.00 | G8 |

170x250: Yellow.

| ds | 17 Hya | 08 55.5 | -07 58 | Hya | 3.9" | 6.8 | 7.0 | A2 A7 |

42x60: Extremely close. *170x250:* A very close pretty much equal pair. Both white. [STF 1295]

| ga | NGC 2655 | 08 55.6 | +78 13 | Cam | 11.0b | 6.6' | 4.8' | S0/a |

HH: *very bright, considerably large, little extended E-W, gradually then suddenly very much brighter middle.*
42x60: Visible. *170x250:* Bright. Sharply very concentrated. Round. Small. No involved stars.

| ga | NGC 2685 | 08 55.6 | +58 44 | UMa | 11.2v | 4.5' | 2.4' | S0+ |

Te: *pretty faint, round, faint star in center.*
170x250: Moderately bright. Sharply pretty concentrated. Edge-on NE-SW. Moderately small. No involved stars. A little ways N is a moderately bright star.

| ga | NGC 2708 | 08 56.1 | -03 21 | Hya | 11.9v | 3.4' | 2.0' | Sb |

WH: *very faint, pretty small, round.*
[See 2695.]

| ga | NGC 2693 | 08 57.0 | +51 20 | UMa | 12.8b | 2.6' | 1.7' | E3 |

HH: *pretty bright, much extended, pretty suddenly much brighter middle.*

170x250: Moderately faint to moderate brightness. Sharply pretty concentrated. Round. Very small. Just off its edge S is a very very faint star [this is *2694 (LR), a stellar galaxy]; the galaxy makes a flat triangle with two moderate brightness stars, one NNE and one a little farther WNW.

| ga | NGC 2713 | 08 57.3 | +02 55 | Hya | 12.7b | 3.6' | 1.5' | Sab |

Ma: pretty bright, irregularly round, much brighter middle.
170x250: Moderate brightness. Evenly moderately concentrated. Slightly elongated WNW-ESE. Moderately small. No involved stars. W is a moderately bright star; S are two moderate brightness stars.

| ga | NGC 2716 | 08 57.6 | +03 05 | Hya | 12.7b | 1.2' | 0.9' | S0+ |

Ma: faint, small, round, much brighter middle.
170x250: Moderately faint. Evenly moderately concentrated. Roughly round. Small. No involved stars. It's just off the N side of a triangular figure of five moderately faint stars.

| ga | NGC 2718 | 08 58.8 | +06 17 | Hya | 12.7p | 2.1' | 2.1' | Sab |

JH: faint, pretty large, round.
170x250: Very faint to faint. Broadly slightly concentrated. Slightly elongated NW-SE. Small. No involved stars. It's on the long side of a small flat triangle of very faint stars.

| ga | NGC 2721 | 08 58.9 | −04 54 | Hya | 12.5 | 2.3' | 1.5' | Sbc |

HH: considerably faint, pretty large, round, very gradually brighter middle.
170x250: Faint to moderately faint. Broadly slightly concentrated. Roughly round. Small. No involved stars.

| ga | NGC 2701 | 08 59.1 | +53 46 | UMa | 12.7b | 2.1' | 1.5' | Sc |

HH: pretty bright, fan-shaped, 11th mag. star attached.
170x250: Moderately faint. Broadly slightly concentrated. Roughly round. Small. Just W of center is a relatively bright involved star; a little ways NW is a wide pair of stars; NNE is another wide pair of stars; each of the two pairs consists of a moderately faint star and a faint star.

| bs | Iota UMa | 08 59.2 | +48 02 | UMa | 3.10 | | 0.23 | A7 |

170x250: White.

| ga | NGC 2712 | 08 59.5 | +44 54 | Lyn | 12.8b | 2.9' | 1.5' | Sb |

JH: pretty bright, large, extended, very gradually brighter middle to an 18th mag. star.
170x250: Faint. Broadly slightly concentrated. Elongated N-S. Small. No involved stars.

| ga | IC 512 | 09 03.9 | +85 30 | Cam | 12.9p | 2.9' | 2.3' | Scd |

170x250: Very very faint. Broadly very slightly concentrated. Roughly round. Small. No involved stars.

| ga | NGC 2749 | 09 05.4 | +18 18 | Cnc | 12.7b | 1.8' | 1.5' | E3 |

dA: pretty faint, small, round, brighter middle to a 15th mag. stellar nucleus.
170x250: Moderate brightness. Evenly pretty concentrated. Round. Very small. Very nearby WSW is a very faint star. [2751 (Ma) and 2752 (Ma) were not seen.]

| ga | NGC 2750 | 09 05.8 | +25 26 | Cnc | 12.6p | 2.2' | 1.9' | Sc |

WH: very faint, considerably large, round, brighter middle to a nucleus, 2 stars W.
170x250: Very faint. Broadly very slightly concentrated. Roughly round. Small. At or near center is either a very faint stellar nucleus or a very faint involved star; the galaxy makes a triangle with two stars SW, bright and moderate brightness.

| ga | NGC 2763 | 09 06.8 | −15 30 | Hya | 12.6b | 2.3' | 2.0' | Scd |

HH: very faint, pretty small, brighter middle, faint star 30" N.
170x250: Faint. Broadly slightly concentrated. Roughly round. Small. Just off its edge N is a faint star.

| ds | STF 1311 | 09 07.4 | +22 59 | Cnc | 7.6" | 6.9 | 7.3 | F5 F5 |

42x60: An almost equal pair. *170x250:* A moderately close pretty much equal pair. Both white.

| ga | NGC 2742 | 09 07.6 | +60 28 | UMa | 12.0b | 3.0' | 1.5' | Sc |

HH: considerably bright, considerably large, extended E-W, easily resolvable.
170x250: Moderate brightness. Broadly very slightly concentrated. Slightly elongated E-W. Moderate size. No involved stars. A little ways SW is a small sharp triangle of stars.

| ga | NGC 2765 | 09 07.6 | +03 23 | Hya | 13.1p | 2.1' | 1.1' | S0 |

HH: very faint, pretty large, extended, gradually brighter middle, easily resolvable.
170x250: Moderately faint to moderate brightness. Broadly slightly concentrated. Elongated WNW-ESE. Small. No involved stars. Nearby NE is a very faint star.

| ga | NGC 2715 | 09 08.1 | +78 05 | Cam | 11.8b | 5.2' | 1.6' | Sc |

By: pretty bright, large, extended.
170x250: Faint. Hardly concentrated. Slightly elongated NNE-SSW. Moderately large. No involved stars. [S is a moderate brightness star.]

| ga | NGC 2756 | 09 09.0 | +53 50 | UMa | 13.2p | 1.7' | 1.1' | Sb |

HH: pretty bright, pretty small, extended, very gradually brighter middle.
170x250: Moderately faint. Broadly slightly concentrated. Round. Small. No involved stars. [ESE are two moderately faint stars.]

| ga | NGC 2770 | 09 09.6 | +33 07 | Lyn | 12.8b | 4.6' | 1.2' | Sc |

HH: faint, large, much extended NNW-SSE, resolvable, 2 stars N.
170x250: Very faint. Hardly concentrated. Edge-on NW-SE. Moderate size. No involved stars. NNE are two moderate brightness stars pretty much parallel to the galaxy; making a shallow arc with these two stars is a faint star NW.

| ga | NGC 2775 | 09 10.3 | +07 02 | Cnc | 11.0b | 4.2' | 3.4' | Sab |

HH: considerably bright, considerably large, round, very gradually then very suddenly much brighter middle, resolvable?
42x60: Visible. ***170x250:*** Bright. Sharply pretty concentrated. Round. Small. No involved stars. E to SE is a long trapezoid of stars.

| ga | UGCA 150 | 09 10.8 | −08 53 | Hya | 11.9b | 5.3' | 1.0' | Sb |

170x250: Moderately faint. Hardly concentrated. Elongated NE-SW. Small. At or near center are a couple of very faint involved stars; at the galaxy's edge W of center is the bright plotted star, which interferes.

| ga | NGC 2781 | 09 11.5 | −14 49 | Hya | 12.5b | 3.4' | 1.5' | S0+ |

HH: bright, small, pretty much extended E-W, pretty suddenly much brighter middle.

170x250: Moderately bright; fairly high surface brightness. Sharply very concentrated. Round. Very small. No involved stars. Nearby W is a very very faint star.

| ga | NGC 2768 | 09 11.6 | +60 02 | UMa | 10.8b | 8.1' | 4.2' | E6 |

HH: considerably bright, considerably large, little extended, pretty suddenly much brighter middle to a large bright nucleus.
42x60: Faintly visible. *170x250:* Moderately bright. Evenly moderately concentrated. Elongated E-W. Moderate size. No involved stars. It's among several nearby faint stars; the galaxy makes a triangle with two bright stars, one WNW and another farther N.

| ga | NGC 2776 | 09 12.2 | +44 57 | Lyn | 12.1b | 3.0' | 2.6' | Sc |

JH: pretty bright, large, round, very gradually brighter middle, resolvable.
170x250: Moderate brightness. Broadly to evenly moderately concentrated. Round. Moderately small. No involved stars.

| ga | NGC 2784 | 09 12.3 | −24 10 | Hya | 11.3b | 6.2' | 2.2' | S0 |

HH: bright, large, much extended ENE-WSW, gradually much brighter middle.
170x250: Moderately bright. Sharply pretty concentrated. Elongated ENE-WSW; at first sight it looks round, but with averted vision it has a very faint elongated halo. Moderate size. No involved stars.

| ga | NGC 2732 | 09 13.4 | +79 11 | Cam | 12.9b | 2.1' | 0.7' | S0 |

JH: pretty bright, small, extended NE-SW, star NE.
170x250: Moderate brightness to moderately bright. Sharply pretty concentrated. Edge-on ENE-WSW. Very small. Right off its E end is a moderately faint star.

| ga | NGC 2748 | 09 13.7 | +76 28 | Cam | 12.4b | 3.0' | 1.1' | Sbc |

JH: pretty bright, pretty large, extended, very gradually little brighter middle.
170x250: Moderately faint to moderate brightness. Evenly slightly concentrated. Elongated NE-SW. Small. No involved stars. [WNW to NE is a long curved line of four bright stars.]

| ga | NGC 2782 | 09 14.1 | +40 06 | Lyn | 12.3b | 3.8' | 2.5' | Sa |

HH: considerably bright, round, much brighter middle to a bright nucleus.
170x250: Moderately bright. Sharply fairly concentrated. Roughly round. Small. No involved stars. It makes a sharp triangle with two moderately faint stars S.

| ga | UGC 4841 | 09 14.8 | +74 13 | Cam | 13.0b | 2.8' | 2.2' | Sd |

170x250: Not visible.

| ga | NGC 2789 | 09 15.0 | +29 43 | Cnc | 13.2p | 1.7' | 1.6' | S0/a |

St: pretty faint, small, round, gradually brighter middle.
170x250: Faint. Sharply pretty concentrated. Round. Very small. No involved stars. A little ways WNW is a moderately bright star.

| ga | NGC 2811 | 09 16.2 | −16 18 | Hya | 12.2b | 2.5' | 0.8' | Sa |

HH: pretty bright, pretty small, extended NE-SW, pretty suddenly much brighter middle.
170x250: Moderately bright. Evenly moderately concentrated. Elongated NNE-SSW. Moderate size. No involved stars. A little ways WNW is a moderate brightness star.

| ga | NGC 2815 | 09 16.3 | −23 38 | Hya | 12.8b | 3.4' | 1.1' | Sb |

HH: faint, small, little extended, gradually brighter middle.
170x250: Faint. Broadly moderately concentrated. Elongated N-S. Moderately small. No involved stars. Nearby S is a very faint star; nearby SSE is another very faint star; E is a little group of moderately faint stars in an upside-down cross figure.

| ga | NGC 2804 | 09 16.8 | +20 11 | Cnc | 12.8v | 2.0' | 1.7' | S0 |

JH: very faint, small, round, NW of 2.
170x250: Faint. Evenly moderately concentrated. Round. Small. No involved stars. A little ways N is a faint star; a little farther S is a moderately faint star; E to SSE is a large shallow arc of four moderately bright stars. S of this arc are two more small galaxies aligned NE-SW [*2809 (JH), *2806 (Dr)]. The NE galaxy is faint. The SW galaxy is very faint. [C91-55 was not seen.]

| ga | NGC 2798 | 09 17.4 | +42 00 | Lyn | 13.0b | 2.5' | 0.9' | Sa |

WH: pretty bright, small, stellar.
170x250: Moderate brightness; fairly high surface brightness. Sharply pretty concentrated. Round. Very small. No involved stars. [2799 (LR) and U4904 were not seen.]

| ga | NGC 2835 | 09 17.9 | −22 21 | Hya | 11.0b | 6.6' | 4.3' | Sc |

Ba: faint, 10th mag. star involved E, between 2 9th mag. stars.

170x250: Very faint. Hardly concentrated. Slightly elongated N-S. Moderately large. No involved stars. It's in a curvy WNW-ESE string of moderate brightness stars, including one at its edge E.

ga	MCG-2-24-7	09 18.1	−12 05	Hya	12.9v	0.8'	0.7'	S0−

170x250: Not visible.

ga	IC 529	09 18.5	+73 45	Cam	12.6b	3.6'	1.6'	Sc

170x250: Very faint; low surface brightness. Diffuse. Slightly elongated NW-SE. Large: a quarter of the field in size [moderately small]. No involved stars.

ga	UGC 4883	09 18.5	+74 19	Cam	13.2p	1.2'	0.6'	S

170x250: Moderate brightness; fairly high surface brightness. Concentrated. Round. Tiny. No involved stars.

ds	38 Lyn	09 18.8	+36 48	Lyn	2.5"	3.9	6.6	A1 A4

170x250: The primary has a very close fainter companion. The primary is white. Very nice; one of the nicer double stars. [STF 1334]

ga	NGC 2787	09 19.3	+69 12	UMa	11.8b	3.1'	2.0'	S0+

HH: bright, pretty large, little extended E-W, much brighter middle, resolvable, very faint star involved SE.
170x250: Moderately bright. Evenly to sharply fairly concentrated. Round. Small. Off its edge SSE is a faint star.

ga	NGC 2832	09 19.8	+33 44	Lyn	11.9v	2.4'	1.9'	E+2

LR: faint, very small, round, 3rd of 3.
170x250: Moderate brightness. Sharply pretty concentrated. Roughly round. Small. No involved stars. It makes a triangle with two moderately bright double stars, a close one SSE and a very wide one ESE. Nearby SW is another galaxy [*2830 (HH)]. Extremely faint. Edge-on WNW-ESE. Farther W is a third galaxy [*2825 (JH)]. Very very faint. Very small. [2831 (LR) and 2834 (LR) were not seen; the NGC observers and descriptions of 2830, 2831, and 2832 are obviously confused.]

ga	NGC 2848	09 20.2	−16 31	Hya	12.4b	2.6'	1.6'	Sc

HH: very faint, considerably large, extended NE-SW, gradually little brighter middle, 11th mag. star [NE].

170x250: Faint. Broadly slightly concentrated. Slightly elongated NE-SW. Moderately small. At its NE end is a very faint star with an extremely faint star next to it W; further NE is a moderately faint star; a little ways NNW is a faint star. A little more than a quarter of the field ENE is a very faint galaxy [*2851 (Sw)].

| ga | NGC 2805 | 09 20.3 | +64 06 | UMa | 11.5b | 6.3' | 4.7' | Sd |

WH: very faint, large, round, much brighter middle.
170x250: Extremely faint. Hardly concentrated. Round. Small. No involved stars. [It's among a number of stars.]

| bs | Alpha Lyn | 09 21.1 | +34 23 | Lyn | 3.16 | | 1.52 | M0 |

170x250: Gold.

| ga | NGC 2855 | 09 21.5 | −11 54 | Hya | 12.6b | 2.4' | 2.1' | S0/a |

HH: pretty bright, pretty large, round, gradually much brighter middle to a nucleus.
170x250: Moderately bright. Evenly moderately concentrated. Round. Small. No involved stars. It's right in between a moderately bright star NNE and a moderately faint star SSW.

| ga | NGC 2820 | 09 21.8 | +64 15 | UMa | 12.8v | 5.5' | 0.7' | Sc |

HH: faint, small, extended, 2nd of 2.
170x250: Very faint. Hardly concentrated. A thin edge-on, ENE-WSW. Moderate size. No involved stars. A little less than a quarter of the field W is an even fainter galaxy [*2814 (WH)]. Slightly concentrated. Edge-on N-S. Small. No involved stars. Nearby SSW is a bright star.

| ga | NGC 2841 | 09 22.0 | +50 58 | UMa | 10.1b | 8.1' | 3.5' | Sb |

HH: very bright, large, very much extended NNW-SSE, very suddenly much brighter middle to a 10th mag. stellar nucleus.
42x60: Visible. *170x250:* Bright. Sharply pretty concentrated. Very elongated to edge-on NNW-SSE. Large: a quarter to a third of the field in size. The ends are very faint, both relative to the core and intrinsically. Right at its end NNW of center is the middle star of a slightly curved line of three faint stars, concave NW, across that end of the galaxy. A little farther NW is a pretty bright star.

| ga | NGC 2810 | 09 22.1 | +71 50 | UMa | 13.2b | 1.6' | 1.6' | E |

WH: faint, considerably small, brighter middle.
170x250: Faint to moderately faint. Evenly moderately concentrated. Round. Small. No involved stars. Nearby SSE is a very faint star.

| ga | CGCG 265-8 | 09 22.5 | +50 25 | UMa | 13.0 | 2.2' | 1.1' | S |

170x250: Not visible.

| ga | IC 2469 | 09 23.0 | −32 26 | Pyx | 12.3 | 6.0' | 1.0' | Sab |

170x250: Moderately faint. Evenly moderately concentrated. Edge-on NE-SW. Moderately large. At its SW end is a faint involved star; nearby are two very faint stars, one on either side of center E and W.

| ga | NGC 2865 | 09 23.5 | −23 09 | Hya | 12.6b | 2.4' | 1.7' | E3-4 |

JH: bright, small, round, gradually brighter middle.
170x250: Moderate brightness. Sharply pretty concentrated. Round. Very small. No involved stars. It's in the middle of a large group of moderate brightness and moderately faint stars in an elongated house-shaped figure, with the three brightest stars forming the roof N.

| ga | NGC 2859 | 09 24.3 | +34 30 | LMi | 11.8b | 4.3' | 4.1' | S0+ |

HH: very bright, pretty large, round, suddenly much brighter middle.
170x250: Bright. Sharply very concentrated. Round. Small. No involved stars.

| ga | NGC 2857 | 09 24.6 | +49 21 | UMa | 12.9b | 2.2' | 2.2' | Sc |

LR: very faint, pretty large, 4 stars W.
170x250: In the position are two galaxies aligned NE-SW [*2856 (HH), *2854 (HH)]. The NE galaxy is a tiny bit brighter but otherwise they are essentially identical. Both are faint. Broadly slightly concentrated. Small. No involved stars. The NE galaxy is elongated NW-SE. The SW galaxy is elongated NE-SW. [The two galaxies make a diamond with a bright star WNW and a moderate brightness star ESE.] Half the field NNE, just E of a little trapezoid of moderate brightness and faint stars, is an extremely faint galaxy [2857]. Diffuse. Roughly round. Small. No involved stars.

| ga | NGC 2872 | 09 25.7 | +11 25 | Leo | 12.9b | 1.6' | 1.4' | E2 |

HH: pretty faint, pretty small, round, brighter middle, W of 2.

170x250: Moderate brightness. Sharply pretty concentrated. Round. Very small. No involved stars. Next to it ESE, making a double galaxy with it, is a larger and fainter galaxy [*2874 (HH)]. Moderately faint. Evenly slightly concentrated. Very elongated NE-SW. Moderate size. No involved stars. A little ways SW of the two galaxies is a very faint star; ENE is a faint star.

| ga | NGC 2884 | 09 26.4 | −11 33 | Hya | 13.2 | 2.0' | 1.0' | S0/a |

dA: faint, small, resolvable?
[See 2889.]

| ga | IC 2482 | 09 27.0 | −12 06 | Hya | 12.6 | 2.3' | 1.5' | E+ |

170x250: Moderate brightness. Sharply very concentrated to a substellar nucleus. Round. Very very small. No involved stars. It's in a NW-SE line with three moderate brightness to very faint stars.

| ga | NGC 2889 | 09 27.2 | −11 38 | Hya | 12.4b | 2.1' | 1.8' | Sc |

HH: pretty faint, pretty small, very little extended, very gradually little brighter middle, resolvable.
170x250: Moderate brightness. Evenly slightly concentrated. Round. Small. No involved stars. Nearby SSE is a moderately faint star that makes a triangle with a moderately bright star SW of it and a moderately faint star ESE of it. 3/4 of a field WNW is 2884. Faint. Evenly moderately concentrated. Elongated N-S. Small. No involved stars. Nearby W is a faint star.

| ga | MCG+12-9-48A | 09 27.4 | +74 26 | Dra | 13.1 | 0.8' | 0.5' | E |

170x250: Not visible.

| bs | Alpha Hya | 09 27.6 | −08 39 | Hya | 1.98 | | 1.44 | K3 |

170x250: Yellow gold. A beautiful star.

| ga | NGC 2880 | 09 29.6 | +62 29 | UMa | 12.5b | 2.0' | 1.1' | S0− |

HH: bright, considerably small, round, much brighter middle, among stars.
170x250: Moderately faint. Evenly to sharply fairly concentrated. Round. Very small. No involved stars. It's on the side of a triangular figure of mostly moderately faint stars.

| ds | Zeta 1 Ant | 09 30.8 | −31 53 | Ant | 8.0" | 6.2 | 7.1 | A1 A1 |

42x60: Split. *170x250:* A moderately close slightly unequal pair. Both white. [DUN 78]

| ga | NGC 2902 | 09 30.9 | −14 44 | Hya | 12.2v | 1.4' | 1.1' | S0 |

WH: very faint, very small, stellar.
170x250: Moderate brightness. Evenly moderately concentrated. Round. Very small. Just off its edge NNW is an extremely faint star; a little ways SSE is a small triangle of faint stars; further SSE is the bright plotted star.

| ga | NGC 2907 | 09 31.6 | −16 44 | Hya | 12.7b | 1.8' | 1.0' | Sa |

HH: pretty faint, small, little extended, much brighter SE.
170x250: Moderate brightness. Evenly moderately concentrated. Round. Small. No involved stars. It's inside a sharp triangle of faint stars; SW is a little triangular group of moderate brightness stars.

| ga | NGC 2903 | 09 32.2 | +21 29 | Leo | 9.7b | 12.6' | 6.0' | Sbc |

HH: considerably bright, very large, extended, gradually much brighter middle, resolvable, SW of 2.
12x35: Visible. *42x60:* Easily visible. *170x250:* Moderately bright. Sharply very concentrated to a small core. Very elongated NNE-SSW. Large: a quarter of the field in size. No involved stars. Nearby ESE of center is a very faint star; S is an E-W line of three moderately bright stars and a fainter star.

| bs | Theta UMa | 09 32.9 | +51 40 | UMa | 3.20 | | 0.43 | F6 |

170x250: Soft white.

| ga | UGCA 167 | 09 33.3 | −16 46 | Hya | 12.8 | 1.9' | 1.4' | Sb |

170x250: Very faint to faint. Evenly slightly concentrated. Roughly round. Small. Off its edge W is an extremely faint star; a little ways NNW is a faint star; W is a large triangular figure of moderately faint stars.

| ga | UGCA 168 | 09 33.4 | −33 02 | Ant | 12.8b | 6.1' | 0.8' | Scd |

170x250: Extremely faint. Unconcentrated. Edge-on E-W. Moderate size. No involved stars. Passing near its W end is a NW-SE linear group of stars; farther SW is a relatively bright star.

| ga | NGC 2911 | 09 33.8 | +10 09 | Leo | 12.5b | 4.1' | 3.1' | S0 |

HH: *faint, pretty large, round, gradually brighter middle, W of 2.*
170x250: Faint. Broadly slightly concentrated. Round. Small. No involved stars. NW is a moderate brightness star; SE is another moderate brightness star and a fainter star; making a triangle with these two stars is a tiny extremely concentrated galaxy [*2914 (HH)].

| ga | NGC 2921 | 09 34.5 | −20 55 | Hya | 12.9p | 2.8' | 1.0' | Sa |

HH: *very faint, pretty small, little extended, very gradually little brighter middle, E of 2.*
170x250: Very faint. Broadly slightly concentrated. Slightly elongated E-W. Small. No involved stars. Off its edge WNW is a very faint star. A third of the field NW is another very faint galaxy [*2920 (JH)].

| ga | NGC 2916 | 09 35.0 | +21 42 | Leo | 11.9v | 2.5' | 1.7' | Sb |

HH: *faint, small, very little extended.*
170x250: Faint. Broadly to evenly slightly concentrated. Slightly elongated NNE-SSW. Small. No involved stars.

| ga | NGC 2924 | 09 35.2 | −16 23 | Hya | 13.0p | 1.3' | 1.1' | E+ |

JH: *pretty bright, small, round.*
170x250: Faint. Evenly slightly concentrated. Slightly elongated NE-SW [round]. Small. No involved stars. At its SW end is an extremely faint involved star; nearby SE is a faint star. [I546 was not seen.]

| ga | NGC 2947 | 09 36.1 | −12 26 | Hya | 13.2 | 1.4' | 1.2' | Sbc |

Le: *extremely faint, pretty large, irregularly round, gradually brighter middle.*
170x250: Faint. Broadly very slightly concentrated. Very slightly elongated N-S. Small. No involved stars. It makes a small triangle with two relatively bright stars SE.

| ga | NGC 2935 | 09 36.7 | −21 07 | Hya | 12.1b | 3.8' | 2.9' | Sb |

HH: *pretty bright, pretty small, very little extended, gradually much brighter middle.*
170x250: Moderate brightness. Evenly to sharply pretty concentrated. Round. Very small. Just SSE of center is a faint involved star; S is a NNE-SSW row of three stars.

ga	NGC 2945	09 37.7	−22 02	Hya	13.2p	1.6'	1.4'	S0−

JH: faint, small, round, gradually little brighter middle, 2 or 3 faint stars near.
170x250: Faint. Sharply pretty concentrated. Round. Very very small. No involved stars. It's off the long leg of a small sharp right triangle of faint stars.

ga	NGC 2942	09 39.1	+34 00	LMi	13.2b	2.4'	1.8'	Sc

JH: faint, pretty large, very little extended N-S, very gradually little brighter middle.
170x250: Extremely faint. Diffuse. Moderately small. No involved stars. Nearby W is a very faint star; the galaxy makes a flat triangle with two moderate brightness stars SE; WNW is another moderate brightness star.

ga	UGC 5139	09 40.5	+71 11	UMa	12.6v	4.0'	3.3'	Im

170x250: Not visible.

ga	NGC 2962	09 40.9	+05 10	Hya	13.0b	2.6'	1.9'	S0+

Ma: faint, very small, very little extended, pretty suddenly brighter middle.
170x250: Moderately faint to moderate brightness. Sharply very concentrated to a substellar nucleus. Round. Small. No involved stars. Nearby NNE is an unequal pair of stars.

ga	NGC 2967	09 42.1	+00 20	Sex	12.3b	3.0'	2.7'	Sc

HH: pretty faint, pretty large, round, very gradually little brighter middle.
170x250: Moderately faint. Broadly very slightly concentrated. Round. Moderately small. No involved stars.

ga	NGC 2950	09 42.6	+58 51	UMa	11.8b	2.7'	1.7'	S0

HH: considerably faint, very small, round, very gradually very much brighter middle to a nucleus.
170x250: Bright to very bright; very high surface brightness. Sharply very concentrated. Slightly elongated NW-SE. Moderately small. The outer halo is much fainter than the core. No involved stars. A little ways W is a moderately faint star.

ga	NGC 2974	09 42.6	−03 41	Sex	11.9b	3.4'	2.0'	E4

HH: bright, considerably small, irregularly round, brighter middle, 9th mag. star SW.

170x250: Moderate brightness. Broadly to evenly moderately concentrated. Elongated NE-SW. Small. Off its SW end is a bright star.

| ga | NGC 2964 | 09 42.9 | +31 50 | Leo | 12.0b | 2.9' | 1.5' | Sbc |

HH: bright, very large, little extended, very gradually brighter middle, W of 2.
170x250: Moderately bright. Broadly slightly concentrated. Slightly elongated E-W. Moderately small. No involved stars. A third of the field NE is 2968. Moderately faint. Evenly moderately concentrated. Roughly round. Small. No involved stars. A quarter of the field further NE is another galaxy [*2970 (JH)]. Extremely faint. Extremely small; substellar. Aligned perpendicular to a line between the two fainter galaxies are two stars, moderately bright and moderately faint.

| ga | NGC 2968 | 09 43.2 | +31 55 | Leo | 12.8b | 2.2' | 1.5' | I0 |

HH: pretty bright, pretty large, little extended, very gradually little brighter middle, E of 2.
[See 2964.]

| ga | NGC 2983 | 09 43.7 | −20 28 | Hya | 12.8b | 2.5' | 1.4' | S0+ |

HH: faint, pretty small, round, brighter middle, resolvable, stellar.
170x250: Moderate brightness. Sharply pretty concentrated. Round. Small. No involved stars.

| ga | ESO 434-28 | 09 44.2 | −28 50 | Ant | 13.1b | 1.5' | 1.1' | S0− |

170x250: Moderately faint. Sharply pretty concentrated. Round. Very very small. No involved stars. It makes a triangle with a pair of moderate brightness stars ENE and a moderately bright star SSE.

| ga | NGC 2986 | 09 44.3 | −21 16 | Hya | 11.7b | 3.1' | 2.6' | E2 |

WH: pretty bright, pretty small, irregularly round, much brighter middle.
170x250: Moderately bright. Sharply pretty concentrated. Round. Moderately small. No involved stars. Very nearby SSW is a very faint star that makes a line with two pairs of stars, one E of it and one W of it. [E566-4 was not seen.]

| ga | UGCA 180 | 09 44.8 | −31 49 | Ant | 13.2p | 2.1' | 2.0' | Sm |

170x250: [The observed galaxy is *I2507; UA180 was not seen.] Very faint. Broadly slightly concentrated. Elongated NE-SW. Small. Just off its NE end is a faint star that makes a triangle with two brighter stars, one NNW and one farther NNE.

| ga | NGC 2997 | 09 45.6 | −31 11 | Ant | 10.1b | 9.2' | 7.4' | Sc |

HH: ! very faint, very large, very gradually then very suddenly brighter middle to a nucleus 4" in diameter.
42x60: Very faintly visible. ***170x250:*** Faint. Sharply concentrated to a small brighter core. Round. Large; a third of the field in size. The halo is low surface brightness and fades out very indistinctly. ESE of center is an extremely faint involved star; at its edge SW is a moderate brightness star; at its edge E is a very faint star; further E are two moderate brightness stars. N is what looks like a very very small very very faint companion galaxy [this is a faint triple star].

| ga | NGC 2992 | 09 45.7 | −14 19 | Hya | 13.1b | 4.5' | 1.0' | Sa |

HH: considerably faint, small, round, brighter middle, stellar, W of 2.
170x250: 2992 and 2993 are a wide double galaxy aligned NNW-SSE. Both are moderate brightness and have no involved stars. 2992 is broadly slightly concentrated. Elongated NNE-SSW. Small. 2993 is sharply very concentrated. Round. Very small.

| ga | NGC 2993 | 09 45.8 | −14 22 | Hya | 13.1b | 4.0' | 2.6' | Sa |

HH: considerably faint, small, round, brighter middle, stellar, E of 2.
[See 2992.]

| bs | Epsilon Leo | 09 45.9 | +23 46 | Leo | 2.98 | | 0.80 | G0 |

170x250: Yellow white.

| ga | NGC 2990 | 09 46.3 | +05 42 | Sex | 13.1b | 1.2' | 0.6' | Sc |

WH: faint, pretty small, little extended E-W.
170x250: Moderately faint. Broadly moderately concentrated. Slightly elongated E-W. Small. No involved stars. It's inside a triangle of stars: a moderately faint star nearby NNE, a faint star WSW, and another faint star farther SSE.

| ga | NGC 3001 | 09 46.3 | −30 26 | Ant | 12.7b | 2.8' | 1.8' | Sbc |

JH: faint, small, round, 12th mag. star attached NW.
170x250: Moderately faint. Broadly slightly concentrated. Slightly elongated N-S. Small. NNW of center is a moderate brightness involved star; NNE is a very sharp triangle of stars.

| ga | NGC 2976 | 09 47.3 | +67 55 | UMa | 10.8b | 5.9' | 2.6' | Sc |

HH: bright, very large, much extended NNW-SSE, stars involved.
42x60: Visible. *170x250:* Moderately bright. Unconcentrated. Very elongated NW-SE. Large: a quarter of the field in size. Just off its edge SW is a moderate brightness star; off its NW end is a fainter star.

| vs | R Leo | 09 47.6 | +11 26 | Leo | 5.2 | 10.5 | 1.30 | M8 |

12x35: Mag. 7.5. *170x250:* Orange red. It makes a triangle with two slightly fainter white stars, one W and one SW; half the field NNW is the bright plotted star, which makes a pretty good comparison star.

| ga | NGC 3003 | 09 48.6 | +33 25 | LMi | 12.3b | 5.9' | 1.3' | Sbc |

HH: ! considerably bright, large, very irregularly much extended E-W.
170x250: Moderately faint. Broadly slightly concentrated. A fairly thin edge-on, ENE-WSW. Moderately large to large. No involved stars.

| ga | NGC 2998 | 09 48.7 | +44 04 | UMa | 12.5v | 3.8' | 1.9' | Sc |

HH: pretty faint, pretty large, irregularly round, brighter middle, resolvable.
170x250: Faint. Broadly slightly concentrated. Elongated NE-SW. Small. No involved stars. Nearby NNW are two stars, extremely faint and moderately faint.

| ga | IC 2511 | 09 49.4 | −32 50 | Ant | 13.0p | 2.9' | 0.5' | Sa |

170x250: Very faint. Broadly very slightly concentrated. Very elongated to edge-on NE-SW. Moderately small. No involved stars. Very nearby E is an extremely faint star; the galaxy is surrounded by faint stars. Half the field ESE is a brighter galaxy [*I2514]. Moderately faint; fairly high surface brightness. Slightly elongated ENE-WSW. Very small. No involved stars. It makes a small arc with two moderate brightness stars E.

| ga | NGC 3022 | 09 49.7 | −05 09 | Sex | 13.0 | 1.6' | 1.6' | S0 |

JH: faint, round, very gradually little brighter middle, E of 2.
170x250: Very faint. Broadly slightly concentrated. Round. Small. No involved stars. [M-1-25-44 was not seen.]

| ga | NGC 3023 | 09 49.9 | +00 37 | Sex | 13.0 | 2.9' | 1.4' | Sc |

St: pretty faint, pretty large, irregularly round, little brighter middle, diffuse.

170x250: Very faint. Hardly concentrated. Roughly round. Small. No involved stars. A little ways E is a moderately faint star; a little ways SW are a couple of faint stars; W is a bright star. [3018 (St) was not seen.]

ga	NGC 3020	09 50.1	+12 48	Leo	12.6p	3.1'	1.5'	Scd

HH: extremely faint, pretty small, little extended N-S, resolvable.
170x250: In the position are three galaxies in a large triangle. 3020 is the northernmost. Moderately faint. Broadly slightly concentrated. Slightly elongated E-W. Small. No involved stars. A third of the field ESE is a faint galaxy [*3024 (HH)]. Broadly slightly concentrated. Edge-on NW-SE. Small. Half the field SSW of 3020 is another faint galaxy [*3016 (dA)]. Broadly slightly concentrated. Roughly round. Small. No involved stars. [3019 (LR) was not seen.]

ga	NGC 2985	09 50.4	+72 16	UMa	11.2b	4.5'	3.5'	Sab

HH: very bright, considerably large, round, pretty suddenly much brighter middle, star involved E.
42x60: Visible. ***170x250:*** Fairly bright. Sharply pretty concentrated. Round. Small. No involved stars. Very nearby E is a relatively bright star.

ga	NGC 3021	09 51.0	+33 33	LMi	12.9p	1.6'	0.8'	Sbc

HH: pretty bright, pretty small, very little extended, much brighter middle, 10th mag. star SE.
170x250: Moderately bright to bright; fairly high surface brightness. Broadly slightly concentrated. Slightly elongated WNW-ESE. Small. At its edge NNE is a very very faint involved star; nearby SE is a moderately faint star; SW is a moderately bright star.

vs	Y Hya	09 51.1	−23 01	Hya	6.9	9	3.82	C5

12x35: Mag. 7.0. ***170x250:*** Orange red. A little more than half the field W is a faint galaxy [*E499-9].

ga	NGC 3038	09 51.3	−32 45	Ant	12.4b	2.5'	1.3'	Sb

Sw: pretty bright, pretty small, round.
170x250: Moderately faint. Evenly moderately concentrated. Round. Very small. No involved stars. It's right in the middle of a NE-SW figure of moderately bright to moderate brightness stars resembling Bootes.

| ga | NGC 3032 | 09 52.1 | +29 14 | Leo | 13.2b | 2.1' | 1.6' | S0 |

JH: faint, small, suddenly brighter middle to a 12th mag. star, between 2 bright stars.
170x250: Moderately faint. Sharply extremely concentrated to a faint stellar nucleus. Round. Very very small. No involved stars. It's just E of a line between two relatively bright stars, one N and one S.

| ga | NGC 3041 | 09 53.1 | +16 40 | Leo | 12.3b | 4.3' | 2.3' | Sc |

HH: globular cluster, faint, large, round, very gradually little brighter middle, partially resolved, 2 bright stars SW.
170x250: Faint. Broadly very slightly concentrated. Roughly round. Moderately small. In an arc around its edge W are three moderately bright to very faint stars; further W are two more stars, making a five-star figure altogether.

| ga | NGC 3044 | 09 53.7 | +01 34 | Sex | 12.5b | 4.9' | 0.7' | Sc |

HH: very faint, very large, very much extended WNW-ESE.
170x250: Moderately faint. Broadly very slightly concentrated. A very thin edge-on, WNW-ESE. Moderately large. No involved stars. A nice galaxy.

| ga | NGC 3051 | 09 54.0 | −27 17 | Ant | 12.8p | 2.1' | 1.9' | S0− |

JH: pretty faint, small, round, gradually brighter middle.
170x250: Moderately faint. Sharply very concentrated. Slightly elongated NW-SE. Small. No involved stars.

| ga | IC 575 | 09 54.5 | −06 51 | Sex | 13.2 | 1.6' | 1.1' | Sa |

170x250: Very very faint. Hardly concentrated. Round. Small. No involved stars. It makes a shallow arc with two moderate brightness stars SE.

| ga | NGC 3052 | 09 54.5 | −18 38 | Hya | 12.8b | 2.0' | 1.4' | Sc |

HH: faint, pretty large, round, gradually little brighter middle.
170x250: Faint. Broadly very slightly concentrated. Round. Small. No involved stars. Nearby SSE is an extremely faint star; S is a faint star; ESE is a moderately bright star; W is a pair of moderately faint stars.

| ga | NGC 3054 | 09 54.5 | −25 42 | Hya | 12.6p | 3.8' | 2.3' | Sb |

Pe: pretty bright, large, irregularly oblong.
170x250: Moderately faint. Broadly moderately concentrated. Slightly elongated WNW-ESE. Moderately small. No involved stars.

| ga | NGC 3056 | 09 54.5 | −28 17 | Ant | 12.6b | 1.8' | 1.1' | S0+ |

JH: pretty bright, small, round, very gradually much brighter middle, 11th mag. star attached SSW.
170x250: Moderate brightness; fairly high surface brightness. Sharply very concentrated. Roughly round. Very very small. At its edge N is a moderately faint star; together the galaxy and this star look a little like a double star; nearby ENE is another star; further ENE is either an edge-on galaxy or a little line of faint stars [the latter is the case].

| ga | NGC 3049 | 09 54.8 | +09 16 | Leo | 13.0p | 2.2' | 1.4' | Sab |

St: very faint, very small, faint star very near.
170x250: Very faint. Evenly slightly concentrated. Round. Small. No involved stars. ENE is a fairly bright star.

| ga | IC 2522 | 09 55.2 | −33 08 | Ant | 12.6p | 3.4' | 2.2' | Sc |

170x250: In the position are two galaxies a quarter of the field apart N-S [I2522, *I2523]. The S galaxy is slightly brighter. Faint. Broadly slightly concentrated. Elongated NNE-SSW. Small. No involved stars. Very nearby ESE is a very faint star; a little ways ENE is a faint star; WNW is a pair of faint stars. The N galaxy is very faint. Hardly concentrated. Slightly elongated N-S. Small, but a little larger than the S galaxy. No involved stars. Closely surrounding it are several faint stars; a little ways N is a moderately bright star.

| ga | NGC 3055 | 09 55.3 | +04 16 | Sex | 12.7b | 2.0' | 1.2' | Sc |

HH: faint, pretty large, very little extended, very gradually brighter middle, partially resolved, 7th mag. star E.
170x250: Moderately faint. Broadly slightly concentrated. Slightly elongated ENE-WSW. Small. No involved stars.

| ga | NGC 3031/M81 | 09 55.6 | +69 03 | UMa | 7.9b | 27.1' | 14.2' | Sab |

JH: ! extremely bright, extremely large, extended NNW-SSE, gradually then suddenly very much brighter middle to a bright resolvable nucleus.

12x35: M81 and M82 are easily visible in the same field. M81 is bright. *42x60:* Both galaxies are in the same field. Both are bright. *170x250:* M81 is very bright. Sharply very concentrated to a small bright core, but not to a stellar nucleus. A fat edge-on, NNW-SSE (it's more elongated than in the pictures). Very large: the field width in size. The light is very smooth, and it tapers off drastically toward the edges; the ends fade out very faintly. It looks a little like a miniature M31. S of center just outside the bright part of the galaxy are two relatively bright involved stars; W of these, making a triangle with them, is a very faint involved star.

ga	NGC 3027	09 55.7	+72 12	UMa	12.2b	4.3'	2.0'	Sd

HH: very faint, very large, little extended, resolvable.
170x250: Extremely faint; very low surface brightness. Diffuse. Elongated NW-SE. Large [moderate size]. At its NW end is an extremely faint star.

ga	NGC 3034/M82	09 55.8	+69 40	UMa	9.3b	11.3'	4.2'	I0

WH: very bright, very large, very much extended (ray).
170x250: Very bright; high surface brightness. Evenly moderately concentrated. Edge-on ENE-WSW. Very large: a third to half the field in size. The E end extends a little farther than the W end. Within the halo are a couple of short transverse dust lanes (perpendicular to the axis of the galaxy), obvious but not well defined; one is close to the center and wide. There's other unevenness in the light as well. No involved stars. Going SW from the S edge of the W side are three bright stars; going S from the E end are three more stars in a shallow arc. M82 makes a big contrast with M81: M81 fades out from the center while M82, aside from the dust lanes and the small-scale unevenness, is much more even in light overall. Unusual: a very bright even-light galaxy. It's actually more impressive than M81, due to its exceptional brightness.

ga	ESO 499-23	09 56.4	−26 05	Hya	12.7p	2.1'	1.2'	S0−

170x250: Moderately faint. Sharply very concentrated. Round. Tiny. No involved stars. Very nearby WSW is a moderately bright star; nearby NE are a couple of moderate brightness stars that make a polygonal figure with a few more moderate brightness stars N of them.

ga	NGC 3067	09 58.4	+32 22	Leo	12.8b	2.4'	0.9'	Sab

HH: pretty bright, pretty large, extended E-W, gradually brighter middle, 9th mag. star NE.
170x250: Moderate brightness. Broadly slightly concentrated. Elongated WNW-ESE. Moderately small. No involved stars. ENE is a moderately bright star.

| ga | NGC 3078 | 09 58.4 | −26 55 | Hya | 12.1b | 2.5' | 2.0' | E2-3 |

HH: pretty bright, small, round, much brighter middle.
170x250: Moderately bright. Sharply very concentrated. Round. Very small. No involved stars. Very nearby W is a very very faint star; the galaxy makes an isosceles triangle with two moderately faint stars, one NNE and one NW.

| ga | NGC 3084 | 09 59.1 | −27 07 | Ant | 13.2p | 2.1' | 1.8' | Sab |

JH: very faint, small, round, 13th mag. star attached SE.
170x250: Faint. Evenly moderately concentrated. Roughly round. Very small. At its edge SE is a moderately faint star that interferes; further SE is a slightly brighter star.

| ga | NGC 3087 | 09 59.2 | −34 13 | Ant | 11.6v | 2.1' | 1.9' | E+ |

JH: pretty bright, small, round, pretty much brighter middle, between 2 stars.
170x250: Moderately faint. Sharply pretty concentrated. Round. Very small. No involved stars. Very nearby NE is a faint star; surrounding the galaxy is a group of moderately faint to moderate brightness stars.

| ga | UGC 5364 | 09 59.4 | +30 44 | Leo | 12.6v | 5.3' | 3.4' | Dw I |

170x250: Not visible. [Leo III]

| ga | NGC 3081 | 09 59.5 | −22 49 | Hya | 12.9b | 3.1' | 2.4' | S0/a |

WII: very faint, considerably small, little brighter middle, triangle of faint stars NW.
170x250: Moderately faint. Evenly moderately concentrated. Round. Small. No involved stars. NW are a couple of moderately bright stars; NE are three stars in a shallow arc.

| ga | NGC 3089 | 09 59.6 | −28 19 | Ant | 13.2p | 1.7' | 1.0' | Sb |

JH: pretty faint, pretty small, round, very faint stars involved.
170x250: Very faint. Broadly very slightly concentrated. Roughly round. Small. At its edge are a couple of faint involved stars, one ESE and one S; a little ways E is the bright plotted star.

| ga | IC 2531 | 09 59.9 | −29 37 | Ant | 12.9b | 7.0' | 0.6' | Sc |

170x250: Very faint. Broadly very slightly concentrated. A thin edge-on, ENE-WSW. Large: a quarter to a third of the field in size. The ends are indistinct.

No involved stars. Nearby are a number of very faint stars. One of the thinnest galaxies in the sky. Very ghostly; tantalizing.

| ga | UGC 5373 | 10 00.0 | +05 19 | Sex | 11.9b | 5.1' | 3.5' | Im |

170x250: Exceedingly faint. The very slightest diffuse glow. Moderate size. No involved stars. Passing NE of it and going completely across the field is a shallow NW-SE arc of faint stars.

| ga | NGC 3095 | 10 00.1 | −31 33 | Ant | 12.4p | 3.4' | 1.9' | Sc |

JH: faint, large, extended, very gradually very little brighter middle.
[See 3100.]

| ga | NGC 3091 | 10 00.2 | −19 38 | Hya | 12.1b | 2.9' | 1.8' | E3 |

HH: pretty bright, pretty small, irregularly round, brighter middle, W of 2.
170x250: Moderate brightness. Evenly to sharply pretty concentrated. Round. Small. No involved stars. It makes a straight line with three faint stars, one NW [this is *Hickson 42C, a stellar galaxy] and two SE, and it makes a slightly curved line with two moderately bright stars a little farther ESE. [3096 (JH) was not seen.]

| ga | IC 2533 | 10 00.5 | −31 14 | Ant | 13.0b | 1.8' | 1.3' | S0− |

170x250: Moderately faint. Evenly to sharply pretty concentrated. Round. Very small. No involved stars.

| ga | NGC 3100 | 10 00.7 | −31 39 | Ant | 12.0b | 3.1' | 1.5' | S0 |

JH: pretty bright, pretty small, round, gradually pretty much brighter middle.
170x250: Moderately faint. Evenly moderately concentrated. Round. Small. No involved stars. Nearby E is a pair of moderate brightness stars; nearby S is a moderately faint star; nearby SW is a very faint star. Half the field NW is 3095. Faint. Broadly very slightly concentrated. Slightly elongated NW-SE. Moderately small. At its edge W is a moderate brightness star.

| ga | NGC 3094 | 10 01.4 | +15 46 | Leo | 13.2p | 1.7' | 1.2' | Sa |

Pa: faint, brighter middle, 9th mag. star 0.5' SE.
170x250: Moderate brightness. Evenly moderately concentrated. Round. Very small. Right next to it SE is a bright star.

| ga | NGC 3079 | 10 02.0 | +55 40 | UMa | 11.5b | 8.0' | 1.4' | Sc |

WH: very bright, large, much extended NW-SE.
170x250: Moderate brightness. Broadly slightly concentrated. Edge-on NNW-SSE. Large: a quarter of the field in size. The galaxy is very slightly curved, concave W. At its N tip is a faint involved star; off its W side are a couple of very faint stars almost parallel to the galaxy; farther WSW is another very faint star; SE to SSW is a large triangle of relatively bright stars. Half the field WSW is a very small very faint galaxy [*3073 (WH)].

| ga | NGC 3098 | 10 02.3 | +24 42 | Leo | 12.9b | 2.2' | 0.5' | S0 |

JH: pretty bright, small, much extended E-W, pretty suddenly brighter middle to a nucleus.
170x250: Moderately bright; high surface brightness. Sharply very concentrated to a small bright core. Edge-on E-W. Small. No involved stars. A little ways E is a faint star in line with the axis of the galaxy.

| ga | NGC 3108 | 10 02.5 | −31 40 | Ant | 12.8b | 2.5' | 1.8' | S0+ |

JH: faint, small, round, gradually little brighter middle.
170x250: Faint. Evenly moderately concentrated. Round. Very small. No involved stars. Nearby NE is a faint star; nearby NW is another faint star; a little ways S are a relatively bright star and a faint star.

| ga | NGC 3109 | 10 03.1 | −26 09 | Hya | 10.4b | 21.0' | 3.7' | Sm |

JH: considerably faint, very large, very much extended E-W, little brighter middle.
170x250: Faint; pretty low surface brightness. Hardly concentrated. Very elongated E-W. Very large: a little more than half the field in size. The galaxy widens at the E end. W of center is a moderately faint involved star; there are several other involved stars, all very faint.

| ga | NGC 3077 | 10 03.3 | +68 44 | UMa | 9.9v | 5.5' | 4.0' | I0 |

HH: considerably bright, considerably large, much brighter middle, round with ray.
42x60: Faintly visible. *170x250:* Moderately bright. Broadly to evenly moderately concentrated. Round. Small. No involved stars.

| ga | IC 2537 | 10 03.9 | −27 34 | Ant | 12.8b | 2.6' | 1.7' | Sc |

170x250: Very faint. Hardly concentrated. Round. Moderately small. No involved stars. It's surrounded by several moderately faint stars in a semicircle NNE to E to SW.

| ga | MCG-1-26-13 | 10 04.0 | −06 29 | Sex | 12.7 | 1.2' | 0.4' | S0+ |

170x250: [The observed galaxy is *3110 (St); M-1-26-13 was not seen.] Moderately faint. Broadly slightly concentrated. Slightly elongated NNW-SSE. Small. Just NW of its N end is a faint star; a little ways S is a faint star; a little farther N is another faint star; these three stars make a very slightly curved N-S line.

| ga | NGC 3115 | 10 05.2 | −07 43 | Sex | 9.9b | 7.2' | 2.4' | S0− |

HH: very bright, large, very much extended NE-SW, very gradually then very suddenly much brighter middle to an extended nucleus.
12x35: Visible. *42x60:* Bright. Elongated. *170x250:* Very bright. Sharply very concentrated. Edge-on NE-SW. Moderate size to moderately large. Just off its edge S is a faint star; further S is a moderately bright star. One of the brightest galaxies. [UA200 was not seen.]

| ga | MCG-1-26-21 | 10 05.7 | −07 58 | Sex | 13.2b | 2.0' | 1.8' | S0 |

170x250: Extremely faint. Diffuse. Small. Off its edge S is a faint star that makes a triangle with two more faint stars, one ESE of the galaxy and one farther SSE; SE are three brighter stars in a little arc.

| ga | NGC 3124 | 10 06.7 | −19 13 | Hya | 12.9b | 2.9' | 2.4' | Sbc |

JH: faint, pretty large, round, little brighter middle, double star S.
170x250: Very faint to faint. Diffuse. Roughly round. Moderate size. No involved stars. S is the bright plotted star, which is double.

| bs | Eta Leo | 10 07.3 | +16 45 | Leo | 3.49 | | −0.03 | A0 |

170x250: White.

| ga | UGC 5459 | 10 08.2 | +53 04 | UMa | 13.2b | 4.7' | 0.7' | Sc |

170x250: Very faint. Unconcentrated. A thin edge-on, NW-SE. Small. At its SE end is the bright plotted star, which interferes.

| bs | Alpha Leo | 10 08.4 | +11 58 | Leo | 1.35 | | −0.11 | B7 |

170x250: White. [Regulus]

| ga | UGC 5470 | 10 08.5 | +12 18 | Leo | 11.2b | 9.8' | 7.4' | E3 |

170x250: Possibly visible; it's uncertain because of the interfering glare from Regulus, a field width S. Off its edge NE is a wide E-W pair of moderately faint stars. [This galaxy is tenuous at best; on most occasions there is no hint of it.] [Leo I]

| ga | NGC 3137 | 10 09.1 | −29 03 | Ant | 12.1b | 6.3' | 2.2' | Scd |

JH: very faint, small, little extended.
170x250: Very faint to faint. Hardly concentrated. Slightly elongated N-S. Small to moderately small. At its edge WSW of center is a moderately faint involved star; a little ways NW is another moderately faint star.

| ga | NGC 3145 | 10 10.2 | −12 26 | Hya | 12.5b | 3.0' | 1.5' | Sbc |

HH: faint, pretty large, round, very gradually then suddenly little brighter middle, E of 2.
170x250: Faint to moderately faint. Broadly slightly concentrated. Elongated NNE-SSW. Moderately small. No involved stars. The bright plotted star NE [Lambda Hya] interferes somewhat.

| ga | UGCA 205 | 10 11.0 | −04 41 | Sex | 11.5v | 5.7' | 5.1' | Im |

170x250: Not visible.

| ga | MCG-3-26-30 | 10 11.3 | −17 12 | Hya | 13.2 | 1.7' | 1.2' | S0− |

170x250: Faint. Sharply concentrated. Round. Very very small. No involved stars. Nearby SE is a faint star. The galaxy is so small and sharply concentrated that it resembles the faint stars in the field.

| ga | NGC 3156 | 10 12.7 | +03 07 | Sex | 13.1b | 2.2' | 1.0' | S0 |

HH: faint, considerably small, round, pretty suddenly brighter middle, forms a triangle with bright stars E.
170x250: Moderately faint. Evenly moderately concentrated. Slightly elongated NE-SW. Small. No involved stars. A little ways ESE is a bright star.

| ga | NGC 3162 | 10 13.5 | +22 44 | Leo | 12.2b | 3.0' | 2.4' | Sbc |

HH: pretty faint, considerably large, round, very gradually little brighter middle, resolvable, faint star involved.
170x250: Faint. Hardly concentrated. Roughly round. Moderately small. No involved stars. Just off its edge SE is a very faint star; in the field with the galaxy are a number of relatively bright stars.

| ga | NGC 3158 | 10 13.8 | +38 45 | LMi | 11.9v | 2.0' | 1.8' | E3 |

HH: considerably bright, considerably small, round, pretty suddenly brighter middle, resolvable.
170x250: Moderately faint. Evenly moderately concentrated. Round. Small. No involved stars. [3158 is at the center of a fairly rich group of galaxies; 3152 (LR) and 3160 (LR) were not seen.]

| ga | NGC 3166 | 10 13.8 | +03 25 | Sex | 11.3b | 4.8' | 2.3' | S0/a |

HH: bright, pretty small, round, pretty suddenly much brighter middle, W of 2.
42x60: 3166 and 3169 are visible together. ***170x250:*** 3166 is a little brighter. Bright to very bright. Sharply very concentrated. Round. Moderately small. No involved stars. It makes a triangle with two moderately faint stars, one WSW and one NNW. Half the field ENE is 3169. Bright. Sharply pretty concentrated. Slightly elongated NE-SW. Moderately small. No involved stars. Nearby E is a moderately bright star. [3165 (LR) was not seen.]

| ga | UGC 5522 | 10 14.0 | +07 01 | Leo | 13.2p | 3.0' | 1.7' | Sd |

170x250: Not visible.

| ga | NGC 3169 | 10 14.2 | +03 28 | Sex | 11.1b | 5.4' | 2.7' | Sa |

HH: bright, pretty large, very little extended, pretty gradually much brighter middle, 11th mag. star 80" ENE.
[See 3166.]

| ga | NGC 3175 | 10 14.7 | −28 52 | Ant | 12.1b | 5.0' | 1.3' | Sa |

JH: considerably bright, large, much extended NE-SW, very gradually little brighter middle.
170x250: Faint. Broadly very slightly concentrated. Edge-on NE-SW. Moderately large. No involved stars. It's surrounded by a number of stars, including a moderately bright star NNE.

| ga | IC 2560 | 10 16.3 | −33 33 | Ant | 12.5p | 3.4' | 1.9' | Sb |

170x250: Faint. Broadly slightly concentrated. Elongated NE-SW. Small to moderately small. No involved stars. It's inside a roughly square figure of moderate brightness stars.

| ga | NGC 3177 | 10 16.6 | +21 07 | Leo | 13.0b | 1.4' | 1.1' | Sb |

HH: considerably faint, small, round, pretty suddenly brighter middle.
170x250: Moderate brightness. Sharply pretty concentrated. Round. Very very small. No involved stars. It's inside a very large triangular group of stars.

| bs | Zeta Leo | 10 16.7 | +23 25 | Leo | 3.40 | | 0.35 | F0 |

170x250: Soft white.

| ga | NGC 3147 | 10 16.9 | +73 24 | Dra | 11.4b | 3.9' | 3.4' | Sbc |

HH: very bright, large, round, very gradually then very suddenly very much brighter middle.
170x250: Bright. Sharply very concentrated to a substellar nucleus. Round. Small. No involved stars. It makes a triangle with two stars, one NNE and one ENE.

| bs | Lambda UMa | 10 17.1 | +42 54 | UMa | 3.40 | | 0.08 | A2 |

170x250: White.

| ga | IC 600 | 10 17.2 | −03 29 | Sex | 13.0p | 2.3' | 1.1' | Sdm |

170x250: Very faint. Diffuse. Roughly round. Small. No involved stars. It makes a large triangle with a double star N and the bright plotted star ESE.

| ga | NGC 3185 | 10 17.6 | +21 41 | Leo | 13.0b | 2.3' | 1.5' | Sa |

LR: pretty faint, pretty large, gradually much brighter middle.
[See 3190.]

| ds | STF 1415 | 10 17.9 | +71 03 | UMa | 16.7" | 6.7 | 7.3 | A7 A |

12x35: Extremely close. *42x60:* A moderately close slightly unequal pair.
170x250: A moderately wide very slightly unequal pair. Both white. It's on the hypotenuse of a perfect right triangle formed by surrounding stars – right at the point of intersection between the hypotenuse and an imaginary altitude line coming

down from the opposite corner: a perfectly symmetrical arrangement, except that the components of the double star itself are at a slight angle to the hypotenuse.

| ga | NGC 3190 | 10 18.1 | +21 49 | Leo | 12.1b | 4.4' | 1.2' | Sa |

HH: bright, pretty small, extended, pretty suddenly brighter middle to a nucleus, SW of 2.
170x250: Bright. Sharply very concentrated. Very elongated WNW-ESE. Moderately small. No involved stars. Nearby W is a faint star. A third of the field NE is 3193. Moderately bright; pretty high surface brightness. Sharply very concentrated. Round. Very small. No involved stars. Nearby N is a moderately bright star. A quarter of the field WNW of 3190, in the same direction as its elongation, is an extremely faint galaxy [*3187 (LR)]. Diffuse. Small. A little more than half the field SW of 3190 is 3185. Faint. Broadly slightly concentrated. Roughly round. Small. No involved stars. Next to it W is a very faint star.

| ga | NGC 3184 | 10 18.3 | +41 25 | UMa | 10.4b | 7.4' | 6.9' | Scd |

HH: pretty bright, very large, round, very gradually brighter middle.
42x60: Faintly visible. ***170x250:*** Moderately faint; fairly low surface brightness. Broadly slightly concentrated. Roughly round. Moderately large. At its edge N is a relatively bright star.

| ga | NGC 3193 | 10 18.4 | +21 53 | Leo | 11.8b | 2.0' | 2.0' | E2 |

HH: bright, small, very little extended, pretty suddenly little brighter middle, resolvable, 9th mag. star 75" N, NE of 2.
[See 3190.]

| ga | NGC 3200 | 10 18.6 | −17 59 | Hya | 12.8b | 4.2' | 1.2' | Sc |

Ho: pretty bright, extended NNW-SSE, brighter middle to a nucleus.
170x250: Faint. Broadly slightly concentrated. Edge-on NNW-SSE. Moderately small. At its N end is a very faint involved star; a little ways W is a brighter star that makes a long curved line with three more stars further W.

| ga | NGC 3182 | 10 19.6 | +58 12 | UMa | 13.0p | 1.8' | 1.5' | Sa |

WH: considerably bright, considerably large, irregularly round, very gradually brighter middle.
170x250: Moderately faint. Evenly moderately concentrated. Round. Very small. No involved stars.

| ga | NGC 3203 | 10 19.6 | −26 41 | Hya | 13.1b | 2.9' | 0.5' | S0+ |

JH: pretty bright, small, considerably extended, gradually brighter middle.
170x250: Moderately faint to moderate brightness. Evenly moderately concentrated. A fairly thin edge-on, ENE-WSW. Moderate size. No involved stars. Starting a little ways from its NE end and going SSE is a long string of about six moderately faint to faint stars.

| ga | NGC 3198 | 10 19.9 | +45 33 | UMa | 10.9b | 8.8' | 3.3' | Sc |

HH: pretty bright, very large, much extended NE-SW, very gradually brighter middle.
42x60: Extremely faintly visible. *170x250:* Moderately faint. Broadly slightly concentrated. Very elongated to edge-on NE-SW. Large: a quarter of the field in size. No involved stars. A little ways S are a couple of faint stars; N are a couple of moderately bright stars. Somewhat ghostly.

| bs | Gamma Leo | 10 20.0 | +19 51 | Leo | 2.61 | | 1.15 | K0 |

42x60: Elongated. *170x250:* A very close slightly unequal pair. Colorful: both yellow. An exceptional, very beautiful double; nice and bright. [STF 1424: 4.5", 2.2, 3.5, K0 G7]

| ga | NGC 3223 | 10 21.6 | −34 15 | Ant | 11.8b | 4.1' | 2.5' | Sb |

JH: pretty bright, very large, very little extended, pretty suddenly little brighter middle to a nucleus.
170x250: Moderately faint to moderate brightness. Evenly moderately concentrated. Roughly round. Small. Just off its edge E is a faint star; a little ways ESE is a moderate brightness star.

| ga | NGC 3224 | 10 21.7 | −34 41 | Ant | 12.0v | 1.9' | 1.5' | E+ |

JH: very faint, pretty small, round, very gradually much brighter middle.
170x250: Moderately faint. Evenly to sharply fairly concentrated. Round. Small. No involved stars. Nearby SW is a very faint star; a little ways ENE are a couple of moderately faint stars; W is a long N-S figure of stars, like a very long question mark with a bright star at the end of its long tail.

| ga | NGC 3183 | 10 21.8 | +74 10 | Dra | 12.7b | 2.3' | 1.3' | Sbc |

dA: faint, pretty large, extended, little brighter middle.

170x250: Very faint. Broadly slightly concentrated. Elongated to edge-on NW-SE. Small. Just off its NW end are a couple of very faint stars; nearby E is another very faint star.

ga	NGC 3206	10 21.8	+56 55	UMa	12.6p	2.9'	1.9'	Scd

HH: *pretty bright, considerably large, extended, very gradually little brighter middle.*
170x250: Exceedingly faint. Diffuse. Roughly round. Small. No involved stars.

bs	Mu UMa	10 22.3	+41 29	UMa	3.05		1.59	M0

170x250: Yellow gold.

ga	PGC 30367	10 22.5	−33 08	Ant	13.1	4.0'	2.8'	Im

170x250: Not visible.

ga	NGC 3222	10 22.6	+19 53	Leo	12.8v	1.5'	1.1'	S0

Wi: *faint, little brighter middle, partially resolved.*
170x250: Faint. Sharply extremely concentrated to a substellar nucleus. Round. Very very small. Right next to it SSW is a faint star; a little ways S is a moderate brightness star.

ga	PGC 30392	10 22.8	−33 43	Ant	12.8	2.3'	1.6'	S0

170x250: Not visible.

ga	MCG+2-27-4	10 23.4	+12 50	Leo	13.2	1.6'	1.4'	S0

170x250: Not visible.

ga	NGC 3226	10 23.4	+19 53	Leo	11.4v	3.3'	2.5'	E2

WH: *very faint, considerably large, round, one of double nebula.*
[See 3227.]

ga	NGC 3227	10 23.5	+19 51	Leo	10.3v	5.2'	4.0'	Sa

WH: *very faint, considerably large, round, one of double nebula.*

170x250: 3226 and 3227 are a double galaxy aligned NNW-SSE. 3227 is a little brighter. Moderately bright. Sharply extremely concentrated to a stellar nucleus. Slightly elongated NNW-SSE. Small. No involved stars. Nearby SE is a faint star in line with the two galaxies. 3226 is moderate brightness. Sharply pretty concentrated, but not to a stellar nucleus. Roughly round. Small. No involved stars. A slightly contrasting pair.

ga	CGCG 9-18	10 23.9	–03 11	Sex	13.2p	1.8'	1.0'	Sb

170x250: Very faint. Hardly concentrated. Slightly elongated NNW-SSE. Small. No involved stars. A little ways SE are a couple of faint stars almost in line with the galaxy.

ga	UGC 5612	10 24.1	+70 52	UMa	12.6b	3.4'	2.2'	Sdm

170x250: Very faint. Diffuse. Right on top of it is a fairly bright star.

ga	NGC 3241	10 24.3	–32 28	Ant	13.1p	2.2'	1.5'	Sab

JH: faint, pretty much extended, gradually little brighter middle, 11th mag. star NW.
170x250: Very faint. Broadly slightly concentrated. Roughly round. Small. No involved stars. Nearby WNW is a moderately bright star.

pn	NGC 3242	10 24.8	–18 38	Hya	8.6p	75.0"		

HH: ! *planetary nebula, very bright, little extended NW-SE, 32" in diameter, blue.*
12x35: It looks like a regular star, except that it has a blinking effect. *42x60:* Bright. A fat star or a tiny disk with a pretty strong blinking effect. *170x250:* Extremely bright. No central star. A fairly good disk, except that the outer ring is fainter than the inner part. There's a subtle unevenness in the light; there seems to be an inner torus. Weak blinking effect (interestingly, the blinking effect diminishes as the scope size increases). Slightly oval NW-SE. Moderate size for a planetary. Slightly greenish. It makes a perfect 30-60-90 right triangle with two stars, one S and one ESE. A terrific and interesting planetary; one of the best and brightest.

ga	NGC 3239	10 25.1	+17 09	Leo	11.7b	5.0'	3.6'	Im

HH: very faint, 9th mag. star involved near middle.
170x250: Moderately faint. Broadly slightly concentrated. Roughly round. Moderate size. SW of center is an interfering bright involved star that makes an isosceles triangle with two moderate brightness stars, one W and one NW. [C94-42 was not seen.]

| ga | NGC 3246 | 10 26.7 | +03 51 | Sex | 13.2p | 2.4' | 1.3' | Sdm |

JH: extremely faint, small, round, forms a triangle with 2 stars, 6th mag. star 8' WNW.
170x250: Very very faint. Diffuse. Roughly round. Small. No involved stars.

| ga | NGC 3245 | 10 27.3 | +28 30 | LMi | 11.7b | 3.2' | 1.7' | S0 |

HH: very bright, pretty large, extended, suddenly much brighter middle to an extended nucleus.
170x250: Moderately bright. Sharply pretty concentrated. Elongated N-S. Small. No involved stars.

| ga | IC 2580 | 10 28.3 | −31 31 | Ant | 13.2b | 2.1' | 1.9' | Sc |

[See E436-27.]

| ga | IC 2574 | 10 28.4 | +68 24 | UMa | 10.8b | 13.2' | 5.3' | Sm |

170x250: Exceedingly faint. An unconcentrated glow. Elongated NE-SW. Very large: it stretches across the entire field. No obvious involved stars. Half the field E is a "W" figure of stars opening up W with the brightest star at the N end.

| ga | ESO 436-27 | 10 28.9 | −31 36 | Ant | 12.6b | 2.6' | 1.1' | S0 |

170x250: Moderate brightness. Sharply very concentrated. Round. Very small. No involved stars. It makes a nearly equilateral triangle with a relatively bright star WSW and a moderate brightness star WNW. Half the field NW is I2580. Extremely faint. Diffuse. Roughly round. Moderately small. No involved stars. A little ways W is a faint star; a little ways SE is an extremely faint star; a little ways NNW is another extremely faint star.

| ga | NGC 3254 | 10 29.3 | +29 29 | LMi | 12.4b | 5.0' | 1.5' | Sbc |

HH: considerably bright, large, extended NE-SW, pretty suddenly much brighter middle to a nucleus.
170x250: Faint. Evenly moderately concentrated. Very elongated NE-SW. Moderate size. No involved stars. E are a couple of bright stars.

| ga | IC 2584 | 10 29.9 | −34 54 | Ant | 12.6v | 1.4' | 0.5' | S0 |

170x250: Moderately faint. Sharply very concentrated; nearly stellar. Elongated NW-SE. Very very small. No involved stars. It makes a very shallow arc with two faint stars E, a faint star W, and another faint star further W.

| ga | NGC 3281 | 10 31.9 | −34 51 | Ant | 11.6v | 3.0' | 1.8' | Sab |

JH: extremely faint, pretty large, extended, gradually little brighter middle.
170x250: Faint. Broadly slightly concentrated. Elongated NW-SE. Moderately small. No involved stars. Nearby NW is a faint star; W to S are several moderate brightness stars.

| ga | NGC 3264 | 10 32.3 | +56 04 | UMa | 12.5b | 2.8' | 1.1' | Sdm |

JH: extremely faint, between 2 faint stars.
170x250: Not visible. In the position are some faint stars that might hide the galaxy [correct], or the galaxy might be stellar.

| ga | NGC 3274 | 10 32.3 | +27 40 | Leo | 13.2b | 2.3' | 1.4' | Sd |

HH: faint, pretty large, round, gradually little brighter middle, double star E.
170x250: Faint. Broadly slightly concentrated. Slightly elongated E-W. Small. No involved stars. It's inside a triangle of moderate brightness to moderately faint stars, the easternmost of which has a faint companion.

| ga | NGC 3259 | 10 32.6 | +65 02 | UMa | 12.9p | 2.2' | 1.2' | Sbc |

HH: faint, small, round, gradually brighter middle.
170x250: Very faint. Broadly slightly concentrated. Round. Very small. No involved stars. [NW is a bright star.]

| ga | NGC 3277 | 10 32.9 | +28 30 | LMi | 12.5b | 2.3' | 2.2' | Sab |

HH: considerably bright, considerably small, round, pretty gradually much brighter middle.
170x250: Moderate brightness. Evenly moderately concentrated. Round. Very small. No involved stars.

| ga | NGC 3285 | 10 33.6 | −27 27 | Hya | 13.1b | 2.5' | 1.4' | Sa |

JH: pretty bright, small, little extended, gradually brighter middle, 1st of 9.
170x250: Moderately faint. Evenly moderately concentrated. Round. Small. No involved stars.

| ga | NGC 3287 | 10 34.8 | +21 39 | Leo | 12.9p | 2.1' | 0.9' | Sd |

dA: faint, pretty large, double star SW.

170x250: Very faint to faint. Broadly very slightly concentrated. Very elongated NNE-SSW. Moderate size. No involved stars. NNE is a moderately bright star; S are two faint stars; farther WSW is the bright plotted star, which is double.

| ga | UGCA 212 | 10 35.4 | −24 45 | Hya | 13.2b | 3.4' | 2.7' | Sdm |

170x250: Very very faint. A diffuse glow surrounding a moderately faint star. Moderately small. The involved star makes a large square figure with moderate brightness stars NE to WNW.

| ga | NGC 3294 | 10 36.3 | +37 19 | LMi | 12.2b | 3.7' | 1.7' | Sc |

HH: considerably bright, large, much extended NW-SE, gradually little brighter middle.
170x250: Moderate brightness. Broadly slightly concentrated. Very elongated WNW-ESE. Moderate size. No involved stars.

| ga | NGC 3308 | 10 36.4 | −27 26 | Hya | 12.9b | 1.7' | 1.2' | S0− |

JH: faint, small, round, 4th of 9.
[See 3309.]

| ga | NGC 3300 | 10 36.6 | +14 10 | Leo | 13.1p | 1.9' | 0.9' | S0 |

HH: considerably faint, considerably small, round, pretty much brighter middle, resolvable, among bright stars.
170x250: Moderately faint. Evenly moderately concentrated. Slightly elongated N-S. Small. No involved stars.

| ga | NGC 3309 | 10 36.6 | −27 31 | Hya | 12.6b | 1.8' | 1.5' | E3 |

JH: bright, large, round, W of double nebula, 5th of 9.
170x250: In the position are six galaxies in a NW-SE string a little more than a field width in length. At the NW end is a moderately faint galaxy [3308]. Evenly moderately concentrated. Round. Small. No involved stars. SE are two galaxies next to each other WNW-ESE. The W galaxy [3309] is moderately faint to moderate brightness. Sharply pretty concentrated. Round. Small. At its edge SE is a moderately faint star. The E galaxy [3311] is fainter and more diffuse. Faint. Broadly very slightly concentrated. Round. Small, but a little larger than the W galaxy. No involved stars. Farther ESE is another galaxy [3312]. Moderately faint. Sharply pretty concentrated to a faint stellar nucleus. Slightly elongated N-S. Small. No involved stars. A little ways ESE is a little triangle of moderate brightness to moderately faint stars. SSE is a faint galaxy [*3314A]. Broadly slightly concentrated.

Elongated NW-SE. Moderately small. Near center is a moderately faint involved star; a little ways SSW is a moderate brightness star; W is another moderate brightness star. NE is another galaxy [*3316 (JH)]. Faint. Evenly moderately concentrated. Round. Very small. No involved stars. It makes a small right triangle with two moderately faint stars, one S and one SE; SSE is a moderately bright star. [3307 (JH) was not seen.]

| ga | NGC 3311 | 10 36.7 | –27 31 | Hya | 11.6v | 2.1' | 1.9' | E+2 |

JH: bright, large, round, E of double nebula, 6th of 9.
[See 3309.]

| ga | NGC 3301 | 10 36.9 | +21 52 | Leo | 12.3b | 3.5' | 1.0' | S0/a |

HH: considerably bright, small, little extended, pretty suddenly brighter middle, resolvable.
170x250: Bright. Sharply pretty concentrated. Nearly edge-on NE-SW: a pretty much round core with faint halo extensions. Moderately small. No involved stars. NNW is a 30-60-90 right triangle of moderate brightness stars.

| ga | NGC 3312 | 10 37.0 | –27 33 | Hya | 12.7p | 3.3' | 1.2' | Sb |

JH: considerably faint, extended, gradually brighter middle, 7th of 9.
[See 3309.]

| ga | NGC 3313 | 10 37.4 | –25 19 | Hya | 12.2b | 4.4' | 3.9' | Sab |

Sn: extremely faint, pretty small, irregularly round, gradually brighter middle to a nucleus, 15th mag. star 3" N.
170x250: Moderate brightness; fairly high surface brightness. Sharply very concentrated. Round. Very small. No involved stars. It makes a trapezoid with moderately faint stars NNW to WSW.

| ga | ESO 501-51 | 10 37.5 | –26 19 | Hya | 12.9b | 2.7' | 1.5' | Sa |

170x250: Moderately faint. Sharply pretty concentrated. Slightly elongated WNW-ESE. Small. At its edge ENE is a moderately faint star that makes a flat triangle with two stars, one NE and one farther NNE. [E501-53 was not seen.]

| vs | U Hya | 10 37.6 | –13 23 | Hya | 4.7 | 6.2 | 2.68 | C7 |

12x35: Mag. 5.0. ***170x250:*** Orangish to orange red. A good carbon star; one of the brightest.

| ga | IC 2597 | 10 37.8 | −27 04 | Hya | 12.8b | 2.5' | 1.7' | E+4 |

170x250: In the position are two galaxies aligned NNW-SSE [I2597, *E501-59]. The N galaxy is moderately faint. Sharply pretty concentrated. Round. Small. Just off its edge SE is a faint star; a little ways WNW is a slightly brighter star. The S galaxy is faint. Broadly slightly concentrated. Roughly round. Small. No involved stars.

| ga | NGC 3310 | 10 38.7 | +53 30 | UMa | 11.2b | 3.3' | 3.0' | Sbc |

HH: *planetary nebula?, considerably bright, pretty large, round, very gradually then very suddenly much brighter middle to a nucleus 15" in diameter.*
42x60: Visible. ***170x250:*** Bright. Broadly concentrated. Round. Very small. It looks almost like a planetary. At times it looks like it might have the slightest halo, but otherwise it looks like only the core of a galaxy, peculiarly even in light. No involved stars. N is a shallow arc of three moderate brightness to moderately faint stars.

| ga | NGC 3319 | 10 39.2 | +41 41 | UMa | 11.5b | 6.2' | 3.6' | Scd |

WH: *considerably faint, large, irregularly extended, much brighter S of middle.*
170x250: Very faint. Broadly very slightly concentrated. Elongated NE-SW. Moderately small. No involved stars. NW are two moderate brightness stars.

| ga | NGC 3320 | 10 39.6 | +47 23 | UMa | 13.0p | 2.4' | 1.1' | Scd |

HH: *faint, pretty small, much extended, 10th mag. star NE.*
170x250: Moderately faint. Broadly slightly concentrated. Very elongated NNE-SSW. Small. At its S end is a faint involved star; a little ways N is a relatively bright star.

| ga | NGC 3336 | 10 40.3 | −27 46 | Hya | 13.0p | 1.9' | 1.5' | Sc |

JH: *very faint, pretty large, little extended, gradually little brighter middle.*
170x250: Faint. Broadly very slightly concentrated. Roughly round. Moderately small. No involved stars. Off its edge E is an extremely faint star; a little ways WNW is an ENE-WSW row of three faint stars.

| ga | ESO 376-9 | 10 42.0 | −33 14 | Ant | 12.8v | 1.5' | 0.4' | S0 |

170x250: Faint. Sharply very concentrated; the entire galaxy is substellar. Round. Tiny. No involved stars. Nearby SE are two very faint stars; the galaxy is just S of a line between two moderate brightness stars, one W and one ENE.

| ga | NGC 3338 | 10 42.1 | +13 44 | Leo | 11.6b | 5.8' | 3.5' | Sc |

HH: *faint, considerably large, extended, very gradually brighter middle, resolvable, 7th mag. star W.*
170x250: Moderately faint. Broadly very slightly concentrated. Elongated E-W. Moderately small. No involved stars. It's among a group of five stars WSW to N, including a bright star off the galaxy's W end.

| ds | 35 Sex | 10 43.3 | +04 45 | Sex | 6.8" | 6.3 | 7.4 | K3 K7 |

42x60: A very close unequal pair. **170x250:** A moderately close slightly unequal pair. Slight color contrast: the primary is slightly yellowish; the companion is white. [STF 1466]

| ga | NGC 3344 | 10 43.5 | +24 55 | LMi | 10.5b | 7.3' | 6.4' | Sbc |

HH: *considerably bright, large, gradually brighter middle, star involved, 2 stars E.*
42x60: Visible with an involved star. **170x250:** Moderate brightness. Sharply concentrated to a substellar nucleus. Round. Moderately small to moderate size. The halo fades out very indistinctly. Just SE of center is a very very faint involved star; farther E are two moderate brightness involved stars making a small arc with the nucleus of the galaxy; these brighter stars interfere.

| ga | NGC 3346 | 10 43.7 | +14 52 | Leo | 12.4p | 2.9' | 2.5' | Scd |

HH: *considerably faint, very large, round, very gradually very little brighter middle, easily resolvable.*
170x250: Faint. Hardly concentrated. Round. Moderately small. No involved stars. W are a couple of moderately faint stars.

| ga | NGC 3351/M95 | 10 44.0 | +11 42 | Leo | 10.5b | 7.5' | 5.0' | Sb |

JH: *bright, large, round, pretty gradually much brighter middle to a resolvable nucleus.*
170x250: Bright. Sharply very concentrated to a small core. Slightly elongated WNW-ESE. Moderately large. The halo fades out very indistinctly. No involved stars.

| ga | NGC 3329 | 10 44.7 | +76 48 | Dra | 13.0p | 1.7' | 0.9' | Sb |

JH: *pretty bright, small, little extended, pretty suddenly much brighter middle.*
170x250: Moderately faint. Evenly to sharply pretty concentrated. Round. Very small. No involved stars.

| vs | VY UMa | 10 45.1 | +67 25 | UMa | 6.0 | 6.6 | 2.39 | C5 |

12x35: Mag. 6.0. *170x250:* Orangish.

| ga | NGC 3365 | 10 46.2 | +01 48 | Sex | 13.2b | 4.4' | 0.7' | Scd |

JH: extremely faint, large, extremely extended, very gradually very little brighter middle, a ray.
170x250: Extremely faint. Unconcentrated. Edge-on NNW-SSE. Moderate size. No involved stars.

| ga | NGC 3359 | 10 46.6 | +63 13 | UMa | 11.0b | 7.3' | 4.3' | Sc |

HH: pretty bright, large, extended N-S, gradually little brighter middle.
42x60: Extremely faintly visible. *170x250:* Faint. Broadly slightly concentrated. Slightly elongated N-S. Moderate size. There's some unevenness in the light; the galaxy is irregular in shape. No involved stars.

| ga | NGC 3367 | 10 46.6 | +13 45 | Leo | 12.1b | 2.5' | 2.1' | Sc |

HH: pretty bright, considerably large, irregularly round, very gradually little brighter middle, resolvable, 1st of 3.
170x250: Moderately faint. Broadly slightly concentrated. Round. Small. No involved stars. It makes a triangle with two moderate brightness stars SW.

| ga | NGC 3368/M96 | 10 46.8 | +11 49 | Leo | 10.1b | 7.6' | 5.2' | Sab |

JH: very bright, very large, little extended, very suddenly very much brighter middle, resolvable.
12x35: M95 and M96 are faintly visible. *42x60:* Both are easily visible. M96 is brighter. *170x250:* M96 is very bright. Sharply pretty concentrated. Slightly elongated NW-SE. Moderately large. The halo fades out indistinctly. No involved stars.

| ga | NGC 3370 | 10 47.1 | +17 16 | Leo | 12.3p | 3.1' | 1.7' | Sc |

HH: considerably bright, pretty large, very little extended, gradually brighter middle, resolvable.
170x250: Moderately faint. Broadly slightly concentrated. Round. Small. No involved stars. WSW is a moderate brightness star.

| ga | NGC 3348 | 10 47.2 | +72 50 | UMa | 12.2b | 2.0' | 1.9' | E0 |

HH: bright, small, irregularly little extended, pretty suddenly brighter middle, 11th mag. star WNW.
170x250: Moderately bright to bright. Evenly to sharply pretty concentrated. Round. Small. Very near center is a moderately faint involved star that interferes; nearby W is a moderately bright star; NW is a moderate brightness star.

| ds | STF 1474 | 10 47.6 | −15 16 | Hya | 7.0" | 6.8 | 7.8 | F5 − |

12x35: A wide double. ***42x60:*** The N component of the wide double is a very very close equal pair. ***170x250:*** A close equal pair. Both white.

| ga | NGC 3377 | 10 47.7 | +13 59 | Leo | 11.2b | 5.2' | 2.9' | E5-6 |

HH: very bright, considerably large, round, suddenly very much brighter middle to a bright nucleus.
42x60: Visible. ***170x250:*** Bright. Sharply very concentrated. Elongated NE-SW. Small. No involved stars.

| ga | NGC 3379/M105 | 10 47.8 | +12 34 | Leo | 10.2b | 5.4' | 4.8' | E1 |

HH: very bright, considerably large, round, pretty suddenly brighter middle, resolvable.
12x35: M105 and 3384 are visible. ***42x60:*** Both are fairly bright, especially M105. ***170x250:*** M105, 3384, and 3389 are all in the same field. M105 is very bright. Sharply very concentrated. Round. Moderately small. 3384 is bright. Sharply very concentrated. Slightly elongated NE-SW. Small. 3389 is moderately faint to moderate brightness. Broadly very slightly concentrated. Elongated WNW-ESE. Moderate size; larger than M105 and 3384. 3389 is very different from M105 and 3384, which are brighter and more concentrated. There are no involved stars with any of the galaxies. A very good trio of galaxies.

| ga | NGC 3390 | 10 48.1 | −31 32 | Hya | 12.9b | 3.5' | 0.5' | Sb |

JH: faint, small, pretty much extended N-S.
170x250: Faint. Broadly slightly concentrated. Edge-on N-S. Moderate size. At its N end is a faint involved star; off its S end is a star.

| ga | NGC 3450 | 10 48.1 | −20 50 | Hya | 12.7p | 2.5' | 2.2' | Sb |

JH: very faint, large, round, very gradually little brighter middle, resolvable.
170x250: Faint. Broadly very slightly concentrated. Round. Moderately small. No involved stars. NNE is a moderate brightness star; S is a faint star.

| ga | NGC 3384 | 10 48.3 | +12 37 | Leo | 10.9b | 5.5' | 2.5' | S0– |

HH: very bright, large, round, pretty suddenly much brighter middle, 2nd of 3. [See M105.]

| ga | NGC 3381 | 10 48.4 | +34 42 | LMi | 12.7p | 2.3' | 2.0' | S |

HH: pretty faint, considerably large, irregularly round, very gradually little brighter middle, 1st of 3.
170x250: Faint. Broadly slightly concentrated. Roughly round. Small. No involved stars.

| ga | NGC 3393 | 10 48.4 | –25 09 | Hya | 13.1p | 2.1' | 1.9' | Sa |

JH: faint, small, round, pretty suddenly brighter middle, 2 10th mag. stars E.
170x250: Faint. Sharply moderately concentrated. Round. Tiny. At its edge NW is an extremely faint star; the galaxy makes a shallow arc with two relatively bright stars a little ways E.

| ga | NGC 3389 | 10 48.5 | +12 32 | Leo | 12.4b | 2.7' | 1.3' | Sc |

HH: faint, large, extended E-W, very gradually little brighter middle, 3rd of 3. [See M105.]

| bs | Nu Hya | 10 49.6 | –16 11 | Hya | 3.11 | | 1.25 | K2 |

170x250: Yellow gold.

| ga | NGC 3395 | 10 49.8 | +32 58 | LMi | 12.0v | 1.8' | 1.6' | Scd |

HH: considerably bright, pretty small, irregularly little extended, 1st of 2.
170x250: 3396 and 3395 are a double galaxy, probably interacting, aligned ENE-WSW with their axes pretty much perpendicular to each other. They're pretty similar. Both are very small and have no involved stars. 3395 is moderate brightness. Broadly concentrated. Elongated N-S. 3396 is a little fainter and more concentrated. Pretty elongated E-W.

| ga | NGC 3396 | 10 49.9 | +32 59 | LMi | 12.6p | 4.2' | 1.4' | Im |

HH: pretty bright, pretty small, irregularly little extended, 2nd of 2. [See 3395.]

| ga | NGC 3411 | 10 50.4 | −12 50 | Hya | 12.9b | 2.0' | 2.0' | E+ |

HH: faint, small, round, little brighter middle.
170x250: Moderately faint. Broadly to evenly moderately concentrated. Round. Small. No involved stars. A little ways W is a N-S line of three faint stars.

| ga | NGC 3394 | 10 50.7 | +65 43 | UMa | 13.1p | 1.9' | 1.4' | Sc |

HH: considerably faint, small, little extended, very gradually brighter middle.
170x250: Extremely faint. Diffuse. Roughly round. Small. No involved stars. It's between two moderate brightness stars, one NNW and one S. A quarter of the field NE is another galaxy [*3392 (HH)]. Very faint. Very small.

| ga | NGC 3412 | 10 50.9 | +13 24 | Leo | 11.5b | 3.6' | 2.0' | S0 |

HH: bright, small, little extended NW-SE, suddenly much brighter middle to a nucleus.
170x250: Moderately bright to bright; the nucleus is relatively very bright. Sharply very concentrated. Elongated NNW-SSE. Moderate size. No involved stars. Nearby N is a very faint star.

| ga | NGC 3421 | 10 51.0 | −12 26 | Hya | 12.8 | 2.0' | 1.5' | Sa |

Cm: faint, round, one of two nebulae.
170x250: Very faint. Broadly to evenly slightly concentrated. Roughly round. Small. No involved stars. ENE is a bright star. Just NNE of this star is a very small galaxy [*3422 (Cm)]. Moderately faint. Concentrated. Roughly round. No involved stars.

| ga | NGC 3423 | 10 51.2 | +05 50 | Sex | 11.6b | 3.8' | 3.2' | Scd |

HH: faint, very large, round, very gradually brighter middle, partially resolved.
170x250: Moderately faint. Broadly very slightly concentrated. Round. Moderately small to moderate size. No involved stars. Off its edge NE are a couple of faint stars that make a narrow tapered quadrilateral with two more stars E.

| ga | NGC 3413 | 10 51.3 | +32 45 | LMi | 13.1p | 2.1' | 1.0' | S0 |

WH: faint, small.
[See 3430.]

| ga | NGC 3414 | 10 51.3 | +27 58 | LMi | 12.0b | 3.5' | 2.7' | S0 |

HH: bright, pretty large, round, much brighter middle.
170x250: Moderate brightness. Evenly to sharply fairly concentrated. Round. Small. No involved stars. Half the field NNE is a small very faint galaxy [*3418 (HH)]. SE is a moderate brightness star.

| vs | V Hya | 10 51.6 | −21 14 | Hya | 6.5 | 12 | 4.14 | C6 |

12x35: Mag. 7.5. ***170x250:*** Deep copper red. A very good carbon star; strongly colored.

| ga | NGC 3424 | 10 51.8 | +32 54 | LMi | 13.2p | 3.2' | 0.9' | Sb |

HH: pretty faint, pretty large, little extended, SW of 3.
[See 3430.]

| ga | NGC 3434 | 10 52.0 | +03 47 | Leo | 12.9p | 2.1' | 1.8' | Sb |

HH: faint, pretty small, round, very gradually little brighter middle.
170x250: Moderately faint. Sharply very concentrated to a substellar nucleus. Round. Very small. No involved stars. Nearby NNE is a faint star; ESE is the bright plotted star.

| ga | NGC 3433 | 10 52.1 | +10 09 | Leo | 12.3p | 3.5' | 3.1' | Sc |

HH: very faint, very large, round, very gradually brighter middle.
170x250: Faint. Hardly concentrated. Round. Small. No involved stars.

| ga | NGC 3430 | 10 52.2 | +32 56 | LMi | 11.5v | 4.6' | 2.3' | Sc |

HH: pretty bright, large, irregularly extended, gradually brighter middle, 2nd of 3.
170x250: Moderate brightness. Broadly slightly concentrated. Elongated NNE-SSW. Moderate size. No involved stars. Nearby SSE is a very faint star. A third of the field WSW is 3424. Moderately faint. Unconcentrated. Elongated WNW-ESE. Moderately small. Near center is a faint involved star; nearby SE is a moderately faint star. In the field with both galaxies are several bright to moderate brightness stars. Half the field SW of 3424 is 3413. Faint. Broadly slightly concentrated. Round. Very small. No involved stars.

| ga | NGC 3432 | 10 52.5 | +36 37 | LMi | 11.7b | 6.8' | 1.4' | Sm |

HH: pretty bright, pretty large, very much extended NE-SW, double star involved?

170x250: Moderate brightness. Broadly very slightly concentrated. A thin edge-on, NE-SW. Moderate size. At its SW end is a pair of stars, the E of which is involved; off the galaxy's edge E is a star. A fairly nice galaxy.

| ga | NGC 3437 | 10 52.6 | +22 56 | Leo | 12.8p | 2.5' | 0.8' | Sc |

HH: pretty bright, pretty large, little extended WNW-ESE, gradually brighter middle.
170x250: Moderate brightness. Broadly slightly concentrated. Edge-on WNW-ESE. Moderately small. No involved stars. A little ways WNW is a very faint star in line with the galaxy's axis; SW is a moderately faint star; N, running all the way across the field, is a meandering WNW-ESE line of stars.

| ga | NGC 3449 | 10 52.9 | −32 55 | Ant | 12.2v | 3.3' | 0.9' | Sab |

JH: faint, small, round, 6th–7th mag. star SE.
170x250: Moderately faint. Evenly to sharply pretty concentrated. Slightly elongated NW-SE. Small. No involved stars. NE to N is an elongated group of 5–6 moderately faint stars; SSW is a faint star.

| ga | NGC 3447 | 10 53.4 | +16 46 | Leo | 13.1p | 3.5' | 1.9' | Sm |

JH: extremely faint, very large, very gradually very little brighter middle, bright double star SW.
170x250: Exceedingly faint. The slightest diffuse glow. Moderately small. No involved stars. SW is a moderately bright star that makes a large isosceles triangle with two more moderately bright stars, one SW of it and one W of it. [3447A was not seen.]

| ga | NGC 3403 | 10 53.9 | +73 41 | Dra | 13.0p | 3.0' | 1.1' | Sbc |

HH: pretty faint, large, irregularly extended, very gradually brighter middle.
170x250: Very very faint. Broadly very slightly concentrated. Elongated ENE-WSW. Small. No involved stars.

| ga | NGC 3456 | 10 54.1 | −16 01 | Crt | 13.2 | 1.8' | 1.2' | Sc |

HH: extremely faint, attached to a 12th mag. star E.
170x250: Faint. Broadly slightly concentrated. Roughly round. Small. NE of center is an interfering moderate brightness involved star with an extremely faint star just ESE of it; nearby SE is a moderately faint star. [M-3-28-17 was not seen.]

| ga | NGC 3455 | 10 54.5 | +17 17 | Leo | 12.8p | 2.6' | 2.0' | Sb |

HH: pretty faint, small, extended, gradually brighter middle, resolvable, SE of 2.
170x250: Faint. Hardly concentrated. Slightly elongated ENE-WSW. Small. No involved stars. A little ways N is a moderately bright star. On the other side of this star is an extremely faint galaxy [*3454 (JH)]. Broadly slightly concentrated. Edge-on WNW-ESE. Small. No involved stars.

| ga | NGC 3445 | 10 54.6 | +56 59 | UMa | 12.9b | 1.6' | 1.4' | Sm |

HH: considerably bright, pretty large, irregularly round, very gradually little brighter middle, 10th mag. star 2' NE.
170x250: Faint. Broadly slightly concentrated. Roughly round. Small. No involved stars. A little ways NE is a bright star.

| ga | NGC 3448 | 10 54.7 | +54 18 | UMa | 12.5b | 4.8' | 1.4' | I0 |

HH: bright, pretty large, much extended ENE-WSW, gradually brighter middle.
170x250: Moderate brightness. Broadly slightly concentrated. Edge-on ENE-WSW. Moderately small. No involved stars.

| ga | NGC 3464 | 10 54.7 | −21 03 | Hya | 13.0b | 2.5' | 1.6' | Sc |

Sn: extremely faint, pretty large, extended NW-SE.
170x250: Very faint to faint. Broadly very slightly concentrated. Round. Small. No involved stars.

| ga | NGC 3462 | 10 55.4 | +07 41 | Leo | 13.2p | 1.7' | 1.2' | S0 |

HH: very faint, very small, very little extended, pretty suddenly brighter middle.
170x250: Moderately faint to moderate brightness. Sharply very concentrated. Round. Very small. No involved stars.

| ds | 54 Leo | 10 55.6 | +24 45 | Leo | 6.6" | 4.5 | 6.3 | A1 A2 |

42x60: Split. *170x250:* A close unequal pair. Both white. [STF 1487]

| ga | NGC 3458 | 10 56.0 | +57 07 | UMa | 13.2p | 1.3' | 0.8' | S0 |

WH: very bright, very small, round, stellar.
170x250: Moderate brightness; fairly high surface brightness. Sharply very concentrated. Round. Tiny. No involved stars. One of the more stellar-looking galaxies.

| ga | ESO 569-24 | 10 57.0 | −20 10 | Crt | 12.9 | 1.3' | 1.2' | Sbc |

170x250: Not visible.

| ga | NGC 3483 | 10 59.0 | −28 28 | Hya | 13.1p | 1.8' | 1.2' | S0+ |

JH: pretty faint, small, round, brighter middle, among stars.
170x250: Moderately faint. Evenly moderately concentrated. Round. Very small. No involved stars. It's on the side of a small triangle of moderate brightness to faint stars.

| ga | NGC 3471 | 10 59.1 | +61 31 | UMa | 13.2b | 1.7' | 0.8' | Sa |

WH: very faint, small, round, brighter middle.
170x250: Moderately faint. Evenly moderately concentrated. Round. Very small. No involved stars.

| ga | NGC 3485 | 11 00.0 | +14 50 | Leo | 12.6p | 2.2' | 1.9' | Sb |

HH: faint, large, round, gradually little brighter middle, resolvable.
170x250: Faint. Broadly very slightly concentrated. Roughly round. Moderately small. No involved stars. Nearby W is a moderate brightness star.

| ga | NGC 3489 | 11 00.0 | +13 54 | Leo | 11.1b | 3.5' | 2.0' | S0+ |

HH: very bright, pretty large, little extended E-W, suddenly much brighter middle to a nucleus.
42x60: Visible. *170x250:* Bright. Sharply very concentrated. Slightly elongated ENE-WSW. Moderately small. No involved stars. Near its SW end is a moderately faint star.

| ga | NGC 3486 | 11 00.4 | +28 58 | LMi | 11.1b | 7.1' | 5.2' | Sc |

HH: considerably bright, considerably large, round, gradually much brighter middle.
42x60: Visible. *170x250:* Moderate brightness. Evenly moderately concentrated. Roughly round. Small. No involved stars.

| ga | NGC 3495 | 11 01.3 | +03 37 | Leo | 12.4b | 4.9' | 1.2' | Sd |

HH: very faint, pretty large, much extended.
170x250: Very faint to faint. Hardly concentrated. Edge-on NNE-SSW. Large: a quarter of the field in size. No involved stars. Near each end are a couple of faint stars.

| bs | Beta UMa | 11 01.8 | +56 22 | UMa | 2.37 | | −0.02 | A1 |

170x250: White.

| ga | NGC 3504 | 11 03.2 | +27 58 | LMi | 11.8b | 2.4' | 2.4' | Sab |

HH: bright, large, extended, much brighter middle to a nucleus, partially resolved, W of 2.
170x250: Bright. Either sharply very concentrated to a stellar nucleus or there's a star pretty much at the center. Slightly elongated NNW-SSE. Small. ESE of center is an extremely faint involved star.

| ga | NGC 3506 | 11 03.2 | +11 04 | Leo | 13.2p | 1.2' | 1.0' | Sc |

HH: very faint, considerably small, round, very gradually very little brighter middle.
170x250: Faint. Broadly slightly concentrated. Round. Small. No involved stars. It's off the E side of a kite-like figure of moderately faint stars pointing SSE.

| ga | NGC 3507 | 11 03.4 | +18 08 | Leo | 11.7b | 4.6' | 3.7' | Sb |

HH: considerably faint, pretty large, round, suddenly brighter middle to a faint star, 9th mag. star attached NNE.
170x250: Moderately faint. Broadly concentrated. Roughly round. Small. Right on top of it is a bright star that is in line with another bright star S and a third bright star farther NNE.

| ga | NGC 3511 | 11 03.4 | −23 05 | Crt | 10.9v | 6.2' | 2.1' | Sc |

WH: very faint, very large, much extended.
170x250: Moderately faint. Broadly very slightly concentrated. Edge-on ENE-WSW. Large: a quarter of the field in size. At its E end is a moderately faint star; at its W end is a faint star. A little more than half the field SSE is 3513. Moderately faint. Broadly very slightly concentrated. Round. Moderately small. No involved stars. It's in a small triangle of stars: a faint star W, another faint star SSW, and a moderately faint star E. A nice pair of galaxies in the same field.

| bs | Alpha UMa | 11 03.7 | +61 45 | UMa | 1.79 | | 1.07 | K0 |

170x250: Yellow.

| ga | NGC 3510 | 11 03.7 | +28 53 | LMi | 12.7p | 4.3' | 0.9' | Sm |

HH: faint, large, considerably extended, 7th mag. star 8' NW.
170x250: Very faint. Broadly slightly concentrated. Edge-on NNW-SSE. Small. No involved stars.

| ga | NGC 3513 | 11 03.8 | −23 14 | Crt | 11.4v | 3.0' | 2.3' | Sc |

WH: very faint, very large, much extended.
[See 3511.]

| ga | NGC 3512 | 11 04.0 | +28 02 | LMi | 13.0b | 1.4' | 1.3' | Sc |

HH: faint, pretty small, round, pretty gradually brighter middle, E of 2.
170x250: Moderately faint. Broadly moderately concentrated. Roughly round. Small. No involved stars. [It's surrounded by moderately faint to faint stars.]

| ga | UGC 6132 | 11 04.5 | +38 12 | UMa | 12.8v | 0.8' | 0.7' | S |

170x250: Not visible.

| ga | ESO 570-2 | 11 05.8 | −20 47 | Crt | 12.8 | 1.5' | 0.6' | Sbc |

170x250: Extremely faint. Hardly concentrated. Elongated ENE-WSW. Small. No involved stars.

| ga | NGC 3521 | 11 05.8 | −00 02 | Leo | 9.8b | 11.0' | 7.1' | Sbc |

HH: considerably bright, considerably large, much extended NW-SE, very suddenly much brighter middle to a nucleus.
12x35, 42x60: Visible. *170x250:* Bright. Sharply very concentrated to a small nucleus. Very elongated NNW-SSE. Large: a quarter of the field in size. No involved stars.

| ga | NGC 3516 | 11 06.8 | +72 34 | UMa | 12.5b | 1.9' | 1.5' | S0 |

HH: pretty bright, very small, irregularly round, pretty suddenly much brighter middle to a star.
170x250: Moderately bright to bright; high surface brightness. Sharply very concentrated. Round. Very small. No involved stars. Nearby ESE is a very faint star; the galaxy makes an isosceles triangle with a moderately bright star NE and a moderately faint star NW; farther W is another moderately bright star.

| ga | MCG+11-14-3B | 11 07.2 | +65 06 | UMa | 13.0 | 0.7' | 0.7' | – |

170x250: Exceedingly faint. Broadly slightly concentrated. Round. Very very small. No involved stars. Very nearby SSW is an extremely faint star. [M+11-14-3A, merged with M+11-14-3B, was not seen as a separate object.]

| ga | NGC 3528 | 11 07.3 | –19 28 | Crt | 12.6 | 2.5' | 1.4' | S0 |

JH: faint, small, round, pretty suddenly little brighter middle, W of 2.
170x250: Moderately faint. Evenly moderately concentrated. Roughly round. Small. No involved stars. Making a shallow little arc with it are two faint stars S. A little more than a quarter of the field S is another galaxy [*3529 (JH)]. Faint. Small.

| bs | Psi UMa | 11 09.7 | +44 29 | UMa | 3.00 | | 1.15 | K1 |

170x250: Yellow.

| ga | IC 2627 | 11 09.9 | –23 43 | Crt | 12.6b | 3.1' | 2.7' | Sbc |

170x250: Very faint to faint. Broadly very slightly concentrated. Round. Moderately small. No involved stars. N is a flat E-W triangle of moderate brightness stars.

| ga | NGC 3547 | 11 09.9 | +10 43 | Leo | 13.2b | 1.9' | 0.9' | Sb |

HH: faint, small, little extended, very little brighter middle.
170x250: Moderately faint. Broadly slightly concentrated. Elongated N-S. Small. No involved stars. Nearby E is a very faint galaxy [this is a very faint double star].

| ga | NGC 3549 | 11 10.9 | +53 23 | UMa | 12.8b | 3.5' | 1.1' | Sc |

WH: considerably bright, considerably large, considerably extended NNW-SSE.
170x250: Faint. Broadly slightly concentrated. Nearly edge-on NE-SW. Moderate size. No involved stars. A little ways SE is a short line of three faint to very faint stars parallel to the galaxy.

| ga | NGC 3556/M108 | 11 11.5 | +55 40 | UMa | 10.7b | 8.7' | 2.2' | Scd |

HH: considerably bright, very large, very much extended E-W, pretty much brighter middle, resolvable.
12x35: Very faintly visible. *42x60:* Visible. *170x250:* Moderately bright. Broadly slightly concentrated. Edge-on E-W. Large: a third of the field in size. There's some

unevenness in the light, including a dark band on the W side and vaguer unevenness toward the E end. E of center is an extremely faint involved star; a little farther W of center is a moderately bright involved star; on either side of the galaxy SSE and NNW is a faint star; farther N is another faint star. A nice edge-on galaxy.

| ga | NGC 3571 | 11 11.5 | −18 17 | Crt | 13.0p | 3.0' | 1.0' | Sa |

WH: pretty faint, pretty large, irregular figure, brighter middle.
170x250: Moderately faint. Broadly to evenly moderately concentrated. Elongated E-W. Small. No involved stars.

| ga | IC 676 | 11 12.7 | +09 03 | Leo | 12.8p | 2.4' | 1.7' | S0+ |

170x250: Faint. Broadly slightly concentrated. Slightly elongated NNW-SSE. Very small. No involved stars. It's in the middle of a long arrowhead group of stars pointing SSW.

| ga | NGC 3562 | 11 13.0 | +72 52 | Dra | 13.2b | 1.7' | 1.3' | E |

HH: pretty faint, pretty small, little extended, gradually brighter middle, 15th mag. star 70" NNE.
170x250: Moderately faint. Evenly to sharply pretty concentrated. Round. Very very small. No involved stars. A quarter of the field ENE is another galaxy [*C334-15]. Very very faint. Elongated [NNE-SSW].

| ga | NGC 3585 | 11 13.3 | −26 45 | Hya | 10.9b | 4.6' | 2.5' | E6 |

HH: bright, pretty large, extended, very suddenly much brighter middle to a nucleus, forms a triangle with 2 bright stars.
42x60: Visible. *170x250:* Bright. Sharply pretty concentrated. Slightly elongated WNW-ESE. Moderately small. It has a very faint outer halo. No involved stars.

| ga | UGC 6253 | 11 13.5 | +22 09 | Leo | 11.9v | 10.1' | 9.0' | Dw E |

170x250: Not visible. [Leo II]

| bs | Delta Leo | 11 14.1 | +20 31 | Leo | 2.56 | | 0.12 | A4 |

170x250: White.

| ga | NGC 3583 | 11 14.2 | +48 19 | UMa | 11.9p | 2.8' | 1.8' | Sb |

WH: pretty bright, pretty large, round, very gradually much brighter middle.

170x250: Moderate brightness. Broadly moderately concentrated. Roughly round. Small. No involved stars. Nearby SSE is a very faint star. [3577 (WH) was not seen.]

| bs | Theta Leo | 11 14.2 | +15 25 | Leo | 3.34 | | −0.01 | A2 |

170x250: White.

| ga | NGC 3593 | 11 14.6 | +12 49 | Leo | 11.9b | 5.2' | 1.9' | S0/a |

HH: bright, considerably large, extended E-W, pretty suddenly much brighter middle.
170x250: Moderately bright. Evenly pretty concentrated. Edge-on E-W. Moderately small. No involved stars.

| ga | ESO 377-29 | 11 14.7 | −33 54 | Hya | 12.7v | 1.5' | 0.5' | S0 |

170x250: Faint. Concentrated. Round. Very very small. No involved stars. It's between two nearby stars: a moderately bright star W and a moderately faint star ESE; these stars make a linear group with more stars W.

| pn | NGC 3587/M97 | 11 14.8 | +55 00 | UMa | 12.0p | 3.4' | | |

JH: !! planetary nebula, very bright, very large, round, very very gradually then very suddenly brighter middle.
12x35: Visible. *42x60:* It looks like a moderately faint round galaxy. *170x250:* Bright. No central star. Not a clean disk; not a sharp edge. There's some slight unevenness in the nebula: at times darker little patches are seen. These famous "eyes" are visible only at times with averted vision, and only very vaguely. No blinking effect. Round. Large for a planetary: an eighth of the field in size. It's on the W side of a NNE-SSW quadrilateral of stars. It looks like a ball of gas. Ghostly, even though it's bright overall. An intriguing object. [Owl Nebula]

| ga | NGC 3596 | 11 15.1 | +14 47 | Leo | 12.0b | 3.9' | 3.7' | Sc |

HH: pretty faint, large, round, gradually little brighter middle.
170x250: Moderately faint. Broadly slightly concentrated. Round. Moderately small. No involved stars.

| ga | NGC 3595 | 11 15.4 | +47 26 | UMa | 13.1p | 1.7' | 1.2' | E |

WH: very faint, very small, very little extended, stellar, considerably bright star N.

170x250: Moderate brightness. Sharply pretty concentrated. Round. Very small. No involved stars. A little ways NNE is the bright plotted star.

| ga | NGC 3599 | 11 15.5 | +18 06 | Leo | 12.0v | 2.7' | 2.2' | S0 |

HH: bright, pretty small, round, pretty gradually much brighter middle.
170x250: Moderately faint. Sharply pretty concentrated. Round. Very small. No involved stars.

| ga | NGC 3600 | 11 15.9 | +41 35 | UMa | 12.6p | 4.1' | 0.9' | Sa |

HH: pretty faint, small, little extended N-S, very gradually brighter middle.
170x250: Faint. Broadly to evenly moderately concentrated. Slightly elongated N-S. Small. No involved stars. [A little ways NNW is a wide pair of stars, moderately faint and moderate brightness.]

| ds | STF 1520 | 11 16.1 | +52 46 | UMa | 12.5" | 6.6 | 7.9 | F6 – |

42x60: The primary has a close fainter companion. *170x250:* A moderate distance unequal pair. Both white.

| ga | NGC 3605 | 11 16.8 | +18 01 | Leo | 12.3v | 1.6' | 1.2' | E4-5 |

HH: faint, small, round, SW of 3.
[See 3607.]

| ga | NGC 3607 | 11 16.9 | +18 03 | Leo | 9.9v | 5.5' | 5.0' | S0 |

HH: very bright, large, round, very much brighter middle, 2nd of 3.
42x60: 3608 and 3607 are visible together, aligned N-S. *170x250:* Both are round. Small. No involved stars. 3607 is bright. Sharply pretty concentrated. 3608 is moderately bright. Sharply very concentrated. Near each galaxy is a little group of stars – a nice effect. A little ways SW of 3607 is a very small faint galaxy [3605 (HH)]. Evenly moderately concentrated. Round. No involved stars.

| ga | NGC 3608 | 11 17.0 | +18 08 | Leo | 10.7v | 4.2' | 3.0' | E2 |

HH: bright, pretty large, round, pretty suddenly brighter middle, 3 rd of 3.
[See 3607.]

| ga | NGC 3611 | 11 17.5 | +04 33 | Leo | 12.8b | 2.1' | 1.7' | Sa |

HH: pretty faint, considerably small, irregularly round, pretty suddenly much brighter middle, 10th mag. star 3' NNW.

170x250: Moderate brightness. Sharply very concentrated. Round. Very small. No involved stars. NNW is a bright star.

| ga | NGC 3614A | 11 18.2 | +45 43 | UMa | 12.7 | 0.6' | 0.6' | Sm |

[See 3614.]

| ga | NGC 3614 | 11 18.3 | +45 44 | UMa | 12.3p | 4.5' | 2.5' | Sc |

HH: faint, pretty large, little extended E-W, gradually little brighter middle, resolvable.
170x250: Very very faint; very low surface brightness. Hardly concentrated. Slightly elongated E-W. Moderate size. No involved stars. [3614A was not seen.]

| ga | NGC 3621 | 11 18.3 | −32 48 | Hya | 9.5v | 13.3' | 6.1' | Sd |

HH: considerably bright, very large, extended NNW-SSE, among 4 stars.
42x60: Easily visible. *170x250:* Moderately bright. Broadly very slightly concentrated. Elongated NNW-SSE. Moderately large to large. At its NE end is a little triangle of moderately faint stars; the galaxy makes a quadrilateral with a bright star SSW, another bright star SSE, and a moderate brightness star E.

| ga | NGC 3610 | 11 18.4 | +58 47 | UMa | 11.7b | 2.7' | 2.2' | E5 |

HH: very bright, pretty small, little extended E-W, very suddenly very much brighter middle to a small nucleus.
170x250: Extremely bright. Sharply very concentrated. Round. Small. It has a small very faint outer halo. No involved stars. One of the highest surface brightness galaxies in the sky.

| bs | Nu UMa | 11 18.5 | +33 05 | UMa | 3.47 | | 1.41 | K3 |

170x250: The primary has a moderately close very faint companion. The primary is yellow gold. [STF 1524: 7.1", 3.5, 9.9]

| ga | NGC 3613 | 11 18.6 | +58 00 | UMa | 11.8b | 3.9' | 1.8' | E6 |

HH: very bright, considerably large, much extended NW-SE, suddenly much brighter middle to a nucleus.
170x250: Very bright. Sharply pretty concentrated. Elongated E-W. Moderately small. No involved stars. A little less than a third of the field NE is a faint stellar galaxy [MAC1119+5803] with a faint star right next to it E.

| ga | NGC 3623/M65 | 11 18.9 | +13 05 | Leo | 10.3b | 9.8' | 2.8' | Sa |

JH: bright, very large, much extended NNW-SSE, gradually brighter middle to a bright nucleus.
170x250: Moderately bright. Evenly pretty concentrated. Edge-on N-S. Large: a third of the field in size. No involved stars. Nearby NE of center is a faint star; nearby SSW of center is another faint star. Compared to M66, M65's extremities are a little brighter and its outline is better defined, even though it's not as bright overall.[†]

| ga | NGC 3619 | 11 19.4 | +57 45 | UMa | 12.5b | 2.7' | 2.3' | S0+ |

HH: considerably bright, considerably large, round, very gradually much brighter middle.
170x250: Moderately bright. Sharply very concentrated. Round. Very small. No involved stars.

| ga | NGC 3626 | 11 20.1 | +18 21 | Leo | 11.8b | 3.2' | 2.3' | S0+ |

HH: bright, small, very little extended, suddenly brighter middle.
170x250: Bright. Sharply very concentrated to a substellar nucleus. Round. Small. No involved stars.

| ga | ESO 570-19 | 11 20.2 | −21 28 | Crt | 11.3 | 1.9' | 1.0' | Sc |

170x250: Extremely faint. Evenly slightly concentrated. Round. Very small. No involved stars. Nearby NE is a moderately faint star.

| ga | NGC 3627/M66 | 11 20.2 | +12 59 | Leo | 9.7b | 9.1' | 4.1' | Sb |

JH: bright, very large, much extended NNW-SSE, much brighter middle, 2 stars NW.
12x35: M66 is easily visible. M65 is faint. ***42x60:*** M66 is fairly bright, with a relatively bright involved star and three slightly fainter stars nearby. M65 is easily visible. 3628 is faintly visible. ***170x250:*** M66 is bright. Sharply very concentrated. Elongated N-S overall, but the core is slightly askew: it's elongated NNW-SSE, at an angle to the rest of the galaxy. Large: a third of the field in size. The outer halo is low surface brightness. Two spiral arms are just made out: a short arm on the E side and a long arm on the W side pointing S. No involved stars. Going NW from the galaxy's N end and opening up WNW is a large curved "Y" of bright stars; a little W of the galaxy's S end is a faint star.

| ga | NGC 3628 | 11 20.3 | +13 35 | Leo | 10.3b | 14.8' | 2.9' | Sb |

HH: pretty bright, very large, very much extended WNW-ESE.
170x250: Moderately faint; fairly low surface brightness. Broadly very slightly concentrated. Edge-on WNW-ESE. Very large: half the field in size. The dust lane is visible through most of the length of the galaxy just inside its S edge; best with averted vision. The ends don't taper; instead they fan out, especially the E end. At the galaxy's S edge toward the E end is a faint star; farther ESE is a brighter star. An intriguing galaxy.

| ga | NGC 3630 | 11 20.3 | +02 57 | Leo | 12.9p | 2.8' | 1.1' | S0 |

JH: pretty bright, small, round, suddenly much brighter middle to a nucleus.
170x250: Moderately bright; fairly high surface brightness. Sharply very concentrated. Edge-on NE-SW. Small. No involved stars.

| ga | NGC 3629 | 11 20.5 | +26 57 | Leo | 12.8p | 2.6' | 1.5' | Scd |

HH: considerably faint, large, round, very gradually very little brighter middle.
170x250: Faint. Broadly slightly concentrated. Roughly round. Small. No involved stars. Nearby SE is a faint star.

| ga | NGC 3635 | 11 20.5 | −09 00 | Crt | 12.6 | 1.9' | 0.8' | Sbc |

Le: extremely faint, extremely small, round, brighter middle to a nucleus.
170x250: Very faint. Broadly slightly concentrated. Elongated E-W. Small. Right at its W end is a faint star. [3634 (Le), merged with 3635, was not seen as a separate galaxy.]

| ga | NGC 3631 | 11 21.0 | +53 10 | UMa | 11.0b | 5.0' | 4.7' | Sc |

HH: pretty bright, large, round, suddenly very much brighter middle to a resolvable nucleus.
42x60: Faintly visible. ***170x250:*** Moderate brightness. Sharply fairly concentrated. Round. Moderate size. It has a faint extended halo and an abrupt brighter core. No involved stars.

| ga | NGC 3640 | 11 21.1 | +03 14 | Leo | 11.4b | 4.3' | 3.4' | E3 |

HH: bright, pretty large, round, pretty suddenly brighter middle.
170x250: Bright. Evenly to sharply pretty concentrated. Round. Small. No involved stars. A little ways N is a faint star. A little ways S is a faint stellar galaxy [*3641 (Ma)]. 3640 is right in between the star and the other galaxy.

| ga | NGC 3646 | 11 21.7 | +20 10 | Leo | 11.8b | 3.9' | 2.2' | – |

HH: considerably faint, considerably large, little extended, gradually brighter middle, SW of 2.
170x250: Faint. Broadly slightly concentrated. Elongated NE-SW. Moderate size. No involved stars. Half the field ENE is a very small galaxy [*3649 (HH)]. Faint. Evenly moderately concentrated. At its edge S is a very faint star.

| ga | NGC 3642 | 11 22.3 | +59 04 | UMa | 11.7b | 5.3' | 4.4' | Sbc |

HH: pretty bright, pretty large, round, very gradually brighter middle.
170x250: Moderate brightness to moderately bright. Evenly fairly concentrated. Round. Small. No involved stars. It makes a sharp right triangle with a star NNE and another star NE; a little S of the NNE star is a fainter star.

| ga | MCG-1-29-15 | 11 22.7 | –07 40 | Crt | 12.6 | 2.3' | 1.5' | S0 |

170x250: In the position are three galaxies in a triangle. The two southern galaxies [M-1-29-15, *M-1-29-13] are aligned E-W. Both are very faint. Round. Small. No involved stars. The W galaxy is a little brighter. Evenly moderately concentrated. Nearby S is a faint star. The E galaxy is broadly slightly concentrated. The third galaxy [*MAC1122-0735], NE, is very very faint. Small. It makes a triangle with two bright stars, one N and one E.

| ga | NGC 3652 | 11 22.7 | +37 46 | UMa | 12.9p | 3.1' | 1.6' | Scd |

HH: pretty faint, considerably large, little extended, very gradually brighter middle.
170x250: Faint to moderately faint. Broadly slightly concentrated. Slightly elongated NNW-SSE. Small. No involved stars.

| ga | NGC 3655 | 11 22.9 | +16 35 | Leo | 12.3b | 1.5' | 0.9' | Sc |

HH: pretty bright, pretty small, irregularly round, brighter middle, resolvable.
170x250: Moderate brightness. Evenly moderately concentrated. Slightly elongated NNE-SSW. Small. No involved stars. A little ways ENE is a faint star.

| ga | NGC 3659 | 11 23.8 | +17 49 | Leo | 12.8p | 2.1' | 1.1' | Sm |

HH: considerably faint, small, little extended, resolvable.
170x250: Moderately faint. Broadly slightly concentrated. Slightly elongated NE-SW. Very small. No involved stars.

ga	NGC 3657	11 23.9	+52 55	UMa	13.1p	1.8'	1.6'	Sc

HH: considerably faint, very small, round, stellar.
170x250: Moderate brightness; fairly high surface brightness. Sharply pretty concentrated. Round. Very very small. No involved stars.

ga	NGC 3658	11 24.0	+38 33	UMa	13.1b	1.6'	1.4'	S0

WH: considerably bright, small, round, suddenly very much brighter middle to a nucleus.
170x250: Moderate brightness. Sharply pretty concentrated. Round. Very small. No involved stars. [It's surrounded by moderately faint stars.]

ga	NGC 3664	11 24.4	+03 19	Leo	13.2b	1.8'	1.8'	Sm

Te: pretty faint, binuclear.
170x250: Very very faint. Diffuse. Moderately small. No involved stars. SE are two moderately bright stars.

ga	NGC 3666	11 24.4	+11 20	Leo	12.7b	4.3'	1.1'	Sc

HH: faint, extended E-W, bright star E.
170x250: Moderately faint to moderate brightness. Broadly slightly concentrated. Edge-on E-W. Moderately small. No involved stars. A little ways NNE is a faint star.

ga	NGC 3665	11 24.7	+38 45	UMa	11.8b	4.3'	3.3'	S0

HH: considerably bright, considerably large, irregularly round, pretty gradually much brighter middle.
170x250: Moderately bright. Evenly moderately concentrated. Round. Small. No involved stars. [ESE are two stars, moderate brightness and moderately faint.]

ga	NGC 3672	11 25.0	−09 47	Crt	12.1b	4.1'	1.9'	Sc

HH: pretty bright, large, extended N-S, gradually brighter middle.
170x250: Moderately faint. Broadly very slightly concentrated. Elongated N-S. Moderate size to moderately large. No involved stars. Surrounding its N end at some distance is a triangular figure of stars like a rooftop.

ga	NGC 3673	11 25.2	−26 44	Hya	12.3p	3.6'	2.5'	Sb

JH: faint, very large, gradually very little brighter middle, 7th mag. star 6' S.

170x250: Faint. Broadly slightly concentrated. Elongated ENE-WSW. Moderately small. No involved stars. Nearby E are a couple of moderate brightness stars; off its W end is a shallow N-S arc of three faint stars.

ga	NGC 3669	11 25.4	+57 43	UMa	13.1p	2.2'	0.5'	Scd

HH: very faint, pretty large, pretty much extended NW-SE, easily resolvable.
170x250: Faint. Broadly very slightly concentrated. Edge-on NNW-SSE. Moderate size. No involved stars.

ga	NGC 3668	11 25.5	+63 26	UMa	13.1p	1.7'	1.3'	Sbc

HH: faint, pretty small, irregularly round, gradually brighter middle, 9th mag. star NW.
170x250: Moderately faint. Broadly slightly concentrated. Slightly elongated NW-SE. Small. No involved stars. It makes a little triangle with a couple of stars SW, the S of which is fairly bright.

ga	NGC 3675	11 26.1	+43 35	UMa	11.0b	5.8'	3.0'	Sb

HH: very bright, considerably large, very much extended N-S, very suddenly much brighter middle to a nucleus, stars W.
42x60: Fairly easily visible. *170x250:* Bright. Evenly moderately concentrated. Elongated N-S. Moderate size. No involved stars. Nearby SSW is a moderately faint star; a little farther SE is a fainter star; [W is a triangle of moderately faint stars].

ga	NGC 3674	11 26.4	+57 02	UMa	13.2p	1.9'	0.6'	S0

WH: pretty faint, irregular figure.
170x250: Moderately bright; fairly high surface brightness. Sharply very concentrated. Slightly elongated NNE-SSW. Small. No involved stars.

ga	NGC 3681	11 26.5	+16 51	Leo	11.9b	2.0'	2.0'	Sbc

HH: bright, pretty small, round, brighter middle.
170x250: Moderate brightness. Evenly moderately concentrated. Round. Small. No involved stars.

ga	UGC 6446	11 26.7	+53 44	UMa	13.1v	3.6'	2.3'	Sd

170x250: Not visible.

| ga | IC 2764 | 11 27.1 | −28 58 | Hya | 13.1p | 1.8' | 1.4' | S0/a |

170x250: Faint. Evenly slightly concentrated. Roughly round. Small. No involved stars. Just off its edge NNE is a moderate brightness star; nearby WSW is a very very faint star.

| ga | NGC 3684 | 11 27.2 | +17 01 | Leo | 12.0b | 3.0' | 2.0' | Sbc |

JH: pretty bright, pretty large, extended, very gradually brighter middle.
170x250: Moderate brightness. Broadly slightly concentrated. Round. Small. No involved stars.

| ga | NGC 3683 | 11 27.5 | +56 52 | UMa | 13.1p | 1.8' | 1.2' | Sc |

HH: considerably bright, pretty large, extended.
170x250: Moderate brightness. Evenly moderately concentrated. Pretty elongated NW-SE. Moderately small. No involved stars.

| ga | NGC 3686 | 11 27.7 | +17 13 | Leo | 11.9b | 3.2' | 2.4' | Sbc |

HH: pretty bright, large, very little extended, very gradually brighter middle, resolvable.
170x250: Moderate brightness. Broadly very slightly concentrated. Round. Moderately small. At its edge S is a faint star; a little ways N is a brighter star.

| ga | NGC 3687 | 11 28.0 | +29 30 | UMa | 12.8p | 2.4' | 2.4' | Sbc |

HH: pretty bright, pretty small, round, little brighter middle, resolvable.
170x250: Moderately faint. Evenly slightly concentrated. Round. Small. No involved stars.

| ga | NGC 3689 | 11 28.2 | +25 39 | Leo | 13.0b | 1.6' | 1.0' | Sc |

HH: pretty bright, pretty large, little extended, brighter middle.
170x250: Moderate brightness. Broadly moderately concentrated. Slightly elongated E-W. Small. No involved stars.

| ga | NGC 3691 | 11 28.2 | +16 55 | Leo | 12.6b | 1.3' | 0.9' | Sb |

HH: faint, pretty small, little extended, resolvable.
170x250: Faint. Broadly very slightly concentrated. Round. Small. No involved stars.

ga	NGC 3693	11 28.2	−13 11	Crt	13.1	3.2'	0.6'	Sb

HH: considerably faint, small, extended, gradually brighter middle.
170x250: Very faint to faint. Broadly slightly concentrated. Elongated E-W. Small. No involved stars. WSW is a moderate brightness star.

ga	NGC 3692	11 28.4	+09 24	Leo	12.9p	3.1'	0.6'	Sb

WH: faint, much extended, resolvable.
170x250: Moderate brightness. Broadly slightly concentrated. A thin edge-on, E-W. Moderately small. No involved stars. A little ways NE is a moderately faint star.

ga	IC 694	11 28.5	+58 33	UMa	12.1	1.1'	0.9'	Sm

[See 3690.]

ga	NGC 3690	11 28.5	+58 33	UMa	12.0p	1.5'	1.0'	Im

HH: pretty bright, pretty small, very little extended E-W, pretty gradually brighter middle, faint stars near SE.
170x250: 3690 and I694 are a merged double galaxy, NE-SW, like a merged double star. Viewed as a single elongated object it's moderately bright to bright. Sharply pretty concentrated. Small. No involved stars. It makes a shallow arc with a bright star SSE and a fainter star further SSE.

ga	NGC 3683A	11 29.2	+57 07	UMa	12.6p	2.3'	1.6'	Sc

170x250: Faint. Broadly slightly concentrated. Roughly round. Small. No involved stars. It makes a triangle with a relatively bright star nearby NE and a fainter star a little ways SE.

ga	NGC 3705	11 30.1	+09 16	Leo	11.9b	4.9'	2.0'	Sab

HH: pretty faint, pretty large, round, very suddenly much brighter middle, resolvable.
170x250: Moderately bright. Sharply very concentrated to a stellar nucleus. Elongated NW-SE. Moderately small. Right next to the nucleus WNW is an involved star.

ga	NGC 3715	11 31.5	−14 13	Crt	11.9	1.3'	0.8'	Sbc

HH: pretty faint, small, round, very gradually very little brighter middle.

170x250: Moderately faint. Evenly moderately concentrated. Round. Small. No involved stars. It makes a triangle with a moderate brightness star NE and a moderately bright star farther NNW.

| ga | NGC 3717 | 11 31.5 | −30 18 | Hya | 12.2b | 7.4' | 1.5' | Sb |

JH: pretty bright, small, much extended, 13th mag. star attached.
170x250: Moderate brightness. Broadly slightly concentrated. Edge-on NNE-SSW. Moderate size. N of center is a moderate brightness involved star. A little less than half the field SE is another galaxy [*I2913]. Faint. Slightly concentrated. Round. Small. No involved stars.

| ga | UGC 6510 | 11 31.5 | −02 18 | Leo | 13.0b | 1.9' | 1.7' | Scd |

170x250: Not visible.

| ga | IC 2910 | 11 31.9 | −09 43 | Crt | 13.2 | 1.6' | 1.1' | S0+ |

170x250: [The first object observed is *MAC1131-0940A.] Extremely faint. Hardly concentrated. Roughly round. Small. Near center is an extremely faint star [this is *MAC1131-0940B, a stellar companion galaxy]. A quarter of the field SE is a very concentrated galaxy [I2910]. Nearly stellar; it looks like a fuzzy star.

| ds | N Hya | 11 32.3 | −29 16 | Hya | 9.5" | 5.8 | 5.9 | F3 F3 |

42x60: A perfectly equal pair. *170x250:* A moderate distance perfectly equal pair. Both yellow white. A nice double in both scopes. [H 96]

| ga | NGC 3718 | 11 32.6 | +53 04 | UMa | 10.7v | 9.2' | 4.4' | Sa |

HH: pretty bright, very large, round, very gradually little brighter middle.
170x250: Moderately faint. Broadly slightly concentrated. Roughly round. Moderately small. No involved stars. It's on the side of a triangle of stars, the SW of which is a wide double.

| ga | NGC 3726 | 11 33.3 | +47 01 | UMa | 10.9b | 6.1' | 4.2' | Sc |

HH: pretty bright, very large, little extended N-S, very suddenly much brighter middle to a 15th mag. star, 11th mag. star N.
42x60: Faintly visible. *170x250:* Moderately faint. Sharply slightly concentrated to a small brighter core. Slightly elongated N-S. Large: almost a quarter of the field in size. Just off its N end is a relatively bright star.

| ga | NGC 3729 | 11 33.8 | +53 07 | UMa | 11.4v | 3.0' | 2.2' | Sa |

HH: pretty bright, pretty large, little extended N-S, gradually brighter middle, 12th mag. star near.
170x250: Moderate brightness. Broadly moderately concentrated. Slightly elongated NNW-SSE. Small. Just off its S end is a relatively bright star.

| ga | NGC 3732 | 11 34.2 | −09 50 | Crt | 12.5v | 1.2' | 1.2' | S0/a |

HH: faint, small, round, pretty suddenly brighter middle, 14th mag. star SW.
170x250: Moderate brightness. Evenly moderately concentrated. Round. Very small. No involved stars. Nearby SW is a moderate brightness star.

| ds | 90 Leo | 11 34.7 | +16 48 | Leo | 3.6" | 6.0 | 7.3 | B4 B3 |

42x60: Elongated. ***170x250:*** A very close slightly unequal pair. Both white. [STF 1552]

| ga | NGC 3733 | 11 35.0 | +54 51 | UMa | 12.9b | 4.7' | 2.1' | Scd |

WH: extremely faint, small, irregularly round, bright star in field.
170x250: Moderately faint. Sharply pretty concentrated. Round. Very very small. No involved stars. It makes a long evenly spaced shallow arc with three stars ENE.

| ga | NGC 3738 | 11 35.8 | +54 31 | UMa | 11.7v | 2.5' | 1.7' | Im |

WH: pretty bright, pretty large, brighter middle.
170x250: Moderate brightness to moderately bright. Broadly moderately concentrated. Slightly elongated NNW-SSE. Small. No involved stars. It's at the W end of a long arc of fairly bright stars NE to ESE.

| ga | NGC 3735 | 11 36.0 | +70 32 | Dra | 12.5p | 4.2' | 1.0' | Sc |

HH: pretty bright, large, much extended NW-SE, much brighter middle.
170x250: Moderately faint to moderate brightness. Broadly moderately concentrated. Edge-on NW-SE. Moderate size. No involved stars.

| ds | HJ 4455 | 11 36.6 | −33 34 | Hya | 3.4" | 5.8 | 7.9 | K0 − |

170x250: The primary has a very close relatively faint companion. Colorful: both yellow. A neat little double.

| ga | NGC 3756 | 11 36.8 | +54 17 | UMa | 12.1b | 4.2' | 2.2' | Sbc |

HH: pretty faint, large, little extended.
170x250: Moderately faint. Broadly slightly concentrated. Elongated N-S. Moderate size. No involved stars. [NNW is a bright star.]

| ga | NGC 3759 | 11 36.9 | +54 49 | UMa | 13.2v | 1.1' | 1.1' | S0 |

dA: faint, small, irregularly round, 11th mag. star near.
170x250: Faint. Evenly to sharply moderately concentrated. Round. Very very small. No involved stars. [It's among a number of stars.]

| ga | NGC 3769 | 11 37.7 | +47 53 | UMa | 12.6b | 3.3' | 0.9' | Sb |

HH: pretty bright, small, pretty much extended.
170x250: Moderate brightness. Broadly moderately concentrated. Very elongated NNW-SSE. Moderately small. No involved stars. [3769A was not seen.]

| ga | NGC 3773 | 11 38.2 | +12 06 | Leo | 13.0p | 1.1' | 0.9' | S0 |

HH: considerably faint, considerably small, round, pretty suddenly brighter middle.
170x250: Moderately faint. Sharply pretty concentrated. Round. Very small. No involved stars.

| ga | NGC 3789 | 11 38.2 | –09 36 | Crt | 13.0 | 2.4' | 0.5' | S0+ |

Le: extremely faint, very small, extended N-S, gradually brighter middle.
170x250: Faint. Sharply pretty concentrated. Round. Tiny. No involved stars. It makes a triangle with two faint stars, one SE and one E.

| ds | STF 1559 | 11 38.8 | +64 21 | UMa | 1.8" | 6.8 | 7.8 | A5 – |

170x250: The primary has an extremely close relatively faint companion. The primary is white.

| ga | NGC 3782 | 11 39.3 | +46 30 | UMa | 13.1b | 2.3' | 1.5' | Scd |

HH: faint, small, attached to a 15th mag. star, another star in contact.
170x250: Moderate brightness. Evenly moderately concentrated. Slightly elongated N-S. Small. The galaxy is just inside a little arc of three moderately bright stars passing S of it; nearby N is a faint star.

| ga | NGC 3780 | 11 39.4 | +56 16 | UMa | 12.2p | 3.1' | 2.4' | Sc |

HH: pretty faint, large, very little extended, very gradually brighter middle, resolvable.
170x250: Moderate brightness. Broadly slightly concentrated. Roughly round. Moderate size. No involved stars. A little ways ENE is a moderately faint star.

| ga | NGC 3786 | 11 39.7 | +31 54 | UMa | 13.2p | 2.2' | 1.2' | Sa |

JH: pretty bright, pretty large, pretty much extended, gradually brighter middle, W of 2.
170x250: In the position are twin galaxies right next to each other NNE-SSW [*3788 (JH), 3786]. The N galaxy is a little brighter. Both are evenly moderately concentrated. Small. No involved stars. The N galaxy is elongated N-S. The S galaxy is elongated ENE-WSW, nearly making a right angle with the N galaxy. A little ways SE is a moderately bright star. A fairly neat pair. [3793 (Te) was not seen.]

| ga | UGC 6628 | 11 40.1 | +45 56 | UMa | 13.2b | 3.8' | 3.5' | Sm |

170x250: Not visible.

| ga | NGC 3798 | 11 40.2 | +24 41 | Leo | 13.1p | 2.5' | 1.7' | S0 |

HH: faint, considerably small, little extended, stellar, resolvable.
170x250: Moderately faint. Evenly moderately concentrated. Round. Very small. No involved stars.

| ga | NGC 3801 | 11 40.3 | +17 43 | Leo | 12.0v | 3.5' | 1.8' | S0 |

HH: pretty faint, pretty large, round, brighter middle, resolvable, 2nd of 3.
170x250: Moderately faint. Broadly slightly concentrated. Round. Small. No involved stars. A little ways NNE is a second galaxy [*3802 (HH)]. Faint. Broadly slightly concentrated. Elongated E-W. Small; a little smaller than 3801. No involved stars. It makes a small triangle with a star just off its E end and an extremely faint star SSE.

| ga | UGCA 241 | 11 40.6 | −10 05 | Crt | 12.8 | 2.7' | 2.1' | Scd |

170x250: Not visible.

| ga | NGC 3810 | 11 41.0 | +11 28 | Leo | 11.4b | 4.3' | 3.2' | Sc |

HH: bright, large, very little extended.

170x250: Moderately bright. Broadly moderately concentrated. Elongated NNE-SSW. Moderate size. No involved stars.

ga	NGC 3811	11 41.3	+47 41	UMa	12.9b	2.8'	1.8'	Scd

HH: faint, small, very little extended, gradually little brighter middle.
170x250: Moderately faint. Evenly moderately concentrated. Roughly round. Small. No involved stars. [It's between two moderate brightness stars, one W and one E.]

ga	NGC 3813	11 41.3	+36 32	UMa	12.2b	2.2'	1.2'	Sb

HH: considerably bright, pretty large, pretty much extended E-W, brighter middle.
170x250: Moderate brightness. Broadly moderately concentrated. Edge-on E-W. Moderately small. At its E end is a very faint star; nearby SSW is another very faint star; a little ways W is third very faint star.

ga	NGC 3818	11 42.0	−06 09	Vir	12.7b	2.0'	1.2'	E5

HH: faint, pretty small, round, pretty suddenly brighter middle.
170x250: Moderately bright. Sharply very concentrated. Slightly elongated E-W. Very small. No involved stars.

ga	MCG-2-30-16	11 42.2	−10 46	Crt	13.2	1.7'	1.2'	Sab

170x250: Not visible.

ga	NGC 3825	11 42.4	+10 15	Vir	13.0v	2.4'	1.5'	Sa

HH: pretty faint, pretty small, 4th of 4.
170x250: In the position are four galaxies, three of which are in a WNW-ESE row. All are roughly round. Very small. No involved stars. The easternmost galaxy [3825] is moderately faint. Sharply very concentrated. The next galaxy W [*3822 (HH)] is faint. Broadly moderately concentrated. The westernmost galaxy [*3817 (JH)] is very faint. Hardly concentrated. The fourth galaxy [*3819 (JH)], N, is very faint. Evenly slightly concentrated.

ga	NGC 3842	11 44.0	+19 56	Leo	12.8b	1.4'	1.2'	E

HH: faint, small, round, very gradually little brighter middle, 3rd of 5.
170x250: Faint. Evenly moderately concentrated. Round. Small. No involved stars. [3842 is at the center of galaxy cluster AGC 1367; no other member is obviously visible.]

| ga | NGC 3835 | 11 44.1 | +60 07 | UMa | 13.2p | 1.9' | 0.7' | Sab |

JH: pretty bright, extended, gradually brighter middle, 8th mag. double star 5' [SE].
170x250: Moderately faint. Broadly slightly concentrated. Edge-on ENE-WSW. Small. No involved stars.

| ga | NGC 3865 | 11 44.9 | −09 13 | Crt | 12.8p | 2.0' | 1.4' | Sb |

Cm: faint, pretty large, diffuse.
170x250: Faint. Broadly very slightly concentrated. Round. Small. No involved stars. A little more than a third of the field SE is a small faint galaxy [*3866 (Cm)]. Very nearby W is a moderate brightness star.

| ga | NGC 3872 | 11 45.8 | +13 45 | Leo | 12.7b | 2.0' | 1.6' | E5 |

HH: bright, small, round, suddenly much brighter middle to a star.
170x250: Moderately bright. Sharply very concentrated. Slightly elongated NNE-SSW. Small. No involved stars.

| ga | NGC 3877 | 11 46.1 | +47 29 | UMa | 11.8b | 5.8' | 1.2' | Sc |

WH: bright, large, much extended NNE-SSW.
170x250: Moderate brightness. Broadly to evenly moderately concentrated. A thin edge-on, NE-SW. Large: a little less than a quarter of the field in size. The ends fade out indistinctly. No involved stars. ENE to NNW is a large isosceles triangle of stars.

| ga | NGC 3885 | 11 46.8 | −27 55 | Hya | 11.9v | 2.8' | 0.9' | S0/a |

HH: considerably faint, very small, very little extended, brighter middle, very faint star SE.
170x250: Moderate brightness to moderately bright. Evenly to sharply pretty concentrated. Round. Very small. No involved stars. It makes a little triangle with two moderate brightness stars SE.

| ga | NGC 3887 | 11 47.1 | −16 51 | Crt | 11.4b | 3.3' | 2.5' | Sbc |

HH: pretty bright, large, irregularly round, very gradually pretty much brighter middle.
170x250: Moderate brightness. Broadly slightly concentrated. Slightly elongated N-S. Moderately small. At its edge NE is a moderate brightness star; this is one of a triangular figure of stars enclosing the galaxy.

| ga | NGC 3888 | 11 47.6 | +55 58 | UMa | 12.0v | 1.8' | 1.3' | Sc |

HH: pretty bright, small, little extended, pretty gradually brighter middle.
170x250: Moderate brightness. Broadly to evenly moderately concentrated. Roughly round. Small. No involved stars. It makes a small triangle with two moderately faint stars, one WNW and one NNE. [3889 (LR) was not seen.]

| ga | NGC 3892 | 11 48.0 | −10 57 | Crt | 12.5p | 2.9' | 2.4' | S0+ |

HH: pretty bright, pretty large, round, gradually brighter middle, resolvable.
170x250: Moderately bright. Evenly moderately concentrated. Round. Small. No involved stars. It's the E point of a triangular figure it forms with four stars W.

| ga | NGC 3891 | 11 48.1 | +30 21 | UMa | 13.2p | 2.0' | 1.6' | Sbc |

HH: pretty bright, small, brighter middle.
170x250: Moderately faint. Broadly to evenly moderately concentrated. Roughly round. Small. No involved stars.

| ga | NGC 3893 | 11 48.6 | +48 42 | UMa | 11.2b | 4.5' | 2.7' | Sc |

HH: bright, pretty large, round, much brighter middle.
170x250: Bright. Evenly moderately concentrated. Roughly round. Moderate size. At its edge NW is a moderately faint star; [SW is a moderately bright star]. A little less than a quarter of the field SE is another galaxy [*3896 (WH)] with a faint star right next to it N. Slightly concentrated. Very small. A quarter of the field ENE of 3893 is a tiny galaxy [M+8-22-9]; almost stellar.

| ga | NGC 3894 | 11 48.8 | +59 24 | UMa | 12.6b | 2.8' | 1.7' | E4-5 |

HH: bright, pretty large, irregularly round, pretty gradually much brighter middle, W of 2.
170x250: Moderate brightness. Evenly to sharply fairly concentrated. Round. Small. No involved stars. A little ways ENE is another galaxy [*3895 (HH)]. Moderately faint. Evenly moderately concentrated. Roughly round. Small. No involved stars.

| ga | UGCA 247 | 11 48.8 | −28 17 | Hya | 13.0p | 3.0' | 2.4' | Sd |

170x250: Exceedingly faint. A very slight diffuse glow. [Moderate size.] Next to it E is a little triangle of moderate brightness stars. SE is a fourth star.

| bs | Beta Leo | 11 49.1 | +14 34 | Leo | 2.14 | | 0.09 | A3 |

170x250: White.

| ga | NGC 3898 | 11 49.2 | +56 05 | UMa | 11.6b | 4.3' | 2.5' | Sab |

HH: bright, pretty large, little extended, suddenly very much brighter middle.
170x250: Bright. Sharply pretty concentrated. Slightly elongated WNW-ESE. Small. No involved stars.

| ga | NGC 3900 | 11 49.2 | +27 01 | Leo | 12.2b | 3.1' | 1.6' | S0+ |

HH: bright, pretty large, very little extended N-S, brighter middle to a nucleus.
170x250: Moderate brightness to moderately bright. Sharply fairly concentrated. Slightly elongated N-S. Small. No involved stars. It's just off the hypotenuse of a large right triangle of stars S to NE.

| ga | NGC 3904 | 11 49.2 | −29 16 | Hya | 11.8b | 2.6' | 1.8' | E2-3 |

HH: pretty bright, small, round, much brighter middle.
170x250: Very bright; high surface brightness. Sharply very concentrated. Roughly round. Small. No involved stars.

| ga | NGC 3907 | 11 49.5 | −01 05 | Vir | 13.1v | 1.1' | 0.7' | S0− |

JH: extremely faint, small, pretty suddenly brighter middle.
170x250: In the position are two galaxies right next to each other E-W [3907, *U6793]. The E galaxy is faint. Sharply very concentrated. Roughly round. Very very small. No involved stars. The W galaxy is very very faint. Broadly slightly concentrated. Elongated ENE-WSW. Very small. No involved stars.

| ga | NGC 3912 | 11 50.1 | +26 28 | Leo | 13.2p | 1.6' | 0.8' | Sb |

HH: faint, pretty large, round, pretty gradually brighter middle.
170x250: Faint to moderately faint. Broadly slightly concentrated. Elongated N-S. Very small. No involved stars.

| ga | NGC 3913 | 11 50.6 | +55 21 | UMa | 13.2b | 2.6' | 2.5' | Sd |

WH: faint, extended.
170x250: Very faint. Hardly concentrated. Roughly round. Small. No involved stars.

| ga | NGC 3917 | 11 50.8 | +51 49 | UMa | 12.5b | 5.1' | 1.2' | Scd |

HH: faint, large, very much extended, very gradually brighter middle.
170x250: Faint. Broadly very slightly concentrated. Edge-on ENE-WSW. Moderately large. No involved stars. Alongside it S are two faint stars aligned at a slight angle to the axis of the galaxy.

| ga | NGC 3923 | 11 51.0 | –28 48 | Hya | 10.8b | 5.8' | 3.8' | E4-5 |

HH: bright, pretty large, little extended, gradually much brighter middle, resolvable, very faint star involved SW.
42x60: Visible. *170x250:* Bright. Evenly to sharply pretty concentrated. Slightly elongated NE-SW. Moderately small to moderate size. Off its SW end is a faint star; further SW are two brighter stars.

| ga | NGC 3921 | 11 51.1 | +55 04 | UMa | 13.1b | 2.1' | 1.2' | S0/a |

HH: pretty faint, small, round, pretty suddenly pretty much brighter middle.
170x250: Moderately faint. Sharply very concentrated. Round. Very small. No involved stars. Nearby ENE is a star. A quarter of the field NNW is a second galaxy [*3916 (HH)]. Very faint. Slightly concentrated. Elongated to edge-on NE-SW. No involved stars. [M+9-19-213 was not seen.]

| ga | NGC 3928 | 11 51.8 | +48 40 | UMa | 13.2b | 1.5' | 1.5' | Sb |

HH: pretty faint, small, round, pretty suddenly pretty much brighter middle.
170x250: Moderate brightness to moderately bright. Sharply pretty concentrated. Round. Very small. No involved stars. Making a short curved line with it are a star SE and another star a little farther NNW.

| ga | NGC 3930 | 11 51.8 | +38 00 | UMa | 13.1p | 4.2' | 2.8' | Sc |

HH: extremely faint, considerably large, irregular figure, gradually little brighter middle.
170x250: Extremely faint. A slight glow; pretty much unconcentrated. Roughly round. Moderately small. No involved stars.

| ga | NGC 3936 | 11 52.3 | –26 54 | Hya | 12.8p | 3.9' | 0.6' | Sbc |

JH: very faint, considerably large, very much extended ENE-WSW.
170x250: Very faint to faint. Broadly very slightly concentrated. A fairly thin edge-on, ENE-WSW. Moderate size. No involved stars.

| ga | NGC 3938 | 11 52.8 | +44 07 | UMa | 10.9b | 5.4' | 4.5' | Sc |

HH: bright, very large, round, brighter middle to a pretty bright nucleus, easily resolvable.
42x60: Visible. ***170x250:*** Moderate brightness. Evenly moderately concentrated. Round. Moderate size. No involved stars.

| ga | NGC 3941 | 11 52.9 | +36 59 | UMa | 11.3b | 3.7' | 2.3' | S0 |

HH: very bright, pretty large, round, suddenly much brighter middle to a 9th mag. star.
170x250: Bright. Sharply very concentrated to a substellar nucleus. Elongated N-S. Small. No involved stars. Nearby E is a very faint star.

| ga | NGC 3945 | 11 53.2 | +60 40 | UMa | 11.8b | 5.2' | 3.4' | S0+ |

HH: bright, pretty large, round, gradually much brighter middle, resolvable, star E.
170x250: Moderately bright. Evenly moderately concentrated. Roughly round. Small. No involved stars. It makes a very small triangle with two moderately faint stars, one SW and one WNW.

| ga | UGCA 250 | 11 53.4 | −28 33 | Hya | 13.2p | 4.3' | 0.6' | Sd |

170x250: Extremely faint. Hardly concentrated. Edge-on ENE-WSW. Moderate size. No involved stars. NE is a faint star; NW is a pair of faint stars.

| ga | NGC 3949 | 11 53.7 | +47 51 | UMa | 11.5b | 3.3' | 2.4' | Sbc |

HH: considerably bright, pretty large, pretty much extended, very gradually brighter middle.
170x250: Bright. Broadly to evenly moderately concentrated. Slightly elongated NW-SE. Moderately small. No involved stars.

| bs | Gamma UMa | 11 53.8 | +53 41 | UMa | 2.44 | | 0.00 | A0 |

170x250: White.

| ga | NGC 3953 | 11 53.8 | +52 19 | UMa | 10.8b | 6.9' | 3.4' | Sbc |

HH: considerably bright, large, extended N-S, very suddenly brighter middle to a large resolvable nucleus.

42x60: Visible. *170x250:* Moderately bright. Evenly moderately concentrated. Elongated NNE-SSW. Moderate size. It has a faint outer halo. Just off its edge W is a faint star; the galaxy makes a triangle with two moderate brightness stars, one ENE and one SE.

ga	NGC 3955	11 54.0	−23 09	Crt	12.6b	2.9'	0.9'	S0/a

WH: considerably faint, small, extended N-S, little brighter S.
170x250: Moderate brightness. Broadly moderately concentrated. Edge-on NNW-SSE. Moderately small. No involved stars.

ga	NGC 3956	11 54.0	−20 33	Crt	12.8p	3.3'	0.9'	Sc

HH: considerably faint, pretty large, pretty much extended ENE-WSW.
170x250: Very faint to faint. Broadly very slightly concentrated. Edge-on ENE-WSW. Moderate size. No involved stars.

ga	NGC 3957	11 54.0	−19 34	Crt	12.8p	3.0'	0.6'	S0+

WH: faint, small, extended, resolvable.
170x250: Moderate brightness. Broadly slightly concentrated. Edge-on N-S. Moderately small. No involved stars.

ga	NGC 3962	11 54.7	−13 58	Crt	11.6b	3.0'	2.2'	E1

HH: considerably bright, pretty large, irregularly round, gradually much brighter middle, forms a triangle with two stars.
170x250: Bright. Evenly to sharply fairly concentrated. Round. Small. No involved stars. It makes a triangle with two bright stars S; a faint star E makes the figure a trapezoid.

ga	NGC 3963	11 55.0	+58 29	UMa	12.5b	2.7'	2.4'	Sbc

HH: pretty faint, considerably large, round, very gradually then suddenly brighter middle.
170x250: Moderately faint. Broadly slightly concentrated. Roughly round. Moderately small. No involved stars. Half the field SSW is another galaxy [*3958 (HH)]. Faint to moderately faint. Evenly moderately concentrated. Elongated NE-SW. Small. No involved stars. Very nearby N is an extremely faint star.

ga	NGC 3967	11 55.2	−07 50	Crt	11.5	1.5'	1.1'	S0−

Te: very faint, small, faint star close W.

170x250: In the position are two galaxies half the field apart WNW-ESE [*3959 (Te), 3967]. Both are faint. Round. Very small. No involved stars. The NW galaxy is broadly moderately concentrated. It's between two very nearby stars: a faint star NW and a moderately faint star SE. The SE galaxy is evenly moderately concentrated. Nearby E is a faint star.

ga	NGC 3968	11 55.5	+11 58	Leo	12.6p	2.7'	1.9'	Sbc

HH: pretty bright, large, irregularly round, brighter middle, 10th mag. star 5' [ENE].
170x250: Moderately faint. Broadly moderately concentrated overall, but it has a very faint substellar nucleus. Round. Small. No involved stars. It makes a quadrilateral with three moderately bright stars ENE to N; near the two northern stars is a fourth fainter star. [3973 (LR) was not seen.]

ga	UGC 6903	11 55.6	+01 14	Vir	13.0p	2.6'	2.3'	Scd

170x250: Exceedingly faint. The very slightest unconcentrated glow. Round. Moderate size. No involved stars. It's on the side of a large triangle of stars, the nearest and faintest of which is a little ways ESE.

ga	NGC 3972	11 55.8	+55 19	UMa	13.0b	4.2'	0.9'	Sbc

WH: pretty bright, extended.
170x250: Faint. Broadly slightly concentrated. Edge-on WNW-ESE. Moderate size. No involved stars. [It makes a shallow arc with two moderately faint stars a little ways SSW.] A little less than a third of the field NE is another galaxy [*3977 (WH)]. Very faint. Slightly concentrated. Small.

ga	NGC 3976	11 56.0	+06 44	Vir	12.3p	3.8'	1.2'	Sb

HH: bright, pretty large, considerably extended NNE-SSW, very suddenly much brighter middle to a nucleus.
170x250: Moderate brightness. Sharply pretty concentrated. Edge-on ENE-WSW. Moderately small. No involved stars. It makes a triangle with two faint stars, one SSE and one ESE.

ga	NGC 3981	11 56.1	–19 53	Crt	12.1p	5.2'	2.3'	Sbc

WH: very faint, pretty large, irregular figure.
170x250: Moderately faint. Broadly slightly concentrated. A fat edge-on, NNE-SSW. Moderately small. No involved stars. Nearby E, in line with the galaxy, are two stars, moderately faint and faint.

| ga | NGC 3982 | 11 56.5 | +55 07 | UMa | 11.8p | 2.3' | 2.0' | Sb |

HH: bright, pretty large, round, gradually then suddenly brighter middle, disc.
170x250: Moderately bright. Broadly moderately concentrated. Round. Small. No involved stars. SSW is a wide pair of stars.

| ga | UGC 6917 | 11 56.5 | +50 25 | UMa | 13.1b | 3.8' | 2.3' | Sm |

170x250: Not visible.

| ga | NGC 3985 | 11 56.7 | +48 20 | UMa | 13.1b | 1.3' | 0.8' | Sm |

HH: very faint, considerably small, another nebula suspected.
170x250: Moderate brightness. Evenly moderately concentrated. Roughly round. Very small. No involved stars.

| ga | NGC 3987 | 11 57.3 | +25 11 | Leo | 12.9v | 3.2' | 0.5' | Sb |

LR: faint, much extended.
[See 4005.]

| ga | UGC 6930 | 11 57.3 | +49 17 | UMa | 12.7b | 4.3' | 3.5' | Sd |

170x250: Very very faint. Hardly concentrated. Roughly round. Moderate size. No involved stars. Running past its edge W is a NNW-SSE line of three stars; the middle star is fairly bright.

| ga | NGC 3992/M109 | 11 57.6 | +53 22 | UMa | 10.6b | 7.6' | 4.6' | Sbc |

HH: considerably bright, very large, pretty much extended, suddenly brighter middle to a bright resolvable nucleus.
42x60: Visible. **170x250:** Moderate brightness. Sharply fairly concentrated. Elongated ENE-WSW. Large: a quarter of the field in size. It has a very faint halo and a small, brighter core. Alongside its N edge are three stars in a slightly curved line concave away from the galaxy; the middle star is involved; with averted vision the halo almost reaches the other two stars.

| ga | NGC 3998 | 11 57.9 | +55 27 | UMa | 11.6b | 3.0' | 2.4' | S0 |

HH: considerably bright, pretty small, round, very gradually then suddenly much brighter middle.
170x250: Bright; pretty high surface brightness. Sharply very concentrated. Roughly round. Small. No involved stars. A sixth of the field W is another galaxy

[*3990 (HH)]. Moderate brightness. Sharply pretty concentrated. Round. Very small. No involved stars. The two galaxies make a large irregular diamond with two bright stars, one NNW and one S; W of the N star is another bright star.

| ga | NGC 4005 | 11 58.2 | +25 07 | Leo | 13.1v | 1.2' | 0.7' | S |

Sv: pretty faint, very small, much brighter middle, 7th mag. star 2' NW.
170x250: Faint. Evenly moderately concentrated. Round. Very small. No involved stars. A little ways WNW is the bright plotted star. Half the field ESE is a small faint galaxy [*4015 (WH)]. Further ESE is a small very very faint galaxy [*4023 (Dr)]. 3/4 of a field ENE of 4005 is a small faint galaxy [*4022 (Dr)]. It makes a sharp triangle with two moderately faint stars W. 2/3 of a field WNW of 4005 is 3987. Very faint. Elongated ENE-WSW. Small. No involved stars. It makes a triangle with a moderately bright star N and a moderate brightness star a little farther E. Half the field NE is a very faint galaxy [*3997 (JH)] between two moderately faint stars. Between 3987 and this last galaxy is a small extremely faint galaxy [*3993 (LR)]. It makes a little arc with two faint stars NNW. [3989 (LR), 3999 (LR), 4000 (LR), and 4011 (Dr) were not seen.]

| ga | NGC 4008 | 11 58.3 | +28 11 | Leo | 13.0b | 2.4' | 1.2' | E5 |

HH: pretty bright, pretty small, round, pretty suddenly brighter middle, resolvable.
170x250: Moderate brightness. Sharply fairly concentrated. Round. Very small. No involved stars.

| ga | NGC 4013 | 11 58.5 | +43 56 | UMa | 12.2b | 5.2' | 1.3' | Sb |

HH: bright, considerably large, much extended ENE-WSW, very suddenly very much brighter middle to a 10th mag. star.
170x250: Bright to very bright. Right at its center is either a moderate brightness star or a relatively very bright stellar nucleus; if the latter, the galaxy is sharply extremely concentrated. Edge-on ENE-WSW. Moderately small to moderate size. It has a relatively faint halo; the ends fade out very indistinctly. No (other) involved stars. The nucleus is one of the most stellar among all galaxies, and the galaxy is very stellar-looking overall. [The central object is an involved star.]

| ga | NGC 4024 | 11 58.5 | –18 20 | Crv | 11.7v | 2.0' | 1.6' | S0– |

WH: faint, very small, irregular figure, brighter middle.
170x250: Moderately bright. Sharply pretty concentrated. Round. Small. No involved stars. WSW is a large "Y" figure of stars opening up W.

| ga | IC 749 | 11 58.6 | +42 44 | UMa | 12.9b | 2.3' | 1.8' | Scd |

[See I750.]

| ga | NGC 4010 | 11 58.6 | +47 15 | UMa | 13.2b | 4.3' | 0.8' | Sd |

JH: faint, pretty large, much extended, very gradually little brighter middle.
170x250: Faint. Hardly concentrated. A thin edge-on, ENE-WSW. Moderate size. No involved stars. [NNE is a moderate brightness star.]

| ga | NGC 4017 | 11 58.8 | +27 27 | Com | 13.0b | 1.8' | 1.3' | Sbc |

HH: faint, large, extended, gradually brighter E of middle.
170x250: Faint. Broadly slightly concentrated. Slightly elongated E-W. Small. No involved stars. [4016 (LR) was not seen.]

| ga | IC 750 | 11 58.9 | +42 43 | UMa | 12.9b | 2.6' | 1.1' | Sab |

170x250: I750 and I749 are next to each other E-W. I750 is smaller and brighter. Moderate brightness. Evenly moderately concentrated. Edge-on NE-SW. Small. No involved stars. I749 is hardly concentrated. Roughly round. Moderate size. No involved stars. SW of I749 is a bright star that makes a flat triangle with the two galaxies. The galaxies contrast somewhat.

| ga | UGC 6983 | 11 59.1 | +52 42 | UMa | 13.1b | 3.4' | 2.3' | Scd |

170x250: Exceedingly faint. Diffuse. Slightly elongated E-W. Moderate size. No involved stars.

| ga | NGC 4026 | 11 59.4 | +50 57 | UMa | 11.7b | 5.2' | 1.4' | S0 |

HH: very bright, considerably large, much extended NNW-SSE, very suddenly very much brighter middle to a bright nucleus.
170x250: Bright; very high surface brightness. Sharply pretty concentrated. Edge-on N-S. Moderately small. The ends are very faint. No involved stars. SSE is a triangle of moderately faint and faint stars. A very nice little galaxy.

| ga | NGC 4027 | 11 59.5 | –19 15 | Crv | 11.7b | 2.8' | 2.5' | Sdm |

HH: globular cluster, pretty faint, pretty large, round, partially resolved, 16th mag. stars.
170x250: Moderate brightness. Broadly moderately concentrated. Roughly round. Moderately small. At its edge NE is a faint star. [4027A was not seen.]

Chapter 12

12 to 18 Hours: SPRING

| ga | NGC 4030 | 12 00.4 | −01 06 | Vir | 11.4p | 4.6' | 3.2' | Sbc |

HH: considerably bright, large, very little extended, pretty suddenly much brighter middle, bright stars near.
170x250: Moderately bright to bright. Broadly moderately concentrated. Round. Moderately small. No involved stars. It makes a little linear figure with three moderately bright stars: one NNW and two SSW.

| ga | NGC 4035 | 12 00.5 | −15 56 | Crv | 13.2 | 1.2' | 1.0' | Sbc |

HH: extremely faint, pretty large, 9th mag. star NE.
170x250: Very very faint. Hardly concentrated. Roughly round. Small. No involved stars. NNE is the bright plotted star.

| ga | NGC 4032 | 12 00.6 | +20 04 | Com | 12.8b | 1.7' | 1.7' | Im |

HH: pretty faint, pretty large, round, gradually brighter middle, 12th mag. star NE.
170x250: Moderately faint. Broadly slightly concentrated. Round. Small. No involved stars. It makes a triangle with two moderately faint stars, one NNE and another farther NNW.

| ga | NGC 4033 | 12 00.6 | −17 50 | Crv | 12.6b | 2.5' | 1.0' | E6 |

WH: pretty bright, small, little extended, brighter middle.
170x250: Moderately bright. Sharply pretty concentrated. Slightly elongated NE-SW. Small. No involved stars.

| ga | NGC 4037 | 12 01.4 | +13 24 | Com | 12.7p | 2.5' | 2.0' | Sb |

HH: extremely faint, pretty large, round, resolvable.
170x250: Very very faint. Diffuse. Moderately small. No involved stars. E is a bright star.

| ga | NGC 4036 | 12 01.5 | +61 53 | UMa | 11.6b | 4.2' | 1.6' | S0− |

HH: very bright, very large, extended.
170x250: Moderately bright. Evenly pretty concentrated. Very elongated to edge-on E-W. Small. No involved stars.

| ga | NGC 4038 | 12 01.9 | −18 52 | Crv | 10.9p | 3.7' | 1.7' | Sm |

HH: pretty bright, considerably large, round, very gradually brighter middle.
42x60: 4038 and 4039 are visible as one object. ***170x250:*** Regarded as one object: Moderate brightness. Broadly slightly concentrated. Two-lobed N-S, in a fetus shape. Moderate size. No involved stars. Individually: 4038, the N galaxy, is larger and brighter. Elongated E-W. Very nearby WNW is a faint star. 4039 is elongated ENE-WSW. [Ring-Tail Galaxy][†]

| ga | NGC 4039 | 12 01.9 | −18 53 | Crv | 11.1p | 4.0' | 2.2' | Sm |

HH: pretty faint, pretty large.
[See 4038.]

| ga | NGC 4041 | 12 02.2 | +62 08 | UMa | 11.9b | 2.6' | 2.4' | Sbc |

HH: bright, considerably large, round, gradually then pretty suddenly very much brighter middle to a resolvable nucleus.
170x250: Moderate brightness. Evenly moderately concentrated. Round. Very small. No involved stars.

| ga | UGCA 266 | 12 02.4 | −14 31 | Crv | 12.9 | 0.9' | 0.5' | Im |

170x250: Not visible.

| ga | NGC 4045 | 12 02.7 | +01 58 | Vir | 11.9v | 3.2' | 1.3' | Sa |

HH: pretty faint, large, round, suddenly brighter middle, star SE.
170x250: Moderate brightness. Evenly moderately concentrated. Round. Small. No involved stars. Nearby ESE is a moderate brightness star that makes a large triangular figure with three more stars farther E. A little ways S is a very very faint star with a tiny very very faint galaxy right next to it NE [*4045A].

| ga | NGC 4047 | 12 02.9 | +48 38 | UMa | 13.0p | 1.8' | 1.6' | Sb |

HH: pretty bright, pretty small, round.
170x250: Moderate brightness. Broadly slightly concentrated. Round. Very small. No involved stars. WSW is a moderately bright star.

| ga | NGC 4050 | 12 02.9 | −16 22 | Crv | 13.1b | 3.4' | 2.3' | Sab |

WH: faint, considerably large, irregularly round, little brighter middle.
170x250: Very faint. Broadly very slightly concentrated. Elongated E-W. Moderate size. No involved stars.

| ga | NGC 4051 | 12 03.2 | +44 31 | UMa | 10.8b | 5.2' | 4.6' | Sbc |

HH: bright, very large, extended, very gradually then very suddenly much brighter middle to an 11th mag. star.
42x60: Visible with a star at its edge. ***170x250:*** Moderate brightness; somewhat low surface brightness. Sharply extremely concentrated to a stellar nucleus. Elongated NW-SE. Moderately large. No involved stars. A little ways WSW is a moderately bright star; a little farther ENE is a faint star.

| ga | NGC 4068 | 12 04.0 | +52 35 | UMa | 13.0p | 2.8' | 1.7' | Im |

WH: pretty faint, small, stellar.
170x250: Pretty much stellar: it looks like an ordinary moderately faint star with the slightest elongated halo around it. Very nearby SSW is an extremely faint star; W to NNW are four evenly spaced stars in a shallow arc. [The object is a very faint galaxy with a star on top of it.]

| ga | NGC 4062 | 12 04.1 | +31 53 | UMa | 11.9b | 4.0' | 1.7' | Sc |

HH: pretty bright, very large, much extended E-W, very gradually brighter middle.
170x250: Moderate brightness. Broadly slightly concentrated. Elongated E-W. Moderately large. No involved stars.

| ga | NGC 4065 | 12 04.1 | +20 14 | Com | 12.6v | 1.3' | 1.2' | E |

JH: pretty faint, round.
170x250: In the position is a widespread group of galaxies taking up 3/4 of the field. The brightest two [*4061 (HH), 4065] are a close pair at the SW corner of the group. Both are moderately faint. Sharply fairly concentrated. Round. Very small. No involved stars. They make a sharp triangular figure with two faint stars S. ESE is another galaxy [*4076 (HH)]. Faint. Broadly moderately concentrated. Round. Small. No involved stars. N is a very faint galaxy [*4074 (HH)]. Evenly moderately concentrated. Round. Very small. No involved stars. A little ways WNW is a moderate brightness star. WNW is a moderately faint galaxy [*4066 (JH)]. Evenly to sharply moderately concentrated. Round. Small. No involved stars. A little ways WSW is a small extremely faint galaxy [*4060 (Ma)]. NNE is a moderately faint galaxy [*4070 (HH)]. Evenly to sharply moderately concentrated. Round. Very small. No involved stars. S are a couple of faint stars. [4072 (LR) was not seen.]

| ga | NGC 4064 | 12 04.2 | +18 26 | Com | 12.2b | 4.3' | 1.7' | Sa |

dA: bright, extended, gradually brighter middle.
170x250: Moderately faint. Broadly slightly concentrated. Slightly elongated NNW-SSE. Small. No involved stars. Nearby SW is a faint star; a little ways E is another faint star.

| ds | 2 Com | 12 04.3 | +21 28 | Com | 3.7" | 5.9 | 7.4 | F0 A9 |

42x60: Slightly elongated. ***170x250:*** A very close unequal pair. Both yellow white. [STF 1596]

| ga | NGC 4073 | 12 04.5 | +01 53 | Vir | 11.4v | 3.4' | 2.3' | E+ |

HH: faint, pretty small, round, pretty gradually brighter middle, NW of 2.
170x250: Moderately faint. Evenly moderately concentrated. Slightly elongated E-W. Small. No involved stars. A little ways SSE is a moderately faint star. A third of the field SSE are two more galaxies aligned NW-SE [*4139 (dA), *4077 (HH)]. Both are faint. The SE galaxy is elongated N-S. Small. The NW galaxy is smaller and fainter.

| ga | NGC 4079 | 12 04.8 | −02 22 | Vir | 13.2p | 2.2' | 1.5' | Sbc |

JH: faint, large, round, 10th mag. star 1' N.
170x250: Very faint to faint. Broadly very slightly concentrated. Round. Small. No involved stars. Nearby N is a moderately faint star; NNW are two brighter stars.

| ga | NGC 4085 | 12 05.4 | +50 21 | UMa | 13.0b | 2.8' | 0.9' | Sc |

WH: bright, pretty large, pretty much extended, very suddenly brighter middle.
170x250: Moderately faint. Broadly slightly concentrated. Edge-on ENE-WSW. Small. No involved stars. [WNW is a moderate brightness star.]

| ga | NGC 4087 | 12 05.6 | −26 31 | Hya | 13.1p | 2.1' | 1.7' | S0− |

HH: pretty bright, small, round, brighter middle.
170x250: Moderately faint. Evenly to sharply fairly concentrated. Round. Very small. No involved stars.

| ga | NGC 4088 | 12 05.6 | +50 32 | UMa | 11.2b | 5.3' | 2.1' | Sbc |

WH: bright, considerably large, pretty much extended [NE-SW], little brighter middle.
42x60: Visible. *170x250:* Moderate brightness. Broadly slightly concentrated. Very elongated NE-SW; a little irregular in shape. Large: a little less than a quarter of the field in size. No involved stars.

| ga | IC 2995 | 12 05.8 | −27 56 | Hya | 12.9p | 3.2' | 0.9' | Sc |

170x250: Very faint. Hardly concentrated. Nearly edge-on WNW-ESE. Moderately small. No involved stars. Very nearby WSW of center is a very faint star; E is a long shallow N-S arc of bright stars.

| ga | NGC 4094 | 12 05.9 | −14 31 | Crv | 12.5p | 4.2' | 1.4' | Scd |

JH: extremely faint, large, pretty much extended, very gradually brighter middle, 2 11th mag. stars near.
170x250: Very faint. Broadly very slightly concentrated. Elongated ENE-WSW. Moderate size. No involved stars. It makes a small triangle with two moderately bright stars, one N and one E.

| ga | NGC 4096 | 12 06.0 | +47 28 | UMa | 11.5b | 7.4' | 1.7' | Sc |

HH: pretty bright, very large, much extended NNE-SSW.
170x250: Moderate brightness. Broadly slightly concentrated. Edge-on NNE-SSW. Large: a quarter of the field in size. No involved stars. NW is a wide unequal double star.

| ga | UGCA 270 | 12 06.1 | −22 50 | Crv | 13.0p | 2.8' | 2.0' | Sm |

170x250: Not visible.

| ga | NGC 4100 | 12 06.2 | +49 34 | UMa | 11.9b | 5.4' | 2.0' | Sbc |

HH: pretty bright, very large, very much extended NNW-SSE, very gradually very little brighter middle.
170x250: Moderate brightness. Hardly concentrated. Edge-on NNW-SSE. Moderate size. No involved stars.

| ga | NGC 4102 | 12 06.4 | +52 42 | UMa | 12.0b | 3.2' | 1.7' | Sb |

HH: bright, pretty small, round, brighter middle to a bright resolvable nucleus, 12th mag. star very near SW.
170x250: Moderate brightness to moderately bright. Sharply pretty concentrated. Elongated NE-SW. Moderately small. Off its edge W is a moderate brightness star; in line with the galaxy and this star are two faint stars ENE.

| ga | NGC 4104 | 12 06.6 | +28 10 | Com | 13.1b | 3.2' | 1.6' | S0 |

HH: pretty bright, pretty small, little extended, brighter middle.
170x250: Moderately faint. Broadly slightly concentrated. Roughly round. Small. No involved stars. A little ways SW is a very small extremely faint galaxy [*C158-23].

| ga | NGC 4105 | 12 06.7 | −29 45 | Hya | 11.6b | 2.7' | 2.0' | E3 |

HH: pretty faint, pretty small, round, pretty suddenly brighter middle, resolvable, W of 2.
170x250: 4105 and 4106 are a nice double galaxy aligned WNW-ESE. 4105 is very slightly brighter. Both are moderately bright. Sharply very concentrated. Round. Small. No involved stars. They make a sharp triangle with a moderately bright star S.

| ga | NGC 4108 | 12 06.7 | +67 09 | Dra | 13.0p | 1.7' | 1.4' | Sc |

JH: bright, small, round, gradually brighter middle.
170x250: Moderately faint. Broadly to evenly moderately concentrated. Round. Small. No involved stars. [A little ways SE is a moderate brightness star.] [4108B was not seen.]

| ga | NGC 4106 | 12 06.8 | −29 46 | Hya | 12.4b | 1.6' | 1.2' | S0+ |

HH: pretty faint, pretty small, round, pretty gradually brighter middle, E of 2. [See 4105.]

| ga | NGC 4111 | 12 07.1 | +43 04 | CVn | 11.6b | 5.2' | 1.2' | S0+ |

HH: very bright, pretty small, much extended NNW-SSE.
170x250: Bright. Sharply pretty concentrated. A thin edge-on, NNW-SSE. Moderate size. No involved stars. Half the field ENE, past a bright star with a companion, is a small very faint galaxy [*4117 (HH)]. A quarter of the field SSW is a very small very very faint galaxy [*4109 (LR)]. Nearby NNW is a faint star.

| ga | NGC 4116 | 12 07.6 | +02 41 | Vir | 12.4b | 3.8' | 2.1' | Sdm |

LR: very faint, extended (hook shape), SW of 2.
170x250: Very faint. Broadly very slightly concentrated. Roughly round. Small. No involved stars. It's in a linear figure with faint stars E and WSW.

| ga | IC 3010 | 12 08.0 | −30 20 | Hya | 13.2b | 1.9' | 1.7' | S0+ |

170x250: Faint. Evenly slightly concentrated. Round. Small. No involved stars. A little ways WSW is a faint star; going SSW from the galaxy is a long line of faint stars.

| ga | NGC 4125 | 12 08.1 | +65 10 | Dra | 10.7b | 5.7' | 3.1' | E6 |

Hi: pretty bright, pretty large, considerably extended, much brighter middle.
42x60: Visible. ***170x250:*** Moderately bright. Evenly moderately concentrated. Slightly elongated E-W. Small. No involved stars. A little ways ESE is a bright star. [4121 (dA) was not seen.]

| ga | NGC 4123 | 12 08.2 | +02 52 | Vir | 12.0b | 4.3' | 3.1' | Sc |

HH: considerably faint, very large, extended E-W, brighter middle to a 16th mag. star.
170x250: Faint. Broadly very slightly concentrated. Slightly elongated WNW-ESE. Moderately small. No involved stars. It makes a triangle with two faint stars S.

| ga | NGC 4124 | 12 08.2 | +10 22 | Vir | 12.2b | 4.2' | 1.3' | S0+ |

HH: pretty bright, pretty large, much extended WNW-ESE, brighter middle, resolvable.
170x250: Moderately faint. Broadly slightly concentrated. Edge-on WNW-ESE. Moderate size. No involved stars. A little ways NW is a faint star.

| ga | NGC 4128 | 12 08.5 | +68 46 | Dra | 12.9b | 2.6' | 0.8' | S0 |

WH: considerably bright, little extended, brighter middle.
170x250: Moderate brightness. Evenly to sharply fairly concentrated. Round. Very very small. No involved stars. [NNW and SSE are several moderate brightness stars.]

| ga | NGC 4133 | 12 08.6 | +74 54 | Dra | 13.1p | 1.8' | 1.3' | Sb |

HH: pretty bright, considerably large, round, gradually much brighter middle.
170x250: Faint. Broadly slightly concentrated. Roughly round. Small. No involved stars. [A little ways E is a moderately faint star; a little ways N is another moderately faint star; WNW is a moderate brightness star.]

| ga | NGC 4129 | 12 08.9 | −09 02 | Vir | 13.1b | 2.3' | 0.5' | Sab |

HH: faint, pretty large, pretty much extended E-W, very gradually little brighter middle.
170x250: Faint. Broadly slightly concentrated. Edge-on E-W. Moderately small. No involved stars. NW is a moderate brightness star; going SW from this star are more stars in a long crooked line.

| ga | IC 3015 | 12 09.0 | −31 31 | Hya | 13.1p | 2.8' | 0.6' | Sbc |

170x250: Moderately faint to moderate brightness. Broadly slightly concentrated. Elongated NNW-SSE. Small. Right at its SE end is a faint star.

| ga | NGC 4136 | 12 09.3 | +29 55 | Com | 11.7p | 4.0' | 4.0' | Sc |

HH: faint, very large, very gradually much brighter middle.
170x250: Faint; low surface brightness. Hardly concentrated. Roughly round. Moderately small. No involved stars.

| ga | NGC 4138 | 12 09.5 | +43 41 | CVn | 12.2b | 3.0' | 2.4' | S0+ |

WH: bright, pretty large, little extended, very gradually brighter middle, star NW.
170x250: Moderately bright. Evenly moderately concentrated. Round. Small. No involved stars. A little ways NNW is a moderate brightness star.

| ga | NGC 4143 | 12 09.6 | +42 32 | CVn | 11.7b | 2.4' | 1.8' | S0 |

HH: considerably bright, round, very gradually then very suddenly brighter middle to a nucleus.
170x250: Moderately bright. Sharply very concentrated to a substellar nucleus. Slightly elongated NW-SE. Small. No involved stars.

| ga | NGC 4144 | 12 10.0 | +46 27 | UMa | 12.1b | 6.1' | 1.3' | Scd |

HH: pretty faint, considerably large, very much extended WNW-ESE, very gradually brighter middle.
170x250: Moderate brightness. Broadly slightly concentrated. Edge-on WNW-ESE. Moderately large. No involved stars. Just S of its E end are a faint star and a faint double star.

| ga | NGC 4145 | 12 10.0 | +39 53 | CVn | 11.8b | 5.8' | 4.2' | Sd |

HH: bright, very large, very gradually little brighter middle.
170x250: Very faint to faint; low surface brightness. Hardly concentrated. Roughly round. Moderately large. No involved stars. Off its edge SSW is a very faint star.

| bs | Epsilon Crv | 12 10.1 | −22 37 | Crv | 3.00 | | 1.33 | K2 |

170x250: Yellow gold.

| gc | NGC 4147 | 12 10.1 | +18 33 | Com | 10.3 | 4.4' | −6.02 | 6 |

HH: globular cluster, very bright, pretty large, round, gradually brighter middle, well resolved.
170x250: Granular. Evenly concentrated. Very small. Fairly regular.

| ga | IC 764 | 12 10.2 | −29 44 | Hya | 12.8p | 4.8' | 1.6' | Sc |

170x250: Very very faint. Hardly concentrated. Elongated N-S. Moderate size. No involved stars. ENE is a shallow arc of faint stars; WNW is a faint star.

| ga | NGC 4151 | 12 10.5 | +39 24 | CVn | 11.5b | 6.5' | 5.0' | Sab |

HH: very bright, small, round, very suddenly much brighter middle to a bright nucleus, W of 2.
170x250: Moderately bright. Sharply extremely concentrated to a substellar nucleus. The faint halo is slightly elongated NW-SE. Moderately small. The galaxy makes a small triangular figure with nearby stars: a very faint star WSW, another very faint star NNW, and a moderately faint star farther N. A little less than a third of the field NE is another galaxy [*4156 (HH)]. Faint. Hardly concentrated. Round. Small. No involved stars.

| ga | NGC 4150 | 12 10.6 | +30 24 | Com | 12.4b | 2.3' | 1.5' | S0 |

HH: bright, small, round, pretty gradually much brighter middle.
170x250: Moderately bright. Sharply very concentrated. Round. Very small. No involved stars.

| ga | NGC 4152 | 12 10.6 | +16 02 | Com | 12.7b | 2.5' | 2.0' | Sc |

HH: pretty bright, pretty large, round, pretty gradually much brighter middle, resolvable.
170x250: Faint. Broadly to evenly slightly concentrated. Round. Small. No involved stars.

| ga | NGC 4157 | 12 11.1 | +50 29 | UMa | 12.2b | 7.7' | 1.3' | Sb |

HH: pretty faint, considerably large, very much extended ENE-WSW.
170x250: Moderately faint. Broadly slightly concentrated. A fairly thin edge-on, ENE-WSW. Large: a quarter of the field in size. No involved stars. [SW is a small triangle of stars; farther NNW is the bright plotted star.]

| ga | NGC 4158 | 12 11.2 | +20 10 | Com | 12.9p | 1.9' | 1.6' | Sb |

HH: faint, pretty small, little extended, brighter middle, pretty bright star [SE].
170x250: Moderately faint. Broadly moderately concentrated. Roughly round. Small. No involved stars. [Nearby ESE is a moderately bright star.]

| ga | NGC 4162 | 12 11.9 | +24 07 | Com | 12.9b | 2.3' | 1.3' | Sbc |

HH: bright, large, irregularly extended, brighter middle.
170x250: Moderate brightness. Broadly to evenly moderately concentrated. Round. Small. No involved stars. It's right in between a bright star WSW and a fainter star ENE.

| ga | NGC 4169 | 12 12.2 | +29 10 | Com | 13.2b | 2.3' | 1.2' | S0 |

HH: faint, small, 1st of 4.
170x250: In the position are four galaxies in a small group. The brightest, 4169, is moderate brightness. Sharply pretty concentrated. Round. Very small. NNE is an extremely faint galaxy [*4170 (dA)]. Unconcentrated. Edge-on NW-SE. Moderately large. SE of 4169 is a very small nearly stellar galaxy [*4174 (HH)]. A little farther ESE of 4169 is a very faint galaxy [*4175 (HH)]. Broadly slightly concentrated. Edge-on NW-SE. Small. There are no involved stars with any of the galaxies.

| ga | NGC 4168 | 12 12.3 | +13 12 | Vir | 12.1b | 3.0' | 2.6' | E2 |

HH: pretty bright, pretty large, irregular figure, pretty suddenly brighter middle, resolvable, star involved.
170x250: Moderate brightness. Evenly moderately concentrated. Round. Small. No involved stars. NW is a moderately bright star. Between 4168 and this star, making a shallow arc with them, is an exceedingly faint galaxy [*4165 (dA)]. Diffuse. Small. [4164 (Te) was not seen.]

| ga | NGC 4178 | 12 12.8 | +10 51 | Vir | 11.9b | 5.1' | 1.8' | Sdm |

JH: very faint, very large, extended NE-SW, 7th mag. star E.
170x250: Very faint to faint. Hardly concentrated. Edge-on NNE-SSW. Large: a quarter of the field in size. No involved stars.

| ga | NGC 4179 | 12 12.9 | +01 18 | Vir | 11.9b | 4.0' | 1.1' | S0 |

HH: pretty bright, pretty small, pretty much extended NW-SE, brighter middle to a nucleus.
170x250: Moderately bright. Evenly to sharply pretty concentrated. Edge-on NW-SE. Moderately small. No involved stars. It's in a long shallow arc with several stars, moderate brightness and fainter, mostly NE.

| ga | NGC 4183 | 12 13.3 | +43 41 | CVn | 12.9b | 6.3' | 0.8' | Scd |

HH: very faint, considerably large, much extended N-S.
170x250: Very faint. Hardly concentrated. A thin edge-on, NNW-SSE. Moderate size. Toward its S end is an extremely faint involved star; a little ways S is a faint star; a little ways WSW is a very faint star.

| ga | NGC 4185 | 12 13.4 | +28 30 | Com | 12.9p | 2.6' | 1.9' | Sbc |

HH: considerably faint, large, round, gradually brighter middle.
170x250: Faint. Broadly very slightly concentrated. Roughly round. Small. No involved stars. A little ways SE is a very faint star; the galaxy makes a large isosceles triangle with two moderate brightness stars, one E and one NNE.

| ga | NGC 4189 | 12 13.8 | +13 25 | Com | 11.7v | 2.5' | 2.0' | Scd |

HH: faint, large, little extended, very gradually little brighter middle, resolvable.
170x250: Faint. Broadly very slightly concentrated. Round. Moderately small to moderate size. No involved stars. A little ways ENE is a faint star; NNE is a very faint star; SSE are a couple more very faint stars.

| ga | NGC 4192/M98 | 12 13.8 | +14 54 | Com | 11.0b | 9.8' | 2.7' | Sab |

JH: bright, very large, very much extended NNW-SSE, very suddenly very much brighter middle.
42x60: Very faintly visible. *170x250:* Moderately faint. Sharply slightly concentrated to a tiny nucleus. Edge-on NNW-SSE. Large: a third of the field in size. No involved stars. A little ways N of center is a faint star.

| ga | NGC 4193 | 12 13.9 | +13 10 | Vir | 12.3v | 2.3' | 1.1' | Sc |

HH: very faint, pretty large, extended, very gradually brighter middle.
170x250: Faint. Broadly slightly concentrated. Elongated E-W. Moderately small. No involved stars. A little ways NW is a very faint star.

| ga | NGC 4194 | 12 14.2 | +54 31 | UMa | 12.5v | 2.8' | 2.1' | Im |

HH: pretty bright, very small, very suddenly brighter middle to a 12th mag. star.
170x250: Moderately bright; fairly high surface brightness. Sharply pretty concentrated. Round. Very very small. No involved stars. [It makes a triangle with two stars a little ways N; SSW is another star.]

| ga | NGC 4203 | 12 15.1 | +33 11 | Com | 11.8b | 3.8' | 3.8' | S0– |

HH: very bright, small, round, pretty suddenly much brighter middle.
170x250: Moderately bright. Sharply very concentrated. Round. Small. It has a very faint outer halo. No involved stars.

| ga | NGC 4204 | 12 15.2 | +20 39 | Com | 12.9p | 3.6' | 2.8' | Sdm |

HH: very faint, considerably large, irregularly round, very gradually brighter middle.
170x250: Extremely faint. A slight unconcentrated glow. Elongated NW-SE. Moderate size. At its SE end is an extremely faint star.

| ga | NGC 4206 | 12 15.3 | +13 01 | Vir | 12.8b | 6.6' | 1.1' | Sbc |

WH: faint, very much extended.
170x250: Very faint. Hardly concentrated. Edge-on N-S. Moderately large. No involved stars.

| bs | Delta UMa | 12 15.4 | +57 01 | UMa | 3.31 | | 0.08 | A3 |

170x250: White.

| ga | NGC 4214 | 12 15.6 | +36 19 | CVn | 10.2b | 7.4' | 6.5' | Im |

HH: considerably bright, considerably large, irregularly extended, binuclear.
42x60: Visible. *170x250:* Moderately bright. Evenly moderately concentrated. Slightly elongated NW-SE. Moderately large. It has an extremely faint outer halo. No involved stars. Passing near it E to NNW is a line of widely spaced very faint stars; SE is a moderate brightness star.

| ga | NGC 4212 | 12 15.7 | +13 54 | Com | 11.8b | 3.8' | 2.1' | Sc |

HH: bright, large, extended E-W, gradually then suddenly brighter middle, resolvable.
170x250: Moderately faint. Broadly very slightly concentrated. Slightly elongated ENE-WSW. Small. No involved stars. A little ways S is a moderate brightness star.

| bs | Gamma Crv | 12 15.8 | −17 32 | Crv | 2.59 | | −0.11 | B8 |

170x250: White.

| ga | NGC 4217 | 12 15.8 | +47 05 | CVn | 11.1v | 5.7' | 1.6' | Sb |

HH: pretty faint, large, much extended NE-SW, double star N, W of 2.
170x250: Faint. Broadly very slightly concentrated. Elongated NE-SW. Moderate size. At its edge N of center is a moderate brightness star; a little ways NNE is a bright star; at its SW end is a very faint star; a little ways S is a moderately faint star; the two brighter stars N interfere. A little less than half the field ESE is a small very faint galaxy [*4226 (JH)].

| ga | NGC 4215 | 12 15.9 | +06 24 | Vir | 13.0b | 1.8' | 0.6' | S0+ |

HH: bright, pretty small, extended, suddenly brighter middle to an 11th mag. star.
170x250: Moderate brightness. Evenly to sharply fairly concentrated. Slightly elongated N-S. Small. No involved stars. A little ways NW is a moderate brightness star.

| ga | NGC 4216 | 12 15.9 | +13 08 | Vir | 11.0b | 8.7' | 1.7' | Sb |

HH: very bright, very large, very much extended NNE-SSW, suddenly brighter middle to a nucleus.
42x60: Visible. *170x250:* Bright. Sharply pretty concentrated to a small core. A thin edge-on, NNE-SSW. Large: a third of the field in size. The E edge is a little sharper; this would be where the dust lane is. At this edge, E of center, right opposite the core, is an extremely faint star.

| ga | NGC 4220 | 12 16.2 | +47 53 | CVn | 12.3b | 3.9' | 1.3' | S0+ |

HH: considerably bright, pretty large, pretty much extended NW-SE, pretty suddenly brighter middle.
170x250: Moderate brightness. Broadly slightly concentrated. Very elongated to edge-on NW-SE. Moderately small. No involved stars. Nearby WSW is a very faint star.

| ga | NGC 4224 | 12 16.6 | +07 27 | Vir | 12.9b | 2.5' | 0.9' | Sa |

HH: pretty bright, pretty small, little extended, gradually brighter middle, resolvable.
170x250: Moderately faint. Evenly moderately concentrated. Slightly elongated ENE-WSW. Small. No involved stars. Nearby N is a faint star; nearby WSW is another faint star.

| ga | NGC 4236 | 12 16.7 | +69 27 | Dra | 10.1b | 22.0' | 7.2' | Sdm |

HH: very faint, extremely large, much extended NNW-SSE, very gradually brighter middle.
170x250: Very faint; low surface brightness. Unconcentrated. Edge-on NNW-SSE. Very large: a narrow glow all the way across the field. No conspicuous involved stars. Near its S end is a bright star.

| ga | NGC 4233 | 12 17.1 | +07 37 | Vir | 12.9b | 2.3' | 0.8' | S0 |

HH: pretty faint, round, very suddenly brighter middle to a small nucleus.
170x250: Moderate brightness. Sharply pretty concentrated. Slightly elongated N-S. Very small. The outer halo is very faint. No involved stars. A little ways NE is a moderate brightness star.

| ga | NGC 4235 | 12 17.2 | +07 11 | Vir | 12.6b | 4.1' | 0.8' | Sa |

HH: pretty bright, pretty large, pretty much extended, little brighter middle, W of 2.
170x250: Moderate brightness. Evenly moderately concentrated. Pretty elongated NE-SW. Moderately small. No involved stars. Nearby N is a faint star; a little ways NW is a moderately faint star; a little ways ENE is a very faint star; the galaxy and these stars form a narrow E-W diamond.

| ga | NGC 4237 | 12 17.2 | +15 19 | Com | 11.6v | 2.5' | 1.7' | Sbc |

HH: pretty bright, pretty large, little extended, very gradually brighter middle, resolvable.
170x250: Moderately faint to moderate brightness. Broadly moderately concentrated. Slightly elongated WNW-ESE. Small. No involved stars.

| ga | NGC 4241 | 12 17.4 | +06 41 | Vir | 13.0b | 2.5' | 1.3' | S0+ |

HH: very faint, large, very gradually brighter middle, 7th mag. star S.
170x250: Moderately faint. Evenly moderately concentrated. Roughly round. Small. No involved stars. A little ways SSW is a moderate brightness star; S is a relatively bright star.

| ga | NGC 4250 | 12 17.4 | +70 48 | Dra | 12.8p | 2.7' | 2.1' | S0+ |

HH: pretty bright, small, round, pretty gradually brighter middle.
170x250: Moderately faint to moderate brightness. Evenly to sharply fairly concentrated. Round. Very very small. No involved stars.

| ga | NGC 4242 | 12 17.5 | +45 37 | CVn | 11.4b | 5.8' | 3.7' | Sdm |

HH: very faint, considerably large, irregularly round, very gradually brighter middle, resolvable.
170x250: Very faint. Diffuse. Roughly round. Moderate size. At its edge S is a faint star; E is a moderate brightness star.

| ga | NGC 4244 | 12 17.5 | +37 48 | CVn | 10.3v | 17.7' | 1.9' | Scd |

HH: pretty bright, very large, extremely extended NE-SW, very gradually brighter middle.
42x60: Very faintly visible. ***170x250:*** Moderate brightness; fairly low surface brightness. Broadly very slightly concentrated. A thin edge-on, NE-SW. Very large: half the field in size. The ends fade out indistinctly. No involved stars. Crossing its NE end are four stars in a long crooked WNW-ESE line; occasionally with averted vision the galaxy reaches and extends a little past the nearest of these stars.†

| ga | NGC 4245 | 12 17.6 | +29 36 | Com | 12.3b | 3.0' | 2.6' | S0/a |

HH: considerably bright, pretty large, very little extended, suddenly much brighter middle, resolvable.
170x250: Moderate brightness to moderately bright. Sharply pretty concentrated. Round. Small. No involved stars. It makes a long shallow arc with five moderately bright and moderate brightness stars W.

| ga | NGC 4248 | 12 17.8 | +47 24 | CVn | 12.5v | 3.1' | 1.1' | I0 |

HH: very faint, small, pretty much extended, pretty suddenly brighter middle.
170x250: Very faint. Hardly concentrated. Elongated WNW-ESE. Moderately small. At its W end is a faint involved star.

| ga | NGC 4251 | 12 18.1 | +28 10 | Com | 11.6b | 3.6' | 1.4' | S0 |

HH: very bright, small, extended, very suddenly very much brighter middle to a nucleus, 6th-7th mag. star E.
170x250: Very bright. Sharply very concentrated to a very bright core. A fat edge-on, E-W. Moderately small. The ends are very indistinct. No involved stars. A little ways ESE is a faint star.

| ga | NGC 4256 | 12 18.7 | +65 53 | Dra | 12.7p | 4.5' | 0.7' | Sb |

HH: pretty bright, large, considerably extended NE-SW, brighter middle to a bright nucleus.
170x250: Moderately faint to moderate brightness. Evenly to sharply fairly concentrated. A thin edge-on, NE-SW. Moderate size. No involved stars.

| ga | NGC 4254/M99 | 12 18.8 | +14 25 | Com | 10.4b | 5.4' | 4.7' | Sc |

HH: !! bright, large, round, gradually brighter middle, resolvable, (Le) 3-branched spiral.
42x60: Visible. ***170x250:*** Moderately bright. Evenly moderately concentrated. Round. Moderate size. Barely visible is a spiral arm coming off its edge S and turning W. No involved stars. Off its edge SE is a faint star.

| ga | NGC 4258/M106 | 12 19.0 | +47 18 | CVn | 9.1b | 18.8' | 7.3' | Sb |

HH: very bright, very large, very much extended N-S, suddenly brighter middle to a bright nucleus.
12x35: Visible. ***42x60:*** Bright. ***170x250:*** Very bright. Sharply very concentrated to a very small bright core. Elongated NNW-SSE. Very large: a little more than half the field in size. The outer halo is very indistinct. Relatively close to center on the N side is a broad spiral arm that extends straight out and points toward an involved star right at the galaxy's NW end. Another spiral arm might be visible on the S side, but this is less certain. One of the nicest galaxies.

| ga | NGC 4260 | 12 19.4 | +06 05 | Vir | 11.8v | 3.9' | 1.7' | Sa |

HH: pretty bright, extended, pretty suddenly brighter middle.
170x250: Moderate brightness. Evenly moderately concentrated. Slightly elongated NE-SW. Small. No involved stars. Making a shallow arc with it are an extremely faint star just off its edge NNE and a very faint star a little ways NE.

| ga | NGC 4261 | 12 19.4 | +05 49 | Vir | 10.4v | 4.3' | 3.5' | E2-3 |

HH: faint, pretty small, round, gradually brighter middle.

42x60: Visible. *170x250:* Bright. Sharply pretty concentrated. Round. Small. No involved stars. A sixth of the field ENE is 4264. Faint. Evenly moderately concentrated. Round. Very small. No involved stars.

ga	NGC 4262	12 19.5	+14 52	Com	11.5v	2.0'	1.8'	S0–

HH: bright, small, round, resolvable.
170x250: Moderately bright; very high surface brightness. Sharply extremely concentrated to a substellar nucleus. Round. Very very small. No involved stars.

ga	NGC 4264	12 19.6	+05 50	Vir	12.8v	1.1'	1.0'	S0+

HH: faint, pretty small, round, gradually brighter middle.
[See 4261.]

ga	NGC 4263	12 19.7	–12 13	Crv	12.6	1.1'	0.5'	Sb

WH: very faint, pretty large, irregular figure.
170x250: Very faint. Broadly slightly concentrated. Roughly round. Small. No involved stars. SE is a little trio of moderately faint stars.

ga	NGC 4267	12 19.8	+12 47	Vir	11.9b	3.2'	2.9'	S0–

HH: pretty bright, very small, round, very suddenly much brighter middle.
170x250: Moderately bright. Sharply very concentrated. Round. Small. It has an extremely faint extended halo. No involved stars.

ga	NGC 4268	12 19.8	+05 17	Vir	12.8v	2.0'	0.9'	S0/a

Sf: pretty faint, small, 2nd of 6 nebulae.
[See 4281.]

ga	NGC 4270	12 19.8	+05 27	Vir	13.1b	2.3'	1.0'	S0

JH: no description.
[See 4281.]

ga	NGC 4274	12 19.8	+29 36	Com	11.3b	6.8'	2.5'	Sab

HH: very bright, very large, extended E-W, much brighter middle to a nucleus.
170x250: Moderately bright. Evenly fairly concentrated. Very elongated WNW-ESE. Large: a quarter of the field in size. The ends fade out very very indistinctly. No involved stars.

| ga | NGC 4273 | 12 19.9 | +05 20 | Vir | 11.9v | 2.4' | 1.5' | Sc |

HH: bright, large, extended, gradually brighter middle.
[See 4281.]

| ga | NGC 4278 | 12 20.1 | +29 16 | Com | 11.1b | 4.0' | 4.0' | E1-2 |

HH: very bright, pretty large, round, much brighter middle, resolvable, W of 2.
42x60: 4278 is easily visible. ***170x250:*** 4283 and 4278 are next to each other ENE-WSW. 4278 is bright to very bright. Sharply pretty concentrated. Round. Small. No involved stars. 4283 is a slightly smaller and fainter version of 4278. Moderate brightness. Sharply very concentrated. Round. Very small. No involved stars. [4286 (WH) was not seen.]

| ga | NGC 4283 | 12 20.3 | +29 18 | Com | 13.0b | 1.5' | 1.5' | E0 |

HH: bright, small, round, brighter middle, 2nd of 3.
[See 4278.]

| ga | NGC 4291 | 12 20.3 | +75 22 | Dra | 12.4b | 1.9' | 1.5' | E |

HH: pretty bright, very small, round, little brighter middle, 3 stars E.
170x250: Moderately bright. Sharply very concentrated. Round. Very small. Just off its edge W is a faint star; the galaxy makes a small quadrilateral with a moderately bright star E and two moderate brightness stars SE. A third of the field ESE is 4319. Moderately faint. Evenly moderately concentrated. Round. Very small. No involved stars. Very nearby S is an extremely faint star [this is *MKN205, a stellar galaxy]; the galaxy makes a perfect isosceles triangle with two moderately faint stars, one E and one SE.

| ga | NGC 4281 | 12 20.4 | +05 23 | Vir | 12.3b | 3.2' | 1.6' | S0+ |

JH: bright, pretty large, little extended, brighter middle, 4th of 4.
170x250: 4268, 4270, 4273, 4281, and a fifth galaxy [*4259 (JH)] are in a diamond figure the field width in size. Going clockwise from the western point of the diamond: The fifth galaxy is very faint. Slightly concentrated. Very small. 4270 is moderately faint. Evenly moderately concentrated. Slightly elongated WNW-ESE. Small. 4281 is moderately bright. Evenly moderately concentrated. Slightly elongated E-W. Small. 4273 is moderate brightness. Evenly moderately concentrated. Slightly elongated N-S. Small. 4268 is moderately faint. Evenly moderately concentrated. Slightly elongated NE-SW. Very small. Very nearby WNW is a very faint star. There are no involved stars with any of the galaxies. One of the nicer groups of galaxies. [4266 (Ma) and 4277 (HH) were not seen.]

| ga | NGC 4290 | 12 20.8 | +58 05 | UMa | 12.7p | 2.3' | 1.5' | Sab |

HH: pretty bright, large, round, gradually much brighter middle.
170x250: Moderately faint. Broadly slightly concentrated. Slightly elongated NNE-SSW. Small. No involved stars. [4284 (WH) was not seen.]

| ga | NGC 4293 | 12 21.2 | +18 22 | Com | 11.3b | 5.6' | 2.5' | S0/a |

HH: faint, very large, extended, little brighter middle, resolvable.
170x250: Faint. Broadly very slightly concentrated. Edge-on ENE-WSW. Moderate size. No involved stars. Nearby SW of center is an extremely faint star; near the galaxy's E end is a triangle of stars; a little ways N is a star.

| ga | NGC 4292 | 12 21.3 | +04 35 | Vir | 12.2v | 1.7' | 1.2' | S0 |

JH: faint, small, round, very gradually little brighter middle, bright star 60" NNW.
170x250: Very faint. Broadly slightly concentrated. Round. Very small. No involved stars. Nearby NNW is a moderate brightness star.

| ga | NGC 4294 | 12 21.3 | +11 30 | Vir | 12.5b | 3.2' | 1.2' | Scd |

HH: faint, large, much extended NW-SE, binuclear, W of 2.
170x250: Faint. Broadly very slightly concentrated. Elongated NNW-SSE. Moderately small. No involved stars. A third of the field E is 4299. Very faint. Broadly very slightly concentrated. Round. Small. No involved stars. Nearby WSW are two very faint stars; nearby SE is another very faint star; further SE is a large group of moderate brightness stars.

| ga | NGC 4298 | 12 21.5 | +14 36 | Com | 11.3v | 3.0' | 1.8' | Sc |

HH: faint, large, extended N-S, very gradually brighter middle, W of 2.
170x250: 4302 and 4298 are right next to each other E-W. 4298 is moderately faint. Broadly slightly concentrated. Slightly elongated NW-SE. Moderately small. At its edge ENE is a moderately faint star. 4302 is faint. Hardly concentrated. A thin edge-on, N-S. Large: nearly a quarter of the field in size. At its N end is a very faint star; a little ways NE is a faint star; the three stars associated with the two galaxies form an ENE-WSW arc.

| ga | NGC 4299 | 12 21.7 | +11 30 | Vir | 12.9b | 1.7' | 1.5' | Sdm |

HH: faint, large, little extended, very gradually brighter middle, E of 2.
[See 4294.]

| ga | NGC 4302 | 12 21.7 | +14 35 | Com | 11.6v | 5.8' | 0.7' | Sc |

HH: large, very much extended N-S, E of 2.
[See 4298.]

| ga | NGC 4319 | 12 21.7 | +75 19 | Dra | 12.8p | 2.9' | 2.3' | Sab |

HH: pretty bright, pretty small, very little extended, suddenly brighter middle.
[See 4291.]

| ga | NGC 4303/M61 | 12 21.9 | +04 28 | Vir | 10.2b | 6.5' | 5.7' | Sbc |

HH: very bright, very large, very suddenly brighter middle to a star, bi-nuclear.
42x60: Visible. ***170x250:*** Moderately bright. Hardly concentrated, except that it has a moderately faint substellar nucleus. Round. Moderate size. There's some vague unevenness in the light. At its edge W is a faint star; a little ways SW is a moderately faint star. Half the field NE is a small very faint galaxy [*4303A]. A little ways W is a moderately faint star.

| ga | NGC 4307 | 12 22.1 | +09 02 | Vir | 12.8b | 3.6' | 0.7' | Sb |

Pe: pretty faint, large, much extended, 3 knots.
170x250: Faint to moderately faint. Broadly slightly concentrated. Edge-on NNE-SSW. Moderate size. No involved stars.

| ga | NGC 4304 | 12 22.2 | −33 29 | Hya | 12.3p | 3.2' | 3.0' | Sbc |

JH: very faint, very large, round, very gradually very little brighter middle, resolvable.
170x250: Faint. Broadly very slightly concentrated. Slightly elongated E-W. Moderately small. No involved stars.

| ga | NGC 4310 | 12 22.4 | +29 12 | Com | 13.2p | 2.2' | 1.1' | S0+ |

HH: bright, considerably large, little extended, NW of 2.
170x250: Very faint. Broadly slightly concentrated. Roughly round. Small. No involved stars. S are a couple of stars, the N of which has a very wide companion.

| ga | NGC 4312 | 12 22.5 | +15 32 | Com | 12.5b | 4.6' | 1.2' | Sab |

HH: pretty bright, considerably large, extended, gradually brighter middle.

170x250: Faint. Broadly very slightly concentrated. Elongated N-S. Moderately small. No involved stars. A little ways ESE is a wide double star.

| ga | NGC 4313 | 12 22.6 | +11 48 | Vir | 12.5b | 4.0' | 1.0' | Sab |

HH: very faint, large, extended NW-SE, resolvable.
170x250: Moderately faint. Broadly slightly concentrated. Very elongated NW-SE. Moderate size. No involved stars.

| ga | NGC 4314 | 12 22.6 | +29 53 | Com | 11.4b | 4.3' | 3.6' | Sa |

HH: considerably bright, large, extended NNW-SSE, suddenly brighter middle, star NW.
170x250: Bright. Evenly pretty concentrated. Elongated NNW-SSE. Moderate size. Halfway to its SE end is an extremely faint involved star; just off its NW end is a moderately faint star; nearby NE is a very very faint star.

| ga | NGC 4332 | 12 22.8 | +65 50 | Dra | 13.1p | 2.1' | 1.4' | Sa |

HH: pretty faint, small, very little extended, very gradually brighter middle.
170x250: Faint. Broadly slightly concentrated. Slightly elongated NW-SE. Small. No involved stars. [A little ways ENE is a moderate brightness star.]

| ga | NGC 4321/M100 | 12 22.9 | +15 49 | Com | 10.1b | 7.5' | 6.3' | Sbc |

HH: !! pretty faint, very large, round, very gradually then pretty suddenly brighter middle to a resolvable nucleus, (Le) 2-branched spiral.
42x60: Visible. ***170x250:*** Moderately bright. Sharply very concentrated. Round. Moderately large. It has a very pronounced two-phase structure: a pretty low surface brightness halo and a small, much brighter core. No involved stars. At its edge SE is a star; at its edge WSW is another star. [4322 (Te) was not seen.]

| ga | NGC 4324 | 12 23.1 | +05 15 | Vir | 11.6v | 3.1' | 1.3' | S0+ |

dA: pretty bright, round or little extended, brighter middle.
170x250: Moderately bright. Evenly to sharply fairly concentrated. Slightly elongated NE-SW. Small. No involved stars. Nearby WSW is a faint star; SSW is another faint star.

| ga | NGC 4329 | 12 23.3 | −12 33 | Crv | 12.5 | 2.4' | 1.3' | E+ |

JH: very faint, very small, round, brighter middle to a nucleus.
170x250: Faint. Evenly moderately concentrated. Roughly round. Very small. No involved stars. ESE is a moderately faint star; SW of this star is either another star or a very small very faint galaxy [it's a faint double star].

| ga | NGC 4330 | 12 23.3 | +11 22 | Vir | 13.1b | 4.4' | 0.8' | Scd |

LR: very faint, large, much extended.
170x250: Extremely faint. Hardly concentrated. Edge-on ENE-WSW. Moderately large to large. The ends are extremely indistinct. No involved stars. A little ways SE of center is a moderately faint star; WSW is a bright star; NNE is another bright star. Very ghostly.

| ga | NGC 4346 | 12 23.5 | +46 59 | CVn | 12.1b | 3.7' | 1.3' | S0 |

HH: very faint, small, much extended E-W, very suddenly much brighter middle to a bright nucleus.
170x250: Moderately bright. Sharply pretty concentrated. Elongated WNW-ESE. Moderately small. No involved stars. It makes a triangle with two moderately faint stars, one S and one SE.

| ga | NGC 4339 | 12 23.6 | +06 04 | Vir | 12.3b | 2.2' | 2.2' | E0 |

HH: bright, pretty large, round, brighter middle, 3rd of 3.
170x250: Moderately faint. Evenly moderately concentrated. Round. Very small. No involved stars. Nearby S is a moderate brightness star. A quarter of the field SW is a small faint galaxy [*4333 (HH)]. Farther W is another small faint galaxy [*4326 (HH)]. These two galaxies make a nearly isosceles triangle with a moderate brightness star SW of the midpoint between them.

| ga | NGC 4340 | 12 23.6 | +16 43 | Com | 12.1b | 3.5' | 2.7' | S0+ |

HH: pretty bright, small, round, pretty suddenly brighter middle.
[See 4350.]

| ga | IC 3253 | 12 23.7 | −34 37 | Cen | 12.6p | 2.8' | 1.1' | Sc |

170x250: Faint. Broadly very slightly concentrated. Elongated NNE-SSW. Moderate size. No involved stars. Off its N end is an extremely faint star; NW is a moderate brightness star; ENE is a pair of moderate brightness stars.

| ga | NGC 4343 | 12 23.7 | +06 57 | Vir | 13.1b | 2.9' | 0.9' | Sb |

HH: pretty faint, small, extended, double?
170x250: Moderately faint. Evenly moderately concentrated. Slightly elongated NW-SE. Small. No involved stars. A third of the field N is a fairly bright galaxy [*4342 (WH)]. Extremely concentrated; almost stellar; it looks like a fuzzy star or a star with a halo. A quarter of the field NE is a very faint galaxy [*4341 (WH)]. Slightly concentrated. Small. A quarter of the field SE is an extremely faint galaxy [*I3267]. Diffuse. Small.

| ga | NGC 4350 | 12 24.0 | +16 41 | Com | 11.9b | 3.0' | 1.4' | S0 |

HH: considerably bright, very small, much extended, very suddenly brighter middle.
170x250: Bright. Sharply very concentrated. Very elongated NNE-SSW. Small. No involved stars. A third of the field WNW is 4340. Moderate brightness. Sharply very concentrated. Roughly round. Very small. No involved stars.

| ga | NGC 4351 | 12 24.0 | +12 12 | Vir | 13.0b | 1.9' | 1.2' | Sab |

dA: faint, pretty large, irregularly round, brighter middle.
170x250: Very faint. Broadly very slightly concentrated. Roughly round. Small. No involved stars. A little ways NNW is an extremely faint star.

| ga | NGC 4357 | 12 24.0 | +48 46 | CVn | 13.2p | 3.6' | 1.2' | Sbc |

Bi: faint, pretty small, gradually brighter middle.
170x250: Faint. Broadly slightly concentrated. Elongated ENE-WSW. Small. No involved stars.

| ga | UGC 7490 | 12 24.4 | +70 20 | Dra | 13.1b | 3.3' | 3.3' | Sm |

170x250: Not visible.

| pn | NGC 4361 | 12 24.5 | –18 47 | Crv | 10.3p | 118" | | |

HH: very bright, large, round, very suddenly much brighter middle to a nucleus, resolvable.
42x60: Visible. ***170x250:*** Pretty bright. It has an obvious moderately faint central star. Not a clean disk or edge at all. The nebula is even in light. No blinking effect. Slightly oval NNE-SSW. Large for a planetary. It looks more like a bright broadly concentrated galaxy than a planetary. Interesting, nice.

| ga | NGC 4365 | 12 24.5 | +07 19 | Vir | 10.5b | 6.9' | 4.9' | E3 |

HH: considerably bright, pretty large, very little extended, gradually little then suddenly much brighter middle.
42x60: Easily visible. ***170x250:*** Very bright. Sharply very concentrated. Round. Moderately small. No involved stars. NW is a moderately bright star; farther SSE is another moderately bright star. A quarter of the field ENE is a faint galaxy [*4366 (WH)].

| ga | NGC 4386 | 12 24.5 | +75 31 | Dra | 12.7b | 2.4' | 1.3' | S0 |

HH: bright, considerably large, little extended, pretty suddenly much brighter middle.
170x250: Moderate brightness. Evenly to sharply fairly concentrated. Slightly elongated NW-SE. Small. No involved stars. [E to NW is a crooked line of four moderate brightness to faint stars.]

| ga | NGC 4369 | 12 24.6 | +39 22 | CVn | 12.3b | 2.1' | 2.1' | Sa |

HH: considerably bright, small, round, much brighter middle to a nucleus, resolvable.
170x250: Moderate brightness. Sharply pretty concentrated. Round. Very small. No involved stars. It makes a shallow arc with a moderately faint star S and a pair of moderately faint stars further S.

| ga | NGC 4371 | 12 24.9 | +11 42 | Vir | 11.8b | 4.0' | 2.2' | S0+ |

HH: bright, pretty small, round, gradually brighter middle.
170x250: Bright. Sharply pretty concentrated. Round. Small. No involved stars.

| oc | MEL 111 | 12 25.1 | +26 07 | Com | 1.8 | 275' | | |

6x30: A fair number of moderately bright stars, very uniform in brightness, with relatively few fainter stars. Elongated NNW-SSE. At the S end is a triangular group of four stars, including 20, 23, and 26 Com. The cluster slightly overflows the field. Somewhat sparse to moderate density. [CR256; Coma Star Cluster]

| ga | NGC 4374/M84 | 12 25.1 | +12 53 | Vir | 10.1b | 6.4' | 5.5' | E1 |

JH: very bright, pretty large, round, pretty suddenly brighter middle, resolvable.
42x60: The brighter members of Markarian's Chain – M84, M86, 4435, 4438, 4473, and 4477 – are all visible. ***170x250:*** M84 is bright. Evenly to sharply pretty concentrated. Round. Moderately small to moderate size. No involved stars.

| ga | NGC 4377 | 12 25.2 | +14 45 | Com | 12.8b | 1.7' | 1.3' | S0– |

HH: bright, small, round, suddenly much brighter middle.
170x250: Moderately bright. Sharply very concentrated. Round. Very small. No involved stars.

| ga | NGC 4379 | 12 25.2 | +15 36 | Com | 12.6b | 2.0' | 1.6' | S0– |

HH: pretty small, round, pretty suddenly brighter middle to a nucleus.
170x250: Moderate brightness. Sharply very concentrated. Round. Very small. No involved stars.

| ga | NGC 4378 | 12 25.3 | +04 55 | Vir | 12.6b | 2.8' | 2.6' | Sa |

HH: bright, small, 8th–9th mag. star 3' SE.
170x250: Moderate brightness. Evenly moderately concentrated. Round. Very small. No involved stars. A little ways NNW is a moderately faint star; the galaxy makes a flat triangle with two bright stars, one N and one ESE.

| vs | SS Vir | 12 25.3 | +00 48 | Vir | 6.0 | 9.6 | 3.97 | C5 |

12x35: Mag. 7.5. ***170x250:*** Light copper red.

| ga | NGC 4380 | 12 25.4 | +10 01 | Vir | 12.7b | 3.4' | 1.8' | Sb |

JH: very faint, pretty large, round, little brighter middle.
170x250: Faint. Broadly slightly concentrated. Roughly round. Moderately small. No involved stars. A little ways S is a faint star.

| ga | NGC 4382/M85 | 12 25.4 | +18 11 | Com | 9.1v | 7.1' | 5.5' | S0+ |

JH: very bright, pretty large, round, brighter middle, star NW.
42x60: Visible. ***170x250:*** Moderately bright. Sharply very concentrated. Roughly round. Moderately small. At its edge NNE is a faint involved star. A little less than half the field E is 4394. Moderately faint. Sharply pretty concentrated. Slightly elongated NW-SE. Small. No involved stars.

| ga | NGC 4383 | 12 25.4 | +16 28 | Com | 12.7b | 2.0' | 0.9' | Sa |

Sf: extremely small, stellar or nebulous 11th–12th mag. star.
170x250: Moderately faint. Evenly moderately concentrated. Round. Very small. No involved stars. A little ways WSW is a moderately faint star.

| ga | NGC 4389 | 12 25.6 | +45 41 | CVn | 12.5b | 2.6' | 1.8' | Sbc |

HH: pretty bright, pretty large, irregularly extended, very gradually little brighter middle.
170x250: Faint. Broadly slightly concentrated. Elongated WNW-ESE. Moderately small. Off its edge SE is a very faint star; S are several bright stars; N are a number of stars in a triangular figure. Half the field NNW, beyond the triangular figure, is a small very faint galaxy [*4392 (HH)].

| ga | NGC 4385 | 12 25.7 | +00 34 | Vir | 13.2b | 2.2' | 1.3' | S0+ |

Ma: very faint, very small, almost stellar.
170x250: Faint. Broadly slightly concentrated. Slightly elongated E-W. Small. No involved stars. S is a moderately bright star; further S is a moderate brightness star.

| ga | NGC 4387 | 12 25.7 | +12 48 | Vir | 13.0b | 1.7' | 1.0' | E |

HH: N of 2.
170x250: Moderate brightness. Evenly moderately concentrated. Round. Very small. No involved stars. Nearby NNW is a moderately faint star.

| ga | NGC 4388 | 12 25.8 | +12 39 | Vir | 11.8b | 7.6' | 1.4' | Sb |

HH: S of 2, extended.
170x250: Moderate brightness. Broadly moderately concentrated. Edge-on E-W. Large: almost a quarter of the field in size. No involved stars. Nearby NE of center is a very faint star; S is a moderate brightness star.

| ga | NGC 4393 | 12 25.8 | +27 33 | Com | 12.7p | 4.4' | 3.4' | Sd |

HH: very faint, very large, irregular figure, bright star W.
170x250: Extremely faint. Diffuse. Moderate size. No involved stars. SSW to NW is a long bent "L" figure of moderate brightness stars.

| ga | NGC 4395 | 12 25.8 | +33 32 | CVn | 10.6b | 13.3' | 11.0' | Sm |

HH: extremely faint, very large, NW of double nebula.
170x250: Extremely faint. An unconcentrated glow. Elongated NW-SE. Very large: half the field in size. No involved stars near center.

| ga | NGC 4394 | 12 25.9 | +18 12 | Com | 10.8v | 3.7' | 3.3' | Sb |

HH: pretty bright, little extended, brighter middle.
[See M85.]

| ga | NGC 4396 | 12 26.0 | +15 40 | Com | 12.5v | 3.5' | 1.0' | Sd |

dA: very faint, pretty large, much extended.
170x250: Very faint. Hardly concentrated. Elongated NW-SE. Moderately small. No involved stars. Nearby N is a moderate brightness star.

| ga | NGC 4402 | 12 26.1 | +13 06 | Vir | 12.6b | 4.7' | 1.0' | Sb |

Au: faint, large, much extended E-W.
170x250: Faint. Hardly concentrated. Edge-on E-W. Large: a quarter of the field in size. No involved stars.

12 to 18 Hours: SPRING

ga	NGC 4405	12 26.1	+16 10	Com	13.0p	1.8'	1.2'	S0/a

HH: pretty faint, small, round, very suddenly brighter middle, resolvable.
170x250: Faint to moderately faint. Broadly slightly concentrated. Round. Small. No involved stars.

ga	NGC 4406/M86	12 26.2	+12 56	Vir	9.8b	8.9'	5.7'	E3

JH: very bright, large, round, gradually brighter middle to a nucleus, resolvable.
170x250: Bright; slightly lower surface brightness than M84. Sharply pretty concentrated. Slightly elongated NW-SE. Large: a quarter of the field in size; larger than M84. It has an extremely faint outer halo. No involved stars.

ga	NGC 4404	12 26.3	−07 40	Vir	12.0	1.1'	0.8'	S0−

WH: very faint, very small, extended.
170x250: In the position is a close double galaxy aligned ENE-WSW [4404, *4403 (WH)]. The two components are very similar. Both are moderately faint. Evenly moderately concentrated. Very small. No involved stars. The E galaxy is roughly round. The W galaxy is slightly elongated NNE-SSW. S of the two galaxies is an arc of moderate brightness stars.

ga	NGC 4414	12 26.4	+31 13	Com	11.0b	4.3'	3.1'	Sc

HH: very bright, large, extended, gradually then very suddenly much brighter middle to a star.
42x60: Visible. ***170x250:*** Bright. Sharply pretty concentrated. Slightly elongated NNW-SSE. Moderately small. No involved stars.

ga	NGC 4413	12 26.5	+12 36	Vir	12.3v	2.3'	1.4'	Sab

HH: considerably faint, small, gradually brighter middle, 2 stars N and NW.
170x250: Very faint to faint. Broadly slightly concentrated. Round. Small. No involved stars. It makes a shallow arc with two moderate brightness stars N; S are two faint stars.

ga	NGC 4412	12 26.6	+03 57	Vir	13.2b	1.4'	1.2'	Sb

HH: faint, pretty large, round, gradually brighter middle, resolvable.
170x250: Faint. Broadly very slightly concentrated. Round. Small. No involved stars. It makes a triangle with two faint stars, one N and one ESE.

ga	NGC 4415	12 26.6	+08 26	Vir	13.1b	1.3'	1.1'	S0/a

WH: extremely faint.
170x250: Very faint. Broadly very slightly concentrated. Round. Very small. No involved stars. A little ways NNW is a faint star.

ga	NGC 4411B	12 26.8	+08 53	Vir	12.9b	2.5'	2.5'	Scd

170x250: [The observed galaxy is *4411A; 4411B was not seen.] A very faint halo with a faint involved star. Round. Small. It's on the side of a triangle of moderately faint stars: one NW, one ENE, and one SE. Half the field N is a brighter galaxy [*4410A and *4410B merged]. Faint. Broadly slightly concentrated. Slightly elongated E-W. Small. No involved stars.

ga	NGC 4416	12 26.8	+07 55	Vir	13.1b	1.7'	1.5'	Scd

JH: very faint, large, round, star 5' SW.
170x250: Extremely faint. Broadly very slightly concentrated. Roughly round. Small. No involved stars. WSW is the bright plotted star.

ga	NGC 4417	12 26.8	+09 35	Vir	12.0b	3.4'	1.3'	S0

WH: faint, pretty large, extended, little brighter W.
170x250: Bright. Sharply very concentrated. Elongated NE-SW. Moderately small. No involved stars. Nearby W is a very faint star.

ga	NGC 4419	12 26.9	+15 02	Com	12.1b	3.3'	1.1'	Sa

HH: bright, pretty much extended NW-SE, suddenly brighter middle.
170x250: Moderately bright to bright. Evenly moderately concentrated. Elongated NW-SE. Moderately small. No involved stars. A little ways S is a moderately faint star.

ga	NGC 4420	12 27.0	+02 29	Vir	12.9b	2.0'	0.9'	Sbc

HH: faint, pretty large, little extended, resolvable.
170x250: Moderate brightness. Broadly very slightly concentrated. Elongated N-S. Small. No involved stars.

ga	NGC 4421	12 27.0	+15 27	Com	11.6v	2.7'	2.0'	S0/a

HH: pretty bright, pretty large, pretty gradually brighter middle, bright star NW.
170x250: Moderately faint. Evenly moderately concentrated. Round. Very small. No involved stars. A little ways WNW is a moderately bright star.

| ga | NGC 4424 | 12 27.2 | +09 25 | Vir | 12.3b | 3.6' | 1.8' | Sa |

dA: faint, pretty large, irregularly round, brighter middle.
170x250: Moderately faint. Evenly moderately concentrated. Elongated E-W. Moderately small. No involved stars. A little ways N is the easternmost of an E-W arc of moderate brightness to moderately faint stars.

| ga | NGC 4425 | 12 27.2 | +12 44 | Vir | 11.8v | 3.0' | 1.1' | S0+ |

HH: pretty faint, small, round, brighter middle.
170x250: Moderate brightness. Broadly moderately concentrated. Very elongated to edge-on NNE-SSW. Moderately small. No involved stars. Nearby W is a moderately faint star.

| ga | NGC 4429 | 12 27.4 | +11 06 | Vir | 11.0b | 5.6' | 2.5' | S0+ |

HH: bright, large, considerably extended, pretty suddenly brighter middle, 10th mag. star NE.
42x60: Visible. ***170x250:*** Moderately bright. Evenly moderately concentrated. Elongated E-W. Moderately large. No involved stars. A little ways NNE is one of the two bright plotted stars in the field; the other is SSE.

| ga | NGC 4430 | 12 27.4 | +06 15 | Vir | 12.8b | 2.3' | 2.0' | Sb |

HH: considerably faint, large, round, gradually brighter middle.
170x250: Very faint. Hardly concentrated. Roughly round. Moderately small. No involved stars. It makes a large triangle with the two plotted stars, one of which is SW and the other of which is farther SSE. [4432 (Ma) was not seen.]

| ga | NGC 4434 | 12 27.5 | +08 09 | Vir | 12.1v | 1.4' | 1.4' | E |

WH: pretty faint, very small.
170x250: Moderate brightness. Sharply very concentrated. Round. Very small. No involved stars.

| ga | NGC 4435 | 12 27.7 | +13 04 | Vir | 11.7b | 2.7' | 2.0' | S0 |

HH: very bright, considerably large, round, W of 2.
170x250: 4435 and 4438 are next to each other NNW-SSE. They're fairly similar. 4435 is moderately bright. Sharply pretty concentrated. Elongated NNE-SSW. Small. No involved stars. 4438 is also moderately bright, but not quite as bright as 4435. Evenly moderately concentrated. Also elongated NNE-SSW, but not as narrow as 4435. Moderately small. No involved stars.

| ga | UGC 7577 | 12 27.7 | +43 29 | CVn | 12.4v | 4.6' | 2.8' | Im |

170x250: Exceedingly faint. A slight diffuse glow. Moderate size. No involved stars. Tangent to it is a long slightly curved WNW-ESE line of three stars: a faint star at the galaxy's edge SE, a moderately faint star further SE, and a moderately bright star W.

| ga | NGC 4438 | 12 27.8 | +13 00 | Vir | 11.0b | 8.6' | 3.1' | S0/a |

HH: bright, considerably large, very little extended, resolvable, E of 2. [See 4435.]

| ga | NGC 4440 | 12 27.9 | +12 17 | Vir | 12.7b | 2.0' | 1.5' | Sa |

HH: bright, pretty small, round, brighter middle, resolvable.
170x250: Moderate brightness. Sharply pretty concentrated. Round. Very small. No involved stars. W are two small very faint galaxies. The closer galaxy [*4436 (HH)] is elongated WNW-ESE. Off its NW end is a moderate brightness star. The farther galaxy [*4431 (HH)] is elongated N-S. Nearby E is a faint star.

| ga | NGC 4442 | 12 28.1 | +09 48 | Vir | 11.4b | 4.5' | 1.7' | S0 |

HH: very bright, pretty large, round, suddenly much brighter middle.
170x250: Bright. Evenly to sharply pretty concentrated. Slightly elongated E-W. Moderately small. It's among a number of faint stars, including one very nearby SSW of center and another at the galaxy's E end.

| ga | NGC 4448 | 12 28.2 | +28 37 | Com | 12.0b | 4.6' | 1.7' | Sab |

HH: bright, large, extended E-W, suddenly brighter middle.
170x250: Moderately bright. Sharply concentrated. Edge-on E-W. Moderately large. No involved stars.

| ga | NGC 4449 | 12 28.2 | +44 05 | CVn | 10.0b | 6.1' | 4.3' | Im |

HH: very bright, considerably large, much extended NNE-SSW, well resolved, 9th mag. star 5' [E].
42x60: Easily visible. *170x250:* Very bright. Broadly slightly concentrated. Irregularly elongated NE-SW: the galaxy is somewhat rectangular, flaring out (widening) toward the NE end. Moderate size. It has a grainy appearance, as if starting to be resolved. There are either threshold stars or knots involved; its light is not at all smooth like other galaxies. E of center is an extremely faint involved star. A very unusual galaxy.

| ga | NGC 4450 | 12 28.5 | +17 05 | Com | 10.9b | 5.2' | 3.8' | Sab |

HH: bright, large, round, gradually very much brighter middle to a star, resolvable, bright star near.
42x60: Faintly visible. ***170x250:*** Moderate brightness. Evenly moderately concentrated. The core is round, but the galaxy has a very faint outer halo elongated N-S. Moderately small. No involved stars.

| ga | IC 3392 | 12 28.7 | +14 59 | Com | 13.0b | 2.3' | 1.0' | Sb |

170x250: Very faint. Broadly very slightly concentrated. Slightly elongated NE-SW. Small. No involved stars. It makes an arc with two moderate brightness stars WSW.

| ga | NGC 4452 | 12 28.7 | +11 45 | Vir | 12.9b | 2.7' | 0.5' | S0 |

WH: pretty bright, small, very much extended.
170x250: Moderate brightness. Broadly very slightly concentrated. A thin edge-on, NNE-SSW. Moderately small. No involved stars.

| ga | NGC 4455 | 12 28.7 | +22 49 | Com | 12.9p | 2.7' | 0.7' | Sd |

HH: faint, large, extended, gradually brighter middle, 2 bright stars NE.
170x250: Faint. Broadly slightly concentrated. Elongated to edge-on NNE-SSW. Moderate size. No involved stars. A little ways WNW is a faint star; NNE are a couple of moderate brightness stars.

| ga | NGC 4454 | 12 28.8 | −01 56 | Vir | 12.7b | 2.7' | 2.0' | S0/a |

HH: faint, large, round, gradually brighter middle, easily resolvable.
170x250: Faint. Broadly slightly concentrated. Round. Small. No involved stars. It makes a triangle with two bright stars, one NNW and one WSW.

| ga | NGC 4460 | 12 28.8 | +44 51 | CVn | 12.3p | 4.7' | 1.2' | S0+ |

HH: bright, pretty large, extended NE-SW, pretty suddenly brighter middle.
170x250: Moderately faint. Broadly moderately concentrated. Edge-on NE-SW. Moderately small. No involved stars. SSE are a couple of moderately faint stars; farther SE is another moderately faint star.

| ga | NGC 4457 | 12 29.0 | +03 34 | Vir | 11.8b | 2.7' | 2.2' | S0/a |

HH: considerably bright, pretty small, round, suddenly much brighter middle to a nucleus.
170x250: Bright. Sharply extremely concentrated to a substellar nucleus. Round. Small. No involved stars. It makes a triangle with two moderately faint stars, one WSW and another farther S.

| ga | NGC 4458 | 12 29.0 | +13 14 | Vir | 12.0v | 1.5' | 1.5' | E0-1 |

HH: pretty bright, small, round, brighter middle, W of 2.
[See 4461.]

| ga | NGC 4459 | 12 29.0 | +13 58 | Com | 11.3b | 3.5' | 2.6' | S0+ |

HH: pretty bright, pretty large, irregularly round, brighter middle, resolvable, 8th mag. star 2' SE.
170x250: Moderately bright. Sharply pretty concentrated. Round. Very small. No involved stars. A little ways SE is the bright plotted star.

| ga | NGC 4461 | 12 29.0 | +13 11 | Vir | 11.1v | 3.6' | 1.4' | S0+ |

HH: pretty faint, small, round, brighter middle, E of 2.
170x250: Moderately bright. Sharply pretty concentrated. Slightly elongated N-S. Small. No involved stars. A little less than a quarter of the field NNW is 4458. Moderate brightness. Evenly moderately concentrated. Round. Small. No involved stars. Making a triangle with the two galaxies is a moderate brightness star a little ways E of 4458.

| ga | NGC 4462 | 12 29.3 | −23 10 | Crv | 12.8b | 3.2' | 1.2' | Sab |

HH: pretty bright, pretty small, extended NW-SE, very gradually brighter middle.
170x250: Moderately faint. Broadly slightly concentrated. Elongated NW-SE. Small. No involved stars. It makes a quadrilateral with three stars: one N, one ENE, and one ESE.

| ga | NGC 4464 | 12 29.4 | +08 09 | Vir | 12.5v | 1.1' | 1.0' | S |

HH: faint, very small, round, pretty gradually brighter middle.
170x250: Moderate brightness. Sharply very concentrated; the whole galaxy is almost stellar. Round. Tiny. No involved stars. A little ways WNW is a moderately faint star.

| ga | NGC 4469 | 12 29.5 | +08 45 | Vir | 12.2p | 3.8' | 1.2' | S0/a |

WH: *pretty faint, pretty large, much extended, brighter middle, resolvable.*
170x250: Moderate brightness. Broadly moderately concentrated. Edge-on E-W. Large: a quarter of the field in size [moderately large]. The ends fade out very indistinctly. No involved stars.

| ga | NGC 4470 | 12 29.6 | +07 49 | Vir | 13.0b | 1.4' | 0.9' | Sa |

HH: *faint, pretty large, irregularly round, brighter middle.*
170x250: Moderately faint. Broadly slightly concentrated. Round. Small. No involved stars.

| ga | NGC 4472/M49 | 12 29.8 | +07 59 | Vir | 9.4b | 9.3' | 7.0' | E2 |

JH: *very bright, large, round, much brighter middle.*
42x60: Bright. ***170x250:*** Very bright. Evenly to sharply very concentrated. Round. Moderately large. It has a very faint outer halo. E of center is an involved star. [4465 (Bi) and 4467 (Sv) were not seen.]

| ga | NGC 4473 | 12 29.8 | +13 25 | Com | 11.2b | 4.5' | 2.5' | E5 |

WH: *very faint, resolvable.*
170x250: Very bright. Sharply pretty concentrated. Slightly elongated E-W. Small. No involved stars. It makes a large triangle with two moderate brightness stars, one SSW and one SE. It has the highest surface brightness of all the galaxies in Markarian's Chain, followed closely by M84.

| bs | Delta Crv | 12 29.9 | −16 30 | Crv | 2.95 | | −0.05 | B9 |

170x250: The primary has a wide relatively very faint companion. The primary is white. [SHJ 145: 24.3", 3.0, 9.2]

| ga | NGC 4474 | 12 29.9 | +14 04 | Com | 12.4b | 2.4' | 1.4' | S0 |

HH: *pretty faint, round, resolvable.*
170x250: Moderate brightness. Sharply pretty concentrated. Round. Very small. No involved stars. It makes a narrow diamond with three stars E to NNW; farther NW is a brighter star. A third of the field WSW is a very faint galaxy [*4468 (WH)].

| ga | NGC 4476 | 12 30.0 | +12 20 | Vir | 13.0b | 1.7' | 1.1' | S0− |

HH: *faint, small, round, brighter middle, 1st of 3.*
[See 4478.]

| ga | NGC 4477 | 12 30.0 | +13 38 | Com | 11.4b | 3.8' | 3.4' | S0 |

WH: very bright, considerably large.
170x250: Bright. Evenly to sharply pretty concentrated. Roughly round. Small. No involved stars. A little less than a third of the field SE is a faint galaxy [*4479 (WH)].

| ga | NGC 4478 | 12 30.3 | +12 19 | Vir | 12.4b | 1.9' | 1.6' | E2 |

HH: pretty bright, small, round, pretty suddenly brighter middle, 2nd of 3.
170x250: Moderate brightness to moderately bright. Evenly to sharply pretty concentrated. Round. Very small. No involved stars. A quarter of the field WNW is 4476. Moderately faint. Sharply pretty concentrated. Round. Very small. No involved stars. 4478 is like a small version of M87, with a comparable surface brightness; 4476 is a slightly smaller and fainter version of 4478.

| ga | UGCA 282 | 12 30.3 | −08 23 | Vir | 12.9 | 2.3' | 2.2' | Sbc |

170x250: Not visible.

| ga | NGC 4480 | 12 30.4 | +04 14 | Vir | 13.2b | 2.3' | 1.1' | Sc |

HH: pretty faint, pretty small, extended, brighter S.
170x250: Very faint. Broadly very slightly concentrated. Slightly elongated N-S. Small. No involved stars. A little ways NW is a faint star.

| ga | NGC 4485 | 12 30.5 | +41 42 | CVn | 12.3b | 2.6' | 1.9' | Im |

HH: bright, pretty small, irregularly round, W of 2.
[See 4490.]

| ga | NGC 4490 | 12 30.6 | +41 38 | CVn | 10.2b | 6.3' | 2.7' | Sd |

HH: very bright, very large, much extended NW-SE, partially resolved.
42x60: 4490 and 4485 are both visible. *170x250:* 4485 and 4490 are next to each other NNW-SSE. 4490 is bigger and brighter. Bright. Broadly moderately concentrated. Elongated WNW-ESE. Large: a quarter of the field in size. No involved stars. 4485 is moderately faint. Broadly very slightly concentrated. Slightly elongated N-S, almost perpendicular to 4490. Small. No involved stars. A nice pair of galaxies.[†]

| ga | NGC 4483 | 12 30.7 | +09 00 | Vir | 13.1b | 1.5' | 0.8' | S0+ |

dA: pretty bright, pretty small, round, brighter middle.
170x250: Moderate brightness. Evenly moderately concentrated. Round. Very small. No involved stars.

| ga | NGC 4486/M87 | 12 30.8 | +12 23 | Vir | 9.6b | 7.4' | 6.0' | E+0-1 |

JH: very bright, very large, round, much brighter middle.
42x60: Easily visible. ***170x250:*** Bright to very bright. Evenly pretty concentrated. Round. Moderately large. The outer halo is very very faint. No involved stars. [The brightest members of the Virgo Cluster – M49, M60, M84, M85, M86, M87 – are visible in small binoculars; they were not observed in the 35mm in this survey, however, since I used the 60mm rather than the 35mm as the finder within the cluster.]

| ga | NGC 4489 | 12 30.8 | +16 45 | Com | 12.8b | 1.7' | 1.7' | E |

HH: pretty faint, considerably small, round, gradually brighter middle.
170x250: Faint. Sharply moderately concentrated. Round. Very small. No involved stars. N to ENE is a group of four stars in an asymmetrical diamond figure.

| ga | NGC 4488 | 12 30.9 | +08 21 | Vir | 12.2v | 3.8' | 1.5' | S0/a |

HH: very faint, very small, little extended.
170x250: Faint. Evenly slightly concentrated. Roughly round. Small. No involved stars. Nearby SSW is a faint star.

| ga | NGC 4492 | 12 31.0 | +08 04 | Vir | 12.6v | 2.0' | 2.0' | Sa |

HH: pretty faint, pretty large, very gradually little brighter middle, 2 stars near.
170x250: Faint. Evenly slightly concentrated. Round. Very small. It makes a little 30–60–90 right triangle with two moderate brightness stars, one NNE and one SE.

| ga | NGC 4487 | 12 31.1 | –08 03 | Vir | 11.6p | 4.1' | 2.7' | Scd |

WH: faint, very large, easily resolvable.
170x250: Moderately faint. Broadly slightly concentrated. Slightly elongated ENE-WSW. Moderate size. Just off its edge N is a moderate brightness star; a little ways E is another star.

| ga | NGC 4494 | 12 31.4 | +25 46 | Com | 9.8v | 4.8' | 3.6' | E1-2 |

HH: very bright, pretty large, round, very suddenly much brighter middle to a nucleus.
42x60: Visible. ***170x250:*** Moderately bright. Sharply pretty concentrated. Round. Small. No involved stars.

| ga | NGC 4500 | 12 31.4 | +57 57 | UMa | 13.1b | 1.6' | 1.0' | Sa |

HH: bright, considerably small, extended, pretty gradually brighter middle, 9th mag. star 30" E.
170x250: Moderately faint. Evenly pretty concentrated. Round. Very very small. No involved stars. Nearby E is a relatively bright star.

| ga | NGC 4497 | 12 31.5 | +11 37 | Vir | 13.2b | 2.0' | 0.9' | S0+ |

WH: very faint.
170x250: Faint. Broadly slightly concentrated. Roughly round. Small. No involved stars. E, in line with the galaxy, are two stars, moderate brightness and moderately faint.

| ga | NGC 4498 | 12 31.6 | +16 51 | Com | 12.8b | 2.9' | 1.5' | Sd |

WH: very faint, very small.
170x250: Very faint. Broadly very slightly concentrated. Elongated NW-SE. Moderately small. No involved stars.

| ga | NGC 4496 | 12 31.7 | +03 56 | Vir | 11.9b | 4.0' | 3.1' | Sm |

HH: faint, considerably large, binuclear or double nebula.
170x250: Very faint. Broadly very slightly concentrated. Round. Moderately small. No involved stars. Nearby S is a faint star that makes a curved line with brighter stars E of it. [4496B, attached to 4496, was not seen as a separate galaxy.]

| ga | NGC 4501/M88 | 12 32.0 | +14 25 | Com | 10.4b | 7.0' | 3.7' | Sb |

JH: bright, very large, very much extended, W of double nebula.
42x60: Pretty bright. **170x250:** Bright. Evenly to sharply pretty concentrated. Very elongated NW-SE. Large: a quarter of the field in size. At its SE end is an involved star – an unequal double; a little ways N is a moderate brightness star; S is a pair of moderately bright stars and a faint star a little E of them; all of these stars are in a NNW-SSE line with the galaxy.

| ga | NGC 4503 | 12 32.1 | +11 10 | Vir | 12.1b | 3.5' | 1.6' | S0– |

HH: pretty bright, small, round, gradually brighter middle.
170x250: Moderate brightness. Evenly to sharply pretty concentrated. Slightly elongated N-S. Small. No involved stars. It makes an isosceles triangle with a moderate brightness star SSW and a moderately faint star ESE. A little more than a third of the field NE is a small very faint galaxy [*I3470].

| ga | NGC 4504 | 12 32.3 | −07 33 | Vir | 11.9p | 4.3' | 2.6' | Scd |

HH: pretty bright, considerably large, irregularly extended, gradually very little brighter middle, easily resolvable.
170x250: Faint. Broadly slightly concentrated. Roughly round. Moderately small to moderate size. No involved stars. A little ways ESE is a very faint star; the galaxy makes a triangle with two faint stars, one N and one ENE.

| ga | UGC 7690 | 12 32.4 | +42 42 | CVn | 13.1b | 2.1' | 1.6' | Im |

170x250: Very faint. Broadly very slightly concentrated. Roughly round. Small. No involved stars. A little ways NNE is a very faint star; WSW are two moderate brightness stars roughly in line with the galaxy.

| ga | NGC 4517A | 12 32.5 | +00 23 | Vir | 12.5v | 5.1' | 3.4' | Sdm |

170x250: Extremely faint. A slight diffuse glow. Elongated NE-SW. Moderately large. No involved stars. NW to SW are three moderately bright stars in a large flat triangle.

| ga | IC 3476 | 12 32.7 | +14 03 | Com | 13.2b | 2.1' | 1.7' | Im |

170x250: Very faint. Broadly very slightly concentrated. Round. Small. No involved stars. It's just off the N side of a very sharp E-W triangular figure of four relatively bright stars.

| ga | NGC 4517 | 12 32.8 | +00 06 | Vir | 11.1b | 11.2' | 1.5' | Scd |

WH: considerably bright, very large, very much extended E-W, bright star in contact.
42x60: Visible with an interfering star near center. *170x250:* Very faint to faint. Hardly concentrated. A thin edge-on, E-W. Very large: half the field in size. The ends fade out very indistinctly. At its edge NE of center is a relatively bright involved star; nearby are several more stars, including a very faint star just NE of its eastern tip; the galaxy doesn't quite reach that far E. Kind of ghostly.

| ga | NGC 4521 | 12 32.8 | +63 56 | Dra | 13.2p | 2.5' | 0.5' | S0/a |

JH: pretty bright, small, pretty much extended, pretty gradually brighter middle, 9th mag. star involved.
170x250: Moderately faint to moderate brightness. Broadly moderately concentrated. Slightly elongated NNW-SSE. Small. No involved stars. [A little ways NW is a moderately bright star.] [4512 (JH) was not seen.]

| ga | UGC 7699 | 12 32.8 | +37 37 | CVn | 13.2p | 5.1' | 1.1' | Scd |

170x250: Extremely faint. Broadly very slightly concentrated. Edge-on NE-SW. Moderately small. No involved stars. Nearby E is a moderate brightness star; SSW is a faint star.

| pn | IC 3568 | 12 32.9 | +82 33 | Cam | 11.6p | 10.0" | | |

42x60: Visible as a moderately faint star. *170x250:* Pretty bright. No central star. A fairly good disk. Even in light. Moderately strong blinking effect. Round. Very small. Right off its edge W is a faint star.

| ga | UGC 7698 | 12 32.9 | +31 32 | CVn | 13.0b | 6.5' | 4.4' | Im |

170x250: Not visible.

| ga | NGC 4516 | 12 33.1 | +14 34 | Com | 12.8v | 1.8' | 0.8' | Sab |

HH: faint, pretty small, round, resolvable.
170x250: Faint. Broadly slightly concentrated. Slightly elongated N-S. Very small. No involved stars.

| ga | NGC 4519 | 12 33.5 | +08 39 | Vir | 12.3b | 3.5' | 2.4' | Sd |

WH: faint, pretty large, round, brighter middle, resolvable.
170x250: Faint. Broadly very slightly concentrated. Round. Moderately small. No involved stars. NW is a little group of four very faint stars.

| ga | NGC 4522 | 12 33.7 | +09 10 | Vir | 13.0b | 3.7' | 0.9' | Scd |

JH: extremely faint, pretty large, little extended, very little brighter middle.
170x250: Very faint. Hardly concentrated. Very elongated NE-SW. Moderate size. No involved stars.

| ga | NGC 4525 | 12 33.9 | +30 16 | Com | 12.9p | 2.8' | 1.3' | Scd |

HH: faint, pretty large, irregularly round, brighter middle.
170x250: Very faint. Broadly very slightly concentrated. Roughly round. Small. No involved stars. A little ways SE is a very faint star.

| ga | NGC 4526 | 12 34.0 | +07 42 | Vir | 10.7b | 7.2' | 2.3' | S0 |

HH: very bright, very large, much extended WNW-ESE, pretty suddenly much brighter middle, bright star E, 9th mag. star W.
42x60: Visible right in between the two plotted stars. *170x250:* Bright. Evenly to sharply pretty concentrated. Elongated WNW-ESE. Large: a third of the field in size. The ends extend very faintly. No involved stars. Nearby S of center is a moderately faint star; the two bright plotted stars are at the edge of the field WSW and ENE.

| ga | NGC 4527 | 12 34.1 | +02 39 | Vir | 10.4v | 6.9' | 2.4' | Sbc |

HH: pretty bright, large, pretty much extended ENE-WSW, much brighter middle.
170x250: Moderately bright. Evenly moderately concentrated. Edge-on ENE-WSW. Large: a quarter of the field in size. No involved stars.

| ga | NGC 4528 | 12 34.1 | +11 19 | Vir | 13.0b | 1.8' | 1.2' | S0 |

HH: pretty faint, considerably small, round, brighter middle, 9th mag. star E.
170x250: Moderate brightness. Sharply pretty concentrated. Round. Very small. No involved stars. Passing a little W of it is a NW-SE line of four moderately faint stars, roughly equally spaced.

| ga | NGC 4534 | 12 34.1 | +35 31 | CVn | 12.9p | 3.0' | 2.2' | Sdm |

HH: considerably faint, large, little extended, very gradually little brighter middle, resolvable.
170x250: Very faint. Broadly very slightly concentrated. Slightly elongated NW-SE. Small. No involved stars. NW is a faint star; SSE is another faint star.

| ga | NGC 4531 | 12 34.3 | +13 04 | Vir | 12.4b | 3.1' | 2.0' | S0+ |

HH: faint, pretty large, round, very gradually brighter middle.
170x250: Faint. Broadly very slightly concentrated. Round. Small. No involved stars.

| ga | NGC 4532 | 12 34.3 | +06 28 | Vir | 12.3b | 2.8' | 1.0' | Im |

HH: pretty bright, pretty large, pretty much extended, very gradually brighter middle, resolvable.
170x250: Moderate brightness. Broadly slightly concentrated. Somewhat rectangular NNW-SSE: a stumpy edge-on. Moderately small. No involved stars. Off its edge E is an extremely faint star. A little unusual: it looks like the elongated core of a larger and thinner edge-on galaxy, with the halo ends missing.

| ga | NGC 4535 | 12 34.3 | +08 11 | Vir | 9.9v | 7.1' | 5.0' | Sc |

WH: very large, easily resolvable.
42x60: Visible. ***170x250:*** Moderately faint. Broadly slightly concentrated. Round. Moderately large. There are at least three involved stars: a faint star N of center, a very faint star WSW of center, and a faint star at the galaxy's edge S.

| bs | Beta Crv | 12 34.4 | −23 23 | Crv | 2.65 | | 0.89 | G5 |

170x250: Yellow white.

| ga | NGC 4536 | 12 34.5 | +02 11 | Vir | 11.2b | 8.4' | 3.2' | Sbc |

HH: bright, very large, much extended WNW-ESE, suddenly brighter middle, easily resolvable.
42x60: 4536 and 4527 are both visible. ***170x250:*** 4536 is moderate brightness. Sharply slightly concentrated. Elongated WNW-ESE. Moderately large. No involved stars.

| ga | NGC 4539 | 12 34.6 | +18 12 | Com | 12.9b | 3.3' | 1.3' | Sa |

JH: pretty bright, pretty much extended.
170x250: Very faint. Hardly concentrated. Elongated E-W. Small. No involved stars. Nearby SE are a couple of very faint stars.

| ga | NGC 4545 | 12 34.6 | +63 31 | Dra | 13.0p | 2.5' | 1.4' | Scd |

HH: faint, large, irregularly round, very gradually brighter middle, faint star NE.
170x250: Very faint to faint. Broadly very slightly concentrated. Roughly round. Small. No involved stars. Very nearby NE is a very faint star.

| ga | NGC 4540 | 12 34.8 | +15 33 | Com | 12.4b | 1.9' | 1.4' | Scd |

HH: faint, pretty small, brighter middle, resolvable.
170x250: Faint to moderately faint. Broadly very slightly concentrated. Round. Small. NW of center is an involved star; off the galaxy's edge W is another star.

| ds | 24 Com | 12 35.1 | +18 23 | Com | 20.6" | 5.2 | 6.7 | K2 A9 |

42x60: A close slightly unequal pair. ***170x250:*** A moderate distance slightly unequal pair. Color contrast: the primary is yellow; the companion is white. [STF 1657]

| ga | NGC 4548/M91 | 12 35.4 | +14 29 | Com | 11.0b | 5.4' | 4.2' | Sb |

HH: bright, large, little extended, little brighter middle.

42x60: Faintly visible. *170x250:* Moderately bright. Sharply very concentrated to a brighter core. Round. Moderate size. It has a very faint halo. No involved stars.

| ga | NGC 4546 | 12 35.5 | –03 47 | Vir | 11.3b | 3.3' | 1.4' | S0– |

HH: very bright, considerably large, pretty much extended ENE-WSW, very suddenly much brighter middle to a nucleus.
170x250: Very bright. Sharply very concentrated. Elongated ENE-WSW. Moderately small. No involved stars. It's right in between a moderately bright star SE and a very faint star NW. [C14-74 was not seen.]

| ga | NGC 4550 | 12 35.5 | +12 13 | Vir | 12.6b | 3.3' | 0.9' | S0 |

HH: pretty bright, small, very little extended, SW of 2.
170x250: 4551 and 4550 are next to each other NE-SW. They're similar. 4550 is a tiny bit brighter. Moderate brightness to moderately bright. Evenly moderately concentrated. Elongated N-S. Small. No involved stars. SE is a moderate brightness star. 4551 is moderate brightness. Broadly to evenly moderately concentrated. Very slightly elongated ENE-WSW. Small. No involved stars. NW are two faint stars. The two galaxies and the associated stars make a narrow diamond figure.

| ga | NGC 4551 | 12 35.6 | +12 15 | Vir | 13.0b | 1.8' | 1.4' | E |

HH: pretty bright, small, round, brighter middle, NE of 2.
[See 4550.]

| ga | UGCA 289 | 12 35.6 | –07 52 | Vir | 12.3 | 4.1' | 2.1' | Sdm |

170x250: Not visible.

| ga | NGC 4552/M89 | 12 35.7 | +12 33 | Vir | 10.7b | 3.5' | 3.5' | E0-1 |

JH: pretty bright, pretty small, round, gradually much brighter middle.
42x60: Visible. *170x250:* Very bright. Sharply pretty concentrated. Round. Small. No involved stars. A little ways ENE is a faint star.

| ga | NGC 4555 | 12 35.7 | +26 31 | Com | 13.1p | 1.9' | 1.2' | E |

HH: bright, pretty small, irregularly round, very suddenly much brighter middle to a 12th mag. star.
170x250: Moderately faint to moderate brightness. Sharply pretty concentrated. Round. Very small. No involved stars.

| ga | NGC 4559 | 12 36.0 | +27 57 | Com | 10.5b | 10.8' | 4.3' | Scd |

HH: very bright, very large, much extended NNW-SSE, gradually brighter middle, 3 stars E.
42x60: Visible. ***170x250:*** Moderate brightness. Broadly slightly concentrated. Elongated NW-SE. Moderately large. Crossing its SE end is a short arc of three moderate brightness to moderately faint stars.

| ga | NGC 4561 | 12 36.1 | +19 19 | Com | 12.9b | 1.5' | 1.2' | Sdm |

HH: pretty bright, pretty large, very little extended, little brighter middle, resolvable.
170x250: Faint. Broadly slightly concentrated. Round. Small. No involved stars.

| ga | NGC 4565 | 12 36.3 | +25 59 | Com | 10.4b | 15.9' | 1.8' | Sb |

HH: bright, extremely large, extremely extended NW-SE, very suddenly brighter middle to a 10th–11th mag. stellar nucleus.
42x60: Visible. ***170x250:*** Moderately bright. Evenly moderately concentrated. An extremely thin edge-on, NW-SE. Very large: 3/4 of the field in size. The ends are faint. The dust lane is visible with averted vision along the NE side of the galaxy through the middle section, but, given that the galaxy is so thin, the dust lane is too far offset from center for it to be very conspicuous. No involved stars. Nearby NE of center is a moderately faint star; SSW is another moderately faint star.

| ga | NGC 4564 | 12 36.4 | +11 26 | Vir | 12.1b | 3.8' | 1.7' | E |

HH: pretty bright, small, little extended, pretty suddenly brighter middle.
170x250: Bright. Sharply pretty concentrated. Slightly elongated NE-SW. Small. No involved stars. It's between two moderate brightness stars, one NW and one SE.

| ga | NGC 4567 | 12 36.5 | +11 15 | Vir | 12.1b | 3.3' | 2.0' | Sbc |

HH: very faint, large, NW of double nebula.
[See 4568.]

| ga | NGC 4568 | 12 36.6 | +11 14 | Vir | 11.7b | 4.8' | 2.0' | Sbc |

HH: very faint, large, SE of double nebula.
170x250: 4567 and 4568 are in contact (optically if not physically) NNW-SSE; they form a fat "V" opening up SW. They're very similar. Both are moderate brightness. Broadly very slightly concentrated. Moderately small. No involved stars. 4567 is elongated E-W. 4568 is elongated NNE-SSW. E of the two galaxies are a couple of moderate brightness stars. [Siamese Twins]

| ga | NGC 4569/M90 | 12 36.8 | +13 09 | Vir | 10.3b | 9.6' | 4.3' | Sab |

HH: pretty large, brighter middle to a nucleus.
42x60: Visible. ***170x250:*** Bright. Sharply extremely concentrated to a relatively bright stellar nucleus. Very elongated NNE-SSW. Large: a third of the field in size. Its nucleus is one of the most prominent among all galaxies; the halo doesn't concentrate until the nucleus. No involved stars. [I3583 was not seen.]

| ga | NGC 4570 | 12 36.9 | +07 14 | Vir | 11.8b | 5.7' | 1.6' | S0 |

HH: considerably bright, pretty small, much extended N-S, suddenly brighter middle to a resolvable nucleus.
170x250: Bright; pretty high surface brightness. Sharply very concentrated. A fairly thin edge-on, NNW-SSE. Moderate size. The ends fade out indistinctly. No involved stars.

| ga | NGC 4571 | 12 36.9 | +14 13 | Com | 11.8b | 3.6' | 3.1' | Sd |

HH: very faint, large, extended, very gradually brighter middle, considerably bright star attached.
170x250: Very faint. Broadly very slightly concentrated. Round. Small. No involved stars. A little ways NE is a bright star.

| ga | NGC 4589 | 12 37.4 | +74 11 | Dra | 11.7b | 3.1' | 2.5' | E2 |

HH: considerably bright, large, little extended, pretty gradually much brighter middle.
170x250: Moderate brightness. Evenly to sharply fairly concentrated. Round. Small. No involved stars. Very nearby WNW is a faint star; NE is a moderately bright star.

| ga | NGC 4578 | 12 37.5 | +09 33 | Vir | 12.4b | 3.3' | 2.4' | S0 |

HH: pretty faint, pretty small, round, suddenly brighter middle to a nucleus, star NW.
170x250: Moderate brightness to moderately bright. Sharply pretty concentrated. Round. Small. No involved stars. It makes a large flat triangle with a moderately bright star W and a moderate brightness star farther SW.

| ga | NGC 4579/M58 | 12 37.7 | +11 49 | Vir | 9.6v | 5.9' | 4.7' | Sb |

JH: bright, large, irregularly round, very much brighter middle, resolvable.
42x60: Visible. ***170x250:*** Bright. Sharply pretty concentrated. Slightly elongated ENE-WSW. Moderate size. It has a very faint outer halo. No involved stars.

| ga | NGC 4580 | 12 37.8 | +05 22 | Vir | 11.8v | 2.1' | 1.6' | Sa |

HH: pretty bright, large, very gradually brighter middle.
170x250: Faint to moderately faint. Broadly very slightly concentrated. Round. Small. No involved stars. ESE is a moderate brightness star.

| ga | NGC 4586 | 12 38.5 | +04 19 | Vir | 12.6b | 4.0' | 1.3' | Sa |

HH: pretty bright, large, extended, pretty suddenly brighter middle.
170x250: Faint. Broadly slightly concentrated. Elongated WNW-ESE. Moderate size. No involved stars.

| ga | NGC 4592 | 12 39.3 | −00 31 | Vir | 12.2p | 5.7' | 1.4' | Sdm |

HH: faint, large, extended E-W, very gradually brighter middle.
170x250: Faint. Broadly very slightly concentrated. Edge-on E-W. Moderately small. No involved stars. A little ways N is a faint star; SE is another faint star.

| gc | NGC 4590/M68 | 12 39.5 | −26 45 | Hya | 8.2 | 11.0' | −6.81 | 10 |

JH: globular cluster, large, extremely rich, very compressed, irregularly round, well resolved, 12th mag. stars, red.
12x35, 42x60: Easily visible. *170x250:* Mostly resolved. A good number of stars are very distinct, especially among the scattered outliers; in the middle is a mass of closely packed stars. Broadly concentrated. A third of the field in size with the outliers. Somewhat faint. Somewhat irregular. It doesn't have a neat central ball; it looks moderately massive.

| ga | NGC 4593 | 12 39.7 | −05 20 | Vir | 11.7p | 3.9' | 2.8' | Sb |

HH: pretty bright, considerably large, extended, suddenly brighter middle to a stellar nucleus.
170x250: Moderately bright. Sharply very very concentrated to a substellar nucleus. Slightly elongated ENE-WSW. Moderately small. No involved stars.

| ga | NGC 4595 | 12 39.9 | +15 17 | Com | 12.9b | 1.7' | 1.0' | Sb |

HH: pretty faint, pretty large, round, gradually brighter middle.
170x250: Moderately faint. Broadly slightly concentrated. Very slightly elongated WNW-ESE. Small. No involved stars. It makes a triangle with a moderately faint star E and a moderate brightness star SE.

| ga | NGC 4596 | 12 39.9 | +10 10 | Vir | 11.4b | 4.4' | 3.1' | S0+ |

HH: bright, pretty small, round, gradually much brighter middle, resolvable, 3 stars E.
170x250: *Moderately bright. Evenly moderately concentrated. Slightly elongated ENE-WSW. Small. No involved stars. At its edge SSW of center is a moderately faint star; SE are two brighter stars.*

| ga | NGC 4594/M104 | 12 40.0 | −11 37 | Vir | 9.0b | 8.8' | 3.5' | Sa |

HH: ! very bright, very large, extremely extended E-W, very suddenly much brighter middle to a nucleus.
12x35: *Visible.* ***42x60:*** *Bright.* ***170x250:*** *Bright to very bright. Sharply pretty concentrated to a small nucleus. Edge-on E-W (it's more elongated than in the pictures). Large: a quarter of the field in size. The dust lane is obvious across most of the galaxy S of the nucleus and is visible all the way across except at the extremely faint ends. With averted vision the halo extends faintly S of the dust lane. N of center is a faint star. An exceptional galaxy; one of the very best, with probably the most distinctive of all galactic dust lanes. [Sombrero Galaxy]*

| ga | NGC 4605 | 12 40.0 | +61 36 | UMa | 10.9b | 5.7' | 2.1' | Sc |

HH: bright, large, very much extended WNW-ESE, gradually little brighter middle.
42x60: *Visible.* ***170x250:*** *Bright; moderate surface brightness. Broadly slightly concentrated. A fat edge-on, WNW-ESE. Moderately large. The galaxy is asymmetrical: it's slightly curved, concave NE. No involved stars.*

| ga | ESO 506-33 | 12 40.2 | −25 19 | Hya | 13.0p | 1.8' | 0.6' | S0 |

170x250: *Moderately faint to moderate brightness. Sharply pretty concentrated. Elongated N-S. Very small to small. No involved stars.*

| ga | NGC 4597 | 12 40.2 | −05 47 | Vir | 12.6p | 4.0' | 1.8' | Sm |

WH: faint, very large, brighter middle.
170x250: *Extremely faint. Diffuse. Slightly elongated NE-SW. Moderate size. No involved stars. NNW is a little triangle of very faint stars.*

| ga | NGC 4602 | 12 40.6 | −05 07 | Vir | 12.3p | 3.4' | 1.1' | Sbc |

HH: faint, large, extended, very gradually little brighter middle.
170x250: *Moderately faint. Broadly slightly concentrated. Very elongated WNW-ESE. Moderately small. Just off its E end is a very very faint star.*

| ga | NGC 4604 | 12 40.8 | −05 18 | Vir | 13.2 | 1.0' | 0.3' | Im |

Pe: no description.
170x250: Very very faint. Broadly slightly concentrated. Slightly elongated WNW-ESE. Very very small. No involved stars.

| ga | NGC 4606 | 12 41.0 | +11 54 | Vir | 12.7b | 2.7' | 1.4' | Sa |

HH: very faint, pretty small, extended, 2 or 3 very faint stars involved.
170x250: Moderate brightness. Broadly slightly concentrated. Edge-on NE-SW. Moderately small. At its SW end are a couple of involved stars in line with the axis of the galaxy. [4607 (LR) was not seen.]

| ga | NGC 4608 | 12 41.2 | +10 09 | Vir | 12.0b | 3.5' | 2.6' | S0 |

HH: pretty bright, pretty large, round, pretty suddenly brighter middle, resolvable, 12th mag. star 1' NW.
170x250: Moderate brightness to moderately bright; fairly high surface brightness. Evenly moderately concentrated. Round. Small. No involved stars. A little ways WNW are two stars, moderately faint and faint.

| ds | STF 1669 | 12 41.3 | −13 01 | Crv | 5.2" | 6.0 | 6.1 | F5 F3 |

42x60: Split. ***170x250:*** A pretty close almost perfectly equal pair. Both white.

| ga | NGC 4612 | 12 41.5 | +07 18 | Vir | 11.9b | 2.4' | 1.9' | S0 |

HH: pretty bright, small, round, pretty suddenly much brighter middle.
170x250: Moderate brightness. Evenly to sharply fairly concentrated. Round. Very small. No involved stars. Nearby E is a moderate brightness star; this is the first of a long line of bright stars going NNE. A fairly nice scene.

| ga | NGC 4618 | 12 41.5 | +41 09 | CVn | 11.2b | 4.2' | 3.4' | Sm |

HH: bright, large, round, much brighter middle, one of double nebula.
42x60: Faintly visible. ***170x250:*** Moderate brightness to moderately bright. Broadly slightly concentrated. Roughly round. Small. No involved stars. S of center is either a second galaxy – very faint, low surface brightness – or a second concentration within 4618. If it's a second galaxy it's unusual because, unlike most double galaxy components, it's neither a clearly distinct galaxy nor of comparable brightness to its twin. If it's part of the first galaxy it's unusual because it's so detached. [It's a gigantic, peculiar spiral arm.] S of both objects is a moderate brightness star. Half the field NNE is 4625. Moderately faint. Broadly slightly concentrated. Round. Small. No involved stars. Nearby WSW is a moderately faint star.

bs	Gamma Vir	12 41.7	−01 27	Vir	3.50		0.30	F0

170x250: Soft white. [This is a double star that I have split in the past; it is now (Jan. 2008) too close.]

ga	NGC 4648	12 41.7	+74 25	Dra	13.0b	2.1'	1.5'	E3

HH: pretty bright, considerably small, round, gradually brighter middle, double star W.
170x250: Moderately faint; fairly high surface brightness. Sharply very concentrated to a faint stellar nucleus. Round. Very very small. No involved stars. It's right at the mouth of a "Y" figure of moderate brightness stars opening up SE.

ga	NGC 4625	12 41.9	+41 16	CVn	12.9b	2.2'	1.9'	Sm

HH: pretty faint, small, round.
[See 4618.]

ga	NGC 4620	12 42.0	+12 56	Vir	13.2b	1.8'	1.5'	S0

JH: very faint, small, round, very gradually brighter middle.
170x250: Very faint. Broadly slightly concentrated. Round. Small. No involved stars.

ga	NGC 4621/M59	12 42.0	+11 38	Vir	10.6b	5.3'	3.2'	E5

JH: bright, pretty large, little extended, very suddenly very much brighter middle, 2 stars W.
170x250: Bright. Evenly to sharply very concentrated. Round. Moderately small. No involved stars. It makes a triangle with a moderately faint star nearby N and another star NE.

ga	NGC 4627	12 42.0	+32 34	CVn	12.5v	2.6'	1.8'	E4

HH: faint, small, round, NW of 2.
[See 4631.]

ga	NGC 4631	12 42.1	+32 32	CVn	9.8b	15.4'	2.6'	Sd

HH: ! very bright, very large, extremely extended ENE-WSW, brighter middle to a nucleus, bright star near.
12x35: Visible. *42x60:* Easily visible. *170x250:* Bright; fairly high surface brightness. Broadly very slightly concentrated. Edge-on E-W. Very large: 3/4 of the field

in size. The W end extends farther than the E end, which is a little fatter and stops more abruptly. There's some small-scale unevenness in the light; E of center is a slight brightening. Just off its edge NW of center is a moderate brightness star; right at its W end is a very faint star. A very nice galaxy; impressive. A little ways NW of center is a small galaxy [4627]. Very faint; much lower surface brightness than 4631. Very slightly concentrated.

| ga | NGC 4623 | 12 42.2 | +07 40 | Vir | 13.2b | 2.2' | 0.7' | S0+ |

HH: considerably faint, pretty large, extended, pretty suddenly little brighter middle, resolvable.
170x250: Moderately faint. Broadly slightly concentrated. Elongated N-S. Small. No involved stars.

| ga | MCG-1-32-38 | 12 42.3 | −05 47 | Vir | 12.3 | 1.0' | 0.7' | S0− |

170x250: In the position are two galaxies right next to each other NE-SW [*M-1-32-39, M-1-32-38]. Both are extremely faint. Tiny. No involved stars. ESE is a triangle of moderately faint stars.

| ga | NGC 4626 | 12 42.4 | −07 02 | Vir | 12.7 | 1.3' | 0.4' | Sbc |

HH: very faint, considerably small, little extended, gradually little brighter middle.
170x250: [The first galaxy observed is *4628 (HH).] Faint. Evenly moderately concentrated. Edge-on NE-SW. Small. No involved stars. Nearby N are a couple of moderate brightness stars. A quarter of the field S is another galaxy [4626]. Broadly very slightly concentrated. Also edge-on NE-SW. Small. No involved stars. A little ways SSE is a faint star.

| ga | NGC 4630 | 12 42.5 | +03 57 | Vir | 13.2b | 1.8' | 1.2' | Im |

HH: considerably faint, small, round, little brighter middle.
170x250: Faint. Broadly slightly concentrated. Round. Small. No involved stars. ENE to N is a shallow arc of moderate brightness to moderately faint stars.

| ga | NGC 4632 | 12 42.5 | −00 04 | Vir | 12.4p | 3.3' | 1.3' | Sc |

HH: pretty bright, large, extended NE-SW.
170x250: Moderate brightness. Broadly slightly concentrated. Very elongated ENE-WSW. Moderately small. No involved stars. It makes an isosceles triangle with two moderately faint stars, one ENE and one SE.

| ga | NGC 4635 | 12 42.6 | +19 56 | Com | 13.2p | 2.0' | 1.4' | Sd |

JH: very faint, large, very gradually little brighter middle.
170x250: Very faint. Diffuse. Moderate size. No involved stars. SSW is a moderately faint star; W is an arc of moderately faint stars.

| ga | NGC 4634 | 12 42.7 | +14 17 | Com | 13.2b | 2.6' | 0.7' | Scd |

HH: very faint, large, much extended NW-SE, very gradually brighter middle.
170x250: Very faint. Hardly concentrated. A thin edge-on, NNW-SSE. Moderate size. No involved stars. A little less than a quarter of the field NNW is an extremely faint galaxy [*4633 (Sw)]. Right at its edge W is a faint star.

| ga | NGC 4636 | 12 42.8 | +02 41 | Vir | 10.4b | 6.0' | 4.6' | E0-1 |

HH: bright, large, irregularly round, very gradually very much brighter middle, resolvable.
42x60: Visible. ***170x250:*** Moderately bright. Evenly pretty concentrated. Round. Small. No involved stars. It's right in between two moderate brightness stars, one NNW and one SSE.

| ga | NGC 4638 | 12 42.8 | +11 26 | Vir | 12.1b | 2.2' | 1.3' | S0– |

HH: faint, round, gradually brighter middle.
170x250: Bright; very high surface brightness. Evenly pretty concentrated. Slightly elongated NW-SE. Very small. No involved stars. It makes a N-S diamond with two faint stars S and a moderate brightness star further S. [4637 (LR) was not seen.]

| ga | NGC 4639 | 12 42.9 | +13 15 | Vir | 12.2b | 3.2' | 2.3' | Sbc |

HH: pretty bright, small, extended, resolvable, 12th mag. star 1' SE.
170x250: Moderate brightness. Evenly moderately concentrated. Roughly round. Small. Just off its edge SE is a moderately faint star.

| ga | NGC 4643 | 12 43.3 | +01 58 | Vir | 11.7b | 3.6' | 2.2' | S0/a |

HH: considerably bright, pretty small, little extended, much brighter middle.
170x250: Bright. Sharply pretty concentrated. Round. Small. No involved stars. NNW is a sharp triangle of stars.

| ga | NGC 4647 | 12 43.5 | +11 34 | Vir | 11.9b | 2.9' | 2.3' | Sc |

HH: very faint, pretty large, little extended WNW-ESE, NW of double nebula.
[See M60.]

| ga | NGC 4649/M60 | 12 43.7 | +11 32 | Vir | 9.8b | 7.4' | 6.0' | E2 |

JH: very bright, pretty large, round, E of double nebula.
42x60: M60 and M59 are both visible. *170x250:* M60 is bright. Evenly moderately concentrated. Round. Moderate size. It has an extremely faint outer halo. No involved stars. Next to it NW is 4647. Moderately faint. Broadly slightly concentrated. Round. Moderately small; nearly the same size as M60's brighter inner part. No involved stars.

| ga | NGC 4651 | 12 43.7 | +16 23 | Com | 11.4b | 4.0' | 2.6' | Sc |

HH: considerably bright, large, extended E-W, gradually brighter middle, resolvable.
170x250: Moderate brightness. Broadly slightly concentrated. Slightly elongated E-W. Moderately small. No involved stars.

| ga | NGC 4653 | 12 43.9 | −00 33 | Vir | 12.2v | 3.0' | 2.6' | Scd |

WH: very faint, pretty large.
170x250: Very faint. Broadly very slightly concentrated. Roughly round. Small. No involved stars. Just off its edge SSE is a moderately faint star; a little ways SE is a moderate brightness star.

| ga | NGC 4654 | 12 44.0 | +13 07 | Vir | 11.1b | 5.2' | 2.8' | Scd |

HH: faint, very large, pretty much extended, double?, 3 stars near.
42x60: Visible next to the plotted star. *170x250:* Moderate brightness. Broadly very slightly concentrated. Elongated WNW-ESE. Moderately large. No involved stars. It makes a triangular figure with the bright plotted star WNW and two moderate brightness stars N.

| ga | NGC 4656 | 12 44.0 | +32 09 | CVn | 11.0b | 9.1' | 1.7' | Sm |

HH: ! pretty bright, large, very much extended NE-SW, SW of 2.
42x60: Very faintly visible. *170x250:* Moderately faint. Broadly slightly concentrated. Elongated NE-SW. Large: a third of the field in size. The galaxy curves E at its N end. The N half is brighter than the S half, which is extremely faint. The whole galaxy is irregular. No involved stars. [4657, a tuft at 4656's N end, was not seen as a separate object.]

| ga | ESO 381-12 | 12 44.1 | −34 12 | Cen | 12.6v | 0.9' | 0.7' | S |

170x250: Moderate brightness; high surface brightness. Sharply very concentrated. Round. Tiny; nearly stellar. No involved stars. It makes a short crooked line with three stars nearby ESE; further ESE is a bright star.

| ga | MCG-1-33-1 | 12 44.1 | −05 40 | Vir | 12.8b | 3.8' | 3.5' | Sdm |

170x250: Exceedingly faint. A diffuse glow. Moderately small. No involved stars. It makes a large triangle with two small groups of moderately faint stars: a sharp triangle of stars NE and a shallow arc of three stars N.

| ga | NGC 4657 | 12 44.2 | +32 12 | CVn | − | 1.3' | 0.6' | Sm |

HH: ! pretty faint, large, extended E-W, NE of 2.
[See 4656.]

| ga | NGC 4659 | 12 44.5 | +13 29 | Com | 13.1p | 1.7' | 1.2' | S0/a |

HH: faint, considerably small, round, brighter middle, resolvable.
170x250: Moderate brightness. Evenly moderately concentrated. Round. Very small. No involved stars. Nearby SSW is the bright plotted star.

| ga | NGC 4660 | 12 44.5 | +11 11 | Vir | 12.2b | 2.2' | 1.6' | E |

HH: very bright, small, very suddenly very much brighter middle to a nucleus.
170x250: Bright; very high surface brightness. Sharply very concentrated. Round. Very small. No involved stars.

| ga | NGC 4658 | 12 44.6 | −10 05 | Vir | 13.0b | 2.1' | 0.9' | Sbc |

HH: very faint, large, extended, 16th mag. star attached, 9th mag. star W.
170x250: Faint. Broadly slightly concentrated. Slightly elongated N-S. Moderately small. Right at its N end is a very faint star; a little ways W is the bright plotted star; a little ways SSE is a moderately faint star. A little less than half the field SSE is another galaxy [*4663 (Te)]. Moderately faint. Very small. Nearby SE is a star.

| ga | NGC 4665 | 12 45.1 | +03 03 | Vir | 10.5v | 3.8' | 3.1' | S0/a |

HH: bright, pretty large, irregularly round, much brighter middle, 10th mag. star SW.
170x250: Moderately bright to bright. Evenly pretty concentrated. Roughly round. Small. At its edge N is an extremely faint star; nearby SW is a moderately bright star.

| ga | NGC 4666 | 12 45.1 | −00 27 | Vir | 10.7v | 5.7' | 1.5' | Sc |

HH: bright, very large, much extended NE-SW, pretty suddenly brighter middle.
170x250: Moderately bright. Evenly moderately concentrated. Edge-on NE-SW. Moderately large. No involved stars. SE is a very small trio of stars. A sixth of the field E of this trio is a very faint to faint galaxy [*4668 (WH)]. Broadly slightly concentrated. Small.

| vs | Y CVn | 12 45.1 | +45 26 | CVn | 5.0 | 6.4 | 2.54 | C5 |

12x35: Mag. 5.5. ***170x250:*** Orangish. Nice, but not very deeply colored for a carbon star.

| ga | NGC 4670 | 12 45.3 | +27 07 | Com | 13.1b | 1.7' | 1.2' | S0/a |

HH: pretty faint, considerably small, round, brighter middle, resolvable, W of 2.
170x250: Moderate brightness. Sharply very concentrated. Round. Very small. No involved stars. A third of the field SE is a smaller and fainter version of 4670 [*4673 (HH)]. Faint. Tiny.

| ga | MCG-1-33-3 | 12 45.7 | −06 04 | Vir | 13.0b | 3.4' | 2.4' | Sm |

170x250: Not visible.

| ga | NGC 4682 | 12 47.3 | −10 03 | Vir | 12.9p | 2.5' | 1.2' | Scd |

HH: considerably faint, large, extended NE-SW, gradually very little brighter middle.
170x250: Very faint. Hardly concentrated. Slightly elongated E-W. Small. No involved stars. Nearby NE is a faint star; ESE is a triangular figure of faint stars; the galaxy is at the southern edge of a very large ring of mostly moderate brightness stars N of it.

| ga | NGC 4684 | 12 47.3 | −02 43 | Vir | 12.4p | 2.8' | 1.0' | S0+ |

HH: bright, pretty large, pretty much extended NNE-SSW.
170x250: Moderately bright. Broadly to evenly moderately concentrated. Edge-on NNE-SSW. Small. At its N end is a very faint involved star.

| ga | NGC 4688 | 12 47.8 | +04 20 | Vir | 12.6p | 3.1' | 2.7' | Scd |

HH: extremely faint, pretty large, 9th–10th mag. star W.
170x250: Extremely faint. Diffuse. Moderate size. No involved stars. WSW is a moderately bright star.

| ga | NGC 4689 | 12 47.8 | +13 45 | Com | 11.6b | 4.3' | 3.4' | Sbc |

HH: pretty bright, very large, extended, very gradually little brighter middle, resolvable.
170x250: Faint. Broadly very slightly concentrated. Round. Moderately small. No involved stars. N are a couple of moderate brightness stars.

| ga | NGC 4691 | 12 48.2 | −03 19 | Vir | 11.7b | 3.2' | 2.4' | S0/a |

HH: pretty bright, pretty large, extended E-W, much brighter middle.
170x250: Moderately bright. Broadly to evenly moderately concentrated. Edge-on E-W. Small. No involved stars.

| ga | NGC 4694 | 12 48.2 | +10 59 | Vir | 11.4v | 4.2' | 2.5' | S0 |

HH: pretty faint, small, very little extended.
170x250: Moderate brightness. Evenly moderately concentrated. Roughly round. Small. No involved stars. Nearby W is a faint star.

| ga | NGC 4698 | 12 48.4 | +08 29 | Vir | 11.5b | 4.0' | 2.4' | Sab |

WH: considerably bright, pretty large, irregularly round, brighter middle, resolvable.
170x250: Moderately bright. Evenly moderately concentrated. Slightly elongated NNW-SSE. Small. No involved stars. It's between two moderately bright stars, one N and one S.

| ga | NGC 4697 | 12 48.6 | −05 48 | Vir | 10.1b | 7.3' | 4.7' | E6 |

HH: very bright, large, little extended NE-SW, suddenly much brighter middle to a resolvable nucleus.
12x35, 42x60: Visible. ***170x250:*** Bright. Sharply pretty concentrated. Elongated ENE-WSW. Moderate size. Off its edge NW is a moderately faint star that makes an evenly spaced line with two more stars, one NE and one W.

| ga | NGC 4699 | 12 49.0 | −08 39 | Vir | 10.4b | 4.0' | 2.8' | Sb |

HH: very bright, round, very much brighter middle to a resolvable nucleus, resolvable.
12x35: Visible. ***42x60:*** Bright. ***170x250:*** Very bright. Sharply very concentrated. Round. Small. No involved stars. It makes a triangle with two faint stars, one SSW and one E; further E is a moderate brightness star.

| ga | NGC 4700 | 12 49.1 | −11 24 | Vir | 12.6p | 3.0' | 0.5' | Sc |

HH: faint, large, much extended NE-SW, very little brighter middle, bright star W.
170x250: Moderately faint. Broadly very slightly concentrated. A thin edge-on, NE-SW. Moderate size. No involved stars. A little ways W is a moderately bright star; a little ways N is a faint star.

| ga | NGC 4701 | 12 49.2 | +03 23 | Vir | 12.8b | 2.7' | 2.0' | Scd |

WH: faint, small.
170x250: Moderate brightness. Broadly moderately concentrated. Round. Small. No involved stars. It's on the side of a large triangular figure of stars: a moderate brightness star WNW, another moderate brightness star SSW, and a little triangle of faint stars NNE.

| ds | STF 1694 | 12 49.2 | +83 25 | Cam | 21.5" | 5.3 | 5.8 | A1 A0 |

12x35: Nicely split. ***42x60:*** A moderately close very slightly unequal pair. ***170x250:*** A wide pretty much equal pair. Both white.

| ga | MCG-2-33-15 | 12 49.4 | −10 07 | Vir | 12.1p | 4.0' | 3.1' | Sm |

170x250: Not visible.

| ga | NGC 4710 | 12 49.6 | +15 09 | Com | 11.9b | 5.6' | 1.3' | S0+ |

HH: considerably bright, pretty large, very much extended NNE-SSW, suddenly brighter middle to a nucleus.
170x250: Bright. Broadly moderately concentrated. A thin edge-on, NNE-SSW. Moderate size. No involved stars. Nearby E is a moderately faint star.

| ga | NGC 4713 | 12 50.0 | +05 18 | Vir | 12.2b | 2.7' | 1.7' | Sd |

HH: pretty bright, large, very little extended, gradually little brighter middle.
170x250: Faint. Broadly slightly concentrated. Roughly round. Small. No involved stars. SE are a couple of moderate brightness stars; farther SSE is another moderate brightness star.

| ga | NGC 4750 | 12 50.1 | +72 52 | Dra | 12.1p | 2.0' | 1.8' | Sab |

HH: pretty bright, large, round, very gradually then very suddenly brighter middle.
170x250: Moderate brightness. Broadly moderately concentrated. Round. Small. No involved stars.

| ga | NGC 4725 | 12 50.4 | +25 30 | Com | 10.1b | 10.7' | 8.0' | Sab |

HH: very bright, very large, extended, very gradually then very suddenly very much brighter middle to an extremely bright nucleus.

42x60: Visible. *170x250:* Bright. Sharply very concentrated to a small brighter core. Slightly elongated NE-SW. Large: a third of the field in size. The halo is low surface brightness and somewhat uneven; SW of center is a slight brightening. Off the galaxy's edge NNW is a faint star. 2/3 of a field W is a small very faint galaxy [*4712 (JH)].

ga	ESO 507-21	12 50.5	−26 50	Hya	13.2p	1.7'	0.7'	S0−

170x250: Moderately faint. Evenly moderately concentrated. Roughly round. Small. No involved stars. Nearby W is an extremely faint star.

ga	NGC 4736/M94	12 50.9	+41 07	CVn	9.0b	14.3'	12.1'	Sab

JH: very bright, large, irregularly round, very suddenly very much brighter middle to a bright nucleus, resolvable.
12x35: Easily visible. *42x60:* Bright. *170x250:* Very bright. Sharply very concentrated to a small brighter core. Round. Large: a quarter of the field in size. The halo fades out very indistinctly; the outer extremities are extremely low surface brightness. No involved stars.

ga	NGC 4731	12 51.0	−06 23	Vir	11.9b	6.6'	3.2'	Scd

HH: very faint, pretty large, extended.
170x250: Faint. Broadly very slightly concentrated. Slightly elongated NW-SE. Moderately small. At its edge WSW of center is a very faint star.

ga	NGC 4733	12 51.1	+10 54	Vir	11.8v	2.2'	2.2'	E+

HH: considerably faint, pretty large, round, resolvable, 12th mag. star W.
170x250: Moderately faint. Broadly slightly concentrated. Round. Small. Just off its edge W is a faint star.

ga	ESO 507-25	12 51.5	−26 27	Hya	12.6b	2.3'	1.7'	S0−

170x250: Moderate brightness. Evenly fairly concentrated. Round. Small. No involved stars. Nearby E are two faint stars.

ga	NGC 4742	12 51.8	−10 27	Vir	12.1b	2.6'	1.4'	E4

HH: considerably bright, very small, very much brighter middle to a 9th mag. stellar nucleus, 10th mag. star SE.
170x250: Bright. Sharply extremely concentrated to a substellar nucleus. Round. Very small. No involved stars. Nearby SE is a moderate brightness star.

| ga | NGC 4747 | 12 51.8 | +25 46 | Com | 13.0b | 3.7' | 1.1' | Scd |

WH: faint, pretty large, little extended.
170x250: Extremely faint. Unconcentrated. Very elongated NNE-SSW. Moderate size. No involved stars. SSE are two bright stars; the N of these is plotted and has a distant companion.

| ga | ESO 442-26 | 12 52.2 | −29 50 | Hya | 12.6p | 2.6' | 0.7' | S0 |

170x250: Moderate brightness. Sharply pretty concentrated. Elongated N-S. Small. No involved stars.

| ga | NGC 4754 | 12 52.3 | +11 18 | Vir | 11.5b | 5.4' | 2.9' | S0− |

HH: bright, pretty large, round, pretty suddenly brighter middle, W of 2.
[See 4762.]

| ga | NGC 4753 | 12 52.4 | −01 11 | Vir | 10.9b | 6.0' | 2.8' | I0 |

HH: considerably bright, large, very little extended, very gradually little brighter middle.
12x35: Faintly visible. ***42x60:*** Visible. ***170x250:*** Moderately bright. Evenly moderately concentrated. Slightly elongated E-W. Moderately small. No involved stars. A little ways W are two faint stars; NNE is a moderately faint star with a fainter star next to it WSW.

| ga | NGC 4756 | 12 52.9 | −15 24 | Crv | 12.4v | 1.8' | 1.4' | S0 |

HH: very faint, pretty small, resolvable.
170x250: Moderately faint. Broadly slightly concentrated. Slightly elongated NE-SW. Moderately small. No involved stars. Off its NE end is a very faint star. Half the field SW is another galaxy [*I829 and *M-2-33-37 combined]. Faint. Small.

| ga | NGC 4762 | 12 52.9 | +11 13 | Vir | 10.2v | 8.8' | 1.7' | S0 |

HH: pretty bright, very much extended NE-SW, 3 bright stars S, E of 2.
42x60: 4762 and 4754 are both visible. ***170x250:*** 4762 is bright. Sharply pretty concentrated. An extremely thin edge-on, NNE-SSW; there's hardly a thinner galaxy in the sky. Moderately large. The ends are indistinct. Around its SW end is an arc of three bright stars; the galaxy doesn't quite reach the imaginary line of this arc. A little more than half the field WNW is 4754. Bright. Sharply pretty concentrated. Round. Small. No involved stars. In line with it are two fairly bright stars WSW. A nice pair of bright galaxies, contrasting drastically in shape.

| ga | NGC 4760 | 12 53.1 | –10 29 | Vir | 11.4v | 1.9' | 1.8' | E0 |

Wi: pretty bright, round.
170x250: Moderately faint. Broadly to evenly slightly concentrated. Round. Small. No involved stars. It's between two bright stars, one NNE and one SW.

| ga | NGC 4763 | 12 53.4 | –17 00 | Crv | 12.6v | 1.5' | 1.0' | Sa |

WH: very faint, small, little brighter middle.
170x250: Faint. Broadly slightly concentrated. Round. Small. No involved stars. It makes a shallow arc with two moderately faint stars N.

| ga | NGC 4771 | 12 53.4 | +01 16 | Vir | 12.9b | 3.9' | 0.8' | Sd |

HH: faint, pretty large, much extended, 9th mag. star W.
170x250: Faint. Broadly slightly concentrated. Elongated NW-SE. Small. No involved stars. A little ways W is a moderately bright star that makes a shallow arc with two moderately faint stars S.

| ga | NGC 4772 | 12 53.5 | +02 10 | Vir | 12.0b | 3.4' | 1.7' | Sa |

HH: pretty faint, pretty small, round, much brighter middle.
170x250: Moderate brightness to moderately bright. Sharply pretty concentrated. Round. Small. No involved stars.

| ga | NGC 4775 | 12 53.8 | –06 37 | Vir | 11.7p | 2.1' | 1.9' | Sd |

HH: faint, considerably large, round, very gradually little brighter middle, resolvable.
170x250: Moderate brightness. Broadly slightly concentrated. Round. Moderately small. No involved stars. Nearby SW is a faint star; a little ways ENE is another faint star.

| ga | NGC 4779 | 12 53.8 | +09 42 | Vir | 13.2b | 2.1' | 1.7' | Sbc |

HH: very faint, pretty large, round, resolvable.
170x250: Faint. Broadly slightly concentrated. Roughly round. Small. No involved stars.

| bs | Epsilon UMa | 12 54.0 | +55 57 | UMa | 1.77 | | –0.02 | A0 |

170x250: White.

| ga | NGC 4789 | 12 54.3 | +27 04 | Com | 13.1b | 1.9' | 1.5' | S0 |

HH: faint, round, 9th mag. star attached 1' N.
170x250: Faint. Concentrated. Round. Very very small. Right next to it N is a moderately bright star. [4787 (dA) and 4789A were not seen.]

| ga | NGC 4781 | 12 54.4 | –10 32 | Vir | 11.7p | 3.4' | 1.5' | Sd |

WH: considerably bright, very large, much extended.
170x250: Moderate brightness. Broadly very slightly concentrated. Elongated WNW-ESE. Moderately small to moderate size. At its W end is an involved star that makes a line with two more stars WSW; NE is another star; all these stars are moderate brightness to moderately faint. A third of the field SE is a nearly stellar galaxy [*4784 (WH)]. It makes a small triangle with a moderate brightness star W and a moderately bright star NNE.

| ga | NGC 4786 | 12 54.5 | –06 51 | Vir | 12.7b | 1.6' | 1.2' | E+ |

WH: pretty bright, pretty small, much brighter middle, resolvable.
170x250: Moderate brightness. Evenly moderately concentrated. Roughly round. Very small. No involved stars. A little ways N is a pair of stars, moderate brightness and very faint; S are several faint stars.

| ga | NGC 4782 | 12 54.6 | –12 34 | Crv | 12.7p | 1.7' | 1.6' | E0 |

WH: pretty faint, pretty small, round, much brighter middle, W of double nebula.
[See 4783.]

| ga | NGC 4783 | 12 54.6 | –12 33 | Crv | 12.5p | 1.7' | 1.7' | E0 |

WH: pretty faint, pretty small, round, much brighter middle, E of double nebula.
170x250: 4783 and 4782 are twin galaxies touching each other N-S, probably interacting; truly a double galaxy. They're similar. Both are moderate brightness. Evenly moderately concentrated. Round. Small. No involved stars. Half the field ESE is a small faint galaxy [*4794 (WH)]. On either side of center E and W is an involved star.

| ga | NGC 4793 | 12 54.6 | +28 56 | Com | 12.3b | 2.9' | 1.4' | Sc |

HH: pretty bright, pretty small, little extended, 8th mag. star 1' NE.
170x250: Moderate brightness. Broadly slightly concentrated. Slightly elongated NE-SW. Small. No involved stars. Nearby N is a moderately bright star.

| ga | NGC 4800 | 12 54.6 | +46 31 | CVn | 12.3b | 1.5' | 1.1' | Sb |

HH: pretty bright, considerably small, round, pretty suddenly brighter middle, 14th mag. star W.
170x250: Moderately bright. Broadly slightly concentrated. Round. Very small. Just off its edge W is a moderately faint star.

| ga | NGC 4790 | 12 54.9 | −10 14 | Vir | 12.8p | 1.7' | 1.0' | Sc |

WH: pretty faint, pretty small, irregularly round.
170x250: Moderately faint. Broadly slightly concentrated. Slightly elongated E-W. Small to moderately small. No involved stars. Nearby N is a faint star; nearby E is another faint star.

| ga | NGC 4795 | 12 55.0 | +08 03 | Vir | 12.1v | 1.8' | 1.5' | Sa |

HH: pretty faint, pretty large, round, brighter middle, resolvable.
170x250: Moderately faint. Elongated E-W. Small. Near center is an involved star that at first sight makes the galaxy look brighter and more concentrated than it really is [this is *4796 (Ma), a companion stellar galaxy]; nearby NNW is a faint star that makes an isosceles triangle with two brighter stars further NNW. [4791 (Ma) was not seen.]

| ga | UGC 8041 | 12 55.2 | +00 06 | Vir | 12.6p | 3.0' | 1.8' | Sd |

170x250: Very very faint. Diffuse. Elongated N-S. Moderately small. No involved stars.

| ga | NGC 4814 | 12 55.4 | +58 20 | UMa | 12.8b | 3.1' | 2.2' | Sb |

HH: bright, pretty small, very little extended, very gradually brighter middle.
170x250: Moderately faint. Broadly slightly concentrated. Round. Small. No involved stars.

| bs | Delta Vir | 12 55.6 | +03 23 | Vir | 3.38 | | 1.58 | M3 |

170x250: Yellow gold.

| ga | ESO 507-45 | 12 55.6 | −26 49 | Hya | 12.8p | 1.9' | 1.3' | S0 |

170x250: Moderately faint. Evenly moderately concentrated. Round. Very small. No involved stars. Nearby ENE is a moderate brightness star; a little farther SSE is a very faint double star; a little ways WSW is a faint star. [E507-46 was not seen.]

| ga | NGC 4802 | 12 55.8 | –12 03 | Crv | 12.6 | 2.4' | 1.6' | S0 |

Te: very faint, small, 10th mag. star attached.
170x250: Moderately faint. Evenly moderately concentrated. Round. Very small. ESE of center is a moderately bright involved star that nearly overwhelms the galaxy; this is one of a large group of relatively bright field stars mostly NE; the next closest is a little ways W.

| ga | NGC 4808 | 12 55.8 | +04 18 | Vir | 12.4b | 2.7' | 1.1' | Scd |

HH: pretty faint, considerably large, extended NW-SE.
170x250: Moderately faint to moderate brightness. Broadly slightly concentrated. Elongated NW-SE. Small. No involved stars. Nearby WNW is a very faint star.

| bs | Alpha CVn | 12 56.0 | +38 19 | CVn | 2.90 | | –0.12 | A0 |

12x35: Extremely close. ***42x60:*** A moderately close unequal pair. A nice double. ***170x250:*** A somewhat wide unequal pair. Color contrast: the primary is white; the companion is yellowish. Beautiful; a great double star. [STF 1692: 18.8", 2.9, 5.5, A0 F0]

| ds | STF 1695 | 12 56.3 | +54 06 | UMa | 3.7" | 6.0 | 7.9 | A5 – |

170x250: The primary has a very close fainter companion. The primary is white.

| vs | RY Dra | 12 56.4 | +66 00 | Dra | 6.0 | 8.2 | 3.26 | C3 |

12x35: Mag. 6.5. ***170x250:*** Slightly orangish.

| ga | NGC 4826/M64 | 12 56.7 | +21 41 | Com | 9.4b | 10.1' | 5.4' | Sab |

JH: ! very bright, very large, very much extended WNW-ESE, brighter middle to a small bright nucleus = double star?
12x35: Easily visible. ***42x60:*** Bright. ***170x250:*** Bright. Sharply extremely concentrated to a relatively bright nucleus. Elongated WNW-ESE. Very large: half the field in size. The halo is extremely vague, especially toward the edge. The little dust lane is visible hugging the nucleus on the NE side; best with averted vision; not very conspicuous (it's not as large as in the photos). No involved stars. [Black Eye Galaxy]

| ga | NGC 4818 | 12 56.8 | –08 31 | Vir | 12.0b | 5.1' | 1.5' | Sab |

HH: pretty bright, large, pretty much extended N-S, gradually brighter middle.

170x250: Moderately faint. Broadly slightly concentrated. Very elongated N-S. Moderately small. No involved stars. Nearby S is a moderately faint star; a little ways E are two very faint stars; W is a NNW-SSE row of three moderate brightness stars.

| ga | NGC 4825 | 12 57.2 | –13 39 | Vir | 12.7p | 1.8' | 1.1' | S0– |

WH: pretty bright, irregular figure, brighter middle.
170x250: Moderately faint. Evenly moderately concentrated. Round. Small. No involved stars. Nearby W is a faint star. [4820 (Te) was not seen.]

| ga | NGC 4839 | 12 57.4 | +27 29 | Com | 13.0b | 4.0' | 1.9' | E+ |

HH: faint, pretty large, round.
170x250: Very faint. Evenly moderately concentrated. Round. Small. No involved stars. A little ways NE is a moderately faint star; a little ways SW is a faint star; W are a couple of faint stars. [4842A and C160-40 were not seen.]

| ga | NGC 4830 | 12 57.5 | –19 41 | Vir | 13.1b | 1.8' | 1.1' | S0– |

Te: faint, large, stars involved, 8th mag. star 5' SE.
170x250: Faint to moderately faint. Evenly moderately concentrated. Round. Very small. No involved stars. Very nearby SW are two stars, extremely faint and faint; the galaxy makes a quadrilateral with three bright stars SSW to ESE.

| ga | NGC 4845 | 12 58.0 | +01 34 | Vir | 12.1b | 5.0' | 1.3' | Sab |

HH: pretty faint, pretty large, pretty much extended, very gradually brighter middle, star NNE.
170x250: Moderate brightness. Broadly very slightly concentrated. A fat edge-on, ENE-WSW. Moderately large. No involved stars. It's inside a flat isosceles triangle of stars: one moderate brightness star and two moderately faint stars.

| ga | NGC 4861 | 12 59.0 | +34 51 | CVn | 12.9b | 4.2' | 1.5' | Sm |

HH: very faint, pretty large, very much extended NNE-SSW, between 2 stars.
170x250: In the position is a moderate brightness substellar object, apparently a galactic nucleus, with an extremely faint halo extending NNE toward a moderately faint star. A very peculiar one-sided galaxy.

| ga | NGC 4868 | 12 59.1 | +37 18 | CVn | 13.0p | 1.6' | 1.4' | Sab |

WH: pretty bright, small, round, much brighter middle.
170x250: Moderately faint. Broadly slightly concentrated. Round. Small. At its edge WSW is a very faint star; nearby N is a moderately faint star.

| ga | NGC 4855 | 12 59.3 | −13 13 | Vir | 13.1 | 1.7' | 1.2' | S0− |

Te: faint, small, stars involved.
170x250: Moderately faint. Evenly moderately concentrated. Roughly round. Small. No involved stars. Nearby S is a very faint star; N is a moderate brightness star; NW is another moderate brightness star.

| ga | NGC 4856 | 12 59.3 | −15 02 | Vir | 11.5b | 4.2' | 1.1' | S0/a |

HH: bright, round, pretty suddenly much brighter middle, 13th mag. star NW.
170x250: Bright. Evenly moderately concentrated. Elongated NE-SW. Moderately small. E of center is a faint involved star.

| ga | NGC 4866 | 12 59.5 | +14 10 | Vir | 12.1b | 9.2' | 1.3' | S0+ |

HH: bright, pretty large, much extended E-W, suddenly brighter middle to a nucleus, faint star involved.
170x250: Moderately bright. Evenly to sharply fairly concentrated. Edge-on E-W. Moderate size. W of center is a faint involved star.

| ga | NGC 4874 | 12 59.6 | +27 57 | Com | 12.6b | 2.3' | 2.3' | E+0 |

dA: faint.
[See 4889.]

| ga | NGC 4889 | 13 00.1 | +27 58 | Com | 12.5b | 2.8' | 2.2' | E+4 |

HH: pretty bright, pretty much extended, brighter middle, 7th mag. star N, 4th of 5.
170x250: 4889 and 4874 [the principal galaxies of AGC 1656, the Coma Galaxy Cluster] are a little less than half the field apart E-W. 4889 is moderately faint. Evenly moderately concentrated. Round. Small. No involved stars. Very nearby SE is a faint star; S are two brighter stars. A little ways ESE is a faint stellar galaxy [*4898 (dA)]. 4874 is faint. Evenly moderately concentrated. Round. Very small. No involved stars. The bright plotted star N interferes. A little ways SW is a moderate brightness star. A quarter of the field SW is a faint galaxy [*4869 (HH)]. A little more than a quarter of the field WNW is another faint galaxy [*4864 (JH)]. These two galaxies make a triangle with 4874, with the star SW of 4874 on one side of the triangle. The above are the only galaxies readily visible in the immediate area without the aid of a very detailed chart, so no way does this look like an exceptional aggregation of galaxies. A little more than half the field ENE of 4889 are a couple of very faint galaxies [*4908 (dA), *I4051].

12 to 18 Hours: SPRING

| ga | NGC 4880 | 13 00.2 | +12 29 | Vir | 12.4p | 3.1' | 2.4' | S0+ |

HH: considerably faint, pretty large, round, very gradually little brighter middle, resolvable.
170x250: Very faint. Broadly very slightly concentrated. Roughly round. Small. No involved stars. SSW is a triangle of faint stars.

| ga | NGC 4877 | 13 00.4 | −15 17 | Vir | 13.2 | 2.4' | 1.0' | Sab |

WH: pretty bright, pretty large, much brighter middle.
170x250: Faint. Broadly slightly concentrated. Slightly elongated N-S. Small. No involved stars. NW is a bright star. A little less than half the field SW is an extremely faint galaxy [*M-2-33-82].

| ga | NGC 4888 | 13 00.6 | −06 04 | Vir | 12.8 | 1.1' | 0.5' | Sab |

HH: pretty faint, considerably small, extended, pretty suddenly brighter middle, double star NW.
170x250: [The first galaxy observed is *4878 (WH).] Faint. Evenly moderately concentrated. Round. Small. No involved stars. A little ways NW is a moderately bright star; ENE is a moderately bright double star. Just E of the double star is a small very faint galaxy [4888].

| ga | NGC 4900 | 13 00.6 | +02 30 | Vir | 11.9b | 2.2' | 2.2' | Sc |

HH: considerably bright, considerably extended, 10th mag. star attached SE.
170x250: Moderate brightness. Broadly very slightly concentrated. Round. Small. At its edge SE is a moderately bright involved star.

| ga | NGC 4890 | 13 00.7 | −04 36 | Vir | 12.7 | 0.9' | 0.7' | Sm |

HH: considerably faint, small, irregularly round, brighter middle.
170x250: Moderately faint. Broadly very slightly concentrated. Round. Small. No involved stars. It makes a flat triangle with two moderately faint stars, one SW and one NNW.

| ga | NGC 4914 | 13 00.7 | +37 18 | CVn | 12.5b | 3.5' | 1.9' | E+ |

HH: pretty bright, considerably small, round, suddenly much brighter middle, 17th mag. star NW.
170x250: Moderately bright. Sharply very concentrated to a substellar nucleus. Round. Very small. No involved stars. It makes a very large nearly perfect diamond with a bright star NE, another bright star W, and a fainter star NNW.

| ga | NGC 4897 | 13 00.9 | –13 26 | Vir | 12.6p | 2.6' | 2.3' | Sbc |

Te: faint.
170x250: Faint. Broadly slightly concentrated. Roughly round. Small. No involved stars. A little ways NW is a very faint star.

| ga | NGC 4899 | 13 00.9 | –13 56 | Vir | 12.6p | 2.6' | 1.4' | Sc |

WH: no description.
170x250: Very faint to faint. Broadly very slightly concentrated. Elongated NNE-SSW. Moderately small. No involved stars. Nearby NE are two stars, extremely faint and moderately faint.

| ga | ESO 443-24 | 13 01.0 | –32 26 | Cen | 12.9b | 1.8' | 1.4' | S0– |

170x250: Moderately faint. Evenly moderately concentrated. Round. Small. No involved stars. It makes a small triangle with two moderate brightness stars, one NW and one WNW; these stars are at the N end of a long NNE-SSW line of stars. [E443-29 was not seen.]

| ga | NGC 4902 | 13 01.0 | –14 30 | Vir | 11.6b | 3.0' | 2.6' | Sb |

HH: pretty bright, pretty large, irregularly round, stars near.
170x250: Faint to moderately faint. Broadly slightly concentrated. Round. Moderately small. No involved stars. It makes a small triangle with two bright stars, one NW and one WSW, each of which has a wide faint companion, but in opposite directions.

| ga | NGC 4904 | 13 01.0 | –00 01 | Vir | 12.6b | 2.2' | 1.4' | Scd |

WH: pretty bright, pretty small, round, brighter middle.
170x250: Moderately faint. Broadly very slightly concentrated. Round. Small. No involved stars. It makes a small triangle with a moderately faint star NNW and a faint star WSW.

| ga | NGC 4921 | 13 01.4 | +27 53 | Com | 13.0b | 2.0' | 1.7' | Sab |

HH: faint, pretty large, 2nd of 3.
170x250: Faint. Broadly slightly concentrated. Round. Small. No involved stars. ENE is a triangle of moderate brightness stars. Half the field NE is another faint galaxy [*4927 (WH)]. Nearby NE is a moderate brightness star. A little ways SSE of 4921 is a very faint galaxy [*4923 (HH)]. Half the field SW of 4921 is a faint galaxy [*4911 (WH)]. It makes a shallow arc with two moderate brightness stars NW. [This field is within the Coma Cluster, E of the main concentration; 4919 (dA) was not seen.]

| ga | NGC 4915 | 13 01.5 | –04 32 | Vir | 13.0b | 1.6' | 1.3' | E0 |

HH: pretty bright, small, round, brighter middle, stellar?
170x250: Bright; high surface brightness. Sharply very concentrated. Round. Very small. No involved stars.

| bs | Epsilon Vir | 13 02.2 | +10 57 | Vir | 2.83 | | 0.94 | G8 |

170x250: Yellow white.

| ga | NGC 4933 | 13 03.9 | –11 29 | Vir | 12.7p | 1.8' | 1.1' | S0/a |

WH: pretty bright, pretty large, irregularly round.
170x250: Moderately faint. Broadly slightly concentrated. Elongated NE-SW. Moderately small to moderate size. No involved stars. At its SW end is a knot or a tiny galaxy [*4933B]; a little ways NW is a faint star.

| ga | NGC 4939 | 13 04.2 | –10 20 | Vir | 11.9b | 5.8' | 3.7' | Sbc |

HH: pretty bright, large, round, gradually much brighter middle.
170x250: Moderately faint. Broadly to evenly slightly concentrated. Slightly elongated N-S. Moderately small to moderate size. No involved stars.

| ga | NGC 4941 | 13 04.2 | –05 33 | Vir | 11.9b | 3.6' | 1.9' | Sab |

HH: pretty faint, large, extended, gradually brighter middle to a bright nucleus, resolvable.
170x250: Moderately faint. Sharply slightly concentrated. Elongated NNE-SSW. Moderately small. No involved stars. A little ways S is a moderate brightness star that is in a long line of stars all the way across the field E-W.

| ga | NGC 4936 | 13 04.3 | –30 31 | Cen | 11.8p | 2.7' | 2.3' | E0 |

JH: pretty bright, small, round, brighter middle, star E.
170x250: Moderately bright. Evenly moderately concentrated. Roughly round. Small. No involved stars. Nearby E is a moderately bright star. A third of the field NNE is a very small very faint galaxy [*MAC1304-3025]. Half the field NNW is a small faint galaxy [*E443-43]. 2/3 of a field W is a moderately faint galaxy [*I844].

| ga | UGCA 322 | 13 04.5 | –03 34 | Vir | 12.9p | 3.5' | 2.5' | Sdm |

170x250: Not visible.

| ga | NGC 4951 | 13 05.1 | –06 29 | Vir | 12.6p | 3.3' | 1.1' | Scd |

HH: faint, pretty large, little extended, resolvable.
170x250: Moderately faint. Broadly slightly concentrated. Elongated E-W. Small. No involved stars.

| ga | IC 4182 | 13 05.8 | +37 36 | CVn | 11.8b | 7.7' | 5.4' | Sm |

170x250: Not visible.

| ga | NGC 4958 | 13 05.8 | –08 01 | Vir | 11.6b | 4.1' | 1.2' | S0 |

HH: very bright, pretty small, extended N-S, brighter middle to a bright nucleus.
170x250: Bright. Sharply pretty concentrated. Edge-on NNE-SSW. Moderately small. No involved stars. Nearby W is a faint star.

| ga | NGC 4955 | 13 06.1 | –29 45 | Hya | 13.2b | 1.8' | 1.2' | E2 |

JH: faint, considerably small, round, gradually brighter middle.
170x250: Faint. Evenly moderately concentrated. Round. Small. No involved stars. Nearby WSW is a moderate brightness star. [E443-66 was not seen.]

| ga | UGC 8201 | 13 06.4 | +67 42 | Dra | 12.8b | 3.4' | 1.8' | Im |

170x250: Not visible.

| ga | UGCA 324 | 13 06.9 | –28 33 | Hya | 13.1p | 2.1' | 1.9' | Sd |

170x250: Extremely faint. A slight diffuse glow. Roughly round. Moderately small. No involved stars.

| ga | NGC 4965 | 13 07.2 | –28 13 | Hya | 12.8p | 2.8' | 2.2' | Sd |

JH: very faint, very large, considerably extended, very gradually brighter middle.
170x250: Extremely faint. A slight unconcentrated glow. Roughly round. Moderately small to moderate size. No involved stars.

| ga | NGC 4970 | 13 07.6 | –24 00 | Hya | 13.2p | 1.7' | 1.0' | S0 |

WH: very faint, pretty large, irregular figure.
170x250: In the position is a very large triangle of galaxies. The northernmost galaxy [*I4197] is faint to moderately faint. Evenly moderately concentrated. Roughly round. Small. No involved stars. Curving around its E side is a large arc of moderately

bright to moderate brightness stars. A little more than 3/4 of a field SSW is a second galaxy [4970]. Moderate brightness. Evenly moderately concentrated. Round. Small. No involved stars. A little ways ESE is a faint star; a little ways W is a moderately faint star. Half the field WNW is a moderately faint galaxy [*I4180]. Evenly moderately concentrated. Round. Small. No involved stars.

| ga | NGC 4981 | 13 08.8 | −06 46 | Vir | 12.1p | 2.7' | 2.0' | Sbc |

HH: bright, pretty large, round, 9th mag. star SE.
170x250: Moderate brightness. Evenly slightly concentrated. Slightly elongated NNW-SSE. Small. At its S end is a moderately bright star.

| ga | NGC 4984 | 13 09.0 | −15 30 | Vir | 12.3b | 4.2' | 2.6' | S0+ |

HH: bright, pretty large, round, pretty suddenly much brighter middle.
170x250: Bright. Sharply very concentrated. Round. Small. Off its edge SSW is a very faint star; a little ways NNW is a faint star; a little ways ENE are a couple of moderate brightness stars.

| ga | NGC 4999 | 13 09.6 | +01 40 | Vir | 12.6p | 2.4' | 1.9' | Sb |

HH: considerably faint, pretty large, round, little brighter middle, easily resolvable.
170x250: Faint to moderately faint. Broadly very slightly concentrated. Roughly round. Small. Just off its edge E is a very faint star.

| ga | NGC 4995 | 13 09.7 | −07 50 | Vir | 12.0b | 2.4' | 1.5' | Sb |

HH: pretty bright, pretty large, round, very gradually pretty much brighter middle, 8th mag. star NW.
170x250: Moderate brightness. Evenly moderately concentrated. Round. Small. No involved stars. NNW is a bright star.

| ga | UGCA 330 | 13 09.8 | −10 19 | Vir | 12.6 | 2.9' | 2.4' | Sd |

170x250: Not visible.

| ga | NGC 5005 | 13 10.9 | +37 03 | CVn | 10.6b | 6.5' | 2.7' | Sbc |

HH: very bright, very large, very much extended [ENE-WSW], very suddenly brighter middle to a nucleus.
42x60: Easily visible. *170x250:* Bright. Evenly to sharply fairly concentrated. Nearly edge-on ENE-WSW. Moderately large. No involved stars.

| ga | NGC 5012 | 13 11.6 | +22 54 | Com | 12.9p | 2.9' | 1.6' | Sc |

HH: pretty faint, considerably large, extended NNE-SSW, binuclear, 9th mag. star E.
170x250: Moderate brightness. Broadly moderately concentrated. Elongated N-S. Moderate size. A little N of center is an extremely faint involved star.

| ga | NGC 5023 | 13 12.2 | +44 02 | CVn | 12.9b | 6.7' | 0.7' | Scd |

HH: pretty faint, large, much extended NNE-SSW, very little brighter middle.
170x250: Faint. Hardly concentrated. A thin edge-on, NNE-SSW. Moderately large. No involved stars.

| ga | NGC 5015 | 13 12.4 | −04 20 | Vir | 12.9 | 1.8' | 1.4' | Sa |

HH: faint, considerably large, irregularly round, little brighter middle.
170x250: Moderately faint. Evenly slightly concentrated. Slightly elongated ENE-WSW. Small. No involved stars. Very nearby ESE is a very faint star; NW is a moderately bright star.

| ga | MCG-3-34-14 | 13 12.6 | −17 32 | Vir | 12.5 | 2.5' | 0.7' | Sc |

170x250: Faint. Broadly very slightly concentrated. Edge-on NW-SE. Moderately small. No involved stars. A little ways E is a faint star.

| ga | NGC 5020 | 13 12.7 | +12 35 | Vir | 12.5p | 3.4' | 2.9' | Sbc |

HH: considerably faint, considerably large, very little extended, little brighter middle.
170x250: Faint. Evenly to sharply moderately concentrated. Roughly round. Small. No involved stars. A little ways NW is a faint star; N is another faint star.

| ga | NGC 5017 | 13 12.9 | −16 45 | Vir | 12.6v | 1.2' | 1.0' | E+ |

HH: very faint, round, brighter middle.
170x250: Moderately faint to moderate brightness. Sharply very concentrated. Round. Small. No involved stars.

| gc | NGC 5024/M53 | 13 12.9 | +18 10 | Com | 7.7 | 13.0' | −8.62 | 5 |

JH: ! globular cluster, bright, very compressed, irregularly round, very very much brighter middle, 12th mag. and fainter stars.
12x35: Visible. ***42x60:*** Bright. ***170x250:*** Resolved throughout into mostly very faint stars, with a few brighter, more distinct stars, the brightest one being at its

edge NE. Broadly to evenly concentrated. A sixth of the field in size. Fairly regular. It looks like a fairly substantial cluster.

| ga | NGC 5018 | 13 13.0 | −19 31 | Vir | 11.7b | 3.3' | 2.4' | E3 |

HH: *considerably bright, small, round, much brighter middle to a pretty bright nucleus.*
170x250: Moderately bright. Sharply pretty concentrated. Round. Small. No involved stars. A little less than half the field ESE is a faint galaxy [*5022 (Te)]. A thin edge-on, NNE-SSW.

| ds | 54 Vir | 13 13.4 | −18 50 | Vir | 5.4" | 6.8 | 7.3 | B9 − |

42x60: Split. *170x250:* A close equal pair. Both white. [SHJ 151]

| ga | NGC 5033 | 13 13.4 | +36 35 | CVn | 10.8b | 12.4' | 5.0' | Sc |

HH: *very bright, pretty large, extended NNW-SSE, suddenly much brighter middle to a very bright nucleus, star NW.*
170x250: Moderately bright. Evenly to sharply fairly concentrated. Elongated N-S. Moderately small. Off its N end is a faint star.

| ga | NGC 5030 | 13 13.9 | −16 29 | Vir | 13.2 | 1.8' | 1.2' | S0+ |

Ho: *very faint, small.*
170x250: Very faint. Broadly very slightly concentrated. Round. Small. No involved stars. Nearby NW is a moderately faint star; nearby SW is a very faint star.

| ga | UGC 8320 | 13 14.5 | +45 55 | CVn | 12.7b | 3.6' | 1.3' | Im |

170x250: Not visible.

| ga | NGC 5035 | 13 14.8 | −16 29 | Vir | 12.8v | 1.5' | 1.2' | S0+ |

Ho: *faint, small, round, brighter middle to a nucleus.*
[See 5037.]

| ga | NGC 5037 | 13 15.0 | −16 35 | Vir | 12.2v | 3.2' | 0.7' | Sa |

HH: *considerably faint, pretty small, very little extended, brighter middle.*
170x250: Moderately faint to moderate brightness. Broadly moderately concentrated. Edge-on NE-SW. Moderately small. NE of center is a moderate brightness

involved star. A third of the field NNW is 5035. Very faint. Broadly to evenly slightly concentrated. Round. Small. No involved stars. It's between two bright stars, one NE and one SSW.

| ga | NGC 5038 | 13 15.0 | –15 57 | Vir | 13.1 | 1.6' | 0.5' | S0 |

Ho: pretty bright, extended E-W, stellar.
170x250: Moderately faint. Broadly slightly concentrated. Edge-on E-W. Small. No involved stars.

| ga | NGC 5044 | 13 15.4 | –16 23 | Vir | 11.8p | 2.9' | 2.9' | E0 |

HH: pretty bright, pretty large, round, brighter middle.
170x250: Moderately bright. Evenly moderately concentrated. Round. Moderately small. No involved stars.

| ga | NGC 5042 | 13 15.5 | –23 58 | Hya | 12.5p | 4.1' | 2.2' | Sc |

JH: faint, large, round, very gradually very little brighter middle, 9th mag. star W.
170x250: Very faint. Diffuse. Roughly round. Moderately small. No involved stars. It's surrounded by faint stars; a little ways WSW is the bright plotted star.

| ga | NGC 5055/M63 | 13 15.8 | +42 02 | CVn | 9.3b | 13.7' | 7.3' | Sbc |

JH: very bright, large, pretty much extended WNW-ESE, very suddenly much brighter middle to a bright nucleus.
12x35: Fairly bright. ***42x60:*** At its W end is a fairly bright star. ***170x250:*** Bright; moderate surface brightness. Broadly concentrated in general, but it has a moderately faint stellar nucleus. Elongated WNW-ESE. Large: a third of the field in size; it doesn't quite reach the bright star off its W end. It has a very faint outer halo. No involved stars. E is a little trio of stars.

| ga | NGC 5049 | 13 16.0 | –16 23 | Vir | 13.0v | 1.9' | 0.6' | S0 |

WH: considerably faint, small.
170x250: Moderately faint. Sharply pretty concentrated. Round. Very small. No involved stars. A little less than a third of the field NW is a small very faint galaxy [*5046 (Ho)]. A little less than half the field SSW is another galaxy [*5047 (WH)]. Edge-on ENE-WSW. It makes a triangle with two faint stars, one E and one SE.

| gc | NGC 5053 | 13 16.4 | +17 42 | Com | 9.8 | 10.0' | −6.20 | 11 |

HH: cluster, very faint, pretty large, irregularly round, very gradually brighter middle, 15th mag. and fainter stars.
170x250: A number of very faint stars are resolved, with a very faint glow of more stars underneath. No concentration. A third of the field in size; larger than M53. ESE is a bright star; NE is a moderate brightness star. Very irregular. A very loose, insubstantial cluster. Probably the faintest large globular in the sky.

| ga | NGC 5054 | 13 17.0 | −16 38 | Vir | 10.8v | 6.6' | 3.3' | Sbc |

HH: faint, pretty small, irregularly round.
170x250: Moderate brightness. Broadly slightly concentrated overall, but it has a substellar nucleus. Slightly elongated NNW-SSE. Moderately small to moderate size. It's among some faint stars, including a moderately faint star just off its edge NE.

| ga | IC 4214 | 13 17.7 | −32 06 | Cen | 12.3b | 2.8' | 1.5' | Sab |

170x250: Moderate brightness. Evenly moderately concentrated. Roughly round. Small. No involved stars. Nearby SSE is a moderately bright star; further SSE is a moderate brightness star.

| ga | NGC 5061 | 13 18.1 | −26 50 | Hya | 11.3b | 3.5' | 2.9' | E0 |

HH: very bright, small, round, very suddenly much brighter middle, 10th mag. star E.
170x250: Bright. Sharply very concentrated. Round. Small. No involved stars. Just off its edge NE is a very faint star; a little ways E is a bright star.

| ga | NGC 5066 | 13 18.5 | −10 13 | Vir | 13.2 | 0.7' | 0.5' | Sa |

Ma: very faint, very small.
170x250: Extremely faint. Hardly concentrated. Round. Small. No involved stars. It makes a triangle with two very faint stars, one SSE and one E.

| bs | Gamma Hya | 13 18.9 | −23 10 | Hya | 3.00 | | 0.92 | G5 |

170x250: Yellow.

| ga | NGC 5068 | 13 18.9 | –21 02 | Vir | 10.7b | 7.3' | 6.5' | Scd |

WH: faint, large, irregularly round, brighter middle.
42x60: Very faintly visible. ***170x250:*** Faint. Broadly slightly concentrated. Round. Moderately large. At its edge N is a faint star; a little farther W is another faint star.

| ga | NGC 5073 | 13 19.3 | –14 50 | Vir | 13.1 | 3.4' | 0.6' | Sc |

HH: very faint, pretty large, pretty much extended NW-SE.
170x250: Very faint. Hardly concentrated. Edge-on NNW-SSE. Moderate size. No involved stars.

| ga | NGC 5077 | 13 19.5 | –12 39 | Vir | 11.3v | 2.8' | 2.3' | E3-4 |

HH: pretty bright, small, very little extended, suddenly brighter middle, 2nd of 3.
170x250: Moderately bright. Evenly to sharply pretty concentrated. Slightly elongated N-S. Small. No involved stars. Very nearby SE is a very faint star. A sixth of the field SSE is another galaxy [*5079 (HH)]. Very faint. Hardly concentrated. Elongated NE-SW. Small, but a little larger than 5077. These two galaxies make a sharp contrast. A sixth of the field SW is a third galaxy [*5076 (HH)]. Faint. Sharply concentrated. Small. Half the field NW of 5077 is a galaxy [*5072 (dA)] with an involved star. It makes a triangular figure with several faint stars ENE to SSW.

| ga | NGC 5078 | 13 19.8 | –27 24 | Hya | 12.0b | 4.0' | 1.9' | Sa |

HH: pretty bright, pretty small, considerably extended, pretty suddenly brighter middle, 7th–8th mag. star E.
170x250: Moderately bright. Evenly moderately concentrated. Edge-on NNW-SSE. Small to moderately small. No involved stars. A little ways SW is a very very faint galaxy [*I879].

| ga | UGCA 348 | 13 19.9 | –22 16 | Vir | 13.2p | 1.9' | 1.3' | Scd |

170x250: Very faint. Hardly concentrated. Round. Moderately small. No involved stars. A little ways S is a moderate brightness star; a little ways NW is a faint star; a little ways E is another faint star.

| ga | NGC 5084 | 13 20.3 | –21 49 | Vir | 11.6b | 10.2' | 1.7' | S0 |

HH: considerably bright, considerably small, very little extended E-W, brighter E.
170x250: Moderately bright to bright. Evenly moderately concentrated. Edge-on E-W. Moderately small. No involved stars. Nearby N is a very faint star; a little ways E is a triangle of faint stars.

12 to 18 Hours: SPRING   343

| ga | NGC 5085 | 13 20.3 | –24 26 | Hya | 12.0p | 4.0' | 3.5' | Sc |

WH: faint, large, round, very gradually little brighter middle.
170x250: Faint. Broadly slightly concentrated. Round. Moderately small. No involved stars. A little ways SW is a moderately faint star; S is the bright plotted star; NE are two moderately bright stars.

| ga | NGC 5088 | 13 20.3 | –12 34 | Vir | 13.2p | 2.7' | 0.8' | Sbc |

LR: pretty bright, pretty small, round, brighter middle.
170x250: Faint. Broadly slightly concentrated. Elongated N-S. Moderately small. No involved stars.

| ga | NGC 5087 | 13 20.4 | –20 36 | Vir | 12.4b | 2.3' | 1.6' | S0 |

WH: considerably faint, very small, irregular figure.
170x250: Bright. Sharply pretty concentrated. Round. Very small. No involved stars. NNW is a moderately bright star with a faint star next to it W.

| ga | MCG+10-19-63 | 13 21.3 | +57 53 | UMa | 13.1 | 3.5' | 2.4' | S |

170x250: Not visible.

| ga | MCG-3-34-61 | 13 21.8 | –17 20 | Vir | 13.1 | 1.7' | 1.2' | S0– |

170x250: Very faint to faint. Broadly slightly concentrated. Round. Very small. No involved stars. It makes a quadrilateral with three faint stars SSW to WNW; ESE is a moderately bright star.

| ga | NGC 5101 | 13 21.8 | –27 25 | Hya | 11.6b | 5.4' | 4.7' | S0/a |

HH: considerably bright, pretty small, little extended, pretty suddenly brighter middle to a star.
170x250: Moderately bright. Sharply very concentrated. Round. Small. No involved stars. It makes a sharp triangle with a moderate brightness star nearby W and a moderately bright star NNW.

| ga | NGC 5105 | 13 21.8 | –13 12 | Vir | 11.7v | 2.0' | 1.6' | Sc |

Sw: extremely faint, pretty small, little extended.
170x250: Very very faint. Hardly concentrated. Roughly round. Moderately small. No involved stars. S is a crooked E-W line of four moderately faint stars; NNW are two moderate brightness stars.

| ga | NGC 5112 | 13 21.9 | +38 44 | CVn | 12.6b | 4.0' | 2.8' | Scd |

HH: faint, large, irregularly round, very gradually little brighter middle.
170x250: Faint to moderately faint. Broadly slightly concentrated. Elongated WNW-ESE. Moderate size. No involved stars. S of its E end is a star.

| ga | NGC 5111 | 13 22.9 | −12 57 | Vir | 12.7 | 2.0' | 1.8' | S0 |

HH: considerably faint, considerably small, irregularly round, gradually little brighter middle.
170x250: Moderately faint. Evenly moderately concentrated. Round. Small. No involved stars. In line with it WNW are two stars, moderately bright and moderate brightness.

| ga | NGC 5127 | 13 23.8 | +31 33 | CVn | 12.9p | 2.8' | 2.1' | E |

HH: pretty bright, pretty large, round, gradually much brighter middle, double star W.
170x250: Moderately faint. Evenly moderately concentrated. Slightly elongated ENE-WSW. Small. No involved stars. WNW is a little quadrilateral of stars. A sixth of the field S is a very small extremely faint galaxy [MAC1323+3130].

| bs | Zeta UMa | 13 23.9 | +54 56 | UMa | 2.27 | | 0.02 | A2 |

12x35: Elongated. ***42x60:*** A close unequal pair; very nicely split. ***170x250:*** A moderate distance unequal pair. Both white. A very nice double star in both the 60 mm and the main scope. [STF 1744: 14.4", 2.3, 4.0, A2 A1]

| ga | NGC 5119 | 13 24.0 | −12 16 | Vir | 13.0 | 1.3' | 0.4' | S0/a |

JH: pretty bright, small, little extended.
170x250: Moderately faint to moderate brightness. Sharply pretty concentrated. Slightly elongated NNE-SSW. Small. No involved stars. Nearby SSW is an extremely faint star; WNW is a faint star; farther SSE is another faint star.

| ga | NGC 5125 | 13 24.0 | +09 42 | Vir | 13.2p | 1.6' | 1.2' | Sb |

JH: pretty faint, small, round, gradually brighter middle.
170x250: Very faint. Broadly slightly concentrated. Round. Small. No involved stars.

| ga | NGC 5129 | 13 24.2 | +13 58 | Vir | 13.0b | 1.6' | 1.3' | E |

HH: pretty bright, very small, round, gradually much brighter middle, double star E.
170x250: Moderately faint. Evenly moderately concentrated. Roughly round. Small. No involved stars. It makes a right triangle with two bright stars, one E and

one SE. Half the field NNE is another galaxy [*5132 (dA)]. Very faint to faint. Evenly slightly concentrated. Roughly round. Small. No involved stars. It makes a triangle with two faint stars, one E and one NNE.

| ga | IC 4237 | 13 24.5 | −21 08 | Vir | 13.2p | 2.0' | 1.2' | Sb |

[See 5134.]

| ga | UGCA 353 | 13 24.7 | −19 41 | Vir | 13.1p | 3.3' | 1.7' | Sc |

170x250: In the position are two galaxies a little less than half the field apart NW-SE [UA353, *E576-51]. The SE galaxy is very faint. Broadly very slightly concentrated. Slightly elongated E-W. Small. A little ways NE is a very faint star. The NW galaxy is very very faint. Hardly concentrated. Roughly round. Moderately small. No involved stars. A little ways SSE is a bright star; E is another relatively bright star.

| ga | NGC 5124 | 13 24.8 | −30 18 | Cen | 13.1b | 3.0' | 0.9' | E6 |

JH: very faint, small, very little extended.
170x250: Moderately faint. Evenly moderately concentrated. Elongated N-S. Small. No involved stars. Nearby SSE is a second galaxy [*5126 (JH)]. SE of both galaxies is a bright pair of stars.

| ga | NGC 5133 | 13 24.9 | −04 04 | Vir | 12.6 | 1.0' | 0.7' | E/S0 |

St: very faint, very small, irregularly round, brighter middle.
170x250: Faint. Sharply moderately concentrated. Round. Very small. No involved stars. It's in a large "L" figure with five stars.

| bs | Alpha Vir | 13 25.2 | −11 09 | Vir | 0.98 | | −0.23 | B1 |

170x250: White. A brilliant star, with diffraction spikes going all the way across the field. Isolated. [Spica]

| ga | NGC 5134 | 13 25.3 | −21 08 | Vir | 12.1b | 2.9' | 1.7' | Sb |

HH: faint, pretty small, little extended, very gradually brighter middle.
170x250: Moderate brightness. Broadly to evenly moderately concentrated. Elongated NNW-SSE. Moderately small to moderate size. No involved stars. S are three faint stars; NE is another faint star. A little more than half the field W is I4237. Faint. Broadly very slightly concentrated. Round. Moderately small. No involved stars. E is a faint star; WNW is another faint star.

| ga | NGC 5135 | 13 25.7 | −29 50 | Hya | 12.2v | 3.2' | 2.6' | Sab |

JH: pretty bright, small, extended.
170x250: Moderately faint to moderate brightness. Sharply pretty concentrated. Roughly round. Small. No involved stars.

| ga | MCG-2-34-48 | 13 26.3 | −12 36 | Vir | 12.5 | 1.9' | 1.9' | S0 |

170x250: Very very faint. Broadly slightly concentrated. Roughly round. Small. No involved stars. Passing E of it is a long shallow NNE-SSW arc of stars including two wide doubles.

| ga | NGC 5147 | 13 26.3 | +02 06 | Vir | 12.3b | 1.9' | 1.5' | Sdm |

HH: pretty bright, pretty large, very little extended, very suddenly much brighter middle to a 12th mag. star.
170x250: Moderately faint. Broadly slightly concentrated. Round. Small to moderately small. Just W of center is an involved star with a very very faint star next to it SW.

| ga | NGC 5140 | 13 26.4 | −33 52 | Cen | 12.8p | 2.0' | 1.6' | S0− |

JH: very faint, small, round, gradually little brighter middle.
170x250: Faint. Broadly moderately concentrated. Round. Small. No involved stars. Very nearby E is a faint star.

| ga | NGC 5146 | 13 26.6 | −12 19 | Vir | 13.1 | 1.9' | 1.3' | E/S0 |

WH: very faint, very small, stellar.
170x250: Moderately faint. Evenly moderately concentrated. Round. Small. No involved stars. Nearby SSE is a very faint star.

| ga | NGC 5150 | 13 27.6 | −29 33 | Hya | 12.6v | 1.3' | 1.0' | Sbc |

JH: considerably faint, small, round, pretty suddenly little brighter middle, star 2' E.
170x250: Moderately faint. Evenly moderately concentrated. Round. Very small. No involved stars. A little ways ENE is a bright star. A little more than a quarter of the field SE is 5153. Moderately faint. Sharply pretty concentrated. Round. Very small. At its edge W are a couple of faint stars. [5152 (JH), adjoining 5153, was not recognized as a galaxy; it might have been one of the faint "stars" at 5153's edge.]

12 to 18 Hours: SPRING   347

| ga | MCG-2-34-54 | 13 27.9 | –13 25 | Vir | 13.1 | 1.8' | 1.2' | Sbc |

170x250: Very very faint. Broadly slightly concentrated. Round. Small. No involved stars. It's between a couple of moderately faint stars NW and a very faint star SE.

| ga | NGC 5153 | 13 28.0 | –29 37 | Hya | 11.8v | 1.9' | 1.2' | E1 |

JH: pretty faint, small, E of double nebula.
[See 5150.]

| ga | NGC 5173 | 13 28.4 | +46 35 | CVn | 13.1b | 1.2' | 1.0' | E0 |

HH: faint, very small, round, stellar.
170x250: Moderately faint. Sharply fairly concentrated. Round. Very small. No involved stars. It makes a linear figure with two very wide pairs of faint stars, one NE and one SSW. A little less than a third of the field NNW is a larger, very very faint galaxy [*5169 (JH)]. Hardly concentrated. Elongated WNW-ESE.

| ga | NGC 5161 | 13 29.2 | –33 10 | Cen | 12.0b | 5.6' | 2.1' | Sc |

JH: pretty faint, large, very much extended, pretty gradually brighter middle, partially resolved.
170x250: Very faint. Hardly concentrated. Elongated ENE-WSW. Moderately small. No involved stars. NE of its E end is a moderately bright star; a little farther from its W end is a bright star.

| ga | NGC 5172 | 13 29.3 | +17 03 | Com | 12.6b | 3.3' | 1.7' | Sbc |

JH: faint, pretty large, round, gradually brighter middle.
170x250: Faint. Broadly slightly concentrated. Elongated E-W. Moderately small. At its edge NNE of center is a faint star; ENE is a moderately faint star. 3/4 of a field S is another galaxy [*5180 (HH)]. Very faint. Sharply concentrated. Round. Small. Right next to it SE is a star; together the galaxy and the star look almost like a double star.

| ga | NGC 5171 | 13 29.4 | +11 44 | Vir | 12.7v | 1.2' | 0.8' | S0– |

Te: pretty bright, large.
170x250: Faint. Evenly moderately concentrated. Slightly elongated NW-SE [round; a faint star at the galaxy's edge NW makes it look elongated in that direction]. Small. No involved stars. A little ways ENE is a small very faint galaxy [*5179 (Te)]. A little farther NNE is another small very faint galaxy [*5176 (Hw)]. The three galaxies make a small triangle. [5177 (Hw) was not seen.]

| ga | NGC 5174 | 13 29.4 | +11 00 | Vir | 13.2 | 3.1' | 1.7' | Scd |

HH: very faint, pretty large, one of double nebula.
170x250: Very faint to faint. Broadly very slightly concentrated. Elongated NNW-SSE. Moderately small. At its S end is a faint involved star; a little ways NNE is a brighter star.

| ga | NGC 5204 | 13 29.6 | +58 25 | UMa | 11.7b | 5.0' | 3.0' | Sm |

HH: pretty bright, considerably large, irregularly round, gradually much brighter middle, resolvable.
170x250: Moderately faint. Broadly slightly concentrated. Round. Moderate size. No involved stars.

| vs | R Hya | 13 29.7 | −23 17 | Hya | 4.0 | 10.0 | 1.60 | M7 |

12x35: Mag. 8.0. ***170x250:*** Copper red.

| ga | NGC 5170 | 13 29.8 | −17 57 | Vir | 12.1b | 8.4' | 1.0' | Sc |

HH: considerably faint, large, much extended NW-SE, pretty gradually brighter middle.
170x250: Moderately faint. Evenly moderately concentrated. A thin edge-on, NW-SE. Large: a quarter of the field in size. With averted vision the ends extend considerably. No involved stars.

| ga | NGC 5194/M51 | 13 29.9 | +47 11 | CVn | 9.0b | 10.3' | 8.1' | Sbc |

JH: !!! nucleus and ring, (LR) spiral.
12x35: Easily visible along with 5195. ***42x60:*** Both galaxies are obvious. ***170x250:*** Both galaxies are bright. M51 is sharply concentrated to a bright core. Round. Large: a third of the field in size. The halo is strangely uneven in light: the core aside, it's ringlike or even annular. ESE of the core is a bright section, the beginning of a spiral arm; this arm goes N toward 5195, comes back around W, and then makes another half loop, so it looks like one arm that winds completely around [there are two distinct arms, one E to N, the other W to S]. Between the arm(s) and the core is a dark gap. With averted vision the classic spiral arm pattern of the pictures can start to be seen, but it's not clear. SW of center is an involved star. 5195 is sharply concentrated to a substellar nucleus. Round. Small to moderately small. No involved stars. No bridge is visible between the two galaxies. [Whirlpool Galaxy]

| ga | NGC 5195 | 13 30.0 | +47 16 | CVn | 10.5b | 5.8' | 4.6' | I0 |

HH: bright, pretty small, round, very gradually brighter middle, E of 2.
[See M51.]

| ga | NGC 5205 | 13 30.1 | +62 30 | UMa | 13.2p | 3.2' | 1.8' | S |

Sw: very faint, pretty small, round, between 2 very faint stars.
170x250: Faint. Broadly very slightly concentrated. Roughly round. Small. No involved stars.

| ga | NGC 5198 | 13 30.2 | +46 40 | CVn | 12.7b | 2.0' | 2.0' | E1-2 |

WH: pretty bright, pretty large, round, much brighter middle.
170x250: Moderate brightness. Evenly moderately concentrated. Round. Very small. No involved stars. It makes a small quadrilateral with three faint stars W to NNE.

| ga | NGC 5188 | 13 31.3 | −34 47 | Cen | 13.0b | 3.0' | 1.1' | Sb |

JH: faint, pretty large, very little extended, very gradually little brighter middle.
170x250: Very faint to faint. Broadly very slightly concentrated. Slightly elongated E-W. Moderately small. No involved stars. A little ways WNW is a moderate brightness star.

| ga | NGC 5193 | 13 31.9 | −33 14 | Cen | 12.5b | 1.6' | 1.6' | E |

JH: pretty bright, small, round, gradually then pretty suddenly brighter middle.
170x250: Moderately faint to moderate brightness. Broadly to evenly moderately concentrated. Round. Small. No involved stars. Very nearby SW is a tiny very faint galaxy [*5193A]; nearby NNW is a faint star; a little farther ESE is another faint star.

| ga | NGC 5218 | 13 32.1 | +62 45 | UMa | 12.3v | 2.9' | 1.6' | Sb |

HH: pretty bright, pretty large, round, gradually brighter middle.
170x250: Moderately faint. Broadly slightly concentrated. Roughly round. Small. No involved stars. A quarter of the field S is another galaxy [*5216 (HH)]. Faint. Broadly slightly concentrated. Roughly round. Very small. No involved stars.

| ga | NGC 5211 | 13 33.1 | −01 02 | Vir | 13.2p | 2.2' | 1.7' | Sab |

JH: pretty bright, small, round, pretty suddenly much brighter middle.
170x250: Moderately faint. Evenly to sharply moderately concentrated. Roughly round. Very small. No involved stars.

| bs | Zeta Vir | 13 34.7 | −00 35 | Vir | 3.40 | | 0.08 | A3 |

170x250: White.

| ga | NGC 5230 | 13 35.5 | +13 40 | Vir | 12.8p | 2.2' | 1.9' | Sc |

HH: considerably faint, pretty large, round, gradually little brighter middle, 3rd of 3.
170x250: Very faint. Broadly very slightly concentrated. Roughly round. Moderately small. No involved stars. It makes a flat isosceles triangle with two faint stars, one WNW and one S.

| ga | NGC 5220 | 13 35.9 | −33 27 | Cen | 13.1p | 2.5' | 0.8' | Sa |

JH: very faint, small, round, 10th mag. star E.
170x250: Moderately faint. Sharply very concentrated. Elongated E-W. Very small. E of center is a moderately bright involved star that makes a crooked line with three bright stars ESE.

| ga | IC 4296 | 13 36.6 | −33 57 | Cen | 11.6b | 3.4' | 3.2' | E |

170x250: Moderate brightness. Evenly moderately concentrated. Round. Small. No involved stars. A little more than a third of the field SSE is a tiny faint galaxy [*I4299].

| ds | H 69 | 13 36.8 | −26 30 | Hya | 10.2" | 5.9 | 6.8 | A7 A7 |

42x60: Split. ***170x250:*** A moderate distance slightly unequal pair. Both white.

| ga | NGC 5236/M83 | 13 37.0 | 29 52 | Hya | 8.2b | 12.8' | 11.4' | Sc |

JH: !! very bright, very large, extended NE-SW, extremely suddenly brighter middle to a nucleus, (Le) 3-branched spiral.
12x35: Fairly bright. ***42x60:*** Moderately bright, with a faint substellar nucleus.
170x250: Bright. Sharply extremely concentrated. Roughly round. Very large: half the field in size. It has a faint low surface brightness halo, milky and uneven, and a

small very bright nucleus; the two contrast tremendously; one of the greatest such contrasts. Going across the middle of the galaxy NE-SW is a bar. Coming off the NE end of the bar at a sharp angle and going S is a spiral arm. Less distinct is a second arm starting from the SW end and going across the N side. Going straight W is a very slight brightening, an extension – apparently a third arm. The spiral arms as a whole are not obvious or steadily visible; what can be said more definitively is that there's a central bar and that the halo is broadly uneven in light. S of center is a faint involved star; at the galaxy's edge SW is a moderate brightness star; at its edge ENE is a moderately faint star.

| ga | UGC 8614 | 13 37.4 | +07 38 | Boo | 13.2b | 3.4' | 1.8' | Im |

170x250: Not visible.

| ga | NGC 5248 | 13 37.5 | +08 53 | Boo | 11.0b | 6.6' | 5.3' | Sbc |

HH: bright, large, extended NNW-SSE, pretty suddenly brighter middle to a resolvable nucleus.
42x60: Faintly visible. *170x250:* Moderately bright. Sharply pretty concentrated to a small core. Slightly elongated NW-SE. Moderate size to moderately large. At times a broad spiral arm is visible curving N from the NW end. No involved stars. A little ways SSW is a faint star.

| ga | NGC 5247 | 13 38.1 | –17 53 | Vir | 10.5b | 5.6' | 4.8' | Sbc |

HH: !! considerably faint, very large, very gradually then pretty suddenly much brighter middle to a large nucleus, (Le) 2-branched spiral.
42x60: Very faintly visible. *170x250:* Moderately faint. Evenly moderately concentrated to a small, brighter nucleus. Round. Large: a quarter of the field in size. With averted vision two big spiral arms are occasionally visible, one curving S and the other curving N; the S arm is a little better seen. No involved stars.

| ga | MCG-2-35-10 | 13 38.2 | –09 48 | Vir | 12.8b | 2.2' | 1.8' | Sm |

170x250: Very faint. Diffuse. Roughly round. Moderately small. No involved stars. A little ways WNW are a couple of faint stars; a little ways SW is a slightly brighter star; N is a small very sharp triangle of moderate brightness stars.

| ga | MCG-2-35-11 | 13 38.5 | –10 42 | Vir | 11.6 | 1.3' | 1.1' | S0 |

170x250: Faint. Evenly moderately concentrated. Round. Very small. No involved stars. Next to it WNW is a moderate brightness star that makes an isosceles triangle with a moderately faint star NW and a moderate brightness star NNW.

| ga | IC 4310 | 13 39.0 | −25 50 | Hya | 13.2p | 2.6' | 0.7' | S0 |

170x250: Moderate brightness. Sharply concentrated. Round. Very small. Right at its edge E is a moderate brightness star; nearby S is another star.

| ga | ESO 445-2 | 13 39.4 | −30 46 | Cen | 12.8p | 2.4' | 1.9' | S0− |

170x250: Moderately faint. Concentrated. Round. Very very small. Off its edge ESE is a moderate brightness star; this is one of a curvy dragon-like figure of moderate brightness stars E of the galaxy.

| ga | NGC 5254 | 13 39.6 | −11 29 | Vir | 13.1 | 3.1' | 1.5' | Sc |

JH: pretty bright, large, pretty much extended, gradually little brighter middle.
170x250: Very very faint. Hardly concentrated. Slightly elongated NW-SE. Moderate size. No involved stars.

| ga | NGC 5253 | 13 39.9 | −31 38 | Cen | 10.9b | 5.0' | 1.9' | Pec |

HH: bright, pretty large, extended NE-SW, pretty suddenly much brighter middle.
42x60: Visible. *170x250:* Very bright. Broadly to evenly pretty concentrated. Slightly elongated NE-SW. Moderately small. No involved stars. It's at the S edge of a large widespread group of stars N of it.

| ga | NGC 5257 | 13 39.9 | +00 50 | Vir | 12.9v | 2.0' | 1.4' | Sb |

HH: very faint, small, round, brighter middle, W of double nebula.
170x250: 5257 and 5258 are twin galaxies aligned WNW-ESE. Both are very faint to faint. Broadly slightly concentrated. Very small. No involved stars. 5257 is roughly round; 5258 is slightly elongated NE-SW.

| ga | NGC 5258 | 13 40.0 | +00 49 | Vir | 12.9v | 2.4' | 1.8' | Sb |

HH: faint, small, irregularly round, E of double nebula.
[See 5257.]

| ga | UGC 8658 | 13 40.7 | +54 20 | UMa | 13.1b | 2.8' | 1.8' | Sc |

170x250: Exceedingly faint. A slight diffuse glow. Moderately small. No involved stars.

| ga | NGC 5264 | 13 41.6 | −29 54 | Hya | 12.6b | 2.5' | 1.5' | Im |

JH: very faint, pretty large, round, very little brighter middle.

170x250: Extremely faint. Diffuse. Roughly round. Moderately small. No involved stars. It's inside a triangle of faint stars; W is the bright plotted star, which is double.

| ga | NGC 5273 | 13 42.1 | +35 39 | CVn | 12.4b | 2.7' | 2.4' | S0 |

HH: considerably bright, pretty large, round, gradually then pretty suddenly much brighter middle.
170x250: Moderately faint. Evenly moderately concentrated. Round. Small. No involved stars. [5276 (LR) was not seen.]

| gc | NGC 5272/M3 | 13 42.2 | +28 23 | CVn | 6.4 | 18.0' | −8.65 | 6 |

JH: !! globular cluster, extremely bright, very large, very suddenly much brighter middle, 11th mag. and fainter stars.
12x35: Bright. Concentrated. ***42x60:*** Bright. It's inside a right triangle of stars.
170x250: Well resolved throughout, with a lot of very distinct stars, except in the middle, where it's granular. Broadly concentrated, with a very dense central ball of stars. A third of the field in size. It looks like a pretty massive cluster. The main body is pretty regular; the outliers are a little irregular in brightness and distribution.

| ds | 84 Vir | 13 43.1 | +03 32 | Vir | 2.6" | 5.5 | 7.9 | K2 − |

170x250: The primary has a very very close much fainter companion. Both yellow. A delicate double star; nice. [STF 1777]

| ga | MCG-1-35-10 | 13 45.6 | −05 59 | Vir | 13.2 | 1.6' | 1.1' | Sm |

170x250: Faint. Hardly concentrated. Elongated NE-SW. Small. A couple of faint involved stars interfere.

| ga | NGC 5297 | 13 46.4 | +43 52 | CVn | 12.5b | 5.6' | 1.2' | Sc |

HH: considerably bright, large, pretty much extended NW-SE, gradually brighter middle.
170x250: Moderately faint. Broadly slightly concentrated. Edge-on NW-SE. Moderate size. No involved stars. It makes a triangle with two moderate brightness stars, one NW and one ENE. [5296 (LR) was not seen.]

| ga | NGC 5308 | 13 47.0 | +60 58 | UMa | 12.3b | 3.7' | 0.6' | S0− |

HH: bright, pretty large, much extended ENE-WSW, pretty suddenly brighter middle to a bright extended nucleus.

170x250: Moderately bright. Evenly to sharply pretty concentrated. A thin edge-on, ENE-WSW. Small. No involved stars. [It's among a number of stars NW and S.]

| ga | ESO 383-76 | 13 47.5 | −32 51 | Cen | 13.0p | 2.3' | 1.0' | E+5 |

170x250: Extremely faint. Diffuse. Slightly elongated N-S. Moderately small to moderate size. No involved stars. A little ways NE is a moderately faint star; SE is a little WNW-ESE line of faint stars; W is a NE-SW line of three stars, the star at the SW end being the brightest. [E383-81 was not seen.]

| bs | Eta UMa | 13 47.5 | +49 18 | UMa | 1.86 | | −0.19 | B3 |

170x250: White.

| ga | NGC 5292 | 13 47.7 | −30 56 | Cen | 12.8p | 1.7' | 1.4' | Sab |

JH: pretty faint, small, round, 2 stars near.
170x250: Moderately faint. Evenly to sharply fairly concentrated. Roughly round. Small. Very nearby NE is a moderately faint star; very nearby W is a very faint star.

| ga | NGC 5300 | 13 48.3 | +03 57 | Vir | 12.1p | 3.9' | 2.5' | Sc |

HH: very faint, very large, little extended, very gradually brighter middle.
170x250: Very very faint. Hardly concentrated. Roughly round. Moderate size. Just off its edge SSE is a very faint star.

| ga | NGC 5298 | 13 48.6 | −30 25 | Cen | 13.0v | 1.9' | 0.8' | Sb |

JH: faint, small, round, gradually brighter middle.
[See I4329.]

| ga | NGC 5302 | 13 48.8 | −30 30 | Cen | 12.0v | 1.9' | 1.1' | S0+ |

JH: faint, small, round, gradually brighter middle.
[See I4329.]

| ga | IC 4329 | 13 49.1 | −30 17 | Cen | 12.2b | 3.5' | 2.2' | S0− |

170x250: Moderately faint. Evenly moderately concentrated. Roughly round. Small. No involved stars. A sixth of the field ESE is I4329A. Moderately faint. Sharply pretty concentrated. Roughly round. Very small. No involved stars. Half the field SW of I4329 is 5298. Faint. Broadly very slightly concentrated. Slightly elongated ENE-WSW. Small. No involved stars. A third of the field SSE is 5302.

12 to 18 Hours: SPRING   355

Faint. Evenly moderately concentrated. Roughly round. Small. No involved stars. Nearby W is a faint star. A third of the field WSW of 5298 is another galaxy [*E445-35]. Very faint. Hardly concentrated. Roughly round. Small. No involved stars. It's inside a triangle of faint stars, the westernmost of which is a wide double.

| ga | NGC 5306 | 13 49.2 | −07 13 | Vir | 13.1b | 1.4' | 1.0' | S0 |

HH: very faint, very small, round, resolvable.
170x250: Moderately faint. Evenly moderately concentrated. Roughly round. Small. No involved stars. Very nearby NW is a faint star.

| ga | IC 4329A | 13 49.3 | −30 18 | Cen | 13.0v | 2.6' | 0.7' | S0+ |

[See I4329.]

| ga | NGC 5322 | 13 49.3 | +60 11 | UMa | 11.1b | 5.9' | 3.8' | E3-4 |

HH: very bright, pretty large, irregularly round, pretty suddenly much brighter middle.
42x60: Visible. ***170x250:*** Bright. Evenly pretty concentrated. Slightly elongated E-W. Moderately small. No involved stars. Off its E end is a faint star.

| ga | NGC 5313 | 13 49.7 | +39 59 | CVn | 12.8p | 2.3' | 1.3' | Sb |

HH: pretty bright, pretty small, very little extended, gradually little brighter middle, E of 2.
170x250: Moderate brightness. Broadly to evenly moderately concentrated. Elongated NE-SW. Small. No involved stars. Half the field W is another galaxy [*5311 (HH)]. Moderately faint. Sharply pretty concentrated. Round. Very small. No involved stars.

| ga | NGC 5320 | 13 50.3 | +41 22 | CVn | 12.8p | 5.3' | 1.9' | Sc |

HH: considerably faint, pretty large, round, gradually brighter middle.
170x250: Faint. Broadly slightly concentrated. Slightly elongated N-S. Small. No involved stars. A little ways SSW is a faint star; the galaxy makes a triangle with a moderately faint star E and a brighter star NE.

| ga | NGC 5326 | 13 50.8 | +39 34 | CVn | 12.9b | 2.8' | 1.8' | Sa |

HH: considerably faint, small, very little extended, suddenly brighter middle.
170x250: Moderately bright; fairly high surface brightness. Sharply fairly concentrated. Roughly round. Very small. No involved stars.

| ga | ESO 384-2 | 13 51.3 | −33 48 | Cen | 12.5b | 4.8' | 2.6' | Sdm |

170x250: Not visible.

| ds | 3 Cen | 13 51.8 | −33 00 | Cen | 7.0" | 4.5 | 6.0 | B4 B8 |

42x60: An unequal pair. *170x250:* A moderately close unequal pair. Both white. [H 101]

| ga | NGC 5324 | 13 52.1 | −06 03 | Vir | 12.4p | 2.2' | 2.0' | Sc |

HH: considerably faint, large, irregularly round, brighter middle.
170x250: Moderately faint. Broadly very slightly concentrated. Round. Small. No involved stars. It makes a small flat triangle with a faint star SSE and a moderately faint star farther SE.

| ga | NGC 5328 | 13 52.9 | −28 29 | Hya | 12.7b | 1.7' | 1.2' | E1 |

HH: pretty bright, small, round, suddenly little brighter middle.
170x250: Moderate brightness. Sharply pretty concentrated. Round. Very small. No involved stars. Making a short line with it are two stars a little ways SSW. [5330 (Sw) was not seen.]

| ga | NGC 5334 | 13 52.9 | −01 06 | Vir | 12.0p | 4.2' | 3.0' | Sc |

WH: considerably faint, very large, round, little brighter middle, resolvable.
170x250: Very faint. Diffuse. Roughly round. Moderate size. No involved stars. S is a moderate brightness star; N are two faint stars.

| ga | NGC 5350 | 13 53.4 | +40 21 | CVn | 11.3v | 3.3' | 2.4' | Sb |

HH: considerably faint, pretty large, brighter middle, bright star W, 1st of 4. [See 5353.]

| ga | NGC 5351 | 13 53.5 | +37 54 | CVn | 13.0b | 2.9' | 1.5' | Sb |

HH: considerably faint, large, little extended E-W, very gradually brighter middle.
170x250: Faint. Broadly very slightly concentrated. Elongated WNW-ESE. Moderate size. No involved stars. [5349 (LR) was not seen.]

| ga | NGC 5353 | 13 53.5 | +40 16 | CVn | 10.9v | 3.3' | 1.8' | S0 |

HH: pretty bright, small, round, 2nd of 4.
170x250: In the position is a group of galaxies taking up half the field. 5353 is bright. Evenly to sharply fairly concentrated. Elongated NW-SE. Small. No involved stars. Next to it N is a second galaxy [5354]. Moderately bright. Sharply fairly concentrated. Roughly round. Very small. No involved stars. Farther NNW is a larger, more diffuse galaxy [5350]. Moderate brightness. Broadly to evenly moderately concentrated. Round. Moderately small. No involved stars. E of these three galaxies, making a triangle with them, is a moderately faint galaxy [*5355 (HH)]. Concentrated. Round. Very small. No involved stars. SE is a fifth galaxy [*5358 (St)] near a double star. Faint. Small. A nice group of galaxies; one of the best in the sky. The galaxies are bright, especially for a cluster of galaxies, and the bright stars in the field create a nice effect.

| ga | NGC 5354 | 13 53.5 | +40 18 | CVn | 11.3v | 2.8' | 2.2' | S0 |

HH: pretty faint, small, round, 3rd of 4.
[See 5353.]

| bs | Eta Boo | 13 54.7 | +18 23 | Boo | 2.68 | | 0.58 | G0 |

170x250: Soft white.

| ga | NGC 5362 | 13 54.9 | +41 18 | CVn | 13.1p | 2.3' | 0.9' | Sb |

WH: pretty bright, pretty large, extended.
170x250: Faint. Broadly slightly concentrated. Elongated E-W. Small. No involved stars. Off its E end is a faint star.

| ds | STF 1788 | 13 55.0 | −08 04 | Vir | 3.6" | 6.5 | 7.7 | F8 G1 |

42x60: Extremely close. ***170x250:*** A very close slightly unequal pair. Both white.

| ga | NGC 5376 | 13 55.3 | +59 30 | UMa | 12.9p | 2.0' | 1.3' | Sb |

WH: considerably bright, pretty large, very little extended, very gradually much brighter middle.
170x250: Moderately faint. Broadly slightly concentrated. Slightly elongated ENE-WSW. Small. No involved stars.

| ga | NGC 5371 | 13 55.7 | +40 27 | CVn | 10.5v | 5.5' | 4.0' | Sbc |

HH: pretty bright, large, round, brighter middle to a faint nucleus.
170x250: Moderate brightness. Sharply concentrated. Round. Moderate size. It has a fairly low surface brightness halo and a small, brighter core. No involved stars. Nearby W are a couple of faint stars; the galaxy makes a flat triangle with two bright stars, one NE and one farther N.

| ga | NGC 5357 | 13 56.0 | −30 20 | Cen | 13.0b | 1.5' | 1.2' | E |

JH: pretty faint, small, round, gradually little brighter middle, between 2 10th mag. stars.
170x250: Faint. Evenly moderately concentrated. Roughly round. Very small. No involved stars. It's right on the side of a small right triangle of moderate brightness stars.

| ga | NGC 5363 | 13 56.1 | +05 15 | Vir | 11.1b | 4.0' | 2.5' | I0 |

HH: bright, pretty large, round, pretty suddenly brighter middle, 8th mag. star NE.
12x35: Extremely faintly visible. ***42x60:*** Visible. ***170x250:*** Bright to very bright. Sharply very concentrated to a stellar nucleus. Roughly round. Moderately small to moderate size. It has an extremely faint extended halo. No involved stars. A little ways SSE is a faint star.

| ga | NGC 5389 | 13 56.1 | +59 44 | UMa | 12.9b | 6.8' | 1.1' | S0/a |

HH: pretty bright, pretty large, extended, much brighter middle to a nucleus.
170x250: Moderately faint to moderate brightness. Evenly to sharply pretty concentrated. Edge-on N-S. Small. No involved stars. [A little ways SSE is a moderate brightness star; E are two bright stars.] A quarter of the field W is another galaxy [*5379 (WH)]. Very faint. Hardly concentrated. Elongated ENE-WSW. Very small. No involved stars.

| ga | NGC 5364 | 13 56.2 | +05 00 | Vir | 11.2b | 6.7' | 5.4' | Sbc |

HH: considerably faint, large, round, gradually brighter middle.
170x250: Moderately faint. Broadly very slightly concentrated. Elongated NE-SW. Moderate size. Nearby NW are two faint stars aligned parallel to the galaxy; farther NNW is a moderate brightness star. 5364 makes a big contrast with 5363: a good example of a major contrast between neighboring galaxies.

| ga | NGC 5377 | 13 56.3 | +47 14 | CVn | 12.2b | 4.4' | 2.1' | Sa |

HH: bright, large, much extended NE-SW, suddenly much brighter middle to a nucleus.
170x250: Moderate brightness. Sharply fairly concentrated. Nearly edge-on NE-SW. Small. No involved stars.

| ga | NGC 5375 | 13 56.9 | +29 09 | CVn | 12.4p | 3.2' | 2.7' | Sab |

JH: pretty bright, pretty large, round, little brighter middle.
170x250: Moderately faint. Evenly moderately concentrated. Round. Small. No involved stars.

| ga | NGC 5383 | 13 57.1 | +41 50 | CVn | 12.1b | 3.1' | 2.6' | Sb |

HH: considerably bright, considerably large, round, gradually brighter middle.
170x250: Moderately faint. Broadly slightly concentrated. Round. Moderately small. No involved stars. Very nearby E is a faint double star; the galaxy is on the hypotenuse of a right triangle of moderate brightness stars; the S corner of this triangle is a wide pair.

| ga | ESO 384-19 | 13 57.7 | −34 13 | Cen | 12.8v | 1.4' | 1.1' | S0 |

170x250: Faint. Sharply pretty concentrated. Round. Tiny. No involved stars. Off its edge ESE is a very faint star; nearby N is a moderately faint star.

| ga | IC 4351 | 13 57.9 | −29 18 | Hya | 12.6b | 6.8' | 0.9' | Sb |

170x250: Moderately faint. Evenly moderately concentrated. Edge-on NNE-SSW. Small. No involved stars. On either side of center E and WSW, just off the galaxy's edge, is a faint star.

| ga | NGC 5395 | 13 58.6 | +37 25 | CVn | 12.1b | 3.1' | 1.6' | Sb |

HH: considerably faint, considerably large, extended NNE-SSW, little brighter middle, SE of 2.
170x250: Faint. Hardly concentrated. Elongated N-S. Moderate size. No involved stars. Off its S end is a faint star. Just NW of its N end is a companion galaxy [*5394 (HH)]. Faint. Concentrated. Round. Small.

| ga | NGC 5406 | 14 00.3 | +38 54 | CVn | 13.1b | 2.0' | 1.4' | Sbc |

HH: faint, pretty small, round, little brighter middle.
170x250: Faint. Evenly moderately concentrated. Roughly round. Small. No involved stars.

| ga | NGC 5422 | 14 00.7 | +55 09 | UMa | 12.8p | 3.9' | 0.8' | S0 |

HH: pretty bright, small, pretty much extended NE-SW, very suddenly very much brighter middle to a nucleus.
170x250: Moderately bright. Evenly to sharply fairly concentrated. Edge-on NNW-SSE. Small. No involved stars. It makes a quadrilateral with three moderate brightness stars E to S.

| ga | NGC 5430 | 14 00.8 | +59 19 | UMa | 12.7p | 2.2' | 1.1' | Sb |

HH: pretty bright, small, irregularly extended, much brighter middle.
170x250: Moderately faint. Evenly moderately concentrated. Slightly elongated NNW-SSE. Small. No involved stars.

| ga | UGCA 378 | 14 01.1 | −30 19 | Cen | 13.1p | 2.5' | 2.1' | Sd |

170x250: Very faint. Evenly moderately concentrated. Round. Small. No involved stars. A little ways S is a moderately faint star; WSW to NW is a flat triangle of stars; farther SW is a relatively bright star.

| ga | NGC 5397 | 14 01.2 | −33 56 | Cen | 12.7v | 1.4' | 0.9' | S0− |

JH: very faint, small, round, gradually brighter middle.
170x250: Faint. Broadly moderately concentrated. Round. Very small. No involved stars. A little ways E is a little triangle of faint stars.

| ga | NGC 5398 | 14 01.4 | −33 03 | Cen | 12.8p | 2.8' | 1.6' | Sdm |

JH: pretty bright, pretty large, round, very gradually brighter middle.
170x250: Very faint. Broadly very slightly concentrated. Roughly round. Moderately small. No involved stars. It's between two faint stars, one a little ways SW and one a little farther NE; W is a triangular figure of stars including a pair of brighter stars.

| ga | NGC 5443 | 14 02.2 | +55 48 | UMa | 13.1p | 3.2' | 1.2' | Sb |

HH: pretty faint, large, extended.

170x250: Moderately faint. Broadly moderately concentrated. Elongated to edge-on NE-SW. Small. Off each end is a faint star; [E is a moderate brightness star; farther N are two bright stars].

| ga | NGC 5448 | 14 02.8 | +49 10 | UMa | 11.9b | 4.0' | 1.7' | Sa |

HH: pretty bright, considerably large, very much extended E-W, suddenly much brighter middle to a nucleus.
170x250: Faint. Broadly slightly concentrated. Edge-on WNW-ESE. Moderate size. No involved stars.

| ga | NGC 5440 | 14 03.0 | +34 45 | CVn | 13.2p | 4.1' | 1.6' | Sa |

HH: pretty faint, considerably small, little extended, brighter middle, 11th mag. star SW.
170x250: Moderately faint. Evenly moderately concentrated. Elongated NE-SW. Small. Near its SW end is a moderately faint star.

| ga | NGC 5457/M101 | 14 03.2 | +54 20 | UMa | 8.3b | 28.9' | 26.9' | Scd |

JH: pretty bright, very large, irregularly round, gradually then very suddenly much brighter middle to a bright small nucleus.
12x35: Easily visible. *42x60:* Broadly very slightly concentrated. *170x250:* Moderate brightness; low surface brightness except for the very center. Evenly to sharply fairly concentrated to a brighter core. Slightly elongated NE-SW. Very large: 2/3 of the field in size. 2/3 of a field ENE from center is a small companion galaxy [*5471 (dA)]. Nearby W is a faint star. Also ENE from center, halfway between the core and the companion galaxy, is a fainter, more diffuse object, apparently an HII region [5462 (WH)]. Closer to center ESE is another HII region that looks like a faint star with nebulosity around it [5461 (WH)]. Nearly half the field SW is another HII region [5447 (WH)] with a moderately faint star off its edge N. Other than these specific objects there's some vague unevenness in the light but no obvious spiral structure. The only thing that looks like it could be a spiral arm is a brightening that curves S from near center and then E and NE through the first two HII regions [5461, 5462]. The individual HII regions are much more obvious than any spiral arms. Near center N is a relatively bright involved star; much farther from center are a couple more involved stars, one WNW and one NE.

| ga | NGC 5426 | 14 03.4 | –06 04 | Vir | 12.7b | 3.0' | 1.6' | Sc |

HH: pretty faint, considerably large, round, gradually much brighter middle, [SW] of 2.
[See 5427.]

| ga | NGC 5427 | 14 03.4 | −06 01 | Vir | 11.9b | 3.2' | 2.3' | Sc |

HH: pretty faint, considerably large, round, [NE] of 2.
170x250: 5427 and 5426 are a close double galaxy aligned N-S. 5427 is moderately faint. Broadly slightly concentrated. Slightly elongated E-W. Moderately small. No involved stars. 5426 is also moderately faint, but a little fainter than 5427. Broadly slightly concentrated. Elongated N-S. Small. At its N end is a faint star. An interesting pair of galaxies.

| ga | NGC 5434 | 14 03.4 | +09 26 | Boo | 13.2v | 1.8' | 1.8' | Sc |

Te: very faint, large.
170x250: In the position are a couple of galaxies a little more than a quarter of the field apart NNE-SSW [*5424 (Te), *5423 (Te)]. Both are faint. Evenly concentrated. Very small. No involved stars. Nearby S of the N galaxy is a faint star. Half the field ENE is a very faint diffuse galaxy [5434]. Half the field NE of this galaxy is a trio of very small very faint galaxies in a flat triangle [*5436 (Te), *5437 (Te), *5438 (Te)].

| ga | NGC 5444 | 14 03.4 | +35 07 | CVn | 12.8b | 2.4' | 2.0' | E+ |

HH: pretty bright, pretty large, irregularly very little extended, very suddenly much brighter middle.
170x250: Moderate brightness. Sharply fairly concentrated. Round. Small. No involved stars. Nearby WSW is a faint star. A little more than a third of the field SSE is another galaxy [*5445 (HH)]. Faint. Sharply fairly concentrated. Edge-on NNE-SSW. Small. Off its S end is a star.

| ga | NGC 5419 | 14 03.7 | −33 58 | Cen | 11.9b | 4.2' | 3.3' | E |

JH: pretty bright, pretty large, round, gradually pretty much brighter middle.
170x250: Moderate brightness. Evenly moderately concentrated. Round. Small. No involved stars. It's right in between a moderate brightness star SE and a faint star NW. [E384-37 was not seen.]

| ga | NGC 5473 | 14 04.7 | +54 53 | UMa | 12.4b | 2.3' | 1.8' | S0− |

HH: pretty bright, small, round, gradually brighter middle.
170x250: Moderately bright. Sharply pretty concentrated. Round. Small. No involved stars. It's roughly in line with three relatively bright stars: one ENE and two SW; a fourth brighter star is off line further SW.

12 to 18 Hours: SPRING

| ga | NGC 5474 | 14 05.0 | +53 39 | UMa | 11.3b | 4.7' | 4.7' | Sc |

WH: pretty bright, large, brighter middle.
170x250: Faint. Broadly slightly concentrated. Roughly round. Moderate size. NE of center is a very faint involved star.

| ga | IC 4366 | 14 05.2 | −33 45 | Cen | 13.2b | 1.6' | 1.4' | Sc |

170x250: Extremely faint. Hardly concentrated. Roughly round. Small. No involved stars. Nearby S is a bright star; a little ways NNW is another star.

| gc | NGC 5466 | 14 05.5 | +28 32 | Boo | 9.1 | 9.0' | −6.86 | 12 |

HH: cluster, large, very rich, very much compressed, 11th mag. and fainter stars.
42x60: Visible. ***170x250:*** Resolved throughout into a good number of very faint but distinct stars, with an underlying glow of additional stars. Hardly concentrated. The main body of the cluster is a quarter of the field in size; with outliers the cluster is a little more than half the field in size. The more distant outliers are few and very scattered. Irregular. Apparently nearby.

| ga | NGC 5480 | 14 06.4 | +50 43 | UMa | 12.8p | 2.1' | 1.6' | Sc |

WH: faint, pretty small, very gradually brighter middle, NW of 2.
170x250: Moderately faint. Broadly very slightly concentrated. Round. Small. No involved stars. A sixth of the field E is another galaxy [*5481 (WH)]: a little higher surface brightness, a little more concentrated, and a little smaller. Moderately faint. Sharply moderately concentrated. Round. Small. No involved stars.

| bs | Pi Hya | 14 06.4 | −26 40 | Hya | 3.27 | | 1.12 | K2 |

170x250: Yellow.

| ga | ESO 384-53 | 14 06.6 | −34 18 | Cen | 12.9p | 2.2' | 1.8' | Sbc |

170x250: Not visible.

| ga | NGC 5468 | 14 06.6 | −05 27 | Vir | 13.0b | 2.6' | 2.3' | Scd |

HH: faint, large, round, very gradually brighter middle.
170x250: Faint. Broadly very slightly concentrated. Roughly round. Moderately small. No involved stars. SSE is the bright plotted star. [5472 (LR) was not seen.]

| ga | UGC 9024 | 14 06.7 | +22 04 | Boo | 13.0 | 0.9' | 0.7' | S |

170x250: Not visible.

| ga | NGC 5485 | 14 07.2 | +55 00 | UMa | 11.4v | 2.4' | 1.8' | S0 |

WH: considerably bright, round, very gradually brighter middle, E of 2.
170x250: Moderate brightness. Evenly moderately concentrated. Roughly round. Small. No involved stars. A little ways SE is a faint star. [5484 (WH) was not seen.]

| ga | NGC 5476 | 14 08.1 | –06 05 | Vir | 13.0 | 1.4' | 1.0' | Sdm |

HH: faint, pretty small, irregularly round.
170x250: Very faint. Hardly concentrated. Round. Small. No involved stars. A little ways W is a moderately faint star; NNW is a pair of very faint stars.

| ga | NGC 5488 | 14 08.1 | –33 18 | Cen | 12.7p | 3.3' | 0.9' | Sbc |

JH: faint, round, 8th mag. star near S.
170x250: Faint to moderately faint. Broadly slightly concentrated. Elongated NNE-SSW. Small. At its S end is a bright star that interferes; very nearby E is a faint star.

| ga | NGC 5490 | 14 10.0 | +17 32 | Boo | 13.1b | 2.4' | 1.9' | E |

HH: considerably faint, considerably small, round, suddenly brighter middle to a faint double star.
170x250: Moderate brightness. Evenly moderately concentrated. Round. Very small. No involved stars. It's among a large E-W group of moderately faint stars. Half the field N are two extremely faint galaxies [I983, *I982]. Both are roughly round. Very very small. No involved stars. They make a sharp isosceles triangle with a bright star E. [5490C was not seen.]

| ga | IC 983 | 14 10.1 | +17 44 | Boo | 12.5b | 5.3' | 4.6' | Sbc |

[See 5490.]

| ga | NGC 5493 | 14 11.5 | –05 02 | Vir | 12.3b | 1.6' | 1.2' | S0 |

HH: pretty bright, very small, round, pretty suddenly much brighter middle to a star, 18th mag. star involved.
170x250: Moderately bright. Sharply very concentrated. Round. Small. No involved stars.

| ga | NGC 5496 | 14 11.6 | −01 09 | Vir | 12.7p | 4.7' | 0.8' | Sd |

Ho: pretty bright, very large, extended N-S.
170x250: Very faint. Hardly concentrated. A thin edge-on, N-S. Moderate size. No involved stars.

| ga | NGC 5494 | 14 12.4 | −30 38 | Cen | 12.6p | 2.2' | 1.9' | Sc |

JH: pretty bright, large, round, gradually brighter middle, partially resolved.
170x250: Very faint. Broadly very slightly concentrated. Round. Moderately small. It's closely surrounded by faint stars.

| ga | NGC 5506 | 14 13.2 | −03 12 | Vir | 12.8b | 2.8' | 0.8' | Sa |

HH: pretty bright, large, extended NNE-SSW, little brighter middle.
170x250: In the position is a pair of moderately faint galaxies a little less than a quarter of the field apart NNE-SSW [*5507 (HH), 5506]. The S galaxy is broadly very slightly concentrated. Edge-on E-W. Moderate size. No involved stars. The N galaxy is sharply pretty concentrated. Round. Very small. No involved stars. A fairly nice contrast in galaxies.

| ds | Kappa Boo | 14 13.5 | +51 47 | Boo | 14.0" | 4.6 | 6.6 | A8 F1 |

42x60: The primary has a close fainter companion. ***170x250:*** A moderate distance unequal pair. Both white. A nice double. [STF 1821]

| ga | NGC 5523 | 14 14.8 | +25 19 | Boo | 12.8p | 4.6' | 1.2' | Scd |

HH: faint, pretty large, pretty much extended E-W, 10th mag. star NW.
170x250: Faint. Broadly slightly concentrated. A thin edge-on, E-W. Moderately large. No involved stars. N of its W end is a moderate brightness star.

| ga | NGC 5529 | 14 15.6 | +36 13 | Boo | 12.8b | 6.2' | 0.8' | Sc |

HH: considerably faint, pretty large, very much extended WNW-ESE, very gradually very much brighter middle.
170x250: Faint. Broadly very slightly concentrated. An extremely thin edge-on, WNW-ESE. Large: a quarter of the field in size. No involved stars. ESE to NE is a group of stars in a narrow parallelogram figure. A nice and interesting galaxy; one of the thinnest galaxies in the sky. [5527 (LR) was not seen.]

bs	Alpha Boo	14 15.7	+19 10	Boo	−0.04		1.23	K2

170x250: Yellow. A brilliant star, with diffraction spikes all the way across the field. [Arcturus]

ga	NGC 5533	14 16.1	+35 20	Boo	12.7b	4.3'	2.7'	Sab

HH: pretty bright, round, very suddenly much brighter middle, 2 or 3 stars involved.
170x250: Moderate brightness. Sharply fairly concentrated. Roughly round. Small. No involved stars.

ga	NGC 5532	14 16.9	+10 48	Boo	12.9b	1.6'	1.6'	S0

HH: very faint, very small, round, gradually brighter middle, resolvable.
170x250: Moderately faint. Evenly moderately concentrated. Roughly round. Small. No involved stars. A little less than a third of the field NNW is another galaxy [*5531 (dA)]. Very faint. Small. No involved stars.

ga	NGC 5534	14 17.7	−07 25	Vir	13.0b	1.4'	0.8'	Sab

St: pretty faint, stars involved, 12th mag. star NW.
170x250: Moderately faint. Sharply pretty concentrated. Round. Very small. No involved stars. It's in the middle of a long NE-SW line of five moderate brightness stars.

ga	NGC 5557	14 18.4	+36 29	Boo	11.9b	3.6'	3.2'	E1

HH: considerably bright, small, round, very suddenly brighter middle to a star.
170x250: Bright. Sharply pretty concentrated. Roughly round. Small. Just S of center is an exceedingly faint involved star.

ga	NGC 5585	14 19.8	+56 43	UMa	11.2b	6.1'	3.8'	Sd

HH: pretty faint, large, irregularly round, very gradually much brighter middle, resolvable.
170x250: Faint. Broadly slightly concentrated. Roughly round. Moderately small. No involved stars. A little ways SSW is a faint star; the galaxy makes a triangle with two brighter stars, one NE and one farther SE.

ga	NGC 5560	14 20.1	+03 59	Vir	12.4v	4.3'	1.2'	Sb

HH: pretty faint, considerably large, extended, gradually brighter middle.
[See 5566.]

| ga | NGC 5566 | 14 20.3 | +03 56 | Vir | 11.5b | 6.7' | 2.1' | Sab |

HH: bright, pretty large, round, pretty suddenly brighter middle, resolvable, 12th mag. star NE.
170x250: Bright; high surface brightness. Sharply very concentrated to a substellar nucleus. Round. Small. No involved stars. Nearby WSW is a faint star; a little farther E is a moderately faint star. A little less than a third of the field NW is 5560. Very faint to faint. Broadly very slightly concentrated. Elongated WNW-ESE. Small. NNW of center is a very faint star. [5569 (LR) was not seen.]

| ga | NGC 5556 | 14 20.6 | −29 14 | Hya | 12.4p | 4.0' | 3.1' | Sd |

JH: extremely faint, large, faint star involved.
170x250: Extremely faint. The very slightest unconcentrated glow. Roughly round. Moderate size. Several very faint stars are involved; at its edge NW is a moderate brightness star; E is a long NW-SE trapezoid of stars; WNW is a bright star.

| ga | NGC 5582 | 14 20.7 | +39 41 | Boo | 12.5b | 2.8' | 1.7' | E |

HH: pretty bright, pretty small, round, brighter middle to a faint nucleus, star SW.
170x250: Moderate brightness. Sharply pretty concentrated. Round. Very small. No involved stars. It makes a trapezoid with three moderate brightness stars SW to SE.

| ga | NGC 5574 | 14 20.9 | +03 14 | Vir | 12.4v | 2.2' | 1.3' | S0− |

HH: pretty faint, pretty small, little extended, W of 2.
[See 5576.]

| ga | NGC 5576 | 14 21.1 | +03 16 | Vir | 11.0v | 3.9' | 2.6' | E3 |

HH: bright, small, round, very suddenly much brighter middle, E of 2.
170x250: 5576 and 5574 are next to each other NE-SW. 5576 is a little brighter. Moderately bright. Sharply pretty concentrated. Round. Very small. No involved stars. Nearby NW is a moderately faint star. 5574 is moderate brightness. Evenly moderately concentrated. Slightly elongated ENE-WSW. Very small. No involved stars. Half the field N is 5577. Very faint. Hardly concentrated. Elongated NE-SW. Moderate size. No involved stars.

| ga | NGC 5577 | 14 21.2 | +03 26 | Vir | 13.1p | 5.7' | 1.0' | Sbc |

LR: pretty faint, pretty large, very much extended NE-SW.
[See 5576.]

| ga | NGC 5584 | 14 22.4 | −00 23 | Vir | 12.1p | 3.4' | 2.4' | Scd |

Ba: faint, large, much extended, diffuse, gradually little brighter middle.
170x250: Very faint. Diffuse. Round. Moderately small. No involved stars. NNE to SE is a sharp triangle of stars: one moderate brightness star and two moderately faint stars.

| ds | STF 1835 | 14 23.4 | +08 27 | Boo | 6.0" | 5.1 | 6.9 | A0 F3 |

42x60: Split. *170x250:* A moderately close slightly unequal pair. Both white. A nice double.

| ga | UGC 9215 | 14 23.5 | +01 43 | Vir | 13.2p | 2.2' | 1.2' | Sd |

170x250: Very faint. Broadly very slightly concentrated. Roughly round. Small. No involved stars. NE to SE are four moderately bright stars in a large triangular figure.

| ga | NGC 5600 | 14 23.8 | +14 38 | Boo | 12.7b | 1.4' | 1.3' | Sc |

HH: pretty bright, pretty small, gradually brighter middle.
170x250: Moderate brightness. Broadly moderately concentrated. Round. Small. No involved stars.

| ga | NGC 5614 | 14 24.1 | +34 51 | Boo | 11.6v | 2.4' | 2.2' | Sab |

HH: pretty bright, small, round, suddenly much brighter middle.
170x250: Moderate brightness to moderately bright. Evenly moderately concentrated. Round. Very small. No involved stars. A little ways ESE is a moderately faint star. [5613 (LR) and 5615 (LR) were not seen.]

| ga | NGC 5595 | 14 24.2 | −16 43 | Lib | 12.6b | 2.2' | 1.2' | Sc |

HH: faint, pretty large, round, very gradually brighter middle, W of 2.
170x250: 5595 and 5597 are a quarter of the field apart NW-SE. 5595 is faint. Broadly very slightly concentrated. Elongated NE-SW. Moderate size. No involved stars. 5597 is very faint. Evenly slightly concentrated. Roughly round. Moderately small. No involved stars.

| ga | NGC 5597 | 14 24.5 | −16 45 | Lib | 12.6b | 2.1' | 1.6' | Scd |

HH: very faint, large, very little extended, very gradually little brighter middle, E of 2. [See 5595.]

| ga | NGC 5605 | 14 25.1 | –13 09 | Lib | 12.9b | 1.6' | 1.3' | Sc |

HH: very faint, pretty large, round, very gradually brighter middle.
170x250: Very faint. Hardly concentrated. Roughly round. Moderately small. No involved stars. W is a moderate brightness star with a wide very faint companion.

| ga | NGC 5631 | 14 26.6 | +56 35 | UMa | 12.4b | 1.7' | 1.7' | S0 |

HH: bright, small, round, pretty suddenly brighter middle to a resolvable nucleus.
170x250: Moderate brightness. Sharply pretty concentrated. Round. Very small. No involved stars.

| ga | NGC 5633 | 14 27.5 | +46 08 | Boo | 13.1b | 2.1' | 1.3' | Sb |

HH: considerably bright, pretty small, round, pretty gradually little brighter middle.
170x250: Moderately faint. Broadly to evenly moderately concentrated. Roughly round. Very small to small. No involved stars.

| ga | IC 1014 | 14 28.3 | +13 46 | Boo | 13.0p | 2.7' | 2.0' | Sdm |

170x250: Very very faint. Diffuse. Moderately small. No involved stars. 2/3 of a field ENE is a very small very faint galaxy [*U9288]. It's at the E end of an E-W linear figure of moderate brightness to moderately faint stars, the two nearest of which make a small triangle with the galaxy.

| ga | NGC 5629 | 14 28.3 | +25 50 | Boo | 13.0b | 1.8' | 1.8' | S0 |

JH: pretty faint, small, round, gradually brighter middle.
170x250: Faint to moderately faint. Sharply pretty concentrated. Round. Small. No involved stars. A little ways WNW is a slightly smaller and fainter galaxy [*I1017]. Nearby WSW is a moderate brightness star.

| ga | ESO 385-30 | 14 29.3 | –33 27 | Cen | 12.9p | 2.0' | 1.0' | S0 |

170x250: Moderately faint. Evenly moderately concentrated. Slightly elongated NNE-SSW. Small. No involved stars. It makes a triangle with two faint stars, one a little ways ENE and another a little ways SSE.

| ga | NGC 5641 | 14 29.3 | +28 49 | Boo | 13.1p | 2.4' | 1.3' | Sab |

St: pretty bright, pretty small, little extended, much brighter middle, resolvable?
170x250: Faint. Broadly slightly concentrated. Slightly elongated NNW-SSE. Small. No involved stars. A little ways W is a faint star; SSE are a couple more faint stars.

| gc | NGC 5634 | 14 29.6 | −05 59 | Vir | 9.6 | 5.5' | −7.33 | 4 |

HH: globular cluster, very bright, considerably large, round, gradually brighter middle, well resolved, 19th mag. stars, [8th] mag. star SE.
170x250: Not resolved. Broadly concentrated. Small. Pretty bright, even for a globular. It's just inside a sharp isosceles triangle of moderately bright and moderate brightness stars – a nice effect.

| ga | NGC 5638 | 14 29.7 | +03 14 | Vir | 11.2v | 2.7' | 2.4' | E1 |

HH: considerably bright, pretty large, round, SE of 2.
170x250: Moderate brightness to moderately bright. Sharply pretty concentrated. Round. Very small. No involved stars. [5636 (HH) and U9310 were not seen.]

| ga | NGC 5660 | 14 29.8 | +49 37 | Boo | 12.4b | 2.8' | 2.7' | Sc |

HH: pretty bright, large, irregularly round, very gradually brighter middle.
170x250: Moderate brightness. Broadly slightly concentrated. Round. Moderately small. No involved stars. [M+8-26-38 was not seen.]

| ga | NGC 5653 | 14 30.2 | +31 12 | Boo | 12.9b | 1.7' | 1.2' | Sb |

HH: pretty faint, pretty small, round, brighter middle.
170x250: Moderate brightness. Evenly moderately concentrated. Round. Very small. No involved stars. Nearby N is an extremely faint star.

| ga | NGC 5656 | 14 30.4 | +35 19 | Boo | 12.7p | 1.9' | 1.5' | Sab |

HH: pretty faint, pretty large, round, much brighter middle, resolvable.
170x250: Moderate brightness. Evenly moderately concentrated. Slightly elongated NE-SW. Small. No involved stars. Nearby SSE is a faint star; [ESE is a bright star].

| ga | NGC 5667 | 14 30.4 | +59 28 | Dra | 13.2p | 1.7' | 1.1' | Scd |

WH: pretty bright, pretty small, extended N-S.
170x250: Faint. Broadly slightly concentrated. Elongated N-S. Small. Right at its N end is a very faint star.

| ga | NGC 5645 | 14 30.7 | +07 16 | Vir | 13.0b | 2.4' | 1.5' | Sd |

HH: considerably faint, pretty large, irregularly round, gradually brighter middle.
170x250: Faint. Broadly slightly concentrated. Round. Small. No involved stars. It's just outside a sharp isosceles triangle of moderate brightness stars.

| ga | NGC 5673 | 14 31.5 | +49 57 | Boo | 12.9b | 2.5' | 0.6' | Sc |

HH: faint, small, considerably extended, 15th mag. star NW.
[See I1029.]

| bs | Gamma Boo | 14 32.1 | +38 18 | Boo | 3.00 | | 0.23 | A7 |

170x250: White.

| ga | NGC 5678 | 14 32.1 | +57 55 | Dra | 12.1p | 3.3' | 1.6' | Sb |

HH: bright, large, little extended N-S, very gradually much brighter middle.
170x250: Moderate brightness. Broadly moderately concentrated. Elongated N-S. Moderately small. No involved stars. It's between a moderately faint star SSE and a bright star NNW.

| ga | IC 1029 | 14 32.4 | +49 54 | Boo | 12.2b | 2.8' | 0.6' | Sb |

170x250: Moderately faint. Evenly moderately concentrated. Edge-on NNW-SSE. Small. No involved stars. [E is a bright star.] Half the field WNW is 5673. Faint. Broadly slightly concentrated. Edge-on NW-SE. Moderate size. At its NW end is a moderate brightness star. A contrasting pair of edge-on galaxies: a concentrated type (I1029) and an even-light type (5673).

| ga | NGC 5665 | 14 32.4 | +08 04 | Boo | 12.7p | 2.5' | 1.6' | Sc |

HH: pretty bright, pretty large, round, gradually brighter middle, resolvable.
170x250: Moderate brightness. Broadly slightly concentrated. Round. Small. No involved stars.

| ga | NGC 5669 | 14 32.7 | +09 53 | Boo | 12.0p | 4.2' | 3.0' | Scd |

WH: faint, large, round, little brighter middle, resolvable.
170x250: Faint. Broadly very slightly concentrated. Roughly round. Moderate size. No involved stars.

| ga | NGC 5676 | 14 32.8 | +49 27 | Boo | 11.9b | 4.0' | 1.9' | Sbc |

HH: bright, large, extended NE-SW, pretty gradually brighter middle, resolvable.
170x250: Moderately bright. Broadly to evenly moderately concentrated. Elongated NE-SW. Moderate size. No involved stars.

| ga | NGC 5668 | 14 33.4 | +04 27 | Vir | 12.2b | 3.3' | 3.0' | Sd |

HH: faint, pretty small, very little extended, 14th mag. star involved.
170x250: Faint. Broadly slightly concentrated. Roughly round. Small. E of center is a faint involved star.

| ga | IC 4453 | 14 34.5 | −27 31 | Hya | 13.2b | 2.2' | 0.9' | S0 |

170x250: Moderately faint. Sharply very concentrated. Round. Very small. No involved stars. SSW is a bright star; NNW to SW is a long crooked line of faint stars.

| ga | IC 4460 | 14 34.6 | +30 16 | Boo | 13.2 | 0.8' | 0.5' | Sa |

170x250: Not visible.

| ga | NGC 5687 | 14 34.9 | +54 28 | Boo | 12.6b | 2.4' | 1.6' | S0− |

HH: pretty faint, small, irregular figure, resolvable, 10th mag. star E.
170x250: Moderate brightness. Evenly moderately concentrated. Elongated E-W. Small. Going SSW from its W end is a short line of three stars, the closest of which is involved; a little ways S is a brighter star.

| ga | NGC 5689 | 14 35.5 | +48 44 | Boo | 12.8b | 4.0' | 1.1' | S0/a |

HH: considerably bright, small, extended E-W, pretty suddenly much brighter middle.
170x250: Moderately bright. Evenly moderately concentrated. Edge-on E-W. Small. No involved stars.

| ga | MCG+9-24-22 | 14 36.8 | +51 27 | Boo | 12.9 | 2.0' | 1.4' | S |

170x250: [The observed galaxy is *5707 (Sw); M+9-24-22 was not seen.] Moderately faint. Evenly to sharply moderately concentrated. Edge-on NE-SW. Very small. No involved stars.

| ga | NGC 5690 | 14 37.7 | +02 17 | Vir | 12.5b | 3.4' | 1.0' | Sc |

HH: very faint, much extended or binuclear NW-SE, 6th–7th mag. star W.
170x250: Faint. Broadly very slightly concentrated. Edge-on NW-SE. Moderately small. At it SE end is a very faint star; WSW is the bright plotted star, which interferes.

| ga | NGC 5691 | 14 37.9 | −00 23 | Vir | 12.3v | 1.8' | 1.3' | Sa |

HH: pretty bright, pretty small, little extended, gradually brighter middle.

170x250: Moderately faint to moderate brightness. Broadly moderately concentrated. Round. Small. No involved stars.

ga	NGC 5701	14 39.2	+05 21	Vir	11.8b	4.4'	4.2'	S0/a

HH: considerably bright, pretty small, round, much brighter middle, 15th mag. star W.
170x250: Moderate brightness. Sharply fairly concentrated. Round. Very small. No involved stars. It's among several stars, moderate brightness to faint.

gc	NGC 5694	14 39.6	−26 32	Hya	10.2	4.3'	−7.60	7

HH: considerably bright, considerably small, round, pretty suddenly brighter middle, resolvable, star near.
42x60: Visible in a little arc with two very faint stars. *170x250:* Starting to be granular. Evenly to sharply concentrated. Very small. It makes an arc with two moderate brightness stars a little ways SSW.

ga	NGC 5713	14 40.2	−00 17	Vir	11.8b	2.7'	2.4'	Sbc

HH: considerably bright, pretty large, round, pretty suddenly much brighter middle, resolvable.
170x250: Moderate brightness. Broadly moderately concentrated. Round. Moderately small. No involved stars.

ds	Pi Boo	14 40.7	+16 25	Boo	5.6"	4.9	5.8	B9 A6

42x60: Split. *170x250:* A moderately close slightly unequal pair. Both white. [STF 1864]

ga	NGC 5719	14 40.9	−00 19	Vir	13.1p	3.2'	1.1'	Sab

HH: pretty faint, small, little extended, brighter middle.
170x250: Faint to moderately faint. Evenly moderately concentrated. Very elongated E-W. Small. No involved stars. A little ways N is a moderately bright star; SSW is another moderately bright star.

ga	NGC 5728	14 42.4	−17 15	Lib	12.3b	3.1'	1.7'	Sa

HH: pretty faint, pretty large, pretty much extended NE-SW, much brighter middle, 10th mag. star S.
170x250: Moderately faint. Broadly moderately concentrated. Edge-on NE-SW. Moderate size. At its SW end is a very faint involved star.

| ga | NGC 5739 | 14 42.5 | +41 50 | Boo | 13.1p | 2.3' | 2.0' | S0+ |

HH: pretty bright, small, round, suddenly much brighter middle, resolvable, star near.
170x250: Moderately faint. Evenly moderately concentrated. Round. Very small. No involved stars. Very nearby ENE is a faint star; SSE is another faint star; these stars make a small very sharp triangle with the galaxy.

| ga | NGC 5735 | 14 42.6 | +28 43 | Boo | 13.1p | 2.4' | 1.9' | Sbc |

HH: very faint, large, irregularly round, little brighter middle.
170x250: Very faint. Broadly very slightly concentrated. Round. Small. No involved stars.

| ga | ESO 512-18 | 14 43.6 | -24 27 | Lib | 12.9p | 2.5' | 1.4' | S0 |

170x250: Faint. Concentrated. Round. Very very small. No involved stars. Very nearby NNE is a very very faint star; very nearby W is the bright plotted star. [E512-19 was not seen.]

| ga | ESO 512-19 | 14 43.6 | -24 27 | Lib | 13.0p | 2.7' | 1.6' | Sa |

[See E512-18.]

| ga | NGC 5740 | 14 44.4 | +01 40 | Vir | 11.9v | 3.0' | 1.5' | Sb |

HH: pretty bright, large, irregularly round, gradually brighter middle, resolvable.
170x250: Faint. Broadly slightly concentrated. Round. Small. No involved stars. Nearby WNW is a very faint star; N is another very faint star.

| ga | NGC 5746 | 14 44.9 | +01 57 | Vir | 11.3b | 7.5' | 1.3' | Sb |

HH: bright, large, very much extended, brighter middle to a bright nucleus.
170x250: Moderately bright. Evenly moderately concentrated. A thin edge-on, N-S. Large: a third of the field in size. At its S tip is a very faint star; N is a long slightly curved ENE-WSW line of four bright stars – a nice effect. A nice galaxy.

| bs | Epsilon Boo | 14 45.0 | +27 04 | Boo | 2.7 | | 1.00 | K0 |

170x250: The primary has an extremely close much fainter companion. Color contrast: the primary is yellow; the companion is white. [STF 1877: 2.6", 2.5, 4.9, K0 A2]

| ga | NGC 5741 | 14 45.9 | −11 54 | Lib | 12.2 | 1.0' | 1.0' | E |

Le: very faint, very small, round, suddenly brighter middle to a nucleus.
170x250: [The first galaxy observed is *5742 (Le).] Faint. Evenly moderately concentrated. Slightly elongated ENE-WSW. Small. No involved stars. Nearby SSW is a faint star. A little less than half the field SSE is another galaxy [5741]. Faint. Very small; almost stellar.

| ds | 54 Hya | 14 46.0 | −25 27 | Hya | 8.3" | 5.1 | 7.1 | F2 F9 |

42x60: Split. ***170x250:*** A moderately close unequal pair. Both white. [H 97]

| ga | NGC 5750 | 14 46.2 | −00 13 | Vir | 12.5b | 3.0' | 1.5' | S0/a |

HH: pretty faint, pretty small, very little extended, resolvable.
170x250: Moderate brightness. Broadly slightly concentrated. Round. Small to moderately small. No involved stars.

| ga | NGC 5756 | 14 47.6 | −14 51 | Lib | 12.3v | 2.7' | 0.9' | Sbc |

JH: pretty bright, pretty large, pretty much extended, gradually pretty much brighter middle.
170x250: Faint. Broadly slightly concentrated. Elongated NE-SW. Moderately small. No involved stars. A little ways WSW is a faint star; NNW are a couple of moderately faint stars.

| ga | NGC 5757 | 14 47.8 | −19 04 | Lib | 12.7p | 2.0' | 1.6' | Sb |

HH: very faint, small, irregularly round, little brighter middle.
170x250: Moderately faint. Evenly moderately concentrated. Elongated N-S. Small. No involved stars. Nearby NE are a couple of stars, moderately faint and faint.

| ga | MCG-2-38-15 | 14 47.9 | −14 16 | Lib | 13.1 | 2.2' | 1.8' | Scd |

170x250: Not visible.

| ds | STF 1884 | 14 48.4 | +24 22 | Boo | 2.2" | 6.1 | 7.7 | F2 − |

170x250: An extremely close almost equal pair. Both white. A faint double.

| ds | Mu Lib | 14 49.3 | −14 09 | Lib | 2.0" | 5.8 | 6.7 | A1 − |

170x250: An extremely close slightly unequal pair. Both white. [BU 106]

| ds | 39 Boo | 14 49.7 | +48 43 | Boo | 2.8" | 6.2 | 6.9 | F6 F5 |

170x250: A very close pretty much equal pair. There's a small gap between the stars. Both white. [STF 1890]

| bs | Beta UMi | 14 50.7 | +74 09 | UMi | 2.05 | | 1.50 | K4 |

170x250: Gold.

| bs | Alpha 2 Lib | 14 50.9 | −16 02 | Lib | 2.75 | | 0.15 | A3 |

170x250: White.

| ds | Xi Boo | 14 51.4 | +19 06 | Boo | 6.3" | 4.7 | 7.0 | G8 K4 |

42x60: Extremely close. *170x250:* A pretty close unequal pair. Color contrast: the primary is yellow; the companion is dark yellow. An unusual color contrast; a neat double. [STF 1888]

| ga | NGC 5768 | 14 52.1 | −02 31 | Lib | 13.2p | 1.8' | 1.3' | Sc |

WH: *faint, round, brighter middle to a faint nucleus, faint star S.*
170x250: Faint. Broadly very slightly concentrated. Roughly round. Small. At its edge S is a moderately faint star.

| ga | IC 1067 | 14 53.1 | +03 19 | Vir | 13.0p | 2.1' | 1.7' | Sb |

170x250: Faint. Broadly slightly concentrated. Round. Small. No involved stars. A little ways SSW is another galaxy [*I1066] – a slightly smaller and fainter version of I1067. E of both galaxies, making a little figure with them, is an arc of three moderate brightness stars.

| ga | NGC 5770 | 14 53.3 | +03 57 | Vir | 13.2b | 1.7' | 1.2' | S0 |

HH: *considerably faint, small, very little extended, brighter middle, binuclear?*
170x250: Moderate brightness. Sharply very concentrated. Round. Very small. No involved stars.

| ga | NGC 5774 | 14 53.7 | +03 34 | Vir | 12.7b | 3.0' | 2.4' | Sd |

LR: *pretty faint, pretty large, round, NW of 2.*
[See 5775.]

| ga | NGC 5775 | 14 54.0 | +03 32 | Vir | 12.2b | 4.2' | 1.0' | Sc |

HH: faint, pretty small, very much extended NNW-SSE, gradually very little brighter middle.
170x250: 5774 and 5775 are next to each other WNW-ESE. 5775 is brighter. Moderate brightness. Broadly very slightly concentrated. Edge-on NW-SE. Moderate size. No involved stars. Very nearby NE of center is a faint star; nearby S is another faint star. 5774 is very faint. Diffuse. Moderately small. No involved stars. Nearby NE is a faint star.

| ga | NGC 5832 | 14 57.8 | +71 40 | UMi | 12.9p | 3.7' | 2.1' | Sb |

WH: pretty bright, considerably large, irregularly round, brighter W, resolvable.
170x250: Very faint. Hardly concentrated. Roughly round. Small to moderate size. No involved stars.

| ga | NGC 5792 | 14 58.4 | −01 05 | Lib | 12.1b | 7.2' | 1.7' | Sb |

WH: pretty bright, pretty large, round, much brighter middle, considerably bright star attached NW.
170x250: Moderately faint. Broadly very slightly concentrated. Round. Small. No involved stars. Right off its edge WNW is a moderately bright star; NE are two more relatively bright stars.

| ga | NGC 5791 | 14 58.8 | −19 16 | Lib | 12.7b | 2.6' | 1.3' | E6 |

HH: pretty faint, small, round, stellar.
170x250: Moderate brightness. Evenly moderately concentrated. Round. Small. No involved stars. It makes a flat triangle with a moderately faint star ESE and a relatively bright star farther SE. [I1081 was not seen.]

| ga | NGC 5796 | 14 59.4 | −16 37 | Lib | 12.7b | 2.4' | 1.7' | E0-1 |

Te: faint, pretty faint star in center.
170x250: Moderately faint. Evenly to sharply moderately concentrated. Round. Small. No involved stars. [5793 (Le) was not seen.]†

| ga | NGC 5806 | 15 00.0 | +01 53 | Vir | 12.4b | 3.0' | 1.5' | Sb |

HH: considerably bright, considerably large, extended NNW-SSE, suddenly brighter middle to a nucleus.
170x250: Moderate brightness. Evenly moderately concentrated. Slightly elongated N-S. Small. No involved stars.

| ga | NGC 5812 | 15 01.0 | –07 27 | Lib | 12.2b | 2.1' | 1.8' | E0 |

HH: considerably bright, small, round, suddenly very much brighter middle.
170x250: Moderate brightness. Evenly to sharply pretty concentrated. Round. Very small. No involved stars.

| ga | NGC 5813 | 15 01.2 | +01 42 | Vir | 11.5b | 4.1' | 2.9' | E1-2 |

HH: bright, pretty small, round, pretty suddenly much brighter middle.
170x250: Moderately bright. Sharply pretty concentrated. Round. Very small. No involved stars. It's almost exactly in the middle of a nearly square quadrilateral of stars, at the intersection of an imaginary "X" connecting the four corners: one of the most perfect formations among galaxies and associated stars. A quarter of the field SSE is a tiny galaxy [*5814 (JH)].

| bs | Beta Boo | 15 01.9 | +40 23 | Boo | 3.50 | | 0.97 | G8 |

170x250: Yellow.

| ds | 44 Boo | 15 03.8 | +47 39 | Boo | 2.1" | 5.3 | 6.2 | G0 G2 |

170x250: An extremely close slightly unequal pair. Both yellow white. [STF 1909]

| gc | NGC 5824 | 15 04.0 | –33 04 | Lup | 9.0 | 7.4' | –8.32 | 1 |

Ba: pretty bright, small, stellar nucleus.
42x60: Bright. Concentrated. *170x250:* Starting to be granular with averted vision. Sharply concentrated. Very small. Very bright, even for a globular. It looks like a substantial cluster, just far away.

| ga | NGC 5831 | 15 04.1 | +01 13 | Vir | 11.5v | 2.0' | 1.7' | E3 |

HH: pretty bright, small, much brighter middle.
170x250: Moderate brightness. Sharply pretty concentrated. Round. Very small. No involved stars. Nearby NNE is a faint star.

| bs | Sigma Lib | 15 04.1 | –25 16 | Lib | 3.29 | | 1.70 | M3 |

170x250: Yellow gold.

| ga | NGC 5838 | 15 05.4 | +02 05 | Vir | 11.9b | 4.1' | 1.4' | S0– |

WH: pretty bright.

170x250: Bright. Sharply very concentrated. Slightly elongated NE-SW. Small. No involved stars. It's in a shallow arc with four faint stars: three NNE and one SW.

| ga | NGC 5845 | 15 06.0 | +01 37 | Vir | 12.5v | 0.8' | 0.5' | E |

WH: very faint, round, W of 2.
[See 5846.]

| ga | NGC 5846 | 15 06.4 | +01 36 | Vir | 10.0v | 3.5' | 3.5' | E0-1 |

HH: very bright, pretty large, round, pretty suddenly brighter middle to a nucleus, E of 2.
42x60: Visible. *170x250:* Bright. Sharply pretty concentrated. Round. Small. At its edge S is a tiny companion galaxy [*5846A]. A little less than half the field WNW is a pretty much stellar galaxy [5845]. Moderately bright. Half the field further W is another galaxy [*5839 (WH)], a little fainter. Moderate brightness. Very small; substellar.

| ga | NGC 5866/M102 | 15 06.5 | +55 45 | Dra | 10.7b | 6.4' | 2.8' | S0+ |

HH: very bright, considerably large, pretty much extended NW-SE, gradually brighter middle.
42x60: Visible. *170x250:* Very bright. Broadly moderately concentrated. A fat edge-on, NW-SE. Moderate size. The ends fade out indistinctly. No involved stars. It makes a small triangle with two moderately bright stars, one NW and one SW.

| ga | NGC 5850 | 15 07.1 | +01 32 | Vir | 10.7v | 4.6' | 4.1' | Sb |

HH: considerably faint, small, little extended, pretty suddenly brighter middle.
170x250: Moderately faint. Evenly moderately concentrated. Round. Small. No involved stars. It makes a triangle with two moderate brightness stars N.

| ga | NGC 5854 | 15 07.8 | +02 34 | Vir | 12.7b | 2.7' | 0.7' | S0+ |

HH: pretty bright, small, very little extended, little brighter middle, among stars.
170x250: Moderate brightness. Evenly to sharply fairly concentrated. Slightly elongated NE-SW. Small. No involved stars. A little ways ESE is a moderate brightness star; NNW to SW is a large triangle of bright stars.

| ga | NGC 5874 | 15 07.9 | +54 45 | Boo | 13.1b | 2.3' | 1.6' | Sc |

Sw: very faint, pretty large, round, in a triangle of 3 bright stars.
170x250: Very faint. Diffuse. Roughly round. Moderate size. No involved stars.

| ga | UGC 9749 | 15 08.8 | +67 11 | UMi | 11.9p | 30.4' | 19.1' | Dw E |

12x35, 42x60, 170x250: Not visible. [Ursa Minor dwarf]

| ga | NGC 5875 | 15 09.2 | +52 31 | Boo | 13.2p | 2.4' | 1.2' | Sb |

WH: pretty bright, pretty large, little extended.
170x250: Faint. Broadly very slightly concentrated. Elongated NW-SE. Moderately small. No involved stars. It's at the NW end of a long crooked line of bright stars.

| ga | NGC 5861 | 15 09.3 | −11 19 | Lib | 12.3p | 3.0' | 1.6' | Sc |

WH: faint, large, extended, resolvable.
170x250: Faint. Hardly concentrated. Round. Moderate size. No involved stars. It's inside a semicircle of moderate brightness and moderately faint stars S, E, and N.

| ga | UGCA 400 | 15 09.4 | −10 41 | Lib | 13.2 | 1.8' | 1.6' | Scd |

170x250: Not visible.

| ga | NGC 5864 | 15 09.6 | +03 03 | Vir | 12.8p | 2.7' | 0.8' | S0 |

HH: pretty faint, considerably small, irregularly little extended, gradually brighter middle, 14th mag. star E.
170x250: Moderate brightness. Evenly moderately concentrated. Elongated ENE-WSW. Small. At its edge E of center is a faint star; nearby SSW is another faint star.

| ga | NGC 5869 | 15 09.8 | +00 28 | Vir | 12.9p | 2.3' | 1.6' | S0 |

HH: pretty faint, small, extended, pretty suddenly brighter middle.
170x250: Moderate brightness. Sharply pretty concentrated. Round. Very small. No involved stars. It makes a little trapezoid with three faint stars E to SE; a little ways WNW is another faint star; SW is a brighter star. [5865 (WH) was not seen.]

| ga | NGC 5879 | 15 09.8 | +57 00 | Dra | 11.5v | 4.2' | 1.4' | Sbc |

HH: considerably bright, small, extended, much brighter middle to a round nucleus, resolvable.
170x250: Moderately bright to bright; fairly high surface brightness. Evenly moderately concentrated. Elongated N-S. Small. No involved stars.

| ga | NGC 5892 | 15 13.7 | −15 27 | Lib | 12.3p | 3.5' | 2.8' | Sd |

Sn: extremely faint, large, gradually brighter middle.
170x250: Not visible.

| ga | NGC 5878 | 15 13.8 | −14 16 | Lib | 12.4b | 3.5' | 1.4' | Sb |

HH: pretty bright, pretty large, pretty much extended N-S, pretty suddenly much brighter middle, star involved.
170x250: Moderately faint. Evenly moderately concentrated. Elongated N-S. Small. No involved stars. It's surrounded by moderately faint and faint stars ESE to NW.

| ga | NGC 5899 | 15 15.0 | +42 02 | Boo | 12.5b | 3.2' | 1.2' | Sc |

HH: considerably bright, pretty large, pretty much extended, suddenly much brighter middle to a nucleus.
170x250: Moderately faint. Broadly to evenly moderately concentrated. Elongated NNE-SSW. Moderately small. No involved stars. It makes a sharp triangle with two fairly bright stars W.

| ga | NGC 5885 | 15 15.1 | −10 05 | Lib | 12.3b | 3.5' | 3.0' | Sc |

HH: faint, considerably large, round, very gradually brighter middle.
170x250: Very very faint. Diffuse. Roughly round. Moderate size. At its edge NE is a moderately bright star.

| ga | NGC 5905 | 15 15.4 | +55 31 | Dra | 12.5p | 4.7' | 3.6' | Sb |

WH: pretty faint, pretty small, irregularly round.
170x250: Faint. Broadly very slightly concentrated, except that it has a substellar nucleus. Roughly round. Small. Near center are two or three very faint involved stars. Between 5905 and 5908, filling the field, is a very large group of stars in a diamond shape.

| bs | Delta Boo | 15 15.5 | +33 18 | Boo | 3.47 | | 0.95 | G8 |

170x250: Yellow.

| ga | NGC 5907 | 15 15.9 | +56 19 | Dra | 11.1b | 12.9' | 1.3' | Sc |

HH: considerably bright, very large, very much extended NNW-SSE, very gradually then pretty suddenly brighter middle to a nucleus.

42x60: Very faintly visible. *170x250:* Moderately faint. Evenly slightly concentrated. A very thin edge-on, NNW-SSE; it has no central bulge. Very large: half the field in size. The ends are indistinct. No involved stars. Nearby W of center is a faint star.

| gc | PAL 5 | 15 16.1 | −00 07 | Ser | 11.8 | 8.0' | −5.00 | 12 |

170x250: Not visible.

| ga | NGC 5908 | 15 16.7 | +55 24 | Dra | 12.8b | 3.2' | 1.6' | Sb |

WH: pretty faint, pretty small, round.
170x250: Moderately faint to moderate brightness. Evenly moderately concentrated. Elongated NNW-SSE. Small. No involved stars. [It's on the side of a triangle of stars.]

| bs | Beta Lib | 15 17.0 | −09 22 | Lib | 2.61 | | −0.11 | B8 |

170x250: White.

| gc | NGC 5897 | 15 17.4 | −21 01 | Lib | 8.6 | 11.0' | −7.05 | 11 |

HH: globular cluster, pretty faint, large, very irregularly round, very gradually brighter middle, well resolved.
12x35, 42x60: Faintly visible. *170x250:* Resolved; quite a few stars are distinct, and there are a lot more very faint and threshold stars. Hardly concentrated. A third of the field in size. Very faint. Somewhat irregular. One of the loosest, most lightweight globulars.

| ga | NGC 5898 | 15 18.2 | −24 05 | Lib | 12.5b | 1.9' | 1.9' | E0 |

HH: faint, small, round, gradually brighter middle.
170x250: 5903 and 5898 are a third of the field apart ENE-WSW. 5898 is brighter. Moderate brightness. Sharply pretty concentrated. Round. Very small. No involved stars. It makes a polygon with several stars SW to SE. 5903 is moderately faint. Evenly moderately concentrated. Round. Small. No involved stars. Nearby NW is a moderate brightness star; further NW is a fainter star. A sixth of the field S of 5903 is a tiny faint galaxy, almost stellar [*E514-3].

| ga | NGC 5903 | 15 18.6 | −24 04 | Lib | 12.2b | 2.7' | 2.0' | E2 |

HH: considerably faint, small, round, gradually pretty much brighter middle.
[See 5898.]

| gc | NGC 5904/M5 | 15 18.6 | +02 05 | Ser | 5.8 | 23.0' | −8.76 | 5 |

JH: !! globular cluster, very bright, large, extremely compressed middle, 11th to 15th mag. stars.

12x35, 42x60: Very bright. *170x250:* Completely resolved with a lot of distinct stars. Sharply concentrated to a moderate-sized but very very dense, very substantial central ball. 2/3 of the field in size. The main body of the cluster is flatter on the W side, but the scattered distant outliers go out farther there than on the E side, where they end abruptly. Very massive looking. A beautiful globular.

| bs | Gamma UMi | 15 20.7 | +71 50 | UMi | 3.03 | | 0.07 | A3 |

170x250: White.

| ga | IC 4538 | 15 21.2 | −23 39 | Lib | 12.8p | 2.5' | 1.9' | Sc |

170x250: Extremely faint. The very slightest unconcentrated glow. Roughly round. Moderately small. No involved stars. E to N is a shallow arc of three stars; SSE are three fainter stars in line with the galaxy.

| vs | S CrB | 15 21.4 | +31 22 | CrB | 6.0 | 14 | 0.96 | M6 |

12x35: Mag. 9.5. *170x250:* Hardly colored.

| ga | NGC 5915 | 15 21.6 | −13 05 | Lib | 12.8b | 1.7' | 1.2' | Sab |

JH: bright, small, round, gradually little brighter middle, W of 2.
170x250: Moderate brightness. Evenly moderately concentrated. Round. Small. Just off its edge S is a faint star. A quarter of the field SSE is a larger, very faint galaxy [*5916 (JH)]. Hardly concentrated. Elongated NNE-SSW.

| ga | NGC 5919 | 15 21.6 | +07 43 | Ser | 12.4 | 1.0' | 0.6' | − |

Sw: most extremely faint, pretty small, little extended, NW of 2.
170x250: Not visible.

| ga | NGC 5921 | 15 21.9 | +05 04 | Ser | 11.5b | 4.9' | 3.9' | Sbc |

HH: considerably bright, considerably large, irregularly round, very suddenly brighter middle to a 12th mag. star, among stars.
170x250: Moderate brightness. Sharply pretty concentrated. Round. Small. No involved stars. It's among a group of moderate brightness to moderately faint stars, including an arc of four stars going S from just off its edge SW.

| pn | ME 2-1 | 15 22.3 | –23 37 | Lib | 11.5p | 16.0" | | |

170x250: Moderately bright. No central star. A tiny disk; not very well defined. Strong blinking effect. Round. Nearby W is a moderately bright star. [PK342+27.1]

| bs | Iota Dra | 15 24.9 | +58 57 | Dra | 3.29 | | 1.16 | K2 |

170x250: Yellow.

| ga | NGC 5928 | 15 26.1 | +18 04 | Ser | 13.2p | 2.1' | 1.5' | S0 |

HH: pretty bright, considerably small, round, pretty suddenly brighter middle, 7th mag. star N.
170x250: Faint. Evenly moderately concentrated. Round. Very small. No involved stars.

| ga | NGC 5930 | 15 26.1 | +41 40 | Boo | 13.0 | 2.2' | 0.8' | Sa |

HH: pretty faint, pretty small, round, NE of double nebula.
170x250: Moderately faint to moderate brightness. Evenly moderately concentrated. Roughly round. Very small. No involved stars. Attached to it SW is a second galaxy [*5929 (JH)] – an even smaller version of 5930.

| ga | UGCA 408 | 15 26.1 | –22 16 | Lib | 12.5 | 2.9' | 1.5' | Sc |

170x250: Extremely faint. Hardly concentrated. Roughly round. Moderately small. No involved stars. Nearby SSE is a faint star; NE is another faint star; WNW is a bright star.

| ga | NGC 5949 | 15 28.0 | +64 45 | Dra | 12.8p | 2.2' | 1.0' | Sbc |

WH: faint, small, little extended NE-SW, very gradually little brighter middle.
170x250: Pretty faint. Broadly very slightly concentrated. Elongated NW-SE. Moderately small. No involved stars.

| ga | NGC 5936 | 15 30.0 | +12 59 | Ser | 13.1b | 1.4' | 1.2' | Sb |

WH: faint, pretty large, irregularly round, very gradually brighter middle, resolvable.
170x250: Faint. Broadly slightly concentrated. Round. Small. No involved stars.

| ga | NGC 5937 | 15 30.8 | –02 49 | Ser | 13.1p | 1.8' | 1.0' | Sb |

HH: pretty bright, pretty small, round, very gradually brighter middle, 3 stars E.
170x250: Faint. Broadly slightly concentrated. Slightly elongated N-S. Small. No involved stars. E to NNE is an arc of four moderate brightness stars.

| ga | NGC 5963 | 15 33.5 | +56 33 | Dra | 13.1b | 3.3' | 2.5' | S |

WH: pretty faint, pretty small, irregular figure.
170x250: Moderate brightness. Broadly slightly concentrated. Roughly round. Very small. No involved stars. Next to it SSE is a moderate brightness star; a little further SSE is a faint star. Half the field NNE is 5965. Also moderate brightness, but slightly fainter than 5963. Sharply fairly concentrated. Edge-on NE-SW. Moderately small. The ends are very indistinct. No involved stars.

| ga | NGC 5965 | 15 34.0 | +56 41 | Dra | 12.6b | 5.2' | 0.7' | Sb |

HH: considerably faint, considerably large, little extended.
[See 5963.]

| ga | NGC 5953 | 15 34.5 | +15 11 | Ser | 12.9 | 1.6' | 1.3' | Sa |

HH: pretty bright, considerably small, W of double nebula.
170x250: 5954 and 5953 are a very close double galaxy aligned NE-SW. Both are roughly round. Very small. No involved stars. 5953 is slightly brighter. Moderately faint. Concentrated. 5954 is faint. Unconcentrated. SE is a pair of moderately faint stars that is similar in separation and brightness to the pair of galaxies.

| ga | NGC 5954 | 15 34.6 | +15 12 | Ser | 12.9 | 1.2' | 0.5' | Scd |

HH: pretty bright, considerably small, E of double nebula.
[See 5953.]

| bs | Alpha CrB | 15 34.7 | +26 42 | CrB | 2.21 | | 0.00 | A0 |

170x250: White.

| ds | Delta Ser | 15 34.8 | +10 32 | Ser | 4.1" | 4.2 | 5.2 | F0 F0 |

42x60: Extremely close: as close as two stars can be in the 60 mm and still be separated. ***170x250:*** A very close slightly unequal pair. Both white. A very nice double. [STF 1954]

| ga | NGC 5956 | 15 35.0 | +11 45 | Ser | 13.0p | 1.6' | 1.6' | Scd |

dA: faint, small, round, 16th mag. star close E.
170x250: Very faint. Evenly moderately concentrated. Round. Very small. On either side of center ENE and WNW is an extremely faint involved star; a little ways SSE are two stars, moderately faint and faint; NE is a faint star.

| ga | NGC 5957 | 15 35.4 | +12 02 | Ser | 12.5p | 2.8' | 2.6' | Sb |

dA: pretty bright, pretty large, cometic, little brighter middle.
170x250: Very faint. Broadly very slightly concentrated. Round. Small. No involved stars. A little ways NNW is a moderately bright star; SW are three moderate brightness stars.

| ga | NGC 5966 | 15 35.9 | +39 46 | Boo | 13.1b | 1.8' | 1.2' | E |

HH: very faint, small, round, gradually brighter middle, 2 8th mag. stars E.
170x250: Moderately faint. Moderately concentrated. Round. Very small. No involved stars. It makes a triangle with the two bright plotted stars NNE.

| ga | NGC 5962 | 15 36.5 | +16 36 | Ser | 12.0b | 2.9' | 2.0' | Sc |

HH: pretty faint, pretty large, irregularly little extended, gradually brighter middle.
170x250: Moderate brightness. Broadly slightly concentrated. Roughly round. Small. No involved stars. It makes a triangle with two moderately faint stars NNE.

| ga | NGC 5964 | 15 37.6 | +05 58 | Ser | 12.6p | 4.2' | 3.2' | Sd |

JH: globular cluster, very faint, very large, round, very gradually brighter middle, well resolved.
170x250: Extremely faint. Diffuse; amorphous. Moderate size. E and N are several very faint stars.

| ga | NGC 5970 | 15 38.5 | +12 11 | Ser | 12.2b | 2.9' | 1.9' | Sc |

WH: pretty faint, pretty large, round, partially resolved.
170x250: Moderately faint. Broadly slightly concentrated. Slightly elongated E-W. Moderately small. No involved stars. ENE to NNW is a line of stars; NE, beyond this line, is the bright plotted star.

| ga | NGC 5982 | 15 38.7 | +59 21 | Dra | 12.0b | 2.5' | 1.8' | E3 |

HH: considerably bright, small, round, pretty suddenly brighter middle, resolvable.
170x250: 5985, 5982, and a third galaxy [*5981 (LR)] are in a row 3/4 of the field in length, ESE to WNW. Starting at the E end: 5985 is moderately faint. Broadly slightly concentrated. Slightly elongated N-S. Moderately large. No involved stars. 5982 is bright; high surface brightness. Sharply pretty concentrated. Roughly round. Very small. No involved stars. The third galaxy is faint. Broadly very slightly concentrated. A thin edge-on, NW-SE. Moderately small. No involved stars. It points toward a bright star NW. A pretty good contrast in galaxies.

| ds | STF 1962 | 15 38.7 | −08 47 | Lib | 11.8" | 6.5 | 6.6 | F6 F6 |

42x60: An equal pair. *170x250:* A moderate distance perfectly equal pair. Both white.

| ds | Zeta CrB | 15 39.4 | +36 38 | CrB | 6.1" | 5.1 | 6.0 | B7 B7 |

42x60: A very close unequal pair; nicely split. *170x250:* A fairly close slightly unequal pair. Both white. A nice double. [STF 1965]

| ga | NGC 5985 | 15 39.6 | +59 19 | Dra | 11.9b | 5.5' | 2.9' | Sb |

WH: pretty bright, considerably large, irregularly extended, resolvable.
[See 5982.]

| ga | NGC 5968 | 15 39.9 | −30 33 | Lup | 13.1b | 2.1' | 1.9' | Sab |

JH: very faint, large, round, gradually brighter middle, resolvable.
170x250: Very very faint. Broadly very slightly concentrated. Roughly round. Small. No involved stars. Crossing the field nearby N is a WNW-ESE zigzag line of moderately faint stars; SE are a few more stars.

| ga | NGC 5987 | 15 40.0 | +58 04 | Dra | 12.7b | 5.2' | 1.5' | Sb |

WH: pretty faint, considerably small.
170x250: Moderate brightness. Sharply moderately concentrated. Slightly elongated ENE-WSW. Small. No involved stars. Nearby NW is a bright star.

| ga | NGC 5984 | 15 42.9 | +14 13 | Ser | 13.1p | 2.9' | 0.7' | Sd |

WH: pretty bright, small, extended NW-SE, brighter middle.
170x250: Very faint. Broadly very slightly concentrated. Very elongated NW-SE. Small. No involved stars. A little ways NNE is a sharp little triangle of faint stars.

| bs | Alpha Ser | 15 44.3 | +06 25 | Ser | 2.65 | | 1.17 | K2 |

170x250: Yellow.

| ga | NGC 5996 | 15 47.0 | +17 53 | Ser | 13.2b | 3.2' | 2.6' | S |

HH: pretty faint, considerably small, round, resolvable, between 2 double stars.
170x250: Faint. Broadly slightly concentrated. Round. Small. No involved stars. Forming an arc around its E side are a pair of moderately faint stars nearby NNE, a moderate brightness star a little ways SE, and another moderate brightness star S. [5994 (LR) was not seen.]

| vs | V CrB | 15 49.5 | +39 34 | CrB | 6.9 | 12.5 | 2.90 | C6 |

12x35: Mag. 8.5. ***170x250:*** Copper red.

| ga | NGC 6000 | 15 49.8 | −29 23 | Sco | 13.0p | 2.0' | 1.7' | Sbc |

JH: very faint, small, round, suddenly brighter middle.
170x250: Faint to moderately faint. Evenly moderately concentrated. Roughly round. Very small. No involved stars. Very nearby NNW is a faint star; SW is a bright star; farther ESE is another bright star.

| ga | NGC 6004 | 15 50.4 | +18 56 | Ser | 13.1p | 1.8' | 1.6' | Sbc |

St: very faint, pretty large, little extended, little brighter middle.
170x250: Very faint. Broadly very slightly concentrated. Round. Small. No involved stars. Off its edge W are a couple of extremely faint stars; SSW is a triangle of faint stars.

| vs | R Ser | 15 50.7 | +15 08 | Ser | 5.7 | 14 | 1.39 | M7 |

12x35: Not visible.

| ga | NGC 6015 | 15 51.4 | +62 18 | Dra | 11.7b | 5.4' | 2.1' | Scd |

WH: very faint, pretty large, round, very gradually brighter middle.
170x250: Moderately faint. Broadly very slightly concentrated. Slightly elongated NNE-SSW. Moderate size. No involved stars. Just off its S end is a faint star; a little ways W is a moderately bright star; SSW are a couple of moderately faint stars.

| ga | NGC 6012 | 15 54.2 | +14 36 | Ser | 12.7b | 2.0' | 1.4' | Sab |

WH: faint, between 2 bright stars.
170x250: Very faint. Broadly very slightly concentrated. Elongated NNW-SSE. Small. No involved stars. It's off a NNE-SSW line of three bright and moderate brightness stars, the northernmost of which is an unequal double.

| ga | NGC 6014 | 15 56.0 | +05 55 | Ser | 13.2p | 1.7' | 1.5' | S0 |

JH: pretty bright, pretty large, extended.
170x250: Very faint. Broadly slightly concentrated. Round. Small. NNE of center is an extremely faint involved star; E are several more very faint stars.

| ds | Xi Lup | 15 56.9 | −33 58 | Lup | 10.3" | 5.3 | 5.8 | A3 B9 |

42x60: Almost equal. A nice double. ***170x250:*** A moderate distance equal pair. Both white. [PZ 4]

| bs | Pi Sco | 15 58.9 | −26 06 | Sco | 2.89 | | −0.19 | B1 |

170x250: White.

| bs | Delta Sco | 16 00.3 | −22 37 | Sco | 2.30 | | −0.10 | B0 |

170x250: White.

| pn | NGC 6026 | 16 01.4 | −34 32 | Lup | 13.2p | 55.0" | | |

JH: faint, small, round, gradually pretty much brighter middle, triangle of stars NW.
42x60: Very faintly visible. ***170x250:*** Somewhat faint. It has an obvious moderately faint central star and a faint nebula. No disk; no sharp edge. Even in light. No blinking effect. Round. Moderate size for a planetary.

| ga | UGC 10144 | 16 02.3 | +16 20 | Her | 13.1v | 1.0' | 1.0' | E |

170x250: Extremely faint. Evenly slightly concentrated. Round. Very small. No involved stars. A little ways ESE is a moderately bright star; the galaxy and this star are at the center of a very large trapezoidal figure of moderately bright stars. [C108-72 was not seen.]

| pn | NGC 6058 | 16 04.4 | +40 40 | Her | 13.3p | 35.0" | | |

HH: pretty faint, very small, round, stellar.
170x250: Faint. It has a very faint central star. An indistinct edge; the planetary is somewhat fuzzy. Aside from the central star it's even in light. Weak blinking effect. Round. Moderately small for a planetary. It's just inside a large triangle of stars.

| ds | Xi Sco | 16 04.4 | −11 22 | Sco | 7.4" | 4.2 | 7.3 | F5 − |

42x60: Extremely close. ***170x250:*** The primary has a moderately close fainter companion. Both white. A nice double. [STF 1998] SE is a wider, equal double [STF 1999: 11.8", 7.4, 8.1]. A nice scene.

| bs | Beta Sco | 16 05.4 | −19 48 | Sco | 2.62 | | −0.07 | B0 |

42x60: A nice double. ***170x250:*** A moderate distance unequal pair. Both white. An exceptionally nice double; one of the best in the sky. [H 7: 13.7", 2.6, 4.9, B0 B2]

| ds | Kappa Her | 16 08.1 | +17 03 | Her | 27.1" | 5.3 | 6.5 | G8 K1 |

12x35: Nicely split. *42x60:* A moderate distance slightly unequal pair. *170x250:* A wide slightly unequal pair. Colorful: both deep yellow. [STF 2010]

| ds | BSO 11 | 16 09.5 | –32 39 | Sco | 7.7" | 6.7 | 7.4 | K1 F6 |

42x60: A slightly unequal pair. *170x250:* A moderately close almost equal pair. Both white.

| ga | NGC 6070 | 16 10.0 | +00 42 | Ser | 12.5b | 3.5' | 1.8' | Scd |

HH: faint, large, pretty much extended, very gradually brighter middle, resolvable.
170x250: Very faint. Hardly concentrated. Elongated NE-SW. Moderate size. No involved stars. It's in line with a pair of faint stars ENE, aligned perpendicular to the axis of the galaxy, and two faint stars SW.

| pn | IC 4593 | 16 11.7 | +12 04 | Her | 10.9p | 30.0" | | |

12x35: Faintly visible. *42x60:* Visible as a moderately faint star. *170x250:* Very bright; high surface brightness. It has a faint central star. A tiny disk. Very sharply concentrated. Pretty strong blinking effect. Round. NW is a good comparison star that is smaller and has no blinking effect.

| ds | 12 Sco | 16 12.3 | 28 25 | Sco | 3.8" | 5.9 | 7.9 | B9 – |

170x250: The primary has a very very close fainter companion. Both white. [HJ 4839]

| ga | NGC 6086 | 16 12.6 | +29 29 | CrB | 12.7v | 1.3' | 1.0' | E |

Ma: faint, very small, stellar nucleus.
170x250: In the position are a couple of galaxies [6086, *6085 (Ma)] nearly half the field apart N-S in a field of stars. The N galaxy is very faint. Broadly slightly concentrated. Roughly round. Very small. No involved stars. Off its edge NW is a moderately faint star. The S galaxy is very very faint. Evenly moderately concentrated. Round. Very very small. No involved stars. It makes a little arc with two faint stars NW and a moderate brightness star E.

| bs | Delta Oph | 16 14.3 | –03 41 | Oph | 2.74 | | 1.58 | M1 |

170x250: Gold. One of the more strongly colored bright stars.

| ds | Sigma CrB | 16 14.7 | +33 52 | CrB | 6.7" | 5.6 | 6.6 | G0 G1 |

42x60: Nicely split. *170x250:* A fairly close slightly unequal pair. Both yellowish. [STF 2032]

| gc | NGC 6093/M80 | 16 17.0 | −22 59 | Sco | 7.2 | 10.0' | −8.08 | 2 |

JH: !!! globular cluster, very bright, large, very much brighter middle (variable star), well resolved, 14th mag. stars.
12x35: Visible with a number of stars around it. *42x60:* Bright. *170x250:* Resolved except in the very middle. Evenly fairly concentrated. A sixth of the field in size. It has a very dense central ball. Probably an intrinsically massive cluster. (225x: Very regular, very smooth; one of the most regular globulars.)

| ga | MCG-2-41-1 | 16 17.3 | −11 43 | Sco | 13.1 | 2.4' | 1.6' | Sb |

170x250: Not visible.

| bs | Epsilon Oph | 16 18.3 | −04 41 | Oph | 3.24 | | 0.96 | G8 |

170x250: Yellow.

| ga | NGC 6106 | 16 18.8 | +07 24 | Her | 12.8b | 2.5' | 1.3' | Sc |

HH: faint, pretty large, little extended, very gradually brighter middle, resolvable.
170x250: Faint. Broadly slightly concentrated. Slightly elongated NW-SE. Small. No involved stars. Nearby S is an extremely faint star.

| ga | NGC 6127 | 16 19.2 | +57 59 | Dra | 13.0b | 1.4' | 1.4' | E |

Sw: pretty faint, very small, round.
170x250: Moderately faint. Sharply fairly concentrated. Round. Very small. No involved stars. W is a large group of relatively bright stars.

| ds | BSO 12 | 16 19.5 | −30 54 | Sco | 23.3" | 5.4 | 6.9 | F5 G1 |

12x35: Split. *42x60:* A wide unequal pair. *170x250:* A very wide unequal pair. Both white.

| ga | NGC 6140 | 16 21.0 | +65 23 | Dra | 11.8b | 6.3' | 4.5' | Scd |

WH: considerably faint, pretty large, irregularly round.

170x250: Faint. Broadly very slightly concentrated. Roughly round. Moderately small. No involved stars. A little ways NW is a moderately bright star.

| bs | Sigma Sco | 16 21.2 | –25 35 | Sco | 2.88 | | 0.13 | B2 |

170x250: The primary has a wide relatively very faint companion, somewhat like Polaris. The primary is soft white. [H 121: 20.0", 2.9, 8.5]

| ga | NGC 6118 | 16 21.8 | –02 17 | Ser | 12.4b | 4.7' | 2.0' | Scd |

HH: very faint, considerably large, considerably extended NE-SW, resolvable.
170x250: Very faint. Hardly concentrated. Slightly elongated NE-SW. Moderately large. At its NE end is a very faint star; further NE are two faint stars.

| gc | NGC 6121/M4 | 16 23.6 | –26 32 | Sco | 5.9 | 30.0' | –6.80 | 9 |

HH: cluster, 8 or 10 bright stars in line, with 5 stars, well resolved.
12x35: Easily visible. *42x60:* Partly resolved. The ridge of stars in the middle is starting to be visible. *170x250:* Well resolved into perfectly distinct stars throughout, although the very middle also contains unresolved stars underneath. Broadly concentrated. With outliers it takes up the entire field. The outer region is very extensive, with widespread scattered stars. There are more outliers E and NE than elsewhere. The central ridge is distinct, going straight through the middle of the core NNE-SSW. The cluster doesn't look massive: it has a central core but not a massive ball. (225x: Somewhat loose.)

| bs | Eta Dra | 16 24.0 | +61 30 | Dra | 2.74 | | 0.91 | G8 |

170x250: Yellow white.

| ds | H 39 | 16 24.7 | –29 42 | Sco | 4.6" | 5.9 | 6.6 | F7 F7 |

42x60: Extremely close. *170x250:* A very close very slightly unequal pair. Both white.

| ds | Rho Oph | 16 25.6 | –23 27 | Oph | 2.9" | 5.3 | 6.0 | B2 B2 |

12x35, 42x60: Near the double are two stars making a little arc with it. *170x250:* A very very close almost equal pair. Both white. A nice scene with the other stars. [H 19]

| ga | NGC 6155 | 16 26.1 | +48 22 | Her | 13.2p | 1.3' | 0.8' | S |

WH: *faint, pretty small, irregular figure, gradually brighter middle.*
170x250: Faint. Broadly slightly concentrated. Roughly round. Small. No involved stars.

| vs | V Oph | 16 26.7 | −12 26 | Oph | 7.3 | 11.5 | 3.50 | C6 |

12x35: Mag. 8.5. ***170x250:*** Light copper red.

| gc | NGC 6144 | 16 27.3 | −26 02 | Sco | 9.1 | 7.4' | −6.57 | 11 |

HH: *cluster, considerably large, much compressed, gradually brighter middle, well resolved.*
12x35: Very faintly visible. ***42x60:*** Faintly visible. ***170x250:*** Still faint. A number of distinct stars are resolved across the entire cluster, with the light of threshold stars underneath. Very broadly concentrated. A little less than a quarter of the field in size. Irregular. Very scrawny; skimpy.

| ga | NGC 6166 | 16 28.6 | +39 33 | Her | 12.8b | 2.2' | 1.5' | E+2 |

HH: *pretty faint, small, very little extended, very gradually much brighter middle.*
170x250: Faint. Broadly slightly concentrated. Round. Small. No involved stars. [6166 is the principal member of AGC 2199, the archetypal cD galaxy cluster.]

| bs | Alpha Sco | 16 29.4 | −26 25 | Sco | 0.96 | | 1.83 | M1 |

170x250: Deep gold. [Antares]

| ga | NGC 6173 | 16 29.7 | +40 48 | Her | 13.1b | 1.9' | 1.4' | E |

HH: *considerably faint, very small, round, brighter middle.*
170x250: Moderately faint. Sharply fairly concentrated to a substellar nucleus. Roughly round. Small. No involved stars. [6174 (LR) was not seen.]

| bs | Beta Her | 16 30.2 | +21 29 | Her | 2.77 | | 0.94 | G8 |

170x250: Yellow white.

| ga | NGC 6181 | 16 32.3 | +19 49 | Her | 12.5b | 2.5' | 1.1' | Sc |

WH: pretty bright, pretty large, very little extended, pretty gradually much brighter middle.
170x250: Moderate brightness. Broadly to evenly moderately concentrated. Round. Small. No involved stars. It makes a triangle with a faint star nearby SSW and a moderate brightness star a little ways W.

| gc | NGC 6171/M107 | 16 32.5 | –13 03 | Oph | 8.1 | 13.0' | –6.90 | 10 |

HH: globular cluster, large, very rich, very much compressed, round, well resolved.
12x35, 42x60: Visible. *170x250:* Resolved throughout, with a lot of distinct stars. Not very concentrated. It has two fairly distinct zones: a modest core and a halo of outliers that extends farthest on the E side. With the farthest outliers it's half the field in size. It might be obscured: it's not as bright as one would expect given the distinctness of its stars, and it's very strange that the outliers are so distant and scattered. Irregular and unusual. Somewhat loose; straggly. It sits within a cradle-like triangle of relatively bright stars: one ESE, one SSW, and one WNW; with the outliers the cluster reaches these stars and goes a little further on the S and E sides.

| ga | NGC 6217 | 16 32.6 | +78 11 | UMi | 11.8b | 3.0' | 2.4' | Sbc |

WH: bright, considerably large, little extended, suddenly little brighter middle.
170x250: Faint. Hardly concentrated. Elongated NNW-SSE. Moderate size. Right in the middle is a faint involved star; off its edge NW is a fainter star.

| bs | Tau Sco | 16 35.9 | –28 12 | Sco | 2.82 | | –0.25 | B0 |

170x250: White.

| ds | 17 Dra | 16 36.2 | +52 55 | Dra | 3.1" | 5.4 | 6.4 | B9 A1 |

12x35, 42x60: A wide pair [16 Dra, 17 Dra: 1.5']. *170x250:* One of the components of the wide double is a very close slightly unequal pair. All the stars are white. A nice multiple star. [STF 2078]

| bs | Zeta Oph | 16 37.2 | –10 34 | Oph | 2.60 | | –0.02 | O9 |

170x250: White.

| ga | UGC 10502 | 16 37.6 | +72 22 | Dra | 12.9p | 2.3' | 2.0' | Sc |

170x250: Extremely faint. The slightest unconcentrated glow. Roughly round. Moderately small to moderate size. No involved stars. A little ways SSE are three stars: one moderate brightness and two faint; W is the bright plotted star. [U10497 was not seen.]

| bs | Zeta Her | 16 41.3 | +31 36 | Her | 2.81 | | 0.65 | F9 |

170x250: Yellow white.

| gc | NGC 6205/M13 | 16 41.7 | +36 28 | Her | 5.9 | 20.0' | −8.49 | 5 |

JH: !! globular cluster, extremely bright, very rich, very gradually extremely compressed middle, 11th to 20th mag. stars.
12x35: Bright. *42x60:* Granular. *170x250:* Resolved throughout, with distinct stars, but the core also has a lot of underlying stars not well resolved. Broadly concentrated. Half the field in size. It has a central ball, but not a well defined one. Very irregular: there are four lines of stars coming out of the main body – linear appendages – making the cluster look a little like a tick or a bug. It looks massive. (225x: The cluster stops pretty suddenly; it doesn't have very distant outliers like some other major globulars; the outliers it has are relatively close to the core.) [Hercules Cluster]

| ga | NGC 6207 | 16 43.1 | +36 49 | Her | 12.2b | 3.3' | 1.7' | Sc |

HH: pretty bright, pretty large, extended NE-SW, very gradually much brighter middle.
170x250: Moderately faint to moderate brightness. Broadly slightly concentrated. Elongated NNE-SSW. Moderately small. Near center is a moderately faint involved star; nearby SSE to NE are four moderately faint stars in a crooked line.

| ga | NGC 6223 | 16 43.1 | +61 34 | Dra | 12.6p | 3.4' | 2.5' | Pec |

dA: faint, small, round, much brighter middle.
170x250: Moderately faint. Sharply very concentrated to a substellar nucleus. Round. Very small. No involved stars.

| pn | NGC 6210 | 16 44.5 | +23 48 | Her | 9.3p | 30.0" | | |

JH: planetary nebula, very bright, very small, round, disc and border.

12x35: Easily visible as a faint star. *42x60:* It looks like the bright plotted star SE, but it's not quite as bright and it has a strong blinking effect – a good contrast with the star. *170x250:* Bright; high surface brightness. No central star. A small disk; not a very clean edge. Even in light. Strong blinking effect. Slightly elongated E-W. NE is a triangle of stars.

ga	NGC 6236	16 44.6	+70 46	Dra	12.6p	2.9'	1.6'	Scd

Sw: faint, pretty large.
170x250: Extremely faint. Diffuse. Roughly round. Moderately small. No involved stars. WSW to SSE is a sharp triangle of moderate brightness stars.

ga	UGC 10528	16 44.8	+22 31	Her	13.1p	2.3'	1.3'	S0+

170x250: Moderately bright. Sharply extremely concentrated: either it has a completely stellar nucleus or there's a moderately faint involved star right in the middle; the halo itself is faint. Round. Very small.

gc	NGC 6229	16 47.0	+47 32	Her	9.4	4.5'	−8.07	4

WH: very bright, large, round, disc and faint, resolvable border.
12x35: Faintly visible. *42x60:* It makes a triangle with two moderately bright stars, one W and one SW. *170x250:* Starting to be granular with averted vision. Evenly fairly concentrated. Small. Fairly bright. It looks like an intrinsically fairly substantial globular. A nice scene with the two bright stars.

gc	NGC 6218/M12	16 47.2	−01 57	Oph	6.6	16.0'	−7.70	9

JH: !! globular cluster, very bright, very large, irregularly round, gradually much brighter middle, well resolved, 10th mag. and fainter stars.
12x35: Bright. *42x60:* Partly resolved. *170x250:* Well resolved into distinct stars, including a lot of outliers. Broadly concentrated. Half the field in size. Irregular; very irregular in the outer regions. Loose. It has a central ball but the cluster is not very massive looking. Nearby E are several relatively bright field stars; these are the westernmost members of a very large E-W rectangular group of 10–12 stars mostly E of the cluster; the closest one is within the cluster S of center.

ga	NGC 6239	16 50.1	+42 44	Her	12.9b	3.3'	1.2'	Sb

WH: considerably faint, small, extended E-W.
170x250: Faint. Broadly slightly concentrated. Elongated WNW-ESE. Moderately small. No involved stars.

| bs | Epsilon Sco | 16 50.2 | –34 17 | Sco | 2.29 | | 1.15 | K2 |

170x250: Yellow.

| ga | NGC 6240 | 16 53.0 | +02 24 | Oph | 12.9v | 2.5' | 1.2' | 10 |

St: very faint, pretty large, little extended, diffuse.
170x250: Very faint. Broadly slightly concentrated. Elongated N-S. Small. At its edge ENE is a moderately faint involved star; nearby N is a faint star; nearby S is another faint star.

| gc | NGC 6235 | 16 53.4 | –22 11 | Oph | 10.2 | 5.0' | –6.2 | 10 |

HH: pretty bright, considerably large, irregularly round, well resolved, 14th to 16th mag. stars.
42x60: Faintly visible. *170x250:* Partly resolved. Broadly concentrated. Small. Somewhat triangular in shape. Irregular; it doesn't have a normal globular appearance.

| vs | RR Sco | 16 56.6 | –30 35 | Sco | 5.1 | 12.3 | 1.15 | M6 |

12x35: Mag. 6.5. *170x250:* Yellow.

| gc | NGC 6254/M10 | 16 57.1 | –04 06 | Oph | 6.6 | 20.0' | –7.48 | 7 |

JH: ! globular cluster, bright, very large, round, gradually very much brighter middle, well resolved, 10th to 15th mag. stars.
12x35: M10 is a little brighter than M12. *42x60:* M10 is granular. *170x250:* Resolved throughout, with a lot of distinct stars. Broadly concentrated. A lot of distant scattered outliers completely fill out the field. A very interesting cluster: the central region could be a globular by itself, but in addition there are the far outliers. M10 is fairly similar to M12 but it's more regular, not as loose, and it has a more distinctive central ball; it's probably a more massive globular. (225x: The distant outliers are very very distinct. One of the best-resolved globulars.)

| bs | Kappa Oph | 16 57.7 | +09 22 | Oph | 3.20 | | 1.15 | K2 |

170x250: Yellow gold.

| ga | NGC 6269 | 16 58.0 | +27 51 | Her | 12.2v | 2.3' | 1.6' | E |

Ma: faint, small, round.

170x250: Very faint. Evenly moderately concentrated. Roughly round. Very small. No involved stars. Nearby S is a faint star. NW of this star is an extremely faint, almost stellar galaxy [this is a faint star].

gc	NGC 6266/M62	17 01.2	–30 07	Oph	6.6	15.0'	–8.78	4

JH: ! globular cluster, very bright, large, gradually much brighter middle, well resolved, 14th to 16th mag. stars.
12x35, 42x60: Bright. Sharply concentrated. *170x250:* Resolved throughout except the very middle, which is granular; the stars are very faint. Evenly pretty concentrated; it has a very dense center. A quarter of the field in size. Pretty regular. Probably a massive cluster. (225x: The outliers go out quite a ways on the N side; they end more abruptly on the S side. M62 and M19 are fairly similar. M62's stars are a little brighter and more distinct. M62 is a little better resolved and more concentrated; it has a smaller but more distinct central ball. Both clusters are pretty regular and look intrinsically massive.)

pn	IC 4634	17 01.6	–21 49	Oph	10.7p	12.0"		

170x250: Bright. It has either no central star or an extremely faint one. A tiny disk; close to stellar. Fairly strong blinking effect. Round.

gc	NGC 6273/M19	17 02.6	–26 16	Oph	7.2	17.0'	–9.20	8

JH: globular cluster, very bright, large, round, very compressed middle, well resolved, 16th mag. and fainter stars, red.
12x35, 42x60: Similar to M62. *170x250:* Resolved into very very faint stars in the periphery; granular in the middle. Broadly concentrated. A quarter of the field in size. Very dense; it has a very large central ball. Slightly oval N-S. Not quite as impressive as M62, although it's roughly the same size and brightness. (225x: Across its edge N are two moderately faint stars aligned E-W – undoubtedly field stars. The cluster looks very distant.)

gc	NGC 6284	17 04.5	–24 46	Oph	9.0	6.2'	–6.9	9

HH: globular cluster, bright, large, round, gradually compressed middle, well resolved, 16th mag. stars.
12x35, 42x60: Visible. *170x250:* Partly resolved with averted vision. Fairly concentrated. Small. It looks like a fairly substantial globular intrinsically.

| gc | NGC 6287 | 17 05.2 | −22 42 | Oph | 9.2 | 4.8' | −6.7 | 7 |

HH: globular cluster, considerably bright, large, round, gradually pretty much compressed middle, well resolved, 16th mag. stars.
42x60: Faintly visible. *170x250:* Granular. Broadly concentrated. Small. Somewhat faint. Somewhat irregular.

| ds | Mu Dra | 17 05.3 | +54 28 | Dra | 2.0" | 5.7 | 5.7 | F7 F7 |

170x250: An extremely close equal pair. Both white. [STF 2130]

| bs | Zeta Dra | 17 08.8 | +65 42 | Dra | 3.17 | | −0.12 | B6 |

170x250: White.

| gc | NGC 6293 | 17 10.2 | −26 35 | Oph | 8.2 | 8.2' | −7.2 | 4 |

HH: globular cluster, very bright, large, round, pretty suddenly brighter middle, well resolved, 16th mag. stars, faint nebula E.
12x35: Faintly visible. *42x60:* Visible. *170x250:* Partly resolved. Pretty concentrated. Small.

| bs | Eta Oph | 17 10.4 | −15 43 | Oph | 2.43 | | 0.06 | A2 |

170x250: White.

| ga | NGC 6340 | 17 10.4 | +72 18 | Dra | 11.9b | 3.2' | 2.9' | S0/a |

HH: considerably faint, pretty large, round, very gradually much brighter middle.
170x250: Moderate brightness. Sharply moderately concentrated. Roughly round. Small. No involved stars. Nearby NNW is a pair of stars.

| pn | NGC 6309 | 17 14.1 | −12 54 | Oph | 10.8p | 16.0" | | |

Te: bright, small, between 2 stars very near.
42x60: Visible. *170x250:* Moderately bright. No central star. Not a clean disk; an indistinct edge. Slight blinking effect. Elongated NNW-SSE. Small. Right at its N end is a moderately faint star.

| gc | NGC 6304 | 17 14.5 | −29 28 | Oph | 8.4 | 8.0' | −7.08 | 6 |

HH: globular cluster, bright, considerably large, round, suddenly then very gradually little brighter middle, well resolved, 16th–17th mag. stars.

12x35, 42x60: Visible. *170x250:* Granular. Broadly concentrated. Small. It looks like a substantial cluster.

| bs | Alpha Her | 17 14.6 | +14 23 | Her | 3.48 | | 1.44 | M5 |

42x60: Elongated. *170x250:* A close unequal pair. Colorful: the primary is gold; the companion is yellow. A very nice double. [STF 2140: 5.0", 3.5, 5.4, M5 G5]

| bs | Delta Her | 17 15.0 | +24 50 | Her | 3.14 | | 0.08 | A3 |

170x250: The primary has a moderate distance relatively faint companion. The primary is white. [STF 3127: 10.9", 3.1, 8.2]

| bs | Pi Her | 17 15.0 | +36 48 | Her | 3.16 | | 1.44 | K3 |

170x250: Yellow gold.

| ds | 36 Oph | 17 15.3 | −26 36 | Oph | 4.7" | 5.1 | 5.1 | K0 K2 |

42x60: Extremely close. *170x250:* A very close perfectly equal pair. Both yellow white. A nice double. [SHJ 243]

| ga | UGC 10803 | 17 16.1 | +73 26 | Dra | 13.1p | 1.3' | 0.7' | E |

170x250: Moderately faint. Broadly to evenly moderately concentrated. Round. Very small. No involved stars.

| gc | NGC 6316 | 17 16.6 | −28 08 | Oph | 9.0 | 5.4' | −8.0 | 3 |

HH: globular cluster, considerably bright, pretty small, round, gradually very much brighter middle, well resolved, 16th to 17th mag. stars.
12x35: Very faintly visible. *42x60:* Visible. *170x250:* Not resolved. Broadly concentrated. Very small. Fairly bright. Regular. Probably a substantial cluster, just far away. Passing just S of it is a slightly curved WNW-ESE line of three moderate brightness stars.

| gc | NGC 6341/M92 | 17 17.1 | +43 08 | Her | 6.5 | 14.0' | −7.98 | 4 |

WH: globular cluster, very bright, very large, extremely compressed middle, well resolved, faint stars.

12x35: Concentrated. *42x60:* Bright. *170x250:* Resolved throughout. Sharply pretty concentrated. A little less than half the field in size. It has a lot of prominent outliers that go out a good distance. It has a modest central ball. Slightly irregular. (225x: Very bright. There are some very distant scattered outliers, well separated from the main body of the cluster, including a group of half a dozen N. Including all possible outliers, not only the closer and more definite ones, the cluster is the entire field width in size [= 3/4 of a 170x field]. It looks moderately massive.)

| gc | NGC 6325 | 17 18.0 | −23 46 | Oph | 10.7 | 4.1' | −6.00 | 4 |

JH: pretty faint, large, round, partially resolved.
170x250: Not resolved. Broadly concentrated. Small. Faint.

| ds | Omicron Oph | 17 18.0 | −24 17 | Oph | 10.2" | 5.4 | 6.9 | K0 F6 |

42x60: Nicely split. *170x250:* A moderate distance unequal pair. Color contrast: the primary is yellow; the companion is white. [H 25]

| gc | NGC 6333/M9 | 17 19.2 | −18 31 | Oph | 7.9 | 12.0' | −7.4 | 8 |

JH: globular cluster, bright, large, round, extremely compressed middle, well resolved, 14th mag. stars.
12x35, 42x60: Easily visible. *170x250:* Resolved throughout into mainly very faint stars. Broadly concentrated. Small. It looks like a moderately massive cluster.

| ds | BU 126 | 17 19.9 | −17 45 | Oph | 2.4" | 6.3 | 7.4 | A2 − |

170x250: A very very close slightly unequal pair. Both white.

| ga | UGC 10822 | 17 20.1 | +57 54 | Dra | 10.9p | 35.5' | 24.5' | E |

12x35, 42x60, 170x250: Not visible. [Draco dwarf]

| gc | NGC 6342 | 17 21.2 | −19 35 | Oph | 9.9 | 4.4' | −7.6 | 4 |

WH: considerably bright, pretty small, little extended, easily resolvable.
42x60: Very faintly visible. *170x250:* Not resolved. Evenly concentrated. Very small. Just off its edge SSW is a moderate brightness star.

| bs | Theta Oph | 17 22.0 | −24 59 | Oph | 3.27 | | −0.22 | B2 |

170x250: White.

| gc | NGC 6356 | 17 23.6 | −17 49 | Oph | 8.4 | 10.0' | −8.67 | 2 |

HH: globular cluster, very bright, considerably large, very gradually very much brighter middle, well resolved, 20th mag. stars.
12x35, 42x60: Visible. ***170x250:*** Granular to resolved; the stars are very faint and very tightly packed. Broadly to evenly concentrated. Small. Very regular; exceptionally smooth. It looks like an intrinsically substantial cluster.

| ds | Rho Her | 17 23.7 | +37 09 | Her | 4.2" | 4.6 | 5.6 | B9 A0 |

42x60: Extremely close. ***170x250:*** A close slightly unequal pair. Both white. A nice double. [STF 2161]

| gc | NGC 6355 | 17 24.0 | −26 21 | Oph | 9.6 | 4.2' | −7.0 | − |

HH: considerably faint, large, round, gradually brighter middle, well resolved.
42x60: Very faintly visible. ***170x250:*** Starting to be granular. Broadly concentrated. Small. Faint. Slightly irregular. It looks like a somewhat insubstantial cluster.

| ne | NGC 6357 | 17 24.6 | −34 10 | Sco | | 31.8' | | E |

JH: faint, large, extended, very gradually little brighter middle, double star involved.
170x250: In the position is the N-S row of plotted stars. There's some nebulosity around the N star of this row. WNW is a small but more conspicuous nebula just N of a little NNE-SSW wedge-shaped figure of stars. It looks like a round diffuse galaxy.

| ga | NGC 6395 | 17 26.5 | +71 05 | Dra | 13.0p | 2.4' | 0.7' | Scd |

Sw: very faint, pretty large, little extended, double star N.
170x250: Very faint. Hardly concentrated. Elongated NNE-SSW. Small. No involved stars. It makes a shallow little arc with two moderate brightness stars N.

| gc | IC 1257 | 17 27.1 | −07 04 | Oph | 13.1 | 5.0' | − | − |

170x250: Extremely faint. A little glow. Very small. It's just N of a line between two moderate brightness stars, one E and one WSW.

| ga | NGC 6368 | 17 27.2 | +11 32 | Oph | 13.1p | 3.8' | 0.9' | Sb |

Ma: faint, small, extended.
170x250: Very faint. Broadly slightly concentrated. Very elongated NE-SW. Moderate size. At its SW end is a faint star; a little ways SSE is a brighter star.

| gc | NGC 6366 | 17 27.7 | –05 05 | Oph | 10.0 | 13.0' | –5.1 | 11 |

Wi: *faint, large, very little brighter middle.*
12x35: A couple of faint stars. ***42x60:*** Two or three faint stars and a possible glow of more stars. ***170x250:*** Resolved throughout into a number of scattered very very faint but distinct stars. Broadly concentrated. A third of the field in size. Very faint. Irregular. Along its W side is a NW-SE line of four stars; the two southernmost of these are at the edge of the cluster SSW of center.

| oc | HAR 15 | 17 28.5 | –29 28 | Oph | 9.5 | 17.0' | | |

12x35: Faintly visible. ***42x60:*** A handful of distinct stars with more threshold stars. ***170x250:*** A fair number of scattered very mixed brightness stars, moderate brightness to threshold, including a central three-spoked triangular figure. Half the field in size. Moderate density. [CR331]

| pn | NGC 6369 | 17 29.3 | –23 45 | Oph | 12.9p | 38.0" | | |

HH: *!! annular nebula, pretty bright, small, round.*
42x60: Faintly visible. ***170x250:*** Moderate brightness. No central star. A fairly good disk; a clean edge. Annular; a thick torus: it has a small hole in the middle. Slight blinking effect. Round. Moderately small for a planetary.

| ga | NGC 6412 | 17 29.6 | +75 42 | Dra | 12.3b | 2.5' | 2.1' | Sc |

WH: *globular cluster, considerably large, round, very gradually brighter middle, partially resolved.*
170x250: Faint. Diffuse. Roughly round. Moderately small. No involved stars. Off its edge SW is a faint star; a little ways SE is a bright star.

| bs | Beta Dra | 17 30.4 | +52 18 | Dra | 2.79 | | 0.98 | G2 |

170x250: Yellow.

| ds | Nu Dra | 17 32.2 | +55 11 | Dra | 61.7" | 4.9 | 4.9 | A A |

12x35, 42x60: A wide equal pair. ***170x250:*** A very wide perfectly equal pair. Both white. [STF 35]

| ga | NGC 6384 | 17 32.4 | +07 03 | Oph | 11.1b | 7.0' | 4.0' | Sbc |

dA: *pretty bright, small, very little extended.*
42x60: Extremely faintly visible. ***170x250:*** Moderately faint. Evenly moderately concentrated. Slightly elongated NE-SW. Small. No involved stars. It's on the long side of an isosceles triangle of moderately faint stars.

| ga | NGC 6389 | 17 32.7 | +16 24 | Her | 12.8b | 2.8' | 1.8' | Sbc |

WH: faint, small, irregular figure, easily resolvable.
170x250: Very faint. Broadly slightly concentrated. Roughly round. Moderately small. No involved stars. A little ways SE are a couple of moderate brightness stars.

| oc | NGC 6383 | 17 34.8 | −32 33 | Sco | 5.5 | 5.0' | | |

JH: cluster, 6th–7th mag. star and 13th mag. stars.
12x35, 42x60: A bright star and a N-S line of stars W of it. ***170x250:*** The cluster is a circular knot of faint and threshold stars right behind the bright star, which is at the center of a crooked E-W line of a few moderate brightness stars. Very small. Dense.

| bs | Alpha Oph | 17 34.9 | +12 33 | Oph | 2.10 | | 0.13 | A5 |

170x250: White.

| ga | NGC 6411 | 17 35.5 | +60 48 | Dra | 12.8b | 2.3' | 1.8' | E |

dA: very small, gradually brighter middle.
170x250: Moderately faint. Evenly moderately concentrated. Roughly round. Small. No involved stars. Nearby SW is a moderate brightness star.

| oc | TR 27 | 17 36.2 | −33 29 | Sco | 6.7 | 6.0' | | |

42x60: A dense little knot of stars with a few more stars beside it. ***170x250:*** A little triangle of brighter stars and a number of additional stars S and E of it. Somewhat rectangular E-W. A little less than half the field in size. Somewhat sparse. [CR336]

| ga | IC 1265 | 17 36.7 | +42 05 | Her | 12.3v | 2.0' | 0.8' | Sab |

170x250: Not visible.

| ga | NGC 6434 | 17 36.8 | +72 05 | Dra | 13.2p | 2.3' | 0.9' | Sbc |

HH: very faint, very small, round, stellar, 8th mag. star S.
170x250: Faint. Broadly slightly concentrated. Roughly round. Very small. No involved stars. It makes a small triangle with a faint star E and a bright star S.

| oc | TR 28 | 17 36.9 | −32 28 | Sco | 7.7 | 7.0' | | |

12x35: Very faintly visible. ***42x60:*** A couple of faint stars with a hint of a few more fainter stars. ***170x250:*** A number of moderate brightness stars. A third of the field in size. Somewhat sparse. [CR337]

| gc | NGC 6402/M14 | 17 37.6 | −03 15 | Oph | 7.6 | 11.0' | −9.34 | 8 |

JH: ! globular cluster, bright, very large, round, extremely rich, very gradually much brighter middle, well resolved, 15th to 16th mag. stars.
12x35, 42x60: *Visible.* ***170x250:*** *Resolved throughout into very very faint stars with averted vision: a whole lot of tiny pinpoints. Very broadly concentrated. A quarter of the field in size. Very regular. It looks like a massive cluster. (225x: Abnormally faint. Neither the stars of the main body nor the outliers are distinct. A very abnormal globular in general. Undoubtedly it's heavily obscured: it's inside the Great Rift.)*

| gc | NGC 6401 | 17 38.6 | −23 55 | Oph | 9.5 | 4.8' | −7.2 | 8 |

HH: pretty bright, pretty large, round, 12th mag. star involved E.
42x60: *Very faintly visible.* ***170x250:*** *Not resolved. Broadly concentrated. Small. Inside its edge ESE is an involved field star.*

| oc | NGC 6404 | 17 39.6 | −33 14 | Sco | 10.6 | 5.0' | | |

JH: cluster, faint, large, pretty rich, little compressed, 13th to 15th mag. stars.
42x60: *Faintly visible.* ***170x250:*** *A fair number of very faint stars between two brighter stars, one E and one W. Somewhat elongated N-S. A quarter of the field in size. Moderate density.*

| oc | NGC 6405/M6 | 17 40.1 | −32 12 | Sco | 4.2 | 30.0' | | |

JH: cluster, large, irregularly round, little compressed, 7th mag. star and 10th mag. and fainter stars.
12x35: *A nice cluster.* ***42x60:*** *A good number of stars: a few bright stars and more moderate brightness stars. The famous butterfly figure doesn't stand out. Elongated NE-SW; somewhat rectangular. A third of the field in size. Moderately dense. Very nice.* ***170x250:*** *A very nice field of bright stars, overflowing the eyepiece field. At its NE end is a conspicuously yellow star; this is one of the brightest stars in the cluster [SAO 209132 (K3)]. [Butterfly Cluster]*

| pn | PK 3+2.1 | 17 41.9 | −24 42 | Oph | 13.1p | 6.0" | | |

170x250: *Not visible.*

| ds | Psi Dra | 17 41.9 | +72 09 | Dra | 30.1" | 4.9 | 6.1 | F5 G0 |

12x35: *An unequal pair; nicely split.* ***42x60:*** *A slightly wide slightly unequal pair.* ***170x250:*** *A wide very slightly unequal pair. Both yellow white. [STF 2241]*

| bs | Beta Oph | 17 43.5 | +04 34 | Oph | 2.77 | | 1.16 | K2 |

170x250: Yellow gold.

| gc | PAL 6 | 17 43.7 | −26 13 | Oph | 13.6 | 1.2' | − | 11 |

170x250: Not visible.

| oc | NGC 6416 | 17 44.4 | −32 21 | Sco | 5.7 | 18.0' | | |

JH: cluster, very large, rich, little compressed.
12x35: Visible. *42x60:* A cluster of faint stars. *170x250:* A good number of scattered moderate brightness stars. The field width in size. Moderate density. A large group of brighter stars SSW, filling the field, merges with the obvious cluster. This other group is sparse. It's probably a separate, closer cluster; the two groups are clearly different.

| gc | NGC 6426 | 17 44.9 | +03 10 | Oph | 11.2 | 4.2' | −6.10 | 9 |

WH: faint, considerably large, irregular figure.
170x250: Not resolved. Broadly concentrated. A sixth of the field in size. Very faint for a globular; it looks more like a galaxy. Insubstantial.

| oc | IC 4665 | 17 46.3 | +05 43 | Oph | 4.2 | 40.0' | | |

12x35: Some dozen very distinct stars of very uniform brightness in a quadrilateral shape. Large and bright. *42x60:* 16–18 widely spaced bright stars, with few fainter stars. Half the field in size. Sparse. A nice cluster in both the 35 mm and the 60 mm.

| bs | Mu Her | 17 46.5 | +27 43 | Her | 3.41 | | 0.76 | G5 |

170x250: Yellow white.

| oc | NGC 6425 | 17 46.9 | −31 31 | Sco | 7.2 | 7.0' | | |

JH: cluster, pretty small, little rich, little compressed, 10th to 12th mag. stars.
12x35: Visible. *42x60:* A fair number of faint stars. *170x250:* A number of moderate brightness stars, pretty uniform in brightness. Square. Half the field in size. Somewhat sparse.

| pn | PK 359-0.1 | 17 47.9 | −29 59 | Sgr | 13.6p | 15.0" | | |

170x250: Not visible.

12 to 18 Hours: SPRING

| pn | NGC 6439 | 17 48.3 | −16 28 | Sgr | 13.8p | 5.0" | | |

Pi: planetary nebula, stellar = 13th mag.
170x250: At first sight it looks like a regular moderately bright star. Then: it has a faint central star and a tiny high surface brightness nebula. Almost stellar. Moderate blinking effect. WSW is a star that is fairly similar to the planetary in brightness and general appearance.

| gc | NGC 6440 | 17 48.9 | −20 22 | Sgr | 9.7 | 4.4' | −6.75 | 5 |

HH: pretty bright, pretty large, round, brighter middle.
42x60: Visible. ***170x250:*** Not resolved. Broadly concentrated. Very small. Probably a fairly substantial cluster intrinsically.

| pn | NGC 6445 | 17 49.2 | −20 00 | Sgr | 13.2p | 44.0" | | |

HH: pretty bright, pretty small, round, gradually brighter middle, resolvable, 15th mag. star NW.
170x250: Moderate brightness to moderately bright. No central star. Slightly rectangular NNW-SSE, with concentrations at either end – two lobes. Practically no blinking effect. Moderate size for a planetary. Very nearby NW is a moderate brightness star. An interesting planetary.

| ga | NGC 6503 | 17 49.4 | +70 08 | Dra | 10.9b | 7.1' | 2.4' | Scd |

Au: pretty faint, large, much extended, 9th mag. star 4' E.
42x60: Visible. ***170x250:*** Bright. Broadly very slightly concentrated. Edge-on WNW-ESE. Large: a little less than a quarter of the field in size. No involved stars. A little ways NNE is a faint star; SW are several more faint stars; E is the bright plotted star. A nice galaxy.

| oc | NGC 6451 | 17 50.7 | −30 12 | Sco | 8.2 | 7.0' | | |

HH: cluster, pretty large, pretty rich, bifid, 12th mag. and fainter stars.
12x35: Faintly visible. ***42x60:*** A cluster of faint stars. ***170x250:*** A good number of moderately faint to faint stars. Elongated NNE-SSW. A third of the field in size. Pretty dense.

| gc | NGC 6453 | 17 50.9 | −34 36 | Sco | 9.9 | 7.6' | −6.5 | 4 |

JH: considerably large, irregularly round, pretty much brighter middle, resolvable.
42x60: Extremely faintly visible. ***170x250:*** Not resolved, but there are a couple of stars made out across its surface – undoubtedly foreground stars. Broadly concentrated. Small. Faint.

| ga | NGC 6482 | 17 51.8 | +23 04 | Her | 12.4b | 2.0' | 1.7' | E |

JH: ! very faint, small, round, very suddenly very much brighter middle to a very small round nucleus.
170x250: Moderately bright. Sharply extremely concentrated to a substellar nucleus. Round. Very small. No involved stars. It's on the long side of a triangle of moderate brightness stars.

| ga | NGC 6484 | 17 51.8 | +24 29 | Her | 13.1p | 1.9' | 1.6' | Sb |

St: extremely faint, very small, round, much brighter middle.
170x250: Faint. Evenly moderately concentrated. Roughly round. Small. No involved stars. [M+4-42-10 was not seen.]

| ga | NGC 6487 | 17 52.7 | +29 50 | Her | 12.9b | 1.6' | 1.6' | E |

St: faint, small, round, gradually brighter middle.
170x250: Faint. Evenly moderately concentrated. Round. Very small. No involved stars. [6486 (St) was not seen.]

| oc | NGC 6469 | 17 52.9 | −22 20 | Sgr | 8.2 | 12.0' | | |

JH: cluster, pretty rich (in Milky Way).
12x35: Visible. ***42x60:*** A few moderately faint stars and more fainter stars.
170x250: A trapezoidal figure of moderate brightness stars plus a lot of much fainter stars, mainly NNW of the figure. Overall the cluster is the field width in size. Pretty dense.

| oc | NGC 6475/M7 | 17 53.9 | −34 47 | Sco | 3.3 | 80.0' | | |

JH: cluster, very bright, pretty rich, little compressed, 7th to 12th mag. stars.
12x35: A nice cluster, with an irregular "H" figure of stars in the middle; the figure is on its side (E-W). ***42x60:*** A number of bright stars, including those of the "H" and some distant outer members, plus pretty faint stars scattered about. The "H" is a quarter of the field in size; with the outer members the cluster takes up the entire field. Moderate density. ***170x250:*** The "H" just fits inside the main scope field. A very nice field of very bright stars.

| pn | PK 356-4.1 | 17 54.6 | −34 22 | Sco | 13.9p | 2.0" | | |

170x250: Not visible.

| ga | NGC 6495 | 17 54.8 | +18 19 | Her | 13.2p | 2.0' | 1.7' | E |

Ma: faint, small, round.
170x250: Moderately faint. Evenly to sharply fairly concentrated. Round. Very small. No involved stars. [6490 (Ma) was not seen.]

| ne | IC 4670 | 17 55.1 | –21 48 | Sgr | | 5.0' | | – |

170x250: [This is a planetary, not a regular nebula.] Moderately bright. Pretty much stellar; it's like a slightly fat star. Pretty strong blinking effect. [PK7+1.1]†

| ga | NGC 6500 | 17 56.0 | +18 20 | Her | 13.1p | 2.2' | 1.6' | Sab |

WH: very faint, very small, W of 2.
170x250: 6501 and 6500 are next to each other NNE-SSW. Both are moderately faint. Evenly to sharply fairly concentrated. Roughly round. Very small. No involved stars.

| ga | NGC 6501 | 17 56.1 | +18 22 | Her | 13.0p | 1.9' | 1.7' | S0+ |

WH: very faint, very small, E of 2.
[See 6500.]

| bs | Gamma Dra | 17 56.6 | +51 29 | Dra | 2.23 | | 1.52 | K5 |

170x250: Yellow gold.

| oc | NGC 6494/M23 | 17 56.8 | –19 01 | Sgr | 5.5 | 27.0' | | |

JH: cluster, bright, very large, pretty rich, little compressed, 9th–10th mag. star and 11th to 13th mag. stars.
12x35: Resolved. ***42x60:*** A good number of moderately faint stars of uniform brightness. Somewhat rectangular NE-SW. A little less than half the field in size. Moderate density to moderately dense. ***170x250:*** A nice field of moderately bright stars of uniform brightness.

| pn | NGC 6543 | 17 58.6 | +66 37 | Dra | 8.8p | 20.0" | | |

WH: planetary nebula, very bright, pretty small, suddenly brighter middle to a very small nucleus.
12x35: It looks like a regular star. Next to it WNW is a star that is fainter than the planetary. ***42x60:*** Starting to look non-stellar. ***170x250:*** Very bright. It has a central star. A well defined disk with a pretty sharp edge. Even in light. Strong blinking effect. Oval N-S. Moderate size for a planetary. Very slightly blue-greenish. A very good planetary.

| bs | Nu Oph | 17 59.0 | −09 46 | Oph | 3.34 | | 0.99 | K0 |

170x250: Yellow.

| ds | PZ 6 | 17 59.1 | −30 15 | Sgr | 5.5" | 5.2 | 6.9 | K5 K5 |

42x60: Extremely close. *170x250:* A pretty close unequal pair. Colorful: both yellow. It looks like it's in the foreground: it's the brightest star in the field and the only colored one. A fairly nice double and a nice scene, unusually three-dimensional because of the rich field of faint background stars.

| ga | NGC 6509 | 17 59.4 | +06 17 | Oph | 13.1p | 1.5' | 1.1' | Sd |

St: very faint, pretty large, irregularly round, little brighter middle.
170x250: Very faint. Broadly very slightly concentrated. Elongated E-W. Moderately small. Near center is a very faint involved star; toward the E end is another very faint involved star; further E is an extremely faint star; a little ways N is a little trapezoid of faint stars; E is a long NNE-SSW string of stars.

| oc | NGC 6507 | 17 59.6 | −17 23 | Sgr | 9.6 | 6.0' | | |

WH: cluster, pretty small, little rich, little compressed.
42x60: Very faintly visible. *170x250:* A number of moderately faint stars, uniform in brightness, in a NNE-SSW half-disk shape. Half the field in size. Sparse.

| oc | TR 31 | 17 59.8 | −28 10 | Sgr | 9.8 | 8.0' | | |

12x35: Faintly visible. *42x60:* A few stars, mostly N-S. *170x250:* Four evenly spaced moderate brightness stars in a N-S line and another star W making a triangle with them. There are a few more stars plus underlying threshold stars. Not a true cluster: this is a Milky Way field framed by dark nebulosity. The brighter stars are random field stars and the fainter stars are background Milky Way stars. The "object" is neither conspicuous nor notable. [CR357]

Chapter 13

18 to 24 Hours: SUMMER

| ds | 41 Dra | 18 00.2 | +80 00 | Dra | 19.3" | 5.7 | 6.1 | F7 F7 |

12x35: Extremely close. *42x60:* A moderately close equal pair. A nice double. *170x250:* A slightly wide equal pair. Both soft white. [STF 2308]

| ds | 95 Her | 18 01.5 | +21 36 | Her | 6.3" | 5.0 | 5.1 | A5 G8 |

42x60: Very very close; nicely split. *170x250:* A close equal pair. Both yellow white. [STF 2264]

| gc | NGC 6517 | 18 01.8 | −08 58 | Oph | 10.3 | 4.0' | −7.8 | 4 |

HH: pretty bright, pretty large, round, partially resolved.
42x60: Very very faintly visible. *170x250:* Not resolved. Evenly concentrated. Very small. Somewhat faint for a globular; it looks more like a galaxy. It's on the hypotenuse of a large right triangle of moderate brightness and moderately bright stars.

| ne | NGC 6514/M20 | 18 02.3 | −23 02 | Sgr | | 17.0' | | E+R |

HH: !!! very bright, very large, trifid, double star involved.
12x35: Visible. *42x60:* The trifid structure is starting to be seen. *170x250:* Moderately bright. It's not cleanly trifid: there are three sections but the division is not symmetrical. Wandering irregularly through the W side is a broad dark lane. Going ENE from a bright double star in the middle of the nebula is a thinner dark

lane. The double star has a third very faint component. N of the main nebula is a second, slightly fainter nebula surrounding a bright star; the two nebulae are separated by a wide E-W dark lane. The combined nebulosity is a little less than the field width in size. The light is very milky. [Trifid Nebula]

pn	IC 4673	18 03.3	–27 06	Sgr	12.9p	15.0"		

170x250: Very faint. No central star. No clear disk. Slight blinking effect. Round. Very small. Very nearby NNE is a faint star.

oc	NGC 6520	18 03.4	–27 53	Sgr	7.6	6.0'		

HH: cluster, pretty small, rich, little compressed, 9th to 13th mag. stars.
12x35: Visible. *42x60:* Three bright stars in a NW-SE line, a few moderately bright stars, and some fainter stars. *170x250:* The cluster is a group of mostly moderately faint stars concentrated around the two northernmost of the three bright stars. Very small; very compact. Moderately dense. W of the cluster is the famous dark nebula [B86], which looks really cool. Not perfectly black, but a lot darker than the surrounding sky, which is gray by comparison. It's in the shape of a narrow triangle pointing NNE. At its edge NW is a bright star. The dark nebula is a little less than half the field in size; it's much larger than the cluster. The two objects together make a nice scene.

gc	NGC 6522	18 03.6	–30 02	Sgr	8.6	9.4'	–7.04	6

HH: globular cluster, bright, pretty large, round, gradually very much brighter middle, well resolved, 16th to 17th mag. stars.
12x35: Faintly visible. *42x60:* 6522 and 6528 are both visible. *170x250:* 6522 is granular. Evenly concentrated. Very small. At its edge ENE is a faint star. 6528 is fainter. Not resolved. Broadly concentrated. Very small. Just off its edge SSW is a faint star.

ne	NGC 6523/M8	18 03.8	–24 23	Sgr		50.0'		E

JH: !!! very bright, extremely large, extremely irregular figure, with large cluster.
12x35: Bright. *42x60:* Large and bright. In the middle of the multiform object is a NE-SW dark rift, on the E side is a cluster [6530], and on the W side are two bright stars plus bright nebulosity. Surrounding the whole thing is more nebulosity, most notably faint nebulosity N of the major part of the nebula. A little more than a third of the field in size. *170x250:* There's unevenness in the nebulosity, but the light is soft and very milky. The nebula's features are broad, not convoluted or sharply defined like M42. The cluster consists of a number of widespread evenly spaced bright stars. Half the field in size. Somewhat sparse. The rift, going right across the field, is neither very dark nor empty; there are a number of stars "floating in the lagoon." The cluster, the rift, and the bright part of the nebula just fit in a single field. 2/3 of a field WNW is a bright star [7 Sgr] surrounded by nebulosity. [Lagoon Nebula]

| gc | NGC 6535 | 18 03.8 | –00 18 | Ser | 10.6 | 3.4' | –5.8 | 11 |

Hi: pretty faint, very small, very faint nebulous star W.
42x60: Extremely faintly visible. ***170x250:*** Three or four stars are resolved at its edge W; otherwise it's granular. Broadly concentrated. Small. Faint.

| oc | NGC 6530 | 18 04.5 | –24 21 | Sgr | 4.6 | 14.0' | | |

JH: cluster, bright, large, pretty rich, very large nebula W.
[See M8.]

| oc | NGC 6531/M21 | 18 04.6 | –22 29 | Sgr | 5.9 | 13.0' | | |

JH: cluster, pretty rich, little compressed, 9th to 12th mag. stars.
12x35: A bright star and a handful of other stars. ***42x60:*** A brighter star and several fainter stars. ***170x250:*** A fair number of moderately bright stars, including the brighter one. The highest concentration is right around the bright star, just N of which is a small circle of stars. The cluster fills the field. Moderate density.

| gc | NGC 6528 | 18 04.8 | –30 03 | Sgr | 9.5 | 5.0' | –6.90 | 5 |

HH: globular cluster, pretty faint, considerably small, round, gradually brighter middle, well resolved, 16th to 17th mag. stars.
[See 6522.]

| gc | NGC 6539 | 18 04.8 | –07 35 | Ser | 9.6 | 7.9' | –6.1 | 10 |

Bn: no description.
42x60: Very faintly visible. ***170x250:*** Not resolved. Broadly concentrated. Small. Just off its edge NW is a triangle of stars.

| pn | NGC 6537 | 18 05.2 | –19 50 | Sgr | 12.5p | 10.0" | | |

Pi: planetary nebula, bright, small, stellar.
170x250: Moderate brightness. Nearly stellar. It makes a flat triangle with two moderately faint stars WNW that are similar in brightness to the planetary.

| ds | 70 Oph | 18 05.5 | +02 30 | Oph | 4.9" | 4.2 | 6.0 | K0 – |

42x60: Extremely close. ***170x250:*** A very close unequal pair. Color contrast: the primary is yellow; the companion is very dark yellow. A very unusual color combination. A very nice double. [STF 2272]

| bs | Gamma Sgr | 18 05.8 | –30 25 | Sgr | 2.99 | | 1.00 | K0 |

170x250: Yellow.

| ga | NGC 6548 | 18 06.0 | +18 35 | Her | 12.7b | 2.9' | 2.7' | S0 |

WH: considerably faint, small, little extended, resolvable.
170x250: Moderately faint. Evenly to sharply fairly concentrated. Round. Very small. No involved stars. [6549 (Ma) was not seen.]

| gc | NGC 6540 | 18 06.3 | –27 47 | Sgr | 15.0 | 1.5' | – | – |

WH: pretty faint, small, irregularly extended, easily resolvable or cluster.
170x250: Granular. Broadly concentrated. Very small. Faint. Irregular. It hardly looks like a globular; apparently it's a loose type. [All sources except MegaStar list this object as an open cluster; the MegaStar photograph doesn't clarify what it is.]

| oc | NGC 6546 | 18 07.2 | –23 18 | Sgr | 8.0 | 13.0' | | |

JH: cluster, very large, very rich.
42x60: Visible. *170x250:* A number of moderately faint stars plus faint and threshold stars. On the E side are three bright stars in a large triangle – undoubtedly foreground stars. 2/3 of the field in size. Moderate density. Unconcentrated; irregular; a somewhat strange cluster.

| gc | NGC 6544 | 18 07.3 | –25 00 | Sgr | 8.3 | 9.2' | –7.10 | – |

HH: considerably faint, pretty large, irregularly round, resolvable.
12x35, 42x60: Visible. *170x250:* Resolved; the stars are mostly very faint and closely packed. Broadly concentrated. Small. Somewhat irregular. Lightweight. It's in a very rich field of threshold stars from which it's not well separated. 6544 makes an interesting contrast with 6553, which looks more substantial.

| ds | 100 Her | 18 07.8 | +26 06 | Her | 14.3" | 5.9 | 6.0 | A3 A3 |

12x35: Extremely close. *42x60:* An equal pair. A nice little double. *170x250:* A moderate distance equal pair. Both white. [STF 2280]

| ga | NGC 6555 | 18 07.8 | +17 36 | Her | 13.0b | 2.0' | 1.5' | Sc |

WH: faint, large, round, very gradually little brighter middle.
170x250: Faint. Broadly slightly concentrated. Round. Small. No involved stars. It's surrounded by moderately bright stars, including a wide pair SE and an arc of three stars N.

| gc | NGC 6553 | 18 09.3 | –25 54 | Sgr | 8.3 | 9.2' | –8.15 | 11 |

HH: globular cluster, faint, large, little extended, very gradually little brighter middle, partially resolved, 20th mag. stars.
12x35, 42x60: Visible. ***170x250:*** Not resolved. Very broadly concentrated. Small. At its edge NW is a moderately faint star. Pretty regular. Interesting: it's bright enough and large enough that it should be at least partly resolved. It's probably obscured; it's somewhat isolated and seems to be behind dark nebulosity, some of which is visible N and especially S.

| ne | NGC 6559 | 18 10.0 | –24 06 | Sgr | | 8.3' | | E |

JH: very faint, very large, little extended, double star involved.
170x250: A faint nebula around a couple of moderate brightness stars aligned NW-SE. Small. Nearby are a few more stars.

| gc | NGC 6558 | 18 10.3 | –31 46 | Sgr | 8.6 | 4.2' | – | – |

JH: globular cluster, pretty bright, pretty large, round, gradually little brighter middle, well resolved, 16th mag. stars.
170x250: Partly resolved. Evenly concentrated. Small. Somewhat insubstantial. 6558 is smaller and fainter than 6569. Although both of these neighboring globulars are small and distant, they exemplify the two basic types: 6558 is fainter, more irregular, looser, and generally scrawny; 6569 is brighter, more regular, more densely packed, and has a considerable central ball.

| gc | IC 1276 | 18 10.7 | –07 12 | Ser | 10.3 | 8.0' | – | 12 |

170x250: Not resolved. Very very faint. The slightest glow. A sixth of the field in size. Right next to it W is the point star of a sharp isosceles triangle of nearby stars; just E of this point star is another star, this one within the cluster. 6517, 6539, and I1276 are all in the huge dark region of the Milky Way known as the Great Rift and are no doubt heavily obscured.

| pn | NGC 6565 | 18 11.9 | –28 10 | Sgr | 13.2p | 14.0" | | |

Pi: planetary nebula, stellar.
12x35: WSW of the planetary is a small dark nebula [B90]. ***170x250:*** 6565 is moderately faint to moderate brightness. No central star. A tiny disk; not a very clean edge. Moderate blinking effect. Round. It's inside a very small triangle of very faint stars, the southernmost of which is a wide double; W is a NNW-SSE line of stars, including a pair of stars in the middle.

| ga | NGC 6574 | 18 11.9 | +14 58 | Her | 12.8b | 1.4' | 1.0' | Sbc |

Ma: pretty bright, small, round.
170x250: Moderate brightness. Evenly moderately concentrated. Round. Small. Just off its edge S is a very faint star.

| pn | NGC 6563 | 18 12.0 | −33 52 | Sgr | 13.8p | 48.0" | | |

JH: planetary nebula, faint, large, considerably extended, hazy border.
170x250: Moderately faint. No central star. A modest disk; not a very sharp edge. Even in brightness. No blinking effect. Round. Moderate size for a planetary.

| ga | NGC 6577 | 18 12.0 | +21 27 | Her | 12.6v | 1.3' | 1.1' | E |

Ma: very faint, small.
170x250: [The observed galaxy is *6580 (Ma); 6577 was not seen.] Faint. Broadly very slightly concentrated. Roughly round. Moderately small. At its edge NW is a moderate brightness star.

| pn | NGC 6572 | 18 12.1 | +06 50 | Oph | 9.0p | 11.0" | | |

JH: planetary nebula, very bright, very small, round, little hazy.
12x35: It looks like a moderate brightness star. ***42x60:*** It looks like a bright star.
170x250: Extremely bright; very high surface brightness. No central star. A very very small disk; not a very clean edge. Moderate blinking effect. Unusual: with direct vision it looks like a very different sort of planetary compared to how it looks with averted vision: smaller, fainter, and fuzzier. Round. Blue-greenish. It makes a nearly isosceles triangle with two stars E that make good comparison stars. It might have the very highest surface brightness of any planetary in the sky. One of the most strongly colored planetaries.

| oc | NGC 6568 | 18 12.8 | −21 35 | Sgr | 8.6 | 12.0' | | |

HH: cluster, very large, little compressed.
12x35: Faintly visible. ***42x60:*** A number of very faint stars. ***170x250:*** A fair number of moderately faint stars plus very faint stars. Elongated N-S. Somewhat irregular. It fills the field. Moderately dense. It looks as much like a modest Milky Way concentration as like a regular cluster.

| gc | NGC 6569 | 18 13.6 | −31 50 | Sgr | 8.7 | 6.4' | −7.8 | 8 |

HH: globular cluster, considerably bright, large, round, well resolved, 15th mag. and fainter stars.
42x60: Visible. ***170x250:*** Granular. Broadly concentrated. Small. Pretty regular.

18 to 24 Hours: SUMMER

| pn | NGC 6567 | 18 13.7 | −19 04 | Sgr | 11.7p | 12.0" | | |

Pi: planetary nebula, stellar, 11th mag., in a cluster.
170x250: Not visible.

| oc | NGC 6583 | 18 15.8 | −22 07 | Sgr | 10.0 | 4.0' | | |

HH: cluster, pretty rich, pretty compressed, considerably extended, 13th mag. and fainter stars.
42x60: A slight cloud. ***170x250:*** A good number of faint stars. Pretty concentrated. Small. Dense. A distant cluster.

| pn | NGC 6578 | 18 16.3 | −20 27 | Sgr | 13.1p | 9.0" | | |

Pi: planetary nebula, stellar = 13th mag.
170x250: Faint. No central star. Not much of a disk. Moderate blinking effect. Round. Very very small. It's in an E-W zigzag line of stars, including a moderately faint star next to it WSW and a pair of stars nearby SE.

| oc | NGC 6605 | 18 16.3 | −14 58 | Ser | 6.0 | 29.0' | | |

JH: cluster, little rich, little compressed, 10th to 12th mag. stars.
12x35: Visible. ***42x60:*** A few moderately faint stars. ***170x250:*** A number of moderately bright stars of uniform brightness, with one brighter star at the cluster's edge NW. It fills the field. Sparse.

| ne | NGC 6589 | 18 16.9 | −19 46 | Sgr | | 4.0' | | R |

Sw: double star in center of extremely faint, pretty large nebulosity.
[See 6590.]

| oc | M24 | 18 17.0 | −18 36 | Sgr | − | 120' | | |

12x35: A cloud of a good number of distinct bright stars plus uncountable faint and threshold stars. Elongated NE-SW. 2/3 of the field in size. On either side NW and SE are dark nebulae; the two NW [B92, B93] are darker. ***42x60:*** More stars are resolved. The complex of dark nebulae along the NW side is well seen. 6603 is faintly visible as a distinct small cloud within the larger aggregation. ***170x250:*** 6603 is a cluster of a lot of very faint stars, uniform in brightness. Small. Very dense. [M24 = Small Sagittarius Star Cloud]

| ne | NGC 6590 | 18 17.0 | −19 51 | Sgr | | 5.6' | | R |

Sw: double star in center of pretty faint, pretty large, round nebulosity.

42x60: A very faint substellar nebulosity. *170x250:* A moderate distance equal double star with a small nebula surrounding it. A third of the field NNW is another nebula [6589] surrounding a moderately bright star; this nebula is larger than 6590 but not quite as conspicuous. Half the field NE is a group of stars [6595]: a few scattered stars, not an obvious cluster.

| oc | TR 32 | 18 17.2 | –13 20 | Ser | 12.2 | 4.0' | | |

42x60: Faintly visible. *170x250:* A number of faint stars. Small. Moderate density. Probably a Milky Way field rather than a true cluster.

| oc | NGC 6595 | 18 17.3 | –19 44 | Sgr | 7.0 | 11.0' | | |

JH: *faint, pretty large, considerably extended, double star involved.*
[See 6590.]

| pn | PK 38+12.1 | 18 17.6 | +10 09 | Oph | 12.4p | 5.0" | | |

170x250: Not visible.

| oc | NGC 6604 | 18 18.1 | –12 13 | Ser | 6.5 | 6.0' | | |

HH: *cluster, little rich, little compressed.*
42x60: A couple of stars in addition to the plotted star. *170x250:* Surrounding the bright plotted star are a few faint stars and a very small glow of unresolved stars. Quite a number of scattered bright stars fill the field apart from the main concentration; if these are members the cluster is much larger. The compact central part looks a little like 2362 in CMa, but this cluster is much less prominent.

| oc | NGC 6603 | 18 18.5 | –18 24 | Sgr | 11.1 | 4.0' | | |

JH: *! cluster, very rich, very much compressed, round, 15th mag. stars (Milky Way).*
[See M24.]

| oc | NGC 6611/M16 | 18 18.8 | –13 45 | Ser | 6.0 | 6.0' | | |

JH: *cluster, at least 100 bright and faint stars.*
12x35: The cluster is easily visible. *42x60:* The bright cluster stars are in a tall N-S backward "S" figure. The N loop of the "S" contains nebulosity. At the very tip N is a concentration of stars. The entire "S" figure is half the field in size. *170x250:* A fair number of bright stars. Each loop of the "S" fills a field. Sparse: the stars are mostly in the figure; there aren't many more stars besides these and the concentration at the N end. Very obvious but pretty much formless nebulosity fills most of the N field (it looks nothing like the pictures). [Eagle Nebula]

| ga | NGC 6643 | 18 19.8 | +74 34 | Dra | 11.7b | 3.8' | 1.9' | Sc |

Tu: pretty bright, pretty large, extended NE-SW, 2 stars W.
170x250: *Moderately faint. Broadly slightly concentrated. Elongated NE-SW. Moderate size. No involved stars. Nearby W are a couple of moderate brightness stars aligned parallel to the galaxy.*

| oc | NGC 6613/M18 | 18 19.9 | −17 06 | Sgr | 6.9 | 9.0' | | |

JH: cluster, poor, very little compressed.
12x35, 42x60: *A small group of stars.* ***170x250:*** *A number of moderately bright stars, mostly in the shape of an arrowhead pointing NE, plus some fainter stars. A little more than a third of the field in size. Moderate density.*

| ne | NGC 6618/M17 | 18 20.8 | −16 09 | Sgr | | 11.0' | | E |

JH: !!! bright, extremely large, extremely irregular figure, "2"-hooked.
12x35: *Easily visible.* ***42x60:*** *The familiar figure is made out.* ***170x250:*** *The nebula is in the shape of a long bar with a hook at the W end. There's obvious unevenness in the bar. S of the bar is fainter nebulosity; E is a detached, narrow NNE-SSW piece of fainter nebulosity between two pairs of stars. The nebula is greenish-gray. The whole thing is elongated WNW-ESE and takes up the entire field exactly. It looks fishlike: the body of a fish with fins. An awesome sight. [Swan Nebula; MegaStar plots a separate open cluster in the same general location (see below at RA 18 21.2).]*[†]

| bs | Delta Sgr | 18 21.0 | −29 49 | Sgr | 2.70 | | 1.38 | K3 |

170x250: *Yellow gold.*

| oc | NGC 6618 | 18 21.2 | −16 11 | Sgr | 6.0 | 27.0' | | |

42x60, 170x250: *The closest thing to a cluster in the vicinity is a loose group of bright stars just N of M17. 3/4 of the main scope field in size. [The Swan Nebula is described above at RA 18 20.8.]*

| bs | Eta Ser | 18 21.3 | −02 53 | Ser | 3.26 | | 0.94 | K0 |

170x250: *Yellow.*

| pn | NGC 6620 | 18 22.9 | −26 49 | Sgr | 15.0p | 8.0" | | |

Pi: planetary nebula, stellar.
170x250: *Pretty much stellar: a faint fuzzy star with a slight blinking effect.*

| oc | NGC 6625 | 18 23.2 | −12 02 | Sct | 9.0 | 40.0' | | |

JH: cluster, little rich, little compressed, 11th to 12th mag. stars.
12x35, 42x60, 170x250: There's no cluster in the position.

| gc | NGC 6624 | 18 23.7 | −30 22 | Sgr | 8.3 | 8.8' | −7.13 | 6 |

HH: globular cluster, very bright, pretty large, round, well resolved, 16th mag. stars.
12x35, 42x60: Visible in a row with two faint stars NE. ***170x250:*** Resolved into very faint stars with averted vision. Sharply concentrated. Small.

| ga | NGC 6654 | 18 24.1 | +73 11 | Dra | 13.0b | 2.6' | 2.0' | S0/a |

Sw: 12th–13th mag. star in pretty bright, pretty large nebulosity.
170x250: Moderate brightness. Sharply fairly concentrated. Roughly round. Small. No involved stars. Off its edge NW are a couple of very faint stars.

| bs | Epsilon Sgr | 18 24.2 | −34 23 | Sgr | 1.80 | | 0.02 | B9 |

170x250: White.

| gc | NGC 6626/M28 | 18 24.5 | −24 52 | Sgr | 6.9 | 13.8' | −8.1 | 4 |

JH: ! globular cluster, very bright, large, round, gradually extremely compressed middle, well resolved, 14th to 16th mag. stars.
12x35: Easily visible. ***42x60:*** Bright. ***170x250:*** Resolved throughout, except for a granular mass of underlying stars at the very center. Evenly pretty concentrated. A sixth of the field in size. Fairly regular. It looks like an intrinsically fairly massive cluster, but pretty distant.

| oc | TR 33 | 18 24.7 | −19 43 | Sgr | 7.8 | 6.0' | | |

42x60: A little line of three stars with a fourth star off line, plus a hint of more stars. ***170x250:*** About a dozen stars: four bright to moderate brightness stars and several moderately faint stars. A quarter of the field in size. Sparse. Hardly a cluster. [CR378]

| ga | NGC 6632 | 18 25.1 | +27 32 | Her | 12.9b | 3.0' | 1.4' | Sbc |

Ma: faint, small, round, gradually brighter middle.
170x250: Very faint. Broadly very slightly concentrated. Roughly round. Small. No involved stars. It makes a very small triangle with a pair of stars WNW.

| pn | NGC 6629 | 18 25.7 | −23 12 | Sgr | 11.6p | 16.0" | | |

HH: planetary nebula or globular cluster, pretty bright, excessively small, round.
170x250: Moderately bright. No central star. A small disk. Even in brightness. Moderately strong blinking effect. Round. It's in the middle of a very sharp triangle of stars: two faint stars NW and a moderate brightness star SE.

| ds | 59 Ser | 18 27.2 | +00 12 | Ser | 3.9" | 5.3 | 7.6 | G0 F5 |

170x250: The primary has a very very close relatively very faint companion. The primary is yellowish. [STF 2316]

| oc | NGC 6631 | 18 27.2 | −12 01 | Sct | 11.7 | 5.0' | | |

JH: cluster, pretty large, pretty rich, 12th to 15th mag. stars.
42x60: A faint elongated cloud. ***170x250:*** A concentration of stars elongated NW-SE. 3/4 of the field in size. It's surrounded by dark nebulosity. Undoubtedly this is just a rich Milky Way field poking through the ubiquitous dark nebulosity, rather than a true cluster; it doesn't look like a cluster at all.

| oc | NGC 6633 | 18 27.7 | +06 34 | Oph | 4.6 | 27.0' | | |

WH: cluster, little compressed, bright stars.
12x35: Elongated. A nice cluster. ***42x60:*** A fair number of moderate brightness stars. The main concentration is in the shape of a sharp triangle of stars pointing ENE; N are more stars. Elongated NE-SW overall. 2/3 of the field in size. Sparse. ***170x250:*** The widely scattered bright stars of the concentration fill the field. Nice.

| bs | Lambda Sgr | 18 28.0 | −25 25 | Sgr | 2.81 | | 1.04 | K0 |

170x250: Yellow.

| ga | MCG+0-47-1 | 18 29.9 | +03 02 | Ser | 12.8 | 0.9' | 0.7' | – |

170x250: Not visible.

| gc | NGC 6638 | 18 30.9 | −25 30 | Sgr | 9.2 | 7.3' | −6.5 | 6 |

HH: globular cluster, bright, small, round, partially resolved.
42x60: Faintly visible. ***170x250:*** Granular. Broadly concentrated. Very small. Relatively faint.

| gc | NGC 6637/M69 | 18 31.4 | −32 21 | Sgr | 7.7 | 9.8' | −7.90 | 5 |

HH: globular cluster, bright, large, round, well resolved, 14th to 16th mag. stars.
12x35, 42x60: Visible with a star next to it NW. ***170x250:*** Completely resolved, especially with averted vision, but, except for a handful of outliers, the stars aren't very bright or distinct. Broadly to evenly concentrated. Small. It looks moderately massive: it has a central concentration but not an exceptional ball.

| oc | IC 4725/M25 | 18 31.8 | −19 07 | Sgr | 4.6 | 32.0' | | |

12x35: A fair number of stars. Concentrated. ***42x60:*** A good number of stars with a little concentration near the middle. Half the field in size. Moderate density. Irregular. ***170x250:*** The central part of the cluster is a nice field of bright and moderate brightness stars. In the very center is a mini-concentration of seven stars in a half-circle.

| gc | NGC 6642 | 18 31.9 | −23 29 | Sgr | 8.8 | 5.8' | − | − |

HH: globular cluster, pretty bright, pretty large, irregularly round, gradually pretty much brighter middle, well resolved, 16th mag. stars.
12x35: Extremely faintly visible. ***42x60:*** Very faintly visible. ***170x250:*** Partly resolved with averted vision; a few stars are made out. Sharply concentrated. Very small.

| pn | NGC 6644 | 18 32.6 | −25 07 | Sgr | 12.2p | 3.0" | | |

Pi: planetary nebula, stellar.
42x60: Easily visible. ***170x250:*** Bright; high surface brightness. Essentially stellar: a slightly fat moderately bright star. Very strong blinking effect.

| oc | NGC 6645 | 18 32.6 | −16 53 | Sgr | 8.5 | 10.0' | | |

HH: cluster, pretty large, very rich, pretty compressed, 11th to 15th mag. stars.
12x35: Visible. ***42x60:*** A lot of very faint stars. ***170x250:*** A lot of faint stars, fairly uniform in brightness. Near the middle is a circle of stars, blank inside. 2/3 of the field in size. Fairly dense.

| oc | NGC 6649 | 18 33.5 | −10 23 | Sct | 8.9 | 5.0' | | |

JH: cluster, poor, little compressed, pretty small, 9th–10th mag. star and 12th to 13th mag. stars.
12x35: Faintly visible. ***42x60:*** A faint cloud. ***170x250:*** A number of moderately faint stars plus fainter stars. At its edge SSW is a moderately bright star. A quarter of the field in size. Moderate density.

| pn | IC 4732 | 18 33.9 | −22 38 | Sgr | 13.3p | 10.0" | | |

170x250: Not visible.

| ga | NGC 6661 | 18 34.6 | +22 54 | Her | 13.1b | 1.7' | 1.0' | S0/a |

Ma: faint, very small, round, gradually brighter middle.
170x250: Faint. Evenly slightly concentrated. Roughly round. Very small. No involved stars. It's among numerous stars.

| ga | NGC 6690 | 18 34.8 | +70 31 | Dra | 13.1b | 3.8' | 1.2' | Sd |

Sw: pretty faint, large, round, between 2 stars.
170x250: Very faint. Hardly concentrated. Elongated N-S. [Moderately small.] N of center is a very faint involved star; just off its edge WSW is a faint star; a little ways SW is a moderate brightness star; a little ways NNE is another moderate brightness star; all of these stars are in a NE-SW line.

| gc | NGC 6652 | 18 35.8 | −32 59 | Sgr | 8.9 | 6.0' | −7.3 | 6 |

JH: bright, small, little extended, well resolved, 15th mag. stars.
12x35: Visible as a faint star. *42x60:* Small and faint. *170x250:* Partly resolved outside of the center, especially with averted vision. Sharply concentrated. Very small. At its edge E is a moderately faint star; at its edge W, a little farther from center, is a slightly brighter star.

| gc | NGC 6656/M22 | 18 36.4 | −23 54 | Sgr | 5.1 | 32.0' | −8.45 | 7 |

JH: !! globular cluster, very bright, very large, round, very rich, very much compressed, 11th to 15th mag. stars.
12x35: Bright. *42x60:* Starting to be resolved with averted vision. Very broadly concentrated. Pretty regular. *170x250:* Very well resolved throughout into countless very distinct stars. Very broadly concentrated. The outer boundary is unclear: there are a lot of distant outliers, very scattered and widespread; with these the cluster takes up the entire field; the main body, with the near outliers, is half the field in size. Fairly regular. Clearly a massive globular but it doesn't have a well defined ball; it just concentrates continuously toward the center, where there's a mass of stars underlying the well resolved stars. By far the largest globular visible from northern latitudes; a couple of others have outliers reaching the edge of the field but not so easily. (225x: Many of the outliers are so bright that they mix with the field stars in this star-rich region.)

| oc | NGC 6664 | 18 36.7 | −08 13 | Sct | 7.8 | 16.0' | | |

HH: cluster, large, pretty rich, very little compressed.
12x35: Visible. ***42x60:*** A fairly large cluster of very faint but distinct stars. ***170x250:*** A good number of stars, including many faint ones. The more concentrated N part contains a very wide E-W arc of brighter stars. The rest of the cluster S is sparser. The cluster is somewhat empty in the middle, where there's only one fairly bright star. It takes up the entire field. Moderate density.

| bs | Alpha Lyr | 18 36.9 | +38 47 | Lyr | 0.03 | | 0.00 | A0 |

170x250: White. A brilliant star. [Vega]

| ga | NGC 6674 | 18 38.6 | +25 22 | Her | 13.0b | 4.0' | 2.2' | Sb |

Ma: faint, pretty small, irregularly round, brighter middle.
170x250: Very faint. Hardly concentrated. Roughly round. Small. At its edge W is a faint star.

| oc | IC 4756 | 18 39.0 | +05 26 | Ser | 4.6 | 52.0' | | |

12x35: Elongated. A third of the field in size. I4756 is larger and less concentrated than 6633 but its stars are fainter – an interesting contrast. ***42x60:*** A lot of scattered moderate brightness stars. Elongated ENE-WSW. It slightly overflows the field. Moderately dense. ***170x250:*** Several nice fields of bright stars.

| oc | TR 34 | 18 39.8 | −08 28 | Sct | 8.6 | 7.0' | | |

12x35, 42x60, 170x250: There's no cluster in the position. [CR387]

| gc | PAL 8 | 18 41.5 | −19 49 | Sgr | 10.9 | 5.2' | – | 10 |

42x60: Extremely faintly visible. ***170x250:*** Starting to be granular with averted vision. Broadly concentrated. Small. At its edge S is a faint star.

| oc | NGC 6682 | 18 41.6 | −04 46 | Sct | – | 47' | | |

JH: cluster, large, rich, 10th to 18th mag. stars.
12x35: This "object" is a detached piece of the Scutum Star Cloud. E is an intervening lane of dark nebulosity cutting it off from the rest of the star cloud further E. ***170x250:*** Not a true cluster: obviously a detached piece of the Scutum Star Cloud. Elongated E-W. 2 1/2 field widths in size.

18 to 24 Hours: SUMMER

| oc | NGC 6683 | 18 42.2 | −06 16 | Sct | 9.4 | 11.0' | | |

JH: cluster, very rich, very little compressed (in Milky Way).
12x35, 42x60, 170x250: There's no notable cluster in the position. What is seen is a detached little piece of a larger star cloud, surrounded by dark nebulosity. Huge dark clouds pervade the area.

| oc | TR 35 | 18 43.0 | −04 13 | Sct | 9.2 | 9.0' | | |

42x60: Faintly visible. ***170x250:*** A fair number of stars: two or three moderately faint and the rest faint. A third of the field in size. Moderate density. Relatively faint; unimpressive. [CR388]

| gc | NGC 6681/M70 | 18 43.2 | −32 18 | Sgr | 8.1 | 8.0' | −7.32 | 5 |

JH: globular cluster, bright, pretty large, round, gradually brighter middle, 14th to 17th mag. stars.
12x35: Visible. ***42x60:*** Somewhat faint. ***170x250:*** Resolved throughout except for the very center; best with averted vision. Sharply concentrated. Small. Coming out of its edge NE and going NNE is a slightly curved line of three field stars. (225x: M70 is a little smaller, more concentrated, and higher surface brightness than M69.)

| ga | NGC 6701 | 18 43.2 | +60 39 | Dra | 13.0p | 1.5' | 1.2' | Sa |

Sw: pretty bright, pretty small, much extended, faint star close E.
170x250: Faint. Broadly slightly concentrated. Elongated WNW-ESE. Very small. Just off its E end is a moderate brightness star.

| ga | UGC 11337 | 18 43.2 | +18 43 | Her | 13.2 | 1.4' | 1.3' | Sa |

170x250: Not visible.

| ds | Epsilon 1 Lyr | 18 44.3 | +39 40 | Lyr | 2.3" | 5.0 | 6.1 | A4 F1 |

170x250: Both pairs are extremely close. The S pair [Epsilon 2 (STF 2383)] is a little closer and pretty much equal. The N pair [Epsilon 1 (STF 2382)] is unequal. Both pairs are very cleanly and nicely split. They're almost perpendicular to each other: the N pair points toward the S pair. All the stars are soft white. [Double Double]

| ds | Epsilon 2 Lyr | 18 44.3 | +39 40 | Lyr | 2.1" | 5.2 | 5.5 | A8 F0 |

[See Epsilon 1 Lyr.]

| ds | Zeta Lyr | 18 44.8 | +37 36 | Lyr | 43.9" | 4.3 | 5.9 | A F0 |

12x35: Split. *42x60:* A wide slightly unequal pair. *170x250:* A very wide slightly unequal pair. The primary is white; the companion is soft white. [STF 38]

| oc | NGC 6694/M26 | 18 45.2 | –09 23 | Sct | 8.0 | 14.0' | | |

JH: cluster, considerably large, pretty rich, pretty compressed, 12th to 15th mag. stars.
12x35: Visible. *42x60:* One bright star and about three fairly bright stars, plus more faint stars. *170x250:* A compact cluster consisting of the bright star, some half-dozen moderate brightness stars, and some fainter stars. A quarter of the field in size. Moderate density.

| ds | STF 2375 | 18 45.5 | +05 30 | Ser | 2.5" | 6.9 | 7.9 | A1 – |

170x250: A very very close equal pair. Both white.

| bs | Phi Sgr | 18 45.7 | –26 59 | Sgr | 3.20 | | –0.14 | B8 |

170x250: White.

| pn | IC 4776 | 18 45.8 | –33 20 | Sgr | 11.7p | 8.0" | | |

170x250: Not visible.

| ds | 5 Aql | 18 46.5 | –00 58 | Aql | 12.7" | 6.0 | 7.8 | A2 A0 |

12x35: Elongated. *42x60:* The primary has a close faint companion. *170x250:* A moderate distance unequal pair. Both white. A third star, extremely faint, makes a small evenly spaced arc with the double. [STF 2379]

| ga | NGC 6702 | 18 47.0 | +45 42 | Lyr | 13.2b | 1.8' | 1.3' | E |

dA: pretty faint, small, little extended.
[See 6703.]

| ga | NGC 6703 | 18 47.3 | +45 33 | Lyr | 12.3b | 2.3' | 2.3' | S0– |

dA: bright, small, round, much brighter middle.
170x250: 6702 and 6703 are half the field apart NNW-SSE. Both are evenly moderately concentrated. Round. Very small. No involved stars. 6703 is moderate brightness. It makes a shallow arc with two stars S and two stars N. 6702 is faint.

| ga | MCG+1-48-1 | 18 49.0 | +05 27 | Ser | 11.0 | 1.2' | 1.0' | – |

170x250: Not visible.

| pn | PK 51+9.1 | 18 49.7 | +20 50 | Her | 12.2p | 3.0" | | |

170x250: Not visible.

| bs | Beta Lyr | 18 50.1 | +33 21 | Lyr | 3.45 | | 0.00 | B7 |

170x250: White. It's on the side of a nice equilateral triangle of moderately bright stars.

| vs | S Sct | 18 50.3 | –07 54 | Sct | 7.3 | 9 | 2.93 | C5 |

12x35: Mag. 7.5. *170x250:* Deep copper red. A nice carbon star. It's in a fairly rich field of faint stars, giving a sense of foreground and depth.

| oc | NGC 6704 | 18 50.9 | –05 12 | Sct | 9.2 | 5.0' | | |

Wi: cluster, bright, 60 13th mag. stars.
42x60: Faintly visible. *170x250:* A little cluster of a few moderately faint stars plus threshold stars. Somewhat rectangular NE-SW. At its SW end is a bit of a concentration. A third of the field in size. Moderate density. Unimpressive.

| oc | NGC 6705/M11 | 18 51.1 | –06 16 | Sct | 5.8 | 13.0' | | |

JH: ! cluster, very bright, large, irregularly round, rich, one bright star and 11th mag. and fainter stars.
12x35: Bright. *42x60:* Resolved into a lot of stars, with one bright star on the E side. *170x250:* One bright star E of center and a whole lot of moderately faint stars. SSE is a relatively bright wide pair. There aren't a lot of fainter stars. The main body of the cluster is squarish, with sharp sides; two sides are very straight. Slightly elongated E-W. A little more than half the field in size. Very dense. Impressive.

| oc | NGC 6709 | 18 51.5 | +10 20 | Aql | 6.7 | 13.0' | | |

JH: cluster, pretty rich, little compressed, irregular figure.
12x35: Visible. *42x60:* A number of moderately faint and faint stars. *170x250:* A fair number of moderate brightness and fainter stars. The brighter stars give the cluster a somewhat triangular shape. 3/4 of the field in size. Moderate density.

| gc | NGC 6712 | 18 53.1 | −08 42 | Sct | 8.2 | 9.8' | −7.30 | 9 |

HH: *globular cluster, pretty bright, very large, irregular, very gradually little brighter middle, well resolved.*
12x35: Faintly visible. ***42x60:*** Easily visible. ***170x250:*** Resolved pretty much throughout into mostly very faint stars, except in the very middle, where it's more granular than resolved. Very broadly concentrated. A sixth of the field in size. Somewhat triangular. Somewhat irregular.

| pn | NGC 6720/M57 | 18 53.6 | +33 01 | Lyr | 9.7p | 1.8' | | |

JH: *!!! annular nebula, bright, pretty large, considerably extended (in Lyra).*
12x35: Visible. ***42x60:*** Annular: the hole is just detected. ***170x250:*** Bright, but not among the very brightest planetaries. No central star. A fairly clean disk. The hole isn't black; it's just fainter nebulosity. No blinking effect. Slightly oval NE-SW. Moderately large for a planetary. Slightly blue-greenish. Just off its edge E is a faint star; nearby [NNW] are a couple of extremely faint stars. An unusual and interesting planetary. [Ring Nebula]

| ds | STF 2452 | 18 53.6 | +75 47 | Dra | 5.7" | 6.6 | 7.4 | A1 − |

42x60: Extremely close. A faint double. ***170x250:*** A close slightly unequal pair. Both white.

| pn | IC 1295 | 18 54.6 | −08 50 | Sct | 15.0p | 90.0" | | |

170x250: Faint. No central star. No disk. Very slightly concentrated. No blinking effect. Roughly round. Moderately large for a planetary. It looks more like a diffuse galaxy than a planetary.

| oc | NGC 6716 | 18 54.6 | −19 52 | Sgr | 7.5 | 6.0' | | |

JH: *cluster, pretty rich, 9th to 13th mag. stars.*
12x35: Visible. ***42x60:*** A parallelogram-shaped cluster of stars. ***170x250:*** A fair number of moderately bright and moderate brightness stars. The brightest stars are in a very skewed NE-SW parallelogram. Somewhat empty in the middle. Half the field in size. Moderate density.

| gc | NGC 6715/M54 | 18 55.1 | −30 29 | Sgr | 7.7 | 12.0' | −9.41 | 3 |

JH: *globular cluster, very bright, large, round, gradually then suddenly much brighter middle, well resolved, 15th mag. stars.*

12x35: Visible. *42x60:* Sharply concentrated. *170x250:* Granular with averted vision. Evenly to sharply concentrated. Small. At its edge SSE are a couple of faint field stars; a little farther ENE is another faint field star. Pretty bright. Probably a massive globular.

| gc | NGC 6717 | 18 55.1 | −22 42 | Sgr | 8.4 | 5.4' | − | 8 |

HH: faint, small, partially resolved, cluster + nebula.
42x60: Extremely faintly visible. *170x250:* Not resolved. Either sharply concentrated or there's a star in the center. On either side of center WNW and NE is a clump of stars or an individual star. Small. Faint. Very irregular; it doesn't look like a normal globular cluster at all; an unusual object.

| bs | Sigma Sgr | 18 55.3 | −26 17 | Sgr | 2.00 | | −0.20 | B2 |

170x250: White.

| pn | PK 3-14.1 | 18 55.6 | −32 15 | Sgr | 10.9p | 4.0" | | |

170x250: Not visible.

| ds | Theta Ser | 18 56.2 | +04 12 | Ser | 22.6" | 4.5 | 5.4 | A5 A5 |

12x35: Nicely split. *42x60:* A moderately close equal pair. A nice double. *170x250:* A moderately wide equal pair. Both white. [STF 2417]

| bs | Gamma Lyr | 18 58.9 | +32 41 | Lyr | 3.24 | | −0.05 | B9 |

170x250: White.

| oc | NGC 6738 | 19 01.4 | +11 36 | Aql | 8.3 | 15.0' | | |

JH: cluster, poor, little compressed.
12x35: Two or three faint stars. *42x60:* Two or three moderate brightness stars and a few fainter stars. *170x250:* A number of moderately bright stars, some moderate brightness stars, and a few fainter stars. It takes up the whole field. Sparse. Not much of a cluster.

| pn | NGC 6741 | 19 02.6 | −00 27 | Aql | 10.8p | 8.0" | | |

Pi: planetary nebula, stellar.
170x250: Not visible.[†]

| bs | Zeta Sgr | 19 02.6 | −29 52 | Sgr | 2.60 | | 0.08 | A3 |

170x250: White.

| vs | V Aql | 19 04.4 | −05 41 | Aql | 6.6 | 8.1 | 4.19 | C5 |

12x35: Mag. 6.5. *170x250:* Copper red. It stands out in the field among a lot of fainter stars.

| gc | NGC 6749 | 19 05.3 | +01 54 | Aql | 11.1 | 4.0' | – | – |

JH: cluster, large, little compressed, bright and faint stars.
170x250: In the position are a number of faint stars in something of a concentration or a group. A globular might be behind these stars [correct], but it's not distinguishable.

| bs | Zeta Aql | 19 05.4 | +13 51 | Aql | 2.99 | | 0.01 | A0 |

170x250: White.

| pn | PK 3-17.1 | 19 05.6 | −33 11 | Sgr | 13.4p | 5.0" | | |

170x250: Not visible.

| pn | NGC 6751 | 19 05.9 | −05 59 | Aql | 12.5p | 26.0" | | |

Ma: pretty bright, small.
42x60: Very faintly visible. *170x250:* Somewhat faint. No central star. Not a clean disk or edge. Somewhat concentrated. Very slight blinking effect. Round. Moderately small for a planetary. Off its edge E is a faint star.

| bs | Lambda Aql | 19 06.2 | −04 52 | Aql | 3.44 | | −0.09 | B9 |

170x250: White.

| vs | R Aql | 19 06.4 | +08 14 | Aql | 5.3 | 12.0 | 1.60 | M5 |

12x35: Mag. 7.5. *170x250:* Slightly orangish.

| bs | Tau Sgr | 19 06.9 | −27 40 | Sgr | 3.32 | | 1.19 | K1 |

170x250: Yellow gold.

| oc | NGC 6755 | 19 07.8 | +04 14 | Aql | 7.5 | 14.0' | | |

HH: cluster, very large, very rich, pretty compressed, 12th to 14th mag. stars.
42x60: Faintly visible. ***170x250:*** A mostly empty square of very mixed brightness stars – moderately bright, moderately faint, and very faint – with a little concentration on the SW side. With the bright peripheral stars, which probably aren't members, it takes up the whole field; without them it's half the field in size. Sparse. It looks more like an unimpressive Milky Way concentration than a true cluster.

| ga | NGC 6764 | 19 08.3 | +50 56 | Cyg | 12.6b | 2.3' | 1.5' | Sbc |

Sw: pretty faint, pretty large, much extended, several very faint stars involved.
170x250: Faint. Diffuse. Roughly round. Moderate size. Near center are two or three very faint involved stars.

| oc | NGC 6756 | 19 08.7 | +04 41 | Aql | 10.6 | 4.0' | | |

HH: cluster, small, rich, little compressed, 11th to 12th mag. stars.
42x60: Faintly visible. ***170x250:*** A little concentration of faint stars, not well resolved. Very small; compact. Within it is an even smaller, unresolved concentration. Despite its density it doesn't look very much like a cluster. Both 6756 and 6755 look like Milky Way detachments peering out of dark nebulosity; they don't have the character of individual, self-contained star clusters.

| bs | Pi Sgr | 19 09.8 | –21 01 | Sgr | 2.89 | | 0.35 | F3 |

170x250: Soft white.

| pn | NGC 6765 | 19 11.1 | +30 32 | Lyr | 13.1p | 40.0" | | |

Ma: faint, small, extended.
170x250: Very faint. No central star. Ill-defined. Elongated NE-SW. Moderate size for a planetary. It doesn't look at all like a planetary; it looks more like a faint galaxy. SSW is a wide pair of stars; the planetary is inside a large triangle of stars.

| gc | NGC 6760 | 19 11.2 | +01 02 | Aql | 9.1 | 9.6' | –6.80 | 9 |

Hi: pretty bright, pretty large, very gradually little brighter middle.
42x60: Visible. ***170x250:*** Starting to be granular with averted vision. Broadly concentrated. Small. Faint for a globular. Very regular.

| ds | STF 2486 | 19 12.1 | +49 51 | Cyg | 7.5" | 6.6 | 6.8 | G4 G4 |

42x60: Slightly unequal. A faint double. *170x250:* A fairly close equal pair. Colorful: both yellow.

| bs | Delta Dra | 19 12.6 | +67 39 | Dra | 3.07 | | 1.00 | G9 |

170x250: Yellow white.

| pn | NGC 6772 | 19 14.6 | −02 42 | Aql | 14.2p | 84.0" | | |

HH: *very faint, large, round, very very little brighter middle, resolvable.*
170x250: Very faint; ghostly. No central star. No disk or clean edge; diffuse. Round. Moderately large for a planetary.

| pn | IC 4846 | 19 16.5 | −09 02 | Aql | 12.7p | 2.0" | | |

170x250: Not visible.

| oc | NGC 6774 | 19 16.6 | −16 16 | Sgr | − | 20' | | |

JH: *cluster, very large, little compressed.*
12x35, 42x60: A large cluster of mixed brightness stars: some scattered bright stars and more fainter stars. Elongated N-S in the shape of a long rectangle. A little less than half the field in size. Somewhat sparse to moderate density. *170x250:* A nice couple of fields of very widely scattered bright stars and more fainter stars. Sparse.

| gc | NGC 6779/M56 | 19 16.6 | +30 11 | Lyr | 8.3 | 8.8' | −7.35 | 10 |

JH: *globular cluster, bright, large, irregularly round, gradually very much compressed middle, well resolved, 11th to 14th mag. stars.*
12x35: Visible. *42x60:* Somewhat faint. *170x250:* Mostly resolved. A lot of the outliers are very distinct. Broadly concentrated. A little less than a quarter of the field in size. Somewhat faint. Somewhat irregular. It has a central mass but it's not an impressive globular.

| gc | TER 7 | 19 17.7 | −34 39 | Sgr | 12.0 | 1.2' | − | − |

170x250: A small exceedingly faint glow. It's among a number of stars; half the field N are the two plotted stars.

18 to 24 Hours: SUMMER

| gc | PAL 10 | 19 18.0 | +18 34 | Sge | 13.2 | 4.0' | – | 12 |

170x250: Not visible.

| pn | NGC 6778 | 19 18.4 | –01 35 | Aql | 13.3p | 16.0" | | |

Ma: small, extended, ill-defined disc.
42x60: Extremely faintly visible. *170x250:* Moderate brightness. A little disk but not a clean edge. Either it's concentrated or it has a very faint indistinct central star. Slight blinking effect. Round.

| pn | NGC 6781 | 19 18.4 | +06 32 | Aql | 11.8p | 1.8' | | |

HH: planetary nebula, faint, large, round, very suddenly brighter middle, disc, faint star NE.
170x250: Faint; ghostly. No central star. Not a sharp disk. At times with averted vision a small hole is seen in the middle, but not a very dark one; the planetary is not typically annular. The N edge is less distinct than the S edge. Round. Large for a planetary. Right at its edge ENE is a faint star. A nice planetary.

| vs | V1942 Sgr | 19 19.2 | –15 54 | Sgr | 6.7 | 7.1 | 2.31 | C4 |

12x35: Mag. 7.5. *170x250:* Copper red.

| oc | NGC 6791 | 19 20.9 | +37 46 | Lyr | 9.5 | 10.0' | | |

Wi: very faint.
42x60: A very faint cloud. *170x250:* A faint cluster of a lot of extremely faint and threshold stars. A little empty in the middle. Slightly elongated E-W. Half the field in size. Dense.

| ga | NGC 6792 | 19 21.0 | +43 07 | Lyr | 12.9p | 2.2' | 1.2' | Sb |

Lo: faint, extended NNE-SSW, gradually little brighter middle, 9.5 mag. star SE.
170x250: Very faint. Hardly concentrated. Very elongated NNE-SSW. Moderately small. No involved stars. Nearby NW is a fairly bright star.

| vs | UX Dra | 19 21.6 | +76 34 | Dra | 6.2 | 7.0 | 2.86 | C7 |

12x35: Mag. 6.5. *170x250:* Copper red. A nice bright carbon star.

| pn | NGC 6790 | 19 23.2 | +01 30 | Aql | 10.2p | 7.0" | | |

Pi: planetary nebula, bright, extremely small, stellar = 9.5 mag.
170x250: Stellar; it looks like a moderately bright star. Strong blinking effect. Very nearby W is a moderate brightness star.

| bs | Delta Aql | 19 25.5 | +03 06 | Aql | 3.40 | | 0.28 | F0 |

170x250: Soft white.

| oc | CR 399 | 19 26.2 | +20 06 | Vul | 3.6 | 60.0' | | |

12x35: Six stars in a nearly straight E-W line, with a hook consisting of four stars going S from near the middle of the line. The hook includes the three brightest stars of the cluster. A little less than half the field in size. Sparse. A nice group. [Brocchi's Cluster/Coathanger]

| oc | NGC 6800 | 19 27.2 | +25 08 | Vul | – | 5.0' | | |

HH: cluster, very large, pretty rich, very little compressed, 10th mag. and fainter stars.
12x35: A few faint stars. *42x60:* A cluster of a few faint stars with a hole in the middle. *170x250:* A fair number of moderate brightness stars, very uniform in brightness and evenly spaced out, plus some fainter stars. There's a hole in the middle. Elongated E-W. It takes up the whole field. Somewhat sparse. A somewhat unusual cluster.

| oc | NGC 6802 | 19 30.6 | 20 17 | Vul | 8.8 | 5.0' | | |

HH: cluster, large, very compressed, extended N-S, 14th to 18th mag. stars.
170x250: A faint cluster of a lot of very faint stars; a few are a little brighter and more distinct. Rectangular N-S. A quarter of the field in size. Moderately dense. An interesting cluster.

| bs | Beta Cyg | 19 30.7 | +27 57 | Cyg | 3.08 | | 1.13 | K3 |

12x35: Nicely split. *42x60:* A slightly wide unequal pair. *170x250:* A very wide unequal pair. Color contrast: the primary is yellow gold; the companion is white. [STF 43: 34.3", 3.1, 5.1, K3 B8]

| ga | UGC 11453 | 19 31.1 | +54 06 | Cyg | 12.9b | 1.8' | 1.3' | Sb |

170x250: Extremely faint. A slight diffuse glow. Moderate size. Around its edge are three very very faint involved stars; the galaxy is among a lot of brighter stars.

| pn | NGC 6803 | 19 31.3 | +10 03 | Aql | 11.3p | 6.0" | | |

Pi: planetary nebula, stellar.
170x250: Not visible.

| pn | NGC 6804 | 19 31.6 | +09 13 | Aql | 12.2p | 35.0" | | |

HH: considerably bright, small, irregularly round, well resolved.
170x250: Somewhat faint. It has a very faint central star. Not a clean disk or edge. Round. Moderate size for a planetary. Right at its edge NE is a star. It has a diffuse appearance compared to many other planetaries, which are brighter and sharper.

| vs | AQ Sgr | 19 34.3 | −16 22 | Sgr | 6.6 | 7.7 | 3.37 | C5 |

12x35: Mag. 7.5. ***170x250:*** Deep copper red. Pretty bright. NW, making a very large shallow arc with it, are a couple of white plotted stars – good comparison stars. One of the best carbon stars.

| pn | NGC 6807 | 19 34.6 | +05 40 | Aql | 13.8p | 2.0" | | |

Pi: planetary nebula, stellar.
170x250: Not visible.

| pn | PK 64+5.1 | 19 34.8 | +30 30 | Cyg | 9.6p | 35.0" | | |

170x250: Not visible.

| oc | NGC 6811 | 19 37.2 | +46 22 | Cyg | 6.8 | 12.0' | | |

JH: cluster, large, pretty rich, little compressed, 11th to 14th mag. stars.
12x35: Visible. ***42x60:*** Well resolved. There's something of a hole in the middle.
170x250: A fair number of moderate brightness stars, uniform in brightness. There's a void in the middle. Half the field in size. Moderate density.

| pn | PK 52-2.2 | 19 39.2 | +15 56 | Aql | 12.6p | 8.0" | | |

170x250: Not visible.

| gc | NGC 6809/M55 | 19 40.0 | −30 58 | Sgr | 7.0 | 19.0' | −6.85 | 11 |

JH: globular cluster, pretty bright, large, round, very rich, very gradually brighter middle, 12th to 15th mag. stars.

12x35: Easily visible. *42x60:* Granular to starting to be resolved. *170x250:* Resolved, with distinct stars throughout. Very broadly concentrated. With all outliers it's 2/3 of the field in size. Slightly irregular. Slightly loose. Although it concentrates toward the middle it doesn't have a well defined ball. An interesting and unusual globular: clearly very major and fairly massive, yet it's not bright. (225x: It has one of the broadest concentrations of all globulars: the very large core pretty much takes up the entire main body of the cluster, and it doesn't have many outliers.)

oc	NGC 6815	19 40.9	+26 51	Vul	–	–		

JH: *cluster, very large, pretty rich, little compressed, 10th to 15th mag. stars.*
12x35: Faintly visible. *42x60:* A few faint stars. *170x250:* Some 10–12 very widely spaced moderate brightness stars, with no concentration at all. 2/3 of the field in size. Sparse. Hardly a cluster.

vs	TT Cyg	19 41.0	+32 37	Cyg	7.8	9.1	3.14	C5

12x35: Mag. 8.0. *170x250:* Copper red.

oc	NGC 6819	19 41.3	+40 11	Cyg	7.3	5.0'		

JH: *cluster, very large, very rich, 11th to 15th mag. stars.*
12x35: Visible. *42x60:* A little cluster of faint stars. *170x250:* A good number of stars: moderate brightness, faint, and threshold. It's in the shape of an "H" with a low bar. A third of the field in size. Pretty dense.

gc	TER 8	19 41.8	–34 00	Sgr	12.4	3.5'	–	–

170x250: Exceedingly faint. Very small. TER8 is smaller and more concentrated than TER7 [the photographic appearances are the opposite of this; apparently only the core of TER8 was seen].

ga	NGC 6814	19 42.7	–10 19	Aql	12.1b	3.0'	2.7'	Sbc

HH: *pretty faint, pretty large, round, brighter middle, resolvable.*
170x250: Faint. Sharply concentrated to a substellar nucleus. Round. Small. The outer halo is vague. No involved stars. Nearby NW is a moderately faint star.

ga	UGC 11466	19 43.0	+45 18	Cyg	12.7p	2.0'	1.1'	S

170x250: Very very faint. Diffuse. Roughly round. Small. No involved stars. In line with it are a moderate brightness star nearby NNW and two faint stars nearby SE; E is another relatively bright star.

| oc | NGC 6823 | 19 43.1 | +23 18 | Vul | 7.1 | 12.0' | | |

HH: cluster, considerably rich, extended, 11th to 12th mag. stars.
12x35: A handful of stars in a little arc. ***42x60:*** There are more stars with averted vision. ***170x250:*** A sparse cluster of mixed brightness stars, with three or four in a little linear group in the very middle. A quarter of the field in size.

| ga | NGC 6824 | 19 43.7 | +56 06 | Cyg | 13.0b | 1.9' | 1.5' | Sb |

WH: pretty bright, irregular figure, brighter middle.
170x250: Moderately faint. Evenly moderately concentrated. Roughly round. Small. Very nearby S is a faint star; the galaxy makes a semicircle with several stars N, including a bright double.

| pn | NGC 6818 | 19 44.0 | −14 09 | Sgr | 9.9p | 48.0" | | |

HH: planetary nebula, bright, very small, round.
12x35: Visible as a faint star. ***42x60:*** Visible as a moderate brightness star. ***170x250:*** Very very bright. No central star. A perfectly round disk with a very clean edge. Perfectly even in light. Pretty strong blinking effect. Moderate size for a planetary. Blue-greenish. It's inside a small flat triangle of faint stars. A very elegant, cool planetary; superlative; one of the very best and brightest in the sky.

| pn | NGC 6826 | 19 44.8 | +50 31 | Cyg | 9.8p | 38.0" | | |

HH: planetary nebula, bright, pretty large, round, 11th mag. star in middle.
12x35, 42x60: Visible as a faint star. 16 Cyg [39.5", 6.0, 6.1] is nearby W. ***170x250:*** Very bright. It has a conspicuous moderate brightness central star. A good disk but a somewhat soft edge. Even in light. Very strong blinking effect. Round. Moderate size for a planetary. Slightly blue-greenish. Compared to 6543, 6826 is slightly less bright, it has a slightly less well defined disk, and it has a stronger blinking effect. [Blinking Planetary]

| ga | NGC 6822 | 19 44.9 | −14 48 | Sgr | 9.3b | 15.6' | 13.5' | Im |

Ba: very faint, large, extended, diffuse.
42x60: Extremely faintly visible. ***170x250:*** A very very faint diffuse glow. Elongated N-S. Half the field in size. Capping its N end is a half-circle of moderate brightness stars; there are other stars on its periphery but there are no conspicuous involved stars within the galaxy. [Barnard's Galaxy]

| bs | Delta Cyg | 19 45.0 | +45 07 | Cyg | 2.87 | | −0.03 | B9 |

170x250: The primary has a relatively very faint companion right next to it, with a tiny gap in between. The primary is white. A very nice double, one of the nicest with a sharp telescope and good seeing. [STF 2579: 2.5", 2.9, 6.3]

| gc | PAL 11 | 19 45.2 | −08 00 | Aql | 9.8 | 10.0' | − | 11 |

170x250: A very faint glow with some extremely faint stars resolved. A little less than a quarter of the field in size. It's surrounded by a semicircle of stars around its E side; NW is the bright plotted star.

| ds | STF 2578 | 19 45.7 | +36 05 | Cyg | 15.2" | 6.4 | 7.2 | B9 A0 |

12x35: Extremely close. *42x60:* A close slightly unequal pair. A faint double. *170x250:* A somewhat wide almost equal pair. Both white.

| bs | Gamma Aql | 19 46.3 | +10 36 | Aql | 2.72 | | 1.52 | K3 |

170x250: Yellow gold.

| ds | Epsilon Dra | 19 48.2 | +70 16 | Dra | 3.4" | 3.8 | 7.4 | G8 − |

170x250: The primary has a very very close relatively very faint companion. The primary is yellow white. [STF 2603]

| ga | ESO 526-7 | 19 49.6 | −26 25 | Sgr | 11.9 | 3.1' | 2.1' | Sbc |

170x250: Not visible.

| pn | NGC 6833 | 19 49.7 | +48 57 | Cyg | 13.8p | 2.0" | | |

Pi: planetary nebula, stellar.
170x250: Not visible.

| vs | Chi Cyg | 19 50.6 | +32 55 | Cyg | 3.6 | 14.2 | 1.82 | S6 |

12x35: Mag. 5.0. *170x250:* Orangish. Bright; fairly nice.

| bs | Alpha Aql | 19 50.8 | +08 52 | Aql | 0.77 | | 0.22 | A7 |

170x250: White. A brilliant star. [Altair]

| oc | NGC 6830 | 19 51.0 | +23 03 | Vul | 7.9 | 12.0' | | |

HH: cluster, large, pretty rich, pretty compressed, 11th to 12th mag. stars.
12x35: Visible. *42x60:* A little group of stars. *170x250:* 8–10 moderate brightness stars in an irregular "X" figure plus a few fainter stars. A third of the field in size. Sparse.

| oc | NGC 6834 | 19 52.2 | +29 25 | Cyg | 7.8 | 5.0' | | |

HH: cluster, poor, little compressed, 11th to 12th mag. stars.
12x35: Visible. *42x60:* A compact little cluster including a couple of brighter stars. *170x250:* One moderately bright star and 4–5 moderate brightness stars in an ENE-WSW linear group plus more faint stars in a concentration near the center. A third of the field in size. Moderate density.

| oc | HAR 20 | 19 53.2 | +18 20 | Sge | 7.7 | 6.0' | | |

12x35: Visible. *42x60:* An elongated group of stars. *170x250:* A group of some ten moderately bright and moderate brightness stars, elongated E-W. The two brightest stars are at the W end. Half the field in size. Not much of a cluster.

| oc | NGC 6837 | 19 53.4 | +11 41 | Aql | 12.0 | 3.0' | | |

HH: cluster, small, poor.
170x250: A little concentration of stars. It doesn't look like an independent cluster at all.

| gc | NGC 6838/M71 | 19 53.8 | +18 47 | Sge | 8.3 | 7.2' | –5.60 | – |

JH: cluster, very large, very rich, pretty much compressed, 11th to 16th mag. stars.
12x35, 42x60: Visible. *170x250:* Well resolved, with an underlying mass of threshold stars. Broadly concentrated, but it doesn't concentrate smoothly like a typical globular. A sixth of the field in size. No central ball. (225x: Triangular. Irregular. Loose. An unimpressive globular.)

| ds | 57 Aql | 19 54.6 | –08 14 | Aql | 35.6" | 5.8 | 6.5 | B7 B8 |

12x35: Split. *42x60:* A wide slightly unequal pair. *170x250:* A very wide very slightly unequal pair. Both white. [STF 2594]

| pn | NGC 6842 | 19 55.0 | +29 17 | Vul | 13.6p | 57.0" | | |

Ma: faint, pretty large, very little extended.

170x250: Very faint; low surface brightness. No central star. Practically no disk; somewhat diffuse. Round. Moderate size for a planetary. It looks more like a faint galaxy than a planetary.

| ds | Psi Cyg | 19 55.6 | +52 26 | Cyg | 3.0" | 4.9 | 7.4 | A4 F4 |

170x250: The primary has an extremely close fainter companion. The primary is yellowish. [STF 2605]

| vs | RR Sgr | 19 55.9 | –29 11 | Sgr | 5.7 | 14.0 | 1.08 | M5 |

12x35: Mag. 8.5. *170x250:* Not colored.

| ds | STF 2609 | 19 58.6 | +38 06 | Cyg | 2.0" | 6.6 | 7.7 | B5 – |

170x250: An extremely close slightly unequal pair. Both white.

| bs | Gamma Sge | 19 58.8 | +19 29 | Sge | 3.47 | | 1.57 | K5 |

170x250: Gold. One of the more strongly colored bright stars.

| pn | NGC 6853/M27 | 19 59.6 | +22 43 | Vul | 7.6p | 6.7' | | |

JH: !!! very bright, very large, binuclear, irregularly extended (Dumbbell Nebula). *12x35:* Easily visible. *42x60:* The major features are made out. *170x250:* Bright; its surface brightness is not as high as that of the very brightest planetaries, but for such a large object it's very high. The larger outer nebula is elongated WNW-ESE; within this is a brighter NNE-SSW hourglass or two-lobed concentration. The edges are indistinct, especially E-W. No blinking effect. Near but not quite in the center is a faint star [this is the central star]; within the nebula is another faint star or two; off the W corner of the S lobe is a relatively bright star. Extremely large for a planetary: half the field in size. A superlative planetary, without equal, and one of the most outstanding deep-sky objects of any kind. [Dumbbell Nebula]

| pn | NGC 6852 | 20 00.7 | +01 43 | Aql | 12.8p | 28.0" | | |

Ma: *faint nebula, among stars.*
170x250: Very faint. No central star. No disk. Roughly round. Moderately small for a planetary. Ill-defined; like a faint diffuse galaxy. At its edge W is a faint star.

| ga | NGC 6869 | 20 00.7 | +66 13 | Dra | 13.1b | 1.5' | 1.2' | S0 |

Sw: *pretty bright, pretty small, round.*

170x250: Faint. Broadly slightly concentrated. Roughly round. Small. No involved stars. Making a shallow arc with it are two stars SSE.

| ne | NGC 6857 | 20 01.8 | +33 31 | Cyg | | 38.0" | | – |

HH: faint, among Milky Way stars.
170x250: A relatively faint nebulosity with a few faint involved stars. Indistinct. Small. Very unimpressive.

| oc | NGC 6866 | 20 03.7 | +44 00 | Cyg | 7.6 | 10.0' | | |

HH: cluster, large, very rich, considerably compressed.
12x35: A faint little cluster. *42x60:* A few stars are resolved. *170x250:* A fair number of moderate brightness and moderately faint stars, fairly uniform in brightness. In the middle is a small N-S concentration. Overall the cluster is linear E-W. It spans the field. Moderate density.

| oc | NGC 6871 | 20 05.9 | +35 46 | Cyg | 5.2 | 20.0' | | |

JH: cluster, bright and faint stars, double star involved.
12x35, 42x60, 170x250: There's no cluster in the position. In the 35 mm and the 60 mm there's a small NNE-SSW arc of stars consisting of a sharp triangle of stars at the N end, a star in the middle, and two stars at the S end – but this doesn't amount to a cluster.

| gc | NGC 6864/M75 | 20 06.1 | –21 55 | Sgr | 8.6 | 6.8' | –8.30 | 1 |

JH: globular cluster, bright, pretty large, round, very much brighter middle to a bright nucleus, partially resolved.
12x35: Visible. *42x60:* Concentrated. Somewhat faint. *170x250:* Granular with averted vision. Sharply concentrated. Small. Very regular. (225x: It's surrounded by four faint stars in a semicircle S to NNE. M75 is very similar to M54: both are unresolved, pretty concentrated, bright, and small; but M75 is a little more concentrated, not quite as bright, and a little smaller.)

| oc | NGC 6874 | 20 07.8 | +38 14 | Cyg | 7.7 | 7.0' | | |

HH: cluster, poor, little compressed.
42x60: An inconspicuous little concentration of stars. *170x250:* A fair number of moderately faint stars in a Capricornus shape. Half the field in size. Moderate density. An undistinguished cluster.

| oc | IC 1311 | 20 10.3 | +41 13 | Cyg | 13.1 | 9.0' | | |

170x250: A very faint cloud: a few extremely faint to threshold stars and a bit of a glow of more stars. It's surrounded by brighter stars.

| pn | NGC 6884 | 20 10.4 | +46 27 | Cyg | 12.6p | 6.0" | | |

Cp: planetary nebula, stellar.
170x250: Moderate brightness. No central star. A tiny disk. Strong blinking effect. Round.

| pn | NGC 6879 | 20 10.5 | +16 54 | Sge | 13.0p | 5.0" | | |

Pi: planetary nebula, stellar = 10th mag.
170x250: Not visible.

| pn | NGC 6881 | 20 10.8 | +37 25 | Cyg | 14.3p | 4.0" | | |

Pi: planetary nebula, stellar.
170x250: Not visible.

| oc | NGC 6883 | 20 11.3 | +35 51 | Cyg | 8.0 | 14.0' | | |

JH: cluster, pretty rich, double star involved.
12x35, 42x60, 170x250: There's no cluster in the position.

| bs | Theta Aql | 20 11.3 | −00 49 | Aql | 3.23 | | −0.07 | B9 |

170x250: White.

| oc | NGC 6885 | 20 12.0 | +26 29 | Vul | 8.1 | 18.0' | | |

HH: cluster, very bright, very large, rich, little compressed, 6th to 11th mag. stars.
12x35: A few random stars. *42x60:* A ring of mixed brightness stars. Sparse.
170x250: On the E side is the very bright plotted star [20 Cyg], on the S side are a half-dozen scattered bright stars, and on the N side are a larger number of fainter stars. Roughly a ring; fairly empty in the middle. The field width in size. Somewhat sparse.

| ne | NGC 6888 | 20 12.0 | +38 21 | Cyg | | 18.0' | | E |

WH: faint, very large, very much extended, double star attached.

170x250: There's the slightest hint of a nebula around two of the bright plotted stars: a little bit of an arc from one star to the other. [On most occasions nothing is seen.] [Crescent Nebula]

| ds | STF 2644 | 20 12.6 | +00 52 | Aql | 2.7" | 6.9 | 7.2 | B9 – |

170x250: A very very close equal pair. Both white.

| pn | NGC 6886 | 20 12.7 | +19 59 | Sge | 12.2p | 6.0" | | |

Cp: planetary nebula, stellar = 10th mag.
170x250: Not visible.

| vs | RS Cyg | 20 13.4 | +38 44 | Cyg | 6.6 | 9.4 | 3.28 | C8 |

12x35: Mag. 7.5. *170x250:* Only slightly colored. The plotted star next to it N makes a good comparison star: there's clearly a difference in color but it's not large.

| pn | NGC 6891 | 20 15.2 | +12 42 | Del | 11.7p | 21.0" | | |

Cp: planetary nebula, stellar = 9.5 mag.
42x60: Bright. *170x250:* Bright. It has a very faint central star. A disk but not a clean edge. Pretty strong blinking effect. Round. Very small.

| pn | NGC 6894 | 20 16.4 | +30 34 | Cyg | 14.4p | 60.0" | | |

HH: !! annular nebula, faint, small, very very little extended.
170x250: Faint. No central star. Round but not disklike; no clean edge. Annular, especially with averted vision: a very fat donut with a relatively small hole in the middle; otherwise even in light. No blinking effect. Moderate size for a planetary. A ghostly and intriguing object.

| oc | IC 4996 | 20 16.5 | +37 39 | Cyg | 7.3 | 5.0' | | |

170x250: A little arc of three stars with a lot more faint stars surrounding it in a narrow NNE-SSW diamond figure. A quarter of the field in size. Moderately dense.

| vs | RT Cap | 20 17.1 | –21 19 | Cap | 6.5 | 8.1 | 4.02 | C6 |

12x35: Mag. 8.0. *170x250:* Copper red. A pretty good carbon star.

| ds | STF 2671 | 20 18.4 | +55 24 | Cyg | 3.7" | 6.1 | 7.5 | A2 A0 |

42x60: Slightly elongated. *170x250:* A very close unequal pair. Both white. A nice little double.

| vs | U Cyg | 20 19.6 | +47 54 | Cyg | 6.7 | 11 | 3.10 | C7 |

12x35: Mag. 7.5. *170x250:* Copper red. A fairly good carbon star. Right next to it NE is the white plotted star, which is equal in brightness and makes a good comparison star.

| pn | IC 4997 | 20 20.2 | +16 44 | Sge | 11.6p | 2.0" | | |

170x250: Moderately bright. It has a faint central star and the tiniest nebula. Very nearly stellar. Strong blinking effect. It makes a little arc with a moderately bright star SW – similar in brightness to the planetary – and a moderate brightness star farther WSW.

| bs | Beta 1 Cap | 20 21.0 | −14 46 | Cap | 3.08 | | 0.79 | G5 |

170x250: Yellow white. Making an arc with it are a second bright star W [Beta 2] and a fairly bright star SE. A nice scene.

| ga | NGC 6900 | 20 21.6 | −02 34 | Aql | 13.1 | 1.0' | 0.8' | Sb |

Ma: very faint, small, round.
170x250: Extremely faint. The slightest unconcentrated glow. Roughly round. Small. No involved stars. N is an E-W elongated group of stars.

| bs | Gamma Cyg | 20 22.2 | +40 15 | Cyg | 2.20 | | 0.68 | F8 |

170x250: Yellow white.

| ga | MCG-3-52-2 | 20 22.2 | −16 47 | Cap | 12.8 | 2.5' | 2.1' | Scd |

170x250: Not visible. The bright plotted star in the position [just N of the galaxy] is a very close double.

| pn | NGC 6905 | 20 22.4 | +20 06 | Del | 11.9p | 72.0" | | |

HH: !! planetary nebula, bright, pretty small, round, 4 faint stars near.
42x60: Faintly visible. It's among some stars. *170x250:* Moderately bright. No central star. A disk but not with a very clean edge. Even in light. Slight blinking effect.

Round. Moderate size for a planetary. It's just outside a small triangle of stars. A fairly nice planetary.

| oc | NGC 6910 | 20 23.1 | +40 47 | Cyg | 7.4 | 7.0' | | |

HH: cluster, pretty bright, pretty small, poor, pretty compressed, 10th to 12th mag. stars.
12x35: A little arc of three stars. *42x60:* Spanning two of the three stars is a smaller arc of very faint stars. *170x250:* Two bright stars and a handful of moderate brightness stars mainly in an arc between them. A quarter of the field in size. Somewhat sparse.

| ga | ESO 462-15 | 20 23.2 | −27 42 | Sgr | 13.0b | 1.7' | 1.3' | E3 |

170x250: Faint. Evenly moderately concentrated. Roughly round. Very small. No involved stars.

| ga | NGC 6906 | 20 23.6 | +06 26 | Aql | 13.2b | 1.6' | 0.7' | Sbc |

Ma: pretty faint, pretty large, round.
170x250: Faint. Broadly slightly concentrated. Elongated NE-SW. Moderately small. No involved stars.

| ga | NGC 6903 | 20 23.8 | −19 19 | Cap | 12.9p | 2.4' | 2.4' | S0− |

JH: considerably large, extended, brighter middle to a 17th mag. star, 10th mag. star attached.
170x250: Moderate brightness. Concentrated. Roughly round. Small. Near center is a bright involved star that interferes; E is a crooked E-W line of faint stars.

| oc | NGC 6913/M29 | 20 23.9 | +38 32 | Cyg | 6.6 | 6.0' | | |

JH: cluster, poor, little compressed, bright and faint stars.
12x35: A little group of stars. *42x60:* A rectangle of stars with appendages coming out of two of the four corners. A nice little figure. *170x250:* The cluster consists of the four corner stars of the rectangle, two or three more bright stars, and a handful of fainter stars among the brighter ones. A quarter of the field in size. Sparse.

| ga | NGC 6907 | 20 25.1 | −24 48 | Cap | 11.9b | 3.3' | 2.9' | Sbc |

HH: considerably faint, considerably large, very little extended, very gradually little brighter middle, resolvable, 3 stars W.
170x250: Faint. Broadly slightly concentrated. Slightly elongated NE-SW. Moderately small. No involved stars. A little ways E is a moderately bright star.

| ga | IC 5020 | 20 30.6 | −33 29 | Mic | 13.1b | 2.9' | 2.0' | Sbc |

170x250: Faint. Evenly moderately concentrated. Round. Very small. No involved stars. It's at the end of the hook of a N-S question mark-like figure of moderate brightness to faint stars.

| oc | NGC 6939 | 20 31.4 | +60 38 | Cep | 7.8 | 7.0' | | |

HH: cluster, pretty large, extremely rich, pretty compressed middle, 11th to 16th mag. stars.
42x60: Faintly visible. *170x250:* A good number of moderately faint and faint stars. The brightest stars are on the SW side. Slightly triangular in shape. Half the field in size. Somewhat compact; fairly dense.

| ga | NGC 6923 | 20 31.7 | −30 49 | Mic | 12.7b | 2.6' | 1.2' | Sb |

JH: pretty faint, considerably small, round, gradually brighter middle, between 2 stars.
170x250: Very faint to faint. Evenly moderately concentrated. Round. Very small. No involved stars. It makes a bent line with four faint stars: three in a row W and one SE. [E462-31 was not seen.]

| ga | ESO 597-6 | 20 32.3 | −19 43 | Cap | 13.0 | 1.8' | 1.3' | S0− |

170x250: Extremely faint. Slightly concentrated. Round. Very small. No involved stars.

| ga | NGC 6928 | 20 32.8 | +09 55 | Del | 13.2b | 2.2' | 0.6' | Sab |

Ma: pretty bright, pretty large, much extended.
170x250: Faint. Broadly slightly concentrated. Roughly round. Small. NE of center is a very faint involved star. [6927 (Ma) and 6930 (Ma) were not seen.]

| ga | NGC 6926 | 20 33.1 | −02 01 | Aql | 13.2b | 1.9' | 1.3' | Sbc |

HH: very faint, pretty large, extended N-S or binuclear, W of 2.
170x250: Very faint. Hardly concentrated. Elongated N-S. Moderately small. No involved stars. [6929 (JH) was not seen.]

| gc | NGC 6934 | 20 34.2 | +07 24 | Del | 8.9 | 7.1' | −7.34 | 8 |

HH: globular cluster, bright, large, round, well resolved, 16th mag. and fainter stars, 9th mag. star W.
12x35: Visible along with a star. *42x60:* Easily visible. Right next to it W is a moderately faint star. *170x250:* Partly resolved, with a granular middle. Broadly to

evenly concentrated. Small. Very regular, with a well defined central ball. Probably an intrinsically massive cluster. The star W is relatively bright.

| ga | NGC 6925 | 20 34.3 | −31 58 | Mic | 12.1b | 4.4' | 1.2' | Sbc |

JH: considerably bright, large, much extended N-S, pretty suddenly little brighter middle.
170x250: Moderately faint. Sharply moderately concentrated. Edge-on N-S. Moderately large. The ends fade out very indistinctly. No involved stars.

| oc | NGC 6940 | 20 34.6 | +28 19 | Vul | 6.3 | 31.0' | | |

WH: cluster, very bright, very large, very rich, considerably compressed, pretty bright stars.
12x35: A faint cloud of stars, with some members resolved. *42x60:* A faint cluster of a lot of faint stars. Half the field in size. *170x250:* A lot of stars, mostly moderate brightness and pretty uniform in brightness. Elongated E-W. A field and a half in size. Moderately dense. A fairly nice cluster.

| ga | NGC 6946 | 20 34.8 | +60 09 | Cyg | 9.6b | 11.6' | 9.8' | Scd |

HH: very faint, very large, very gradually then very suddenly brighter middle, partially resolved.
12x35: 6946 and open cluster 6939 are both visible. 6939 is a little brighter. *42x60:* Both objects are in the same field. *170x250:* 6946 is faint; low surface brightness. Hardly concentrated. Slightly elongated E-W. Large: a third of the field in size. There's some unevenness in the light. There are a couple of involved stars within the halo and a number of stars on its periphery; S is a triangle of bright stars.

| ga | NGC 6951 | 20 37.2 | +66 06 | Cep | 11.6b | 3.9' | 3.2' | Sbc |

Sw: pretty bright, pretty large, little extended.
170x250: Faint. Evenly moderately concentrated. Slightly elongated E-W. Moderately small. Off its E end is a star.

| ds | 49 Cyg | 20 41.0 | +32 18 | Cyg | 2.8" | 5.7 | 7.8 | G8 G8 |

170x250: The primary has a very very close relatively faint companion. The primary is yellowish. [STF 2716]

| bs | Alpha Cyg | 20 41.4 | +45 16 | Cyg | 1.25 | | 0.09 | A2 |

170x250: White. A brilliant star. [Deneb]

| ga | IC 5041 | 20 43.6 | −29 42 | Mic | 12.6v | 2.6' | 1.4' | Sd |

170x250: Extremely faint. Hardly concentrated. Elongated NE-SW. Moderately small to moderate size. No involved stars.

| ga | NGC 6956 | 20 43.9 | +12 30 | Del | 13.1p | 1.9' | 1.3' | Sb |

HH: very faint, small, stellar, double star attached.
170x250: Moderately faint. Broadly concentrated. Elongated E-W. Small. Near center is a bright involved star that interferes, with a very faint star right next to it E; the galaxy makes a sharp triangle with a couple of stars NNE.

| ga | MKN 509 | 20 44.2 | −10 43 | Aqr | 13.0 | 0.5' | 0.4' | − |

170x250: Not visible.

| bs | Eta Cep | 20 45.3 | +61 50 | Cep | 3.41 | | 0.94 | K0 |

170x250: Yellow.

| bs | Epsilon Cyg | 20 46.2 | +33 58 | Cyg | 2.50 | | 0.99 | K0 |

170x250: Yellow.

| ds | Gamma Del | 20 46.7 | +16 07 | Del | 9.5" | 4.5 | 5.5 | K1 A2 |

42x60: Nicely split. Slightly unequal. *170x250:* A moderately close very slightly unequal pair. Both yellowish. A very nice double. [STF 2727]

| ga | NGC 6962 | 20 47.3 | +00 19 | Aqr | 13.0b | 3.3' | 2.2' | Sab |

HH: considerably faint, small, round, brighter middle.
170x250: In the position are two galaxies next to each other NW-SE [6962, *6964 (HH)]. They're very similar. Both are faint. Sharply fairly concentrated. Roughly round. Very small. No involved stars. Very nearby SE of the SE galaxy is a faint star; [curving SW to SE of both galaxies is an arc of five moderately bright stars]. A little more than a third of the field NNE of both galaxies is another galaxy [6967]. Evenly fairly concentrated. Elongated E-W. Small. Just off its E end is a bright star. [6961 (LR) and 6965 (LR) were not seen.]

| ne | NGC 6960 | 20 47.5 | +30 43 | Cyg | | 60.0' | | E |

HH: !! pretty bright, considerably large, extremely irregular figure, Kappa Cyg involved.

170x250: The nebula is split in two by the bright plotted star [52 Cyg], which has a faint companion. The N part is pretty distinct. Narrow and winding; it ends right at a moderate brightness star, tapering to a sharp point. There are some fairly well defined dense strands, especially on the W edge. The S part is irregular and diffuse. The N part is a field and a half in length. The S part is at least as long, fading out indefinitely in a field of stars at its southern end. [Veil Nebula]

| ga | NGC 6967 | 20 47.6 | +00 24 | Aqr | 13.1v | 1.3' | 0.6' | S0+ |

LR: *extremely faint, very small, 10th mag. star 50" E.*
[See 6962.]

| ga | ESO 597-41 | 20 50.3 | −19 25 | Cap | 12.4 | 2.4' | 1.7' | Sb |

170x250: Not visible.

| ne | NGC 6979 | 20 51.0 | +32 09 | Cyg | | 28.0' | | E |

WH: *faint, small, irregularly extended, resolvable.*
42x60, 170x250: Not visible.

| gc | NGC 6981/M72 | 20 53.5 | −12 32 | Aqr | 9.4 | 6.6' | −6.94 | 9 |

JH: *globular cluster, pretty bright, pretty large, round, gradually much compressed middle, well resolved.*
12x35: Faintly visible. *42x60:* Visible. *170x250:* Partly resolved with averted vision. Broadly concentrated. Small. Very regular; smooth.

| ds | STF 2735 | 20 55.7 | +04 32 | Del | 2.0" | 6.1 | 7.6 | G6 − |

170x250: The primary has an extremely close fainter companion, almost touching it. Colorful: both yellow.

| ne | NGC 6992 | 20 56.4 | +31 43 | Cyg | | 80.0' | | E |

HH: *!! extremely faint, extremely large, extremely extended, extremely irregular figure, bifurcated.*
12x35, 42x60: Faintly visible. It spans the entire field in the 60 mm. *170x250:* There's a good deal of broad irregularity and unevenness in the light throughout. There are some narrow detached pieces of nebula in the N half, which is a little brighter than the S half, but there are no thin strands as in 6960. [Veil Nebula; 6992=6995]

| oc | NGC 6997 | 20 56.5 | +44 38 | Cyg | 10.0 | 8.0' | | |

WH: cluster, poor, little compressed, bright stars.
12x35: Extremely faintly visible. ***42x60:*** A few faint stars. ***170x250:*** A number of moderate brightness to moderately faint stars. Half the field in size. Somewhat sparse. [This cluster is within the North America Nebula.]

| ds | STF 2741 | 20 58.5 | +50 28 | Cyg | 2.2" | 5.9 | 7.2 | B5 – |

170x250: An extremely close slightly unequal pair. Both white.

| ne | NGC 7000 | 20 58.8 | +44 20 | Cyg | | 120' | | E |

HH: faint, excessively large, diffuse nebulosity.
12x35: A very ill-defined brightening. It looks more like a rich Milky Way star field – with a dark nebula intruding into it from the SW – than a regular nebula. Not at all impressive; hardly a distinct object at all. NE is an especially dense Milky Way star cloud. [North America Nebula]

| oc | NGC 6994/M73 | 20 59.0 | –12 38 | Aqr | 8.9 | 2.8' | | |

HH: cluster?, extremely poor, very little compressed, no nebula.
12x35: Visible as a faint star. ***42x60:*** A tiny triangle of faint stars. ***170x250:*** A triangular figure of four stars. The southernmost star is moderate brightness; the other three stars are moderately faint. Very small.

| ds | Epsilon Equ | 20 59.1 | +04 18 | Equ | 9.9" | 6.0 | 7.1 | F5 G0 |

42x60: Nicely split. ***170x250:*** A moderate distance unequal pair. Both white. [STF 2737]

| ga | ESO 401-25 | 20 59.3 | –32 41 | Mic | 11.5 | 1.5' | 1.1' | S0– |

170x250: Extremely faint. Slightly concentrated. Roughly round. Very small. No involved stars. SSE is a wide pair of stars; W is a bright star.

| ne | NGC 7023 | 21 00.5 | +68 10 | Cep | | 14.0' | | R |

WH: extremely faint, 7th mag. star in nebula?
12x35, 42x60: A star with faint diffuse nebulosity around it. ***170x250:*** An obvious diffuse nebula around the bright plotted star. Moderate size. Apparently a reflection nebula.

| pn | NGC 7008 | 21 00.6 | +54 32 | Cyg | 13.3p | 86.0" | | |

HH: considerably bright, large, extended NE-SW, resolvable, double star attached.
12x35: A faint star. ***42x60:*** A moderately faint star with some nebulosity around it. ***170x250:*** In the position is a wide unequal double star with a moderately bright primary and a moderately bright nebula next to it NNW. The nebula has a central star that is actually W of center; E of this star, still within the nebula, is an extremely faint star. Not a clean disk. Slightly annular. No blinking effect. Roughly round. Moderately large for a planetary. Just off its edge W is a faint star. An irregular planetary; an unusual object.

| gc | NGC 7006 | 21 01.5 | +16 11 | Del | 10.6 | 3.6' | −7.52 | 1 |

HH: bright, pretty large, round, gradually brighter middle.
42x60: Faintly visible. ***170x250:*** Not resolved. Broadly concentrated. Small. Faint.

| ga | MCG-2-53-23 | 21 03.0 | −14 15 | Aqr | 13.2 | 1.5' | 1.3' | Sab |

170x250: Not visible.

| ga | NGC 7013 | 21 03.6 | +29 53 | Cyg | 12.4b | 4.4' | 1.4' | S0/a |

HH: pretty bright, considerably small, round, pretty suddenly brighter middle, pretty bright star NW.
170x250: Moderately faint. Evenly moderately concentrated. Slightly elongated NNW-SSE. Small. No involved stars. A little ways NNW, in the direction of the galaxy's elongation, is a relatively bright star.

| ds | 12 Aqr | 21 04.1 | −05 49 | Aqr | 2.3" | 5.9 | 7.3 | G4 A3 |

170x250: The primary has an extremely close faint companion. The primary is yellowish. [STF 2745]

| pn | NGC 7009 | 21 04.2 | −11 22 | Aqr | 8.3p | 70.0" | | |

HH: !!! planetary nebula, very bright, small, elliptical.
12x35: It looks like a moderate brightness star. ***42x60:*** It looks like a fat moderately bright star. ***170x250:*** Very bright. No central star. A good disk. Even in light. Weak blinking effect. Slightly oval ENE-WSW. Moderately small to moderate size for a planetary. Slightly blue-greenish. Occasionally visible with averted vision, beyond the ends of the bright oval disk, are faint pointed nebulous extensions (not well defined ringlike appendages). [Saturn Nebula]

| pn | NGC 7026 | 21 06.3 | +47 51 | Cyg | 12.7p | 40.0" | | |

Bu: planetary nebula, pretty bright, binuclear.
42x60: Visible. *170x250:* Moderate brightness. No central star. Not a clean disk. Even in light. Fairly strong blinking effect. Roughly round. Small. Right next to it ENE is a moderate brightness star; together the two objects look like a double star.

| ds | 61 Cyg | 21 06.9 | +38 45 | Cyg | 30.4" | 5.2 | 6.0 | K5 K7 |

12x35: Nicely split. Slightly unequal. *42x60:* A very slightly unequal pair. *170x250:* A wide almost equal pair. Colorful: both yellow. [STF 2758]

| oc | NGC 7031 | 21 06.9 | +50 51 | Cyg | 9.1 | 15.0' | | |

HH: cluster of triple stars, little compressed.
12x35: Very faintly visible. *42x60:* A very small group of a handful of stars. *170x250:* Some ten stars of mixed brightness: moderate brightness to moderately faint. The main part of the cluster is a small NW-SE linear concentration, with a couple of stars out of line. SW are a couple more bright stars. A quarter of the field in size. Sparse. A messy little cluster.

| pn | NGC 7027 | 21 07.1 | +42 14 | Cyg | 10.4p | 60.0" | | |

St: planetary nebula, stellar = 8.5 mag.
12x35, 42x60: Easily visible as a faint star. *170x250:* Very bright; high surface brightness. No central star. A modest disk; not a very clean edge. Moderate blinking effect. With averted vision the nebula appears even in light; with direct vision it looks smaller and more concentrated. Slightly oblong NW-SE. Very small. Blue-greenish.

| ds | V389 Cyg | 21 08.6 | +30 12 | Cyg | 3.5" | 5.8 | 7.8 | B9 – |

170x250: The primary has a very close faint companion. The primary is white. [STF 2762]

| vs | T Cep | 21 09.5 | +68 29 | Cep | 5.3 | 10.9 | 1.49 | M5 |

12x35: Mag. 6.5. *170x250:* Orangish.

| ds | STF 2769 | 21 10.5 | +22 27 | Vul | 18.1" | 6.9 | 7.7 | A1 A0 |

12x35: Extremely close. *42x60:* A faint star with a fairly close fainter companion. *170x250:* A slightly wide slightly unequal pair. Both white.

| oc | NGC 7039 | 21 11.2 | +45 39 | Cyg | 7.6 | 24.0' | | |

JH: cluster, very large, pretty rich, extended, 10th mag. and fainter stars.
42x60: A number of faint stars. *170x250:* A good number of stars, mostly moderately faint. It spans the field between the two bright plotted stars, which are exactly a field width apart NE-SW. Moderate density. It doesn't stand out well in the rich field.

| oc | IC 1369 | 21 12.1 | +47 44 | Cyg | 8.8 | 4.0' | | |

12x35, 42x60: Extremely faintly visible. *170x250:* A square cluster of faint stars. Small. Moderate density. Not much of a cluster.

| oc | NGC 7044 | 21 12.9 | +42 29 | Cyg | 12.0 | 5.0' | | |

HH: cluster, very faint, pretty large, very rich, very compressed, 15th to 18th mag. stars.
42x60: Extremely faintly visible. *170x250:* A cloud of very faint and threshold stars with a handful of brighter stars overlaying these. Slightly elongated E-W. A quarter of the field in size. Fairly dense. Beyond the E end are some brighter field stars.

| bs | Zeta Cyg | 21 12.9 | +30 13 | Cyg | 3.20 | | 0.99 | G8 |

170x250: Yellow white.

| ga | NGC 7042 | 21 13.8 | +13 34 | Peg | 12.8p | 2.0' | 1.7' | Sb |

WH: very faint, small, round.
170x250: Very faint. Broadly very slightly concentrated. Roughly round. Small. No involved stars. [7043 (Ma) was not seen.]

| pn | NGC 7048 | 21 14.2 | +46 16 | Cyg | 11.3p | 61.0" | | |

St: pretty faint, pretty large, diffuse, irregularly round, very little brighter middle.
170x250: Faint. No central star. Not a clean disk. Even in light. No blinking effect. Roughly round. Moderately large for a planetary. It doesn't look like a typical planetary; it looks more like a small galaxy. SSE is a group of stars, including a moderately bright star at the planetary's edge.

| bs | Alpha Cep | 21 18.6 | +62 35 | Cep | 2.44 | | 0.22 | A7 |

170x250: White.

| ds | STT 437 | 21 20.8 | +32 27 | Cyg | 2.4" | 6.2 | 6.9 | G8 – |

170x250: A very very close almost equal pair. Both slightly yellowish.

| oc | NGC 7058 | 21 21.8 | +50 49 | Cyg | 9.0 | 10.0' | | |

JH: cluster, poor, little compressed.
12x35: Two or three stars. *42x60:* A sharp little triangle of stars. *170x250:* A sharp little triangle of bright stars with a few more fairly bright stars surrounding it in a semicircle [NE to SW]. The whole thing is a third of the field in size. Not really a cluster.

| oc | NGC 7062 | 21 23.2 | +46 23 | Cyg | 8.3 | 6.0' | | |

HH: cluster, pretty small, pretty rich, pretty compressed, 13th mag. and fainter stars.
12x35, 42x60: Faintly visible. *170x250:* A cluster in a diamond or parallelogram shape, slightly elongated E-W, with corner stars of moderate brightness and a fair number of moderately faint stars within the figure. A quarter of the field in size. Moderate density.

| oc | NGC 7067 | 21 24.2 | +48 01 | Cyg | 9.7 | 3.0' | | |

HH: cluster, poor, nebulous?
12x35, 42x60, 170x250: There's no clearly identifiable cluster in the position.

| oc | NGC 7063 | 21 24.5 | +36 30 | Cyg | 7.0 | 7.0' | | |

JH: cluster, poor, 10th mag. and fainter stars.
12x35: A handful of very faint stars. *42x60:* A small compact cluster of a few moderately faint stars. It looks a little like M29 except that the stars aren't nearly as bright. *170x250:* A compact cluster of eight moderate brightness stars of uniform brightness in a NW-SE rectangular figure, with few other stars. Half the field in size. Sparse.

| ga | NGC 7065A | 21 27.0 | –07 01 | Aqr | 13.1v | 1.4' | 1.4' | Sc |

170x250: Not visible.

| bs | Beta Cep | 21 28.7 | +70 34 | Cep | 3.19 | | –0.20 | B2 |

170x250: The primary has a moderate distance much fainter companion. Both white. [STF 2806: 13.3", 3.2, 7.9, B2 A2]

| oc | NGC 7082 | 21 29.4 | +47 05 | Cyg | 7.2 | 24.0' | | |

HH: cluster, large, considerably rich, little compressed, 10th to 13th mag. stars.
12x35, 42x60: An elongated cluster. ***170x250:*** A good number of stars. Elongated E-W. A field and a half in size. Moderate density. It looks more like a Milky Way star field than an actual cluster.

| gc | NGC 7078/M15 | 21 30.0 | +12 10 | Peg | 6.4 | 18.0' | −8.91 | 4 |

JH: ! globular cluster, very bright, very large, irregularly round, very suddenly much brighter middle, well resolved, very faint stars.
12x35, 42x60: Bright. ***170x250:*** Completely resolved except for the very center of the extremely dense core. Sharply concentrated; one of the most concentrated globulars. The core is rather small; the globular doesn't have a big central ball. It has a large field of outliers. With outliers it's half the field in size. Fairly regular. (225x: A lot of the stars outside of the very center are very distinct, although they're not as bright as those of some other major globulars. It has an exceptionally large envelope of very distinct outliers – the most outstanding field of outliers relative to the main body among all globulars. Unlike with globulars in the plane of the Milky Way, it's easily seen that these stars, relatively distant from the main body, are outliers and not field stars because the cluster is in a relatively starless section of the sky. A neat, elegant globular.)

| ga | NGC 7080 | 21 30.0 | +26 43 | Vul | 13.1b | 1.8' | 1.7' | Sb |

Ma: very faint, small, very little extended.
170x250: Very faint. Hardly concentrated. Roughly round. Small. At its edge NE is an involved star; nearby E are a couple more stars; all of these stars interfere.

| oc | NGC 7086 | 21 30.5 | +51 35 | Cyg | 8.4 | 9.0' | | |

HH: cluster, considerably large, very rich, pretty compressed, 11th to 16th mag. stars.
12x35: Visible. ***42x60:*** A little cluster of faint stars. ***170x250:*** A fair number of mixed brightness stars: moderate brightness to moderately faint; most are in a slight concentration on the S side. On the N side is a little group of four or five stars. Half the field in size. Moderate density.

| bs | Beta Aqr | 21 31.6 | −05 34 | Aqr | 2.91 | | 0.83 | G0 |

170x250: Yellow white.

| ga | UGC 11762 | 21 31.9 | +30 08 | Cyg | 13.0 | 0.1' | 0.1' | − |

170x250: Not visible.

| oc | NGC 7092/M39 | 21 32.2 | +48 27 | Cyg | 4.6 | 31.0' | | |

JH: cluster, very large, very poor, very little compressed, 7th to 10th mag. stars.
12x35: A loose group of some 18 distinct stars. *42x60:* A loose cluster of widely spaced moderately bright stars in the shape of a triangle. A little less than half the field in size. Sparse.

| pn | IC 5117 | 21 32.5 | +44 35 | Cyg | 13.3p | 1.0" | | |

170x250: Not visible.

| pn | PK 86-8.1 | 21 33.1 | +39 38 | Cyg | 12.7p | 32.0" | | |

170x250: Moderately bright. No central star. Moderate blinking effect. Tiny: more like a fat fuzzy star than a disk. It makes a triangle with two stars similar in brightness to it, one a little ways W and the other SSE.

| gc | NGC 7089/M2 | 21 33.5 | −00 49 | Aqr | 6.5 | 16.0' | −8.95 | 2 |

JH: !! globular cluster, bright, very large, gradually pretty much brighter middle, well resolved, extremely faint stars.
12x35: Easily visible. *42x60:* Starting to be granular. *170x250:* Resolved throughout into mostly faint stars except for the very middle of the core, which is too dense to be resolved; a lot of stars are distinct. Evenly concentrated to a dense central mass. With outliers it's a third of the field in size. Very regular; very symmetrical and smooth. Probably an intrinsically massive cluster: it seems far away but it's still very bright. Just off the edge of the core E is a slightly brighter star; otherwise the stars are very uniform in brightness. A beautiful, elegant globular; isolated and surrounded by black sky. (225x: The outliers are not very bright.)

| vs | S Cep | 21 35.2 | +78 37 | Cep | 7.4 | 12.9 | 4.09 | C7 |

12x35: Mag. 8.5. *170x250:* Copper red.

| ga | IC 1392 | 21 35.5 | +35 23 | Cyg | 12.5p | 1.5' | 1.2' | S0− |

170x250: Very faint. Slightly concentrated. Round. Very small. No involved stars. It's in a little NNE-SSW line of stars. A quarter of the field SE is another galaxy with an involved star [*U11775].

| ga | UGC 11781 | 21 36.7 | +35 41 | Cyg | 13.1p | 1.4' | 1.0' | S0 |

170x250: Not visible.

| oc | IC 1396 | 21 39.1 | +57 30 | Cep | 3.5 | 49.0' | | |

12x35: A moderately bright star and some half-dozen moderate brightness stars in a big "Y" opening up NNE; the brighter star is in the middle. A few fainter stars extend the group a little beyond the "Y" figure. Half the field in size. A Milky Way aggregation; not really a cluster. *42x60:* The middle star is actually a nice shallow little arc of three stars. The "object" is not a cluster at all.

| gc | NGC 7099/M30 | 21 40.4 | −23 11 | Cap | 7.5 | 12.0' | −7.10 | 5 |

JH: ! globular cluster, bright, large, little extended, gradually pretty much brighter middle, 12th to 16th mag. stars.
12x35: Easily visible. *42x60:* Fairly bright. Pretty concentrated. *170x250:* Resolved throughout, especially with averted vision; a number of stars are quite distinct. Evenly to sharply pretty concentrated. A quarter of the field in size. It has a central mass: a moderate-sized core. Coming out of the globular are two little lines of stars - the brightest stars of the cluster, assuming they're members: one line of three stars comes out of the center and goes NNW; a slightly longer line of four stars comes off the western edge of the main body and goes NW. A few bright field stars surrounding the cluster further add to the interesting scene. Somewhat irregular, but a pretty nice globular.

| oc | NGC 7129 | 21 41.3 | +66 06 | Cep | 11.5 | 8.0' | | |

HH: ! considerably faint, pretty large, gradually brighter middle to a triple star.
12x35: A couple of faint stars. *42x60:* A sharp little triangle of stars with a couple more stars SW. *170x250:* A small triangle of moderate brightness stars plus a number of moderately faint stars. Enveloping the triangle stars is faint nebulosity [7133]. The triangle is very small; altogether the cluster is half the field in size.

| vs | V460 Cyg | 21 42.0 | +35 31 | Cyg | 6.1 | 7.0 | 2.52 | C6 |

12x35: Mag. 6.5. *170x250:* Orangish.

| ne | NGC 7133 | 21 42.8 | +66 06 | Cep | | 9.5' | | R |

Bi: very faint, pretty large.
[See 7129.]

| vs | RV Cyg | 21 43.3 | +38 01 | Cyg | 7.1 | 9.3 | 3.7 | C6 |

12x35: Mag. 7.5. *170x250:* Light copper red.

| oc | NGC 7127 | 21 43.9 | +54 37 | Cyg | – | 2.8' | | |

JH: cluster, small, poor, little compressed.
42x60: Two little groups of stars next to each other. ***170x250:*** Two trapezoidal groups of stars aligned NW-SE, each consisting of some half-dozen moderate brightness stars. Together they're half the field in size. Sparse. Hardly a cluster, but cute.

| oc | NGC 7128 | 21 44.0 | +53 43 | Cyg | 9.7 | 3.1' | | |

HH: cluster, small, pretty rich, ruby star.
12x35: A couple of faint stars. ***42x60:*** A couple of distinct stars within a small cloud of stars. ***170x250:*** A very small compact cluster of 10–12 stars forming a rough circle. The brightest star is at the SE end. Somewhat sparse.

| bs | Epsilon Peg | 21 44.2 | +09 52 | Peg | 2.39 | | 1.53 | K2 |

170x250: Yellow gold.

| oc | NGC 7142 | 21 45.9 | +65 48 | Cep | 9.3 | 4.3' | | |

HH: cluster, considerably large, considerably rich, pretty compressed, 11th to 14th mag. stars.
170x250: A number of faint stars plus threshold stars. Along the N side is a long shallow arc of three brighter stars that probably are not members. Irregular. A little more than half the field in size. Somewhat sparse to moderate density.

| pn | NGC 7139 | 21 46.1 | +63 48 | Cep | 13.3p | 77.0" | | |

WH: very faint, considerably small, round, resolvable.
170x250: Very faint. A diffuse glow. No central star. Roughly round. Moderately large for a planetary. It looks like a faint galaxy. Off its edge SE is a faint star.

| gc | PAL 12 | 21 46.6 | –21 15 | Cap | 11.7 | 2.9' | –4.30 | 12 |

170x250: Not visible.

| bs | Delta Cap | 21 47.0 | –16 07 | Cap | 2.87 | | 0.29 | A |

170x250: Soft white.

| ga | NGC 7137 | 21 48.2 | +22 09 | Peg | 13.1b | 1.6' | 1.6' | Sc |

HH: faint, pretty small, round, very gradually little brighter middle, resolvable.

170x250: Very faint. Hardly concentrated. Roughly round. Small. At its edge WNW is an extremely faint star.

| ga | NGC 7130 | 21 48.3 | −34 57 | PsA | 13.0p | 1.7' | 1.6' | Sa |

JH: pretty bright, small, round, gradually little brighter middle.
170x250: Moderately faint. Evenly moderately concentrated. Round. Very small. No involved stars.

| ga | NGC 7135 | 21 49.8 | −34 52 | PsA | 12.7b | 2.9' | 1.9' | S0− |

JH: pretty bright, pretty large, round, very gradually brighter middle, 14th mag. star attached W.
170x250: Very faint to faint. Broadly slightly concentrated. Round. Small. Just W of center is a very faint involved star; the galaxy is inside a very large narrow WNW-ESE pentagonal figure of moderate brightness stars.

| ds | STF 2840 | 21 52.0 | +55 48 | Cep | 18.1" | 5.5 | 7.3 | B6 A1 |

12x35: Extremely close. *42x60:* A slightly unequal pair. *170x250:* A wide slightly unequal pair. Both white.

| oc | NGC 7160 | 21 53.7 | +62 36 | Cep | 6.1 | 7.0' | | |

HH: cluster, poor, very little compressed.
12x35: A sharp little triangle of stars. *42x60:* There are a few more faint stars. *170x250:* A sharp triangular figure of stars including two bright stars and three or four moderate brightness stars, plus a dozen fainter stars surrounding the figure. A third of the field in size. Sparse.

| ga | NGC 7156 | 21 54.6 | +02 56 | Peg | 13.1b | 1.6' | 1.3' | Scd |

HH: faint, pretty large, round, brighter middle, resolvable.
170x250: Faint. Broadly very slightly concentrated. Roughly round. Moderately small. No involved stars.

| ga | NGC 7154 | 21 55.4 | −34 48 | PsA | 12.9b | 2.4' | 1.8' | Sm |

JH: bright, pretty large, irregularly round, gradually little brighter middle, resolvable.
170x250: Extremely faint. Broadly very slightly concentrated. Round. Small. No involved stars. Nearby WNW is an extremely faint star; a little ways N is a slightly brighter star.

| ga | ESO 404-12 | 21 57.1 | −34 34 | PsA | 12.9p | 2.2' | 1.7' | Sc |

170x250: Moderately faint. Broadly slightly concentrated. Round. Small. No involved stars. A little ways WNW is a moderately faint star; E is a moderate brightness star; NE is another moderate brightness star.

| ga | NGC 7167 | 22 00.5 | −24 37 | Aqr | 13.2p | 1.8' | 1.5' | Sc |

JH: faint, pretty small, round, very gradually little brighter middle, 10th mag. star E.
170x250: Extremely faint. Hardly concentrated. Roughly round. Small. No involved stars. Nearby E is a moderately bright star.

| ga | NGC 7177 | 22 00.7 | +17 44 | Peg | 12.0b | 3.1' | 2.0' | Sb |

HH: pretty bright, pretty small, round, brighter middle to a nucleus, resolvable, star SW.
170x250: Moderate brightness. Evenly moderately concentrated. Roughly round. Small. No involved stars. Nearby SSW is a faint star.

| ga | NGC 7171 | 22 01.0 | −13 16 | Aqr | 12.9b | 2.6' | 1.5' | Sb |

HH: very faint, considerably large, extended NW-SE, very gradually brighter middle.
170x250: Very faint. Hardly concentrated. Slightly elongated NW-SE. Moderately small. No involved stars. NE is an ENE-WSW line of three evenly spaced moderate brightness stars.

| ga | NGC 7172 | 22 02.0 | −31 52 | PsA | 12.9b | 2.3' | 1.2' | Sa |

JH: pretty bright, pretty large, little extended, gradually brighter middle, 1st of 4.
[See 7176.]

| ga | NGC 7173 | 22 02.0 | −31 58 | PsA | 13.0b | 1.3' | 1.0' | E+ |

JH: considerably bright, considerably small, round, suddenly brighter middle to a star, 2nd of 4.
[See 7176.]

| ga | NGC 7176 | 22 02.1 | −31 59 | PsA | 12.3b | 1.2' | 1.2' | E |

JH: bright, pretty large, round, E of double nebula, 4th of 4.
170x250: 7173 and 7176 are a close pair aligned NW-SE. 7176 is moderate brightness. Sharply very concentrated. Very small. It has an extension going WSW, possibly a tiny companion galaxy [*7174 (JH)]. 7173 is moderately faint to moderate brightness.

Sharply very concentrated. Round. Tiny. A little less than half the field N is 7172, which is larger and more diffuse than the other galaxies. Faint. Broadly very slightly concentrated. Slightly elongated E-W. Moderately small. A little ways SW are two moderately faint stars making a slightly curved line with it; a little ways SE is a moderate brightness star. There are no involved stars with any of the galaxies.

| ga | NGC 7183 | 22 02.4 | −18 54 | Aqr | 12.9p | 3.8' | 1.1' | S0+ |

HH: very faint, pretty small, very little extended E-W, little brighter middle.
170x250: Faint. Broadly slightly concentrated. Elongated ENE-WSW. Moderate size. No involved stars.

| ga | NGC 7184 | 22 02.7 | −20 48 | Aqr | 11.7b | 6.0' | 1.4' | Sc |

HH: pretty bright, pretty large, much extended ENE-WSW, between 3 stars, easily resolvable.
170x250: Moderately faint. Evenly moderately concentrated. Edge-on NE-SW. Moderately large. At its NE end is a moderately faint star; a little ways W are two stars.

| ga | NGC 7185 | 22 02.9 | −20 28 | Aqr | 12.5v | 2.4' | 1.5' | S0− |

JH: very faint, pretty large, irregularly round, very gradually little brighter middle, E of 2.
170x250: Very faint. Evenly moderately concentrated. Roughly round. Small. No involved stars. W is a NNW-SSE line of three widely spaced stars. Half the field WSW is a brighter galaxy [*7180 (HH)]. Moderately faint. Evenly moderately concentrated. Slightly elongated ENE-WSW. Small. No involved stars.

| ga | IC 5156 | 22 03.3 | −33 50 | PsA | 13.0b | 2.2' | 0.7' | Sab |

170x250: Moderately faint. Broadly moderately concentrated. Elongated N-S. Small. No involved stars. It makes a triangle with two moderately faint stars NW.

| ga | IC 5157 | 22 03.4 | −34 56 | PsA | 12.7v | 1.6' | 1.4' | E |

170x250: Faint. Evenly moderately concentrated. Round. Very small. No involved stars. E to NNE is a curved line of three moderate brightness stars.

| ds | Xi Cep | 22 03.8 | +64 38 | Cep | 7.8" | 4.4 | 6.5 | A A3 |

42x60: The primary has a close fainter companion. ***170x250:*** The primary has a fairly close fainter companion. The primary is yellow white; the companion is yellow. [STF 2863]

| oc | NGC 7209 | 22 05.2 | +46 30 | Lac | 7.7 | 24.0' | | |

HH: cluster, large, considerably rich, pretty compressed, 9th to 12th mag. stars.
12x35: A lot of faint stars. ***42x60:*** The main body of the cluster is in an arc or a wide "V" shape. The stars are fairly faint. ***170x250:*** A lot of widely scattered moderate brightness stars of very uniform brightness plus some fainter stars. It takes up the whole field. Moderately dense.

| bs | Alpha Aqr | 22 05.8 | –00 19 | Aqr | 2.90 | | 1.04 | G2 |

170x250: Yellow.

| vs | RZ Peg | 22 05.9 | +33 30 | Peg | 7.7 | 13.5 | 2.86 | C9 |

12x35: Mag. 8.5. ***170x250:*** Hardly colored.

| ga | NGC 7217 | 22 07.9 | +31 21 | Peg | 11.0b | 3.9' | 3.2' | Sab |

HH: bright, pretty large, gradually brighter middle, easily resolvable.
42x60: Visible. ***170x250:*** Bright. Sharply pretty concentrated. Round. Small to moderately small. No involved stars.

| ga | UGC 11920 | 22 08.5 | +48 26 | Lac | 12.9p | 2.4' | 1.5' | S0/a |

170x250: Moderately faint. Very concentrated. Round. Tiny. No involved stars. It's surrounded by faint stars; N is the bright plotted star; S is a moderate brightness star.

| ga | NGC 7218 | 22 10.2 | –16 39 | Aqr | 12.7p | 2.7' | 1.3' | Scd |

HH: pretty bright, little extended, resolvable.
170x250: Faint. Broadly slightly concentrated. Elongated NNW-SSE. Moderately small. At its N end is a very faint involved star; nearby E is a slightly brighter star.

| ga | NGC 7223 | 22 10.2 | +41 01 | Lac | 13.0p | 1.6' | 1.1' | Sbc |

HH: extremely faint, pretty small, little extended, resolvable, among 3 stars.
170x250: Very faint. Small. N and S of center are two or three faint involved stars that severely interfere; S is a large triangular group of six moderately bright stars.

| oc | NGC 7226 | 22 10.4 | +55 23 | Cep | 9.6 | 1.8' | | |

Ho: pretty bright, large, in cluster.
170x250: A little cloud of threshold stars: a very small, dense concentration within a rich field of stars. At its edge NW are two moderate brightness stars.

| oc | IC 1434 | 22 10.5 | +52 50 | Lac | 9.0 | 7.0' | | |

12x35: Visible. *42x60:* Three moderate brightness stars and the glow of more stars, mostly N. *170x250:* Four stars in an irregular diamond shape plus a lot of moderately faint and fainter stars. Elongated N-S. 2/3 of the field in size. Very dense.

| ds | STF 2883 | 22 10.6 | +70 08 | Cep | 14.6" | 5.6 | 7.6 | F2 – |

42x60: The primary has a faint companion. *170x250:* The primary has a somewhat wide fainter companion. The primary is yellowish.

| bs | Zeta Cep | 22 10.9 | +58 12 | Cep | 3.35 | | 1.57 | K1 |

170x250: Yellow gold. Nearby (but not a companion) is another fairly bright star.

| ga | NGC 7221 | 22 11.3 | –30 33 | PsA | 12.8b | 2.0' | 1.5' | Sbc |

JH: faint, small, round, gradually brighter middle, resolvable, 2 very faint stars near.
170x250: Very faint to faint. Broadly slightly concentrated. Slightly elongated N-S. Small. NNW of center is an extremely faint involved star; nearby NNE is a faint star; nearby SSE is a very faint star; N are three brighter stars in an E-W arc.

| oc | NGC 7235 | 22 12.6 | +57 17 | Cep | 7.7 | 4.0' | | |

JH: cluster, pretty compressed, ruby 10th mag. star.
12x35, 42x60: Visible. *170x250:* Some half-dozen moderate brightness stars and a few fainter stars. Triangular. Small. Sparse.

| ga | NGC 7225 | 22 13.1 | –26 08 | PsA | 13.2p | 2.0' | 1.0' | S0/a |

JH: pretty faint, small, little extended, brighter middle.
170x250: Very faint. Broadly to evenly slightly concentrated. Slightly elongated NW-SE. Small. No involved stars. A little ways NW is either a faint double star or possibly a tiny very faint galaxy [the former is the case].

| ga | NGC 7229 | 22 14.0 | −29 23 | PsA | 13.1b | 1.8' | 1.5' | Sc |

JH: faint, pretty large, round, very gradually little brighter middle.
170x250: Exceedingly faint. The slightest diffuse glow. Small. No involved stars. Nearby E is a faint star; further E is the bright plotted star.

| ds | 41 Aqr | 22 14.3 | −21 04 | Aqr | 5.1" | 5.6 | 7.1 | K0 F2 |

42x60: Extremely close: as close as can be in the 60 mm and still be split. *170x250:* A very very close unequal pair. Slight color contrast: the primary is yellowish; the companion is slightly bluish by comparison. [H 56]

| oc | NGC 7243 | 22 15.3 | +49 53 | Lac | 6.4 | 21.0' | | |

HH: cluster, large, poor, little compressed, very bright stars.
42x60: A loose cluster with the brighter stars in an ENE-WSW central bar. A third of the field in size. Very irregular. *170x250:* A good number of widely scattered moderately bright stars plus fainter stars. A field and a half in size. Moderately dense.

| oc | NGC 7245 | 22 15.3 | +54 20 | Lac | 9.2 | 5.0' | | |

HH: cluster, compressed, extremely faint stars.
170x250: A fair number of faint stars inside a triangle of fairly bright stars. Very small. Fairly dense. A Milky Way concentration.

| ga | IC 1438 | 22 16.5 | −21 25 | Aqr | 12.6p | 2.4' | 2.0' | Sa |

170x250: Moderately faint. Evenly to sharply fairly concentrated. Round. Very small. No involved stars. W is a nearly square trapezoid of stars; NNW is a little triangle of stars. [I1439 was not seen.]

| oc | IC 1442 | 22 16.5 | +54 03 | Lac | 9.1 | 5.0' | | |

42x60: A number of very faint stars next to two widely spaced bright stars. *170x250:* Mixed brightness stars: the two brighter stars, some ten moderately bright stars, and more moderately faint stars. The brighter stars are on the periphery, mostly E; the fainter stars are concentrated toward the middle. Somewhat triangular overall. Half the field in size. Moderate density.

| ga | UGC 11973 | 22 16.8 | +41 30 | Lac | 12.9p | 3.2' | 0.8' | Sbc |

170x250: Not visible.

| ga | NGC 7250 | 22 18.3 | +40 33 | Lac | 13.2b | 1.7' | 0.7' | Sdm |

WH: very faint, small, much extended NNW-SSE.
170x250: Very faint. Broadly slightly concentrated. Elongated N-S. Small. Off its S end is a moderately bright star.

| oc | NGC 7261 | 22 20.4 | +58 05 | Cep | 8.4 | 5.0' | | |

JH: cluster, large, pretty rich, little compressed.
42x60: One moderate brightness star and a few faint stars. ***170x250:*** Half a dozen moderate brightness stars and about ten faint stars; the brightest star is at the cluster's edge SE. Oval N-S. A third of the field in size. Sparse.

| ga | NGC 7251 | 22 20.5 | –15 46 | Aqr | 12.6 | 1.9' | 1.6' | Sa |

HH: faint, pretty small, round, gradually pretty much brighter middle.
170x250: Very very faint. Broadly slightly concentrated. Round. Small. No involved stars.

| ga | NGC 7252 | 22 20.7 | –24 40 | Aqr | 12.1v | 3.8' | 2.4' | S0 |

HH: faint, small, round, easily resolvable.
170x250: Moderately faint. Evenly to sharply fairly concentrated. Round. Very small. No involved stars.

| ds | STF 2903 | 22 21.8 | +66 42 | Cep | 4.2" | 6.6 | 7.3 | G5 A2 |

170x250: A very close slightly unequal pair. Slight color contrast: the primary is yellowish; the companion is bluish.

| ga | NGC 7265 | 22 22.5 | +36 12 | Lac | 13.2b | 2.4' | 1.9' | S0– |

St: faint, very small, round, much brighter middle.
170x250: Faint. Evenly moderately concentrated. Round. Very small. No involved stars. Nearby ESE is a little linear figure of four moderate brightness stars.

| pn | IC 5217 | 22 23.9 | +50 58 | Lac | 12.6p | 7.0" | | |

170x250: Very high surface brightness. Almost stellar: a slightly fat star. Pretty strong blinking effect. It's in a pointed oval figure with seven moderately bright stars N, E, and S that are similar to it in brightness; the southern point star is the brightest.

| ga | NGC 7267 | 22 24.4 | –33 41 | PsA | 12.9b | 1.6' | 1.3' | Sa |

JH: considerably bright, pretty small, very little extended, gradually little brighter middle, bright triple star SW.
170x250: Faint. Broadly slightly concentrated. Slightly elongated WNW-ESE. Small. No involved stars. WSW is a little right triangle of relatively bright stars, including the bright plotted star.

| oc | NGC 7281 | 22 24.7 | +57 51 | Cep | – | 12.0' | | |

JH: cluster, large, pretty rich, little compressed, 10th to 16th mag. stars.
12x35, 42x60: Mainly a little row of three stars: a perfect line of evenly spaced moderate brightness stars. ***170x250:*** The cluster fills the field with irregularly scattered stars, but it's not much of a cluster.

| ga | NGC 7280 | 22 26.5 | +16 08 | Peg | 13.0b | 2.2' | 1.5' | S0+ |

HH: faint, considerably small, round, gradually brighter middle to a faint star, 3 stars near.
170x250: Moderately faint. Sharply fairly concentrated. Round. Very small. No involved stars. It makes a kite-like figure with three moderately faint stars a little ways NNE, the westernmost of which is double.

| ds | 53 Aqr | 22 26.6 | –16 45 | Aqr | 1.8" | 6.4 | 6.6 | G1 G2 |

170x250: Not split. [SHJ 345]

| oc | NGC 7296 | 22 28.2 | +52 18 | Lac | 9.7 | 4.0' | | |

WH: cluster, irregularly round, little compressed, very faint stars.
42x60: A little cloud of stars right next to a moderate brightness star. ***170x250:*** A number of moderate brightness and fainter stars in a N-S club shape, with the bright plotted star (which probably is not a member) at the S point. A third of the field in size. Moderate density.

| ga | NGC 7292 | 22 28.4 | +30 17 | Peg | 13.0b | 2.1' | 1.6' | Im |

St: extremely faint, small, oval, faint star involved.
170x250: Faint. Broadly very slightly concentrated. Roughly round. Small. No involved stars. Off its edge NW is a little group of faint stars.

| ga | NGC 7284 | 22 28.6 | −24 50 | Aqr | 12.9p | 2.1' | 1.4' | S0 |

HH: considerably faint, considerably small, little extended, resolvable, double star involved.

170x250: 7285 and 7284 are a very close double galaxy, with the components tagged right next to each other ENE-WSW; they almost make one object. 7284 is moderately faint. Sharply pretty concentrated. Round. Very very small. 7285 is a little fainter, more diffuse, and a little larger. Together the two galaxies are moderately small. No involved stars.

| ga | NGC 7285 | 22 28.6 | −24 50 | Aqr | 12.8p | 2.3' | 1.3' | Sa |

Ls: nebulous star.
[See 7284.]

| ds | Zeta Aqr | 22 28.8 | −00 01 | Aqr | 2.3" | 4.3 | 4.5 | F1 F5 |

170x250: An extremely close equal pair. Both white. [STF 2909]

| pn | NGC 7293 | 22 29.6 | −20 47 | Aqr | 7.5p | 16.0' | | |

Ha: ! pretty faint, very large, extended or binuclear.

12x35: Visible. *42x60:* Faint and diffuse. Along its W side is a very narrow NNW-SSE quadrilateral of stars. *170x250:* A diffuse glow. Roughly in the middle are a couple of widely spaced stars aligned E-W; the W one is apparently the central star. The planetary darkens in the middle: the nebulosity there is less bright but not black. The NW sector of the outer ring is a little more tenuous than the rest of the ring. Round. Extremely large for a planetary: a little more than half the field in size. Right at its edge W is a sharp triangle of stars pointing NNW; further W is the very long quadrilateral of stars seen in the 60 mm, the SE corner of which is a double. [Helix Nebula]

| ga | IC 1447 | 22 30.0 | −05 07 | Aqr | 12.0 | 1.4' | 0.7' | Sb |

170x250: Not visible.

| ga | NGC 7302 | 22 32.4 | −14 07 | Aqr | 13.2b | 1.7' | 1.0' | S0− |

HH: faint, pretty small, round, very suddenly brighter middle to a faint small round nucleus.

170x250: Moderately faint to moderate brightness. Sharply very concentrated. Round. Very small. No involved stars.

ga	NGC 7309	22 34.3	–10 21	Aqr	13.0b	1.9'	1.7'	Sc

HH: very faint, pretty large, round, gradually little brighter middle, resolvable.
170x250: Very faint. Hardly concentrated. Round. Small. No involved stars.

ga	NGC 7314	22 35.8	–26 02	PsA	11.6b	5.0'	2.1'	Sbc

JH: considerably bright, large, much extended N-S, very little brighter middle.
170x250: Moderately faint. Broadly very slightly concentrated. Very elongated to nearly edge-on N-S. Moderately large. No involved stars. A little ways W is a star that makes a crooked line with more stars SW. [7313 (Ma) was not seen.]

ds	8 Lac	22 35.9	+39 38	Lac	22.6"	5.7	6.5	B1 B2

12x35: Split. *42x60:* Very slightly unequal. *170x250:* A slightly wide almost equal pair. Both white. The two components of the double star make an arc with two more stars, faint and moderately faint, but these are undoubtedly unrelated background stars. [STF 2922]

ga	NGC 7319	22 36.1	+33 58	Peg	13.1v	1.5'	1.1'	Sbc

St: extremely faint, extremely small.
[See 7320.]

ga	NGC 7320	22 36.1	+33 56	Peg	13.2b	2.3'	1.1'	Sd

St: faint, very small.
170x250: In the position are two galaxies next to each other NW-SE [*7318 (St), 7320]. The NW galaxy is brighter. Both are very faint. Broadly slightly concentrated. Roughly round. Very small. No involved stars. WSW of the two galaxies are some faint stars in a line parallel to them. [7318 consists of 7318A and 7318B; they were seen as one object; 7317 (St) and 7319 were not seen.] [Stephan's Quintet]

ga	NGC 7330	22 36.9	+38 32	Lac	13.2p	1.2'	1.2'	E

St: pretty bright, small, little extended, brighter middle.
170x250: Faint. Evenly moderately concentrated. Round. Very small. No involved stars. Nearby WNW is a moderately faint star.

ga	NGC 7331	22 37.1	+34 25	Peg	9.4v	14.5'	3.7'	Sb

HH: bright, pretty large, pretty much extended NNW-SSE, suddenly much brighter middle.

12x35: Faintly visible. *42x60:* Visible. *170x250:* Bright. Sharply pretty concentrated. Elongated NNW-SSE. Moderately large. The long ends fade out very very indistinctly. No involved stars. A little less than a quarter of the field ENE is a small very faint galaxy [*7335 (HH)]. Looking around for additional galaxies known to be in the area: half the field E is a galaxy [*7340 (LR)]. A little ways NNE are two relatively bright stars in an evenly spaced line with it. A quarter of the field ESE is another galaxy [*7337 (LR)] with faint stars involved.

| ga | NGC 7332 | 22 37.4 | +23 47 | Peg | 12.0b | 4.0' | 1.1' | S0 |

HH: considerably bright, small, much extended NNW-SSE, very suddenly much brighter middle to an 11th mag. star, W of 2.

170x250: 7339 and 7332 are a third of the field apart E-W. 7332 is much brighter. Moderately bright. Sharply very concentrated. A thin edge-on, NNW-SSE. Moderately small. No involved stars. A little ways SSE is a moderately bright star. 7339 is faint. Broadly slightly concentrated. Also a thin edge-on, E-W, pointing toward 7332. Moderately small; a little larger than 7332. No involved stars. A good example of the two types of edge-on galaxies right next to each other: concentrated and even-light; a really neat contrasting pair.

| ga | NGC 7339 | 22 37.8 | +23 47 | Peg | 13.1b | 3.0' | 0.7' | Sbc |

HH: faint, pretty small, much extended E-W, very gradually little brighter middle, E of 2.
[See 7332.]

| pn | NGC 7354 | 22 40.4 | +61 16 | Cep | 12.9p | 36.0" | | |

HH: bright, small, round, pretty gradually very little brighter middle, easily resolvable.

42x60: Very faintly visible. *170x250:* Moderately faint. No central star. Not a very clean disk; not a very well defined edge. Round. Moderately small for a planetary. Very nearby W are three very faint stars in a little arc curving around that side of the planetary.

| ga | NGC 7351 | 22 41.4 | −04 26 | Aqr | 13.1 | 1.7' | 1.2' | S0 |

St: pretty faint, pretty small, round, brighter middle, resolvable.

170x250: Very faint. Broadly slightly concentrated. Elongated to edge-on N-S. Small. No involved stars.

| bs | Zeta Peg | 22 41.5 | +10 49 | Peg | 3.40 | | −0.09 | B8 |

170x250: White.

| ga | NGC 7361 | 22 42.3 | −30 03 | PsA | 12.7b | 3.8' | 0.9' | Sc |

JH: faint, pretty large, very much extended N-S, very gradually very little brighter middle.
170x250: Extremely faint. Hardly concentrated. Edge-on N-S. Moderately large. No involved stars. A little ways S is a moderately faint star.

| pn | IC 1454 | 22 42.6 | +80 26 | Cep | 14.8p | 32.0" | | |

170x250: Very faint. No central star. No disk. Roughly round. Moderately small for a planetary. Right at its edge NE is a pair of stars; nearby SE is another star. It looks more like a galaxy than like a typical planetary.

| bs | Eta Peg | 22 43.0 | +30 13 | Peg | 3.00 | | 0.80 | G8 |

170x250: Yellow.

| ds | STF 2935 | 22 43.1 | −08 19 | Aqr | 2.5" | 6.9 | 7.9 | A5 − |

170x250: The primary has a very very close fainter companion. The primary is white.

| ga | NGC 7371 | 22 46.1 | −11 00 | Aqr | 12.3b | 2.0' | 1.9' | S0/a |

HH: very faint, pretty large, round, little brighter middle.
170x250: Faint. Evenly slightly concentrated. Round. Small. No involved stars. A little ways SE is a pair of faint stars.

| oc | NGC 7380 | 22 47.0 | +58 06 | Cep | 7.2 | 12.0' | | |

HH: cluster, pretty large, pretty rich, little compressed, 9th to 13th mag. stars.
42x60: A little group of stars. ***170x250:*** A fair number of moderately bright stars of uniform brightness plus some fainter stars. Triangular. Half the field in size. Somewhat sparse.

| ga | NGC 7377 | 22 47.8 | −22 18 | Aqr | 12.1b | 2.9' | 2.4' | S0+ |

HH: pretty bright, small, very little extended, very gradually much brighter middle, faint star near.
170x250: Moderate brightness. Evenly moderately concentrated. Round. Small. No involved stars. SW is a group of five stars.

| ga | NGC 7385 | 22 49.9 | +11 36 | Peg | 12.0v | 2.5' | 2.0' | E |

HH: considerably faint, small, round, gradually little brighter middle, 11th mag. star NW.
[See 7386.]

| bs | Mu Peg | 22 50.0 | +24 36 | Peg | 3.48 | | 0.93 | G8 |

170x250: Yellow.

| ga | NGC 7386 | 22 50.0 | +11 41 | Peg | 12.3v | 2.6' | 1.8' | S0 |

HH: considerably faint, small, round, pretty gradually brighter middle, E of 2.
170x250: In the position are four galaxies occupying a quarter of the field. The two brightest, 7386 and 7385, are aligned NNE-SSW. Both are faint. Roughly round. Very small. No involved stars. 7386 is sharply fairly concentrated. 7385 is evenly to sharply fairly concentrated; not quite as concentrated as 7386. Next to it NW is a bright star. E and SE of these two galaxies, making a large quadrilateral with them, are two small extremely faint galaxies [*7387 (LR), *7389 (LR)]. [7383 (LR) was not seen.]

| ga | NGC 7391 | 22 50.6 | -01 32 | Aqr | 13.1b | 2.0' | 1.8' | E |

HH: considerably faint, considerably small, round, suddenly brighter middle to a 13th mag. star, star NW.
170x250: Moderately faint. Sharply pretty concentrated. Round. Very small. No involved stars. Nearby N is a moderately faint star.

| ga | NGC 7392 | 22 51.8 | -20 36 | Aqr | 12.6b | 2.1' | 1.2' | Sbc |

HH: pretty bright, pretty small, little extended WNW-ESE, much brighter middle.
170x250: Moderate brightness. Evenly to sharply fairly concentrated. Nearly edge-on WNW-ESE. Small. No involved stars. A little ways N are two stars, moderately faint and faint; further N is a brighter star; a little ways E is a moderately faint star.

| oc | NGC 7419 | 22 54.3 | +60 50 | Cep | 13.0 | 6.0' | | |

HH: cluster, pretty rich, considerably compressed.
42x60: A very small very faint cloud. *170x250:* A very small compact cluster of a fair number of faint stars. At its edge NW is a moderately bright star; further NW is the bright plotted star. Pretty dense.

| bs | Delta Aqr | 22 54.7 | −15 49 | Aqr | 3.27 | | 0.05 | A3 |

170x250: White.

| ga | NGC 7416 | 22 55.7 | −05 29 | Aqr | 13.2b | 3.1' | 0.8' | Sb |

Ma: faint, pretty large, pretty much extended, very gradually brighter middle.
170x250: Very faint. Broadly very slightly concentrated. Very elongated WNW-ESE. Small. No involved stars. A little ways ENE is a moderately faint star.

| pn | PK 107-2.1 | 22 56.3 | +57 09 | Cep | 14.0p | 8.0" | | |

170x250: Not visible.

| bs | Alpha PsA | 22 57.7 | −29 37 | PsA | 1.16 | | 0.09 | A3 |

170x250: White. [Fomalhaut]

| ga | IC 5271 | 22 58.0 | −33 44 | PsA | 11.6v | 2.6' | 0.8' | Sb |

170x250: Moderately faint. Broadly slightly concentrated. Elongated NW-SE. Moderately small to moderate size. No involved stars. A little ways NNE is a moderately faint star.

| ga | NGC 7443 | 23 00.1 | −12 48 | Aqr | 12.6v | 1.4' | 0.5' | S0+ |

HH: faint, very small, very little extended, suddenly much brighter middle, easily resolvable, [N] of 2.
170x250: In the position are two galaxies close together N-S [7443, *7444 (HH)]. They're pretty similar. The N galaxy is a little brighter. Both are moderately faint. Evenly moderately concentrated. Small. No involved stars. The N galaxy is slightly elongated NE-SW. The S galaxy is slightly elongated N-S. In line with it, nearby SE, are a very faint star and a moderately faint star.

| ga | NGC 7448 | 23 00.1 | +15 58 | Peg | 11.6v | 2.5' | 1.2' | Sbc |

HH: pretty bright, large, extended N-S, very gradually brighter middle.
170x250: Moderate brightness. Broadly moderately concentrated. Slightly elongated N-S. Moderately small. No involved stars. It's in a WNW-ESE line of stars.

| ga | NGC 7457 | 23 01.0 | +30 08 | Peg | 12.1b | 4.3' | 2.3' | S0– |

HH: considerably bright, considerably large, little extended, gradually much brighter middle, resolvable, 2 faint stars N.
170x250: Moderate brightness. Evenly moderately concentrated. Elongated NW-SE. Moderately small. No involved stars. Nearby SSW is a pair of moderately faint stars aligned parallel to the galaxy; ENE is a triangle of moderately bright stars.

| ga | NGC 7454 | 23 01.1 | +16 23 | Peg | 11.8v | 2.0' | 1.4' | E4 |

WH: faint, considerably small, little extended, little brighter middle, pretty bright star W.
170x250: Moderately faint. Evenly moderately concentrated. Round. Small. No involved stars. It makes a very small triangle with two stars NW, moderate brightness and faint.

| ga | NGC 7465 | 23 02.0 | +15 57 | Peg | 12.6v | 2.2' | 1.8' | S0 |

HH: very faint, very small, E of 2.
170x250: In the position are three galaxies in a short WNW-ESE line [*7463 (HH), *7464 (Ma), 7465]. The westernmost galaxy is the most obvious. Faint. Broadly slightly concentrated. Elongated E-W. Moderate size. No involved stars. Right next to it SE is a tiny companion galaxy. A little ways ESE is the third galaxy. Moderate brightness; high surface brightness. Sharply very concentrated to a stellar nucleus. Round. Tiny. No involved stars.

| ga | NGC 7469 | 23 03.3 | +08 52 | Peg | 13.0b | 1.4' | 1.0' | Sa |

HH: very faint, very small, very suddenly much brighter middle to a 12th mag. star.
170x250: Moderate brightness; high surface brightness. Sharply extremely concentrated. Round. Very very small. No involved stars. It's just off the side of a very large right triangle of bright stars. [I5283 was not seen.]

| bs | Beta Peg | 23 03.8 | +28 04 | Peg | 2.42 | | 1.67 | M2 |

170x250: Gold. A deeply colored star; very pretty.

| bs | Alpha Peg | 23 04.8 | +15 12 | Peg | 2.49 | | –0.04 | B9 |

170x250: White.

| ga | NGC 7479 | 23 04.9 | +12 19 | Peg | 11.6b | 4.1' | 3.1' | Sc |

HH: pretty bright, considerably large, much extended NNE-SSW, between 2 stars.
170x250: Moderately faint. Broadly slightly concentrated. Elongated N-S. Moderate size. At its N end is a faint star; very nearby SW of center is a very faint star.

| ga | NGC 7490 | 23 07.4 | +32 22 | Peg | 13.1p | 2.7' | 2.5' | Sbc |

St: very faint, very small, irregularly round, little brighter middle.
170x250: Very faint. Broadly slightly concentrated. Round. Small. No involved stars. Nearby N is a moderately faint star; this star and at least three others form a long shallow E-W arc of stars curving around the galaxy's N side.

| ds | STF 2978 | 23 07.5 | +32 50 | Peg | 8.4" | 6.3 | 7.5 | A4 – |

42x60: An unequal pair. *170x250:* A moderately close unequal pair. The primary is white.

| oc | K 19 | 23 08.3 | +60 31 | Cep | 9.2 | 6.0' | | |

42x60: A couple of faint stars. *170x250:* A couple of moderate brightness stars and a few moderately faint stars. Small. Not really a cluster; just a few stars in a small grouping.

| gc | NGC 7492 | 23 08.4 | –15 37 | Aqr | 11.5 | 4.2' | –5.20 | 12 |

HH: extremely faint, large, between 2 double stars.
170x250: Not resolved. Hardly concentrated. A little less than a quarter of the field in size. Extremely faint; a slight round glow. It's between a moderately faint star WNW and another moderately faint star E, this one with a faint star next to it SSE.

| ga | NGC 7497 | 23 09.1 | +18 10 | Peg | 13.0b | 4.8' | 1.1' | Sd |

HH: very faint, large, pretty much extended NE-SW, little brighter middle.
170x250: Very faint. Broadly very slightly concentrated. Edge-on NE-SW. Moderate size to moderately large. The ends are very indistinct. No involved stars.

| oc | NGC 7510 | 23 11.5 | +60 34 | Cep | 7.9 | 4.0' | | |

HH: cluster, pretty rich, pretty compressed, fan-shaped, pretty bright stars.
12x35, 42x60: A dense little linear group of stars. *170x250:* A dense group of moderately bright stars plus some fainter stars. Triangular overall, with an ENE-WSW linear concentration on the S side. Small.

18 to 24 Hours: SUMMER

| ga | NGC 7507 | 23 12.1 | −28 32 | Scl | 11.4b | 2.7' | 2.6' | E0 |

HH: pretty bright, considerably small, round, pretty suddenly very much brighter middle, 10th mag. star NW.
170x250: Bright. Sharply very concentrated. Round. Small. No involved stars.

| ga | NGC 7515 | 23 12.8 | +12 40 | Peg | 13.2b | 1.7' | 1.6' | S |

HH: faint, considerably small, round, very gradually little brighter middle, resolvable.
170x250: Faint. Broadly slightly concentrated. Round. Small. No involved stars. It makes a triangle with two faint stars, one a little ways SW and one SSE.

| ga | NGC 7513 | 23 13.2 | −28 21 | Scl | 11.4v | 3.1' | 2.0' | Sb |

Ma: very faint, pretty large, extended, gradually brighter middle.
170x250: Very faint. Hardly concentrated. Slightly elongated ENE-WSW. Small. No involved stars. SSE is the bright plotted star along with a second star.

| ne | NGC 7538 | 23 13.5 | +61 31 | Cep | | 9.7' | | E |

WH: very faint, large, 2 pretty bright stars involved.
42x60: Very faintly visible. ***170x250:*** A pretty conspicuous but somewhat faint nebula with two moderate brightness stars involved, one of which may be the central star. Round. Somewhat diffuse. Small.

| ga | UGC 12433 | 23 13.7 | +49 40 | And | 13.2 | 1.8' | 1.8' | S0/a |

170x250: Not visible.

| ga | NGC 7530 | 23 14.2 | −02 46 | Psc | 13.1 | 0.9' | 0.5' | S0/a |

Ma: extremely faint, very small, almost stellar.
170x250: [The observed galaxy is *7532 (Ma); 7530 (Ma) was not seen.] Faint. Sharply concentrated. Slightly elongated NNW-SSE. Very small. No involved stars. A little ways S is a bright star.

| ga | NGC 7541 | 23 14.7 | +04 32 | Psc | 12.4b | 3.5' | 1.2' | Sbc |

HH: bright, large, much extended E-W, much brighter middle, NE of 2.
170x250: Moderately faint. Broadly slightly concentrated. Edge-on E-W. Moderate size. Just off its E end is a moderately faint star. A sixth of the field SW is another

galaxy [*7537 (HH)] – a smaller version of 7541. Faint. Broadly slightly concentrated. Edge-on ENE-WSW, not quite parallel to 7541. Small; half the size of 7541. No involved stars.

| ga | NGC 7550 | 23 15.3 | +18 57 | Peg | 13.2b | 1.3' | 1.2' | S0– |

HH: considerably faint, small, round, SE of 2.
170x250: Moderately faint. Evenly moderately concentrated. Round. Very small. No involved stars. It makes a sharp isosceles triangle with a faint star a little ways WSW and a moderately faint star S. A sixth of the field WNW is a very faint galaxy [*7547 (JH)]. Small. A quarter of the field N is another very faint galaxy [*7549 (LR)]. Small, but a little larger than either of the first two galaxies. Nearby WNW is a moderately bright star. [7558 (Ma) was not seen.]

| ga | NGC 7556 | 23 15.7 | –02 22 | Psc | 12.5 | 2.4' | 1.6' | S0– |

HH: considerably faint, pretty large, round, bright double star E.
170x250: Very faint. Broadly slightly concentrated. Elongated WNW-ESE. Small. SW of center is a faint involved star; at each end of the galaxy is a very faint star [the W "star" is *7554 (Ma), a stellar galaxy].

| ga | NGC 7562 | 23 16.0 | +06 41 | Psc | 12.6b | 2.2' | 1.4' | E2-3 |

HH: considerably bright, pretty small, irregularly round, pretty suddenly brighter middle.
170x250: Moderately faint to moderate brightness. Sharply fairly concentrated. Round. Very small. No involved stars. [7557 (LR) was not seen.]

| ga | NGC 7596 | 23 17.2 | –06 54 | Aqr | 12.9 | 1.0' | 0.5' | S0 |

Le: very faint, pretty small, little extended N-S, little brighter middle to a nucleus.
170x250: Not visible.

| ga | NGC 7585 | 23 18.0 | –04 39 | Aqr | 12.3b | 3.0' | 2.5' | S0+ |

HH: pretty bright, pretty small, irregularly round, gradually brighter middle.
170x250: Moderately bright. Sharply pretty concentrated. Round. Small. No involved stars.

| ds | Omicron Cep | 23 18.6 | +68 07 | Cep | 3.4" | 4.9 | 7.1 | K0 F6 |

170x250: The primary has an extremely close much fainter companion. The primary is light yellow. [STF 3001]

| ga | NGC 7600 | 23 18.9 | −07 34 | Aqr | 12.9b | 2.7' | 1.1' | S0− |

HH: considerably faint, small, round, pretty suddenly pretty much brighter middle.
170x250: Moderate brightness. Sharply fairly concentrated. Elongated ENE-WSW. Small. No involved stars.

| ds | 94 Aqr | 23 19.1 | −13 28 | Aqr | 12.4" | 5.3 | 7.3 | G5 K2 |

42x60: The primary has a faint companion. ***170x250:*** The primary has a moderately close fainter companion. The primary is yellow. [STF 2998]

| ga | NGC 7606 | 23 19.1 | −08 29 | Aqr | 11.5b | 5.7' | 2.2' | Sb |

HH: pretty faint, considerably large, pretty much extended N-S.
170x250: Moderately faint. Evenly moderately concentrated. Very elongated NW-SE. Moderate size. No involved stars.

| ga | NGC 7611 | 23 19.6 | +08 03 | Psc | 12.5v | 1.8' | 0.7' | S0+ |

dA: faint, small, round, forms a triangle with 2 19th mag. stars N.
[See 7619.]

| ga | NGC 7619 | 23 20.2 | +08 12 | Peg | 11.0v | 2.5' | 2.3' | E |

HH: considerably bright, pretty small, round, pretty suddenly brighter middle.
170x250: Moderately bright. Sharply pretty concentrated. Round. Small. No involved stars. A little less than half the field E is 7626. Moderate brightness. Sharply fairly concentrated. Round. Small. Very nearby W is a faint star. A little more than half the field further E is a faint diffuse galaxy [*7631 (LR)]. A little more than half the field NNW of 7626 is a small concentrated galaxy [*7623 (HH)]. 2/3 of a field SW of 7619 is 7611. Moderately faint. Sharply very concentrated to a substellar nucleus. Very very small. No involved stars. [7617 (dA) was not seen.]†

| ga | NGC 7625 | 23 20.5 | +17 13 | Peg | 12.8b | 1.5' | 1.3' | Sa |

HH: pretty bright, considerably small, round, suddenly much brighter middle.
170x250: Moderate brightness to moderately bright. Evenly fairly concentrated. Round. Very small. No involved stars.

| ga | NGC 7626 | 23 20.7 | +08 13 | Peg | 11.1v | 2.2' | 2.0' | E |

HH: considerably bright, pretty small, round, pretty suddenly brighter middle.
[See 7619.]

| ne | NGC 7635 | 23 20.7 | +61 12 | Cas | | 16.0' | | E |

HH: very faint, 9th mag. star involved, little eccentric.
170x250: A small faint nebula surrounding the bright plotted star, making it look foggy. It looks like a reflection nebula [it's classified as an emission nebula]. Round. [Bubble Nebula]

| ga | NGC 7640 | 23 22.1 | +40 50 | And | 11.9b | 11.6' | 1.9' | Sc |

HH: considerably faint, large, much extended N-S, very little brighter middle, resolvable.
170x250: Very faint. Hardly concentrated. Elongated NNW-SSE. Moderately large. SE of center is a faint involved star; the galaxy is inside a triangle of moderate brightness stars; it doesn't quite reach the NW corner star.

| oc | NGC 7654/M52 | 23 24.2 | +61 35 | Cas | 6.9 | 12.0' | | |

JH: cluster, large, rich, much compressed middle, round, 9th to 13th mag. stars.
12x35: Visible. **42x60:** A nice dense cluster. **170x250:** A whole lot of stars: moderate brightness stars of uniform brightness plus faint stars. At its edge W is a brighter star. Slightly elongated E-W. With outer members it takes up the entire field. Very dense. A very nice cluster.

| ga | NGC 7653 | 23 24.8 | +15 16 | Peg | 12.6v | 1.6' | 1.5' | Sb |

JH: very faint, pretty small, round, gradually brighter middle.
170x250: Faint. Evenly moderately concentrated. Roughly round. Very small. No involved stars. A little less than half the field SE is an extremely faint galaxy [*U12590]. Very small.

| pn | NGC 7662 | 23 25.9 | +42 32 | And | 9.2p | 37.0" | | |

HH: !!! planetary nebula, very bright, pretty small, round, blue.
12x35: It looks just like the surrounding 8th magnitude stars. **42x60:** A tiny disk. **170x250:** Very bright. No central star. At times with averted vision there's a bit of a very faint diffuse halo around it, but otherwise it's a perfect even-light disk with a very clean edge – as clean a disk as is seen in a planetary. Moderate blinking effect. Perfectly round and symmetrical; extremely regular. Moderately small for a planetary. Slightly bluish. Nearby ENE is a faint star. A classic planetary; one of the best and brightest.

| pn | PK 111-2.1 | 23 26.3 | +58 10 | Cas | 14.0p | 1.0" | | |

170x250: Not visible.

| ga | NGC 7673 | 23 27.7 | +23 35 | Peg | 13.2b | 1.3' | 1.1' | Sc |

Ma: faint, small, round.
170x250: Faint. Evenly moderately concentrated. Round. Very small. No involved stars. It's between two faint stars, one SE and one NW.

| ga | NGC 7678 | 23 28.5 | +22 25 | Peg | 12.4b | 2.3' | 1.6' | Sc |

HH: very faint, pretty large, very little extended, little brighter middle, among 4 stars.
170x250: Moderately faint. Broadly very slightly concentrated. Slightly elongated N-S. Moderately small. No involved stars. It fits comfortably inside a small sharp triangle of moderate brightness stars.

| ga | UGC 12613 | 23 28.6 | +14 44 | Peg | 12.6v | 8.3' | 2.5' | Dw I |

170x250: Not visible. [Pegasus dwarf]

| ga | UGC 12632 | 23 30.0 | +40 59 | And | 12.8b | 4.5' | 3.7' | Sm |

170x250: Not visible.

| oc | NGC 7686 | 23 30.2 | +49 07 | And | 5.6 | 14.0' | | |

HH: cluster, poor, little compressed, 7th to 11th mag. stars.
42x60: Two bright stars, more fainter stars, and some very faint stars. ***170x250:*** Widely scattered very mixed brightness stars: two bright stars, the E of which is in the middle of the cluster and is yellow, an E-W linear grouping of moderate brightness stars passing N of the bright middle star, and a few faint stars, especially near the middle star. 3/4 of the field in size. Sparse. Somewhat interesting.

| ga | MCG-1-59-27 | 23 30.5 | −02 27 | Psc | 13.0 | 1.8' | 1.2' | Sb |

170x250: Not visible.

| ga | IC 1496 | 23 30.9 | −02 56 | Psc | 13.1 | 1.6' | 1.3' | S0/a |

170x250: In the position are two faint galaxies half the field apart NE-SW [I1496, *I1492]. The SW galaxy is moderately faint. Sharply concentrated. [Round.] Very small. No involved stars. Nearby SSW is a moderately bright star; the galaxy makes a sharp triangle with this star and a second star SE. The NE galaxy is very very faint. Slightly concentrated. Roughly round. Very small. No involved stars. It makes a flat triangle with two moderate brightness stars, one SSW and one ENE.

| ga | NGC 7711 | 23 35.7 | +15 18 | Peg | 13.1b | 2.6' | 1.2' | S0 |

HH: faint, small, round, pretty suddenly brighter middle, stellar.
170x250: Moderately faint to moderate brightness. Sharply pretty concentrated. Round. Very small. No involved stars. It makes a small triangle with two faint stars, one SSE and another ENE.

| ga | NGC 7714 | 23 36.2 | +02 09 | Psc | 13.0b | 2.2' | 1.1' | Sb |

JH: pretty bright, small, round, pretty suddenly brighter middle, 12th mag. star SW.
170x250: Moderately faint. Sharply very concentrated. Round. Tiny. No involved stars. Very nearby SW is a faint star. [7715 (LR) was not seen.]

| ga | UGC 12704 | 23 36.4 | +50 17 | And | 12.9 | 0.8' | 0.6' | E |

170x250: Not visible.

| ga | NGC 7716 | 23 36.5 | +00 17 | Psc | 12.8b | 2.1' | 1.7' | Sb |

JH: faint, pretty large, little extended, gradually brighter middle, 10th mag. star S.
170x250: Moderately faint. Sharply moderately concentrated. Round. Small. No involved stars. A little ways S is a bright star.

| ga | ESO 605-16 | 23 37.1 | −20 27 | Aqr | 13.2b | 1.6' | 1.4' | Sc |

170x250: Very faint. Hardly concentrated. Round. Small. No involved stars. A little ways SE is a moderately bright star.

| ga | MCG-1-60-15 | 23 37.9 | −03 29 | Aqr | 13.1 | 1.3' | 0.5' | S0+ |

170x250: Very very faint. Slightly concentrated. Round. Very very small. No involved stars. It makes a shallow little arc with two moderately faint stars NW; farther N is a brighter star.

| ga | NGC 7720 | 23 38.5 | +27 01 | Peg | 12.3v | 1.8' | 1.3' | E+ |

HH: faint, small, little extended, brighter middle, among stars.
170x250: Faint. Evenly moderately concentrated. Roughly round. Very small. No involved stars. A sixth of the field S is a very small extremely faint galaxy [*C476-90].

| ga | NGC 7721 | 23 38.8 | –06 30 | Aqr | 12.2b | 3.5' | 1.4' | Sc |

HH: pretty faint, considerably large, extended NNE-SSW, very gradually brighter middle.
170x250: Faint. Broadly very slightly concentrated. A fat edge-on, NNE-SSW. Moderately large. No involved stars.

| ga | NGC 7723 | 23 38.9 | –12 57 | Aqr | 11.9b | 3.5' | 2.3' | Sb |

HH: considerably bright, considerably large, extended, gradually much brighter middle, resolvable.
170x250: Moderate brightness. Broadly moderately concentrated, except that it has a very faint substellar nucleus. Slightly elongated NE-SW. Moderately small. No involved stars.

| ga | NGC 7725 | 23 39.2 | –04 32 | Aqr | 13.0 | 0.7' | 0.7' | S0 |

WH: excessively faint.
170x250: Extremely faint. Concentrated. Roughly round. Very very small. No involved stars.

| bs | Gamma Cep | 23 39.3 | +77 37 | Cep | 3.21 | | 1.03 | K1 |

170x250: Yellow.

| ga | NGC 7727 | 23 39.9 | –12 17 | Aqr | 11.5b | 4.7' | 3.5' | Sa |

HH: pretty bright, pretty large, irregularly round, much brighter middle.
170x250: Moderately bright. Evenly to sharply pretty concentrated. Round. Small. No involved stars. It's in a large hexagon of moderately faint stars; the galaxy is slightly off center E.

| ga | NGC 7741 | 23 43.9 | +26 04 | Peg | 11.8b | 4.4' | 2.9' | Scd |

WH: considerably large, round, 10th–11th mag. star NW.
170x250: Very faint. Broadly very slightly concentrated. Roughly round. Moderately small. No involved stars. A little ways NNW is a pair of stars, the brighter of which is moderately bright.

| ga | NGC 7742 | 23 44.3 | +10 46 | Peg | 12.4b | 1.7' | 1.7' | Sb |

HH: considerably bright, considerably small, gradually much brighter middle, resolvable, bright star E.

170x250: Moderately bright. Sharply pretty concentrated. Round. Small. No involved stars. Nearby ESE is a moderate brightness star.

| ga | NGC 7743 | 23 44.4 | +09 56 | Peg | 12.4b | 3.0' | 2.5' | S0+ |

HH: pretty faint, small, round, 15th mag. star SE.
170x250: Moderate brightness to moderately bright; fairly high surface brightness. Sharply pretty concentrated. Round. Very small. No involved stars. Nearby SSE is a faint star.

| ds | 107 Aqr | 23 46.0 | −18 41 | Aqr | 6.8" | 5.7 | 6.7 | F0 F2 |

42x60: The primary has an extremely close fainter companion. *170x250:* A close unequal pair. Both white. [H 24]

| vs | TX Psc | 23 46.4 | +03 29 | Psc | 5.5 | 6.0 | 2.61 | C6 |

12x35: Mag. 5.5. *170x250:* Orange red.

| ga | NGC 7753 | 23 47.1 | +29 29 | Peg | 12.8p | 3.3' | 2.0' | Sbc |

HH: considerably faint, considerably large, very little extended, very gradually little brighter middle, resolvable.
170x250: In the position is a pair of galaxies aligned NE-SW [7753, *7752 (LR)]. The NE galaxy is brighter. Faint. Broadly slightly concentrated. Roughly round. Moderately small. No involved stars. The SW galaxy is very faint. Very small. Between the two galaxies is an extremely faint star.

| ga | NGC 7755 | 23 47.9 | −30 31 | Scl | 12.6b | 3.8' | 2.8' | Sc |

JH: bright, considerably large, round, pretty suddenly much brighter middle.
170x250: Faint. Broadly concentrated except for either a faint substellar nucleus or an involved star at the center. Roughly round. Moderately small. A little ways E is a moderately faint star.

| ga | NGC 7757 | 23 48.8 | +04 10 | Psc | 13.1b | 2.4' | 1.7' | Sc |

JH: very faint, considerably large, round, very gradually little brighter middle, 13th mag. star N.
170x250: Extremely faint. Diffuse. Roughly round. Moderately small. No involved stars. Nearby N are two stars in a short line with the galaxy. [7756 (LR), a tiny knot in the halo of 7757, was not seen as a separate object.]

| oc | NGC 7762 | 23 49.8 | +68 01 | Cep | 10.0 | 11.0' | | |

HH: cluster, pretty rich, pretty compressed, 11th to 15th mag. stars.
42x60: Very faintly visible. ***170x250:*** A fair number of stars: scattered brighter stars plus faint and threshold stars. Elongated NW-SE. It takes up 3/4 of the field. Somewhat sparse. Irregular. More a random Milky Way concentration than a true cluster.

| oc | K 21 | 23 49.9 | +62 42 | Cas | 9.6 | 2.5' | | |

42x60, 170x250: There's no veritable cluster in the position.

| ga | NGC 7768 | 23 51.0 | +27 08 | Peg | 12.2v | 1.4' | 1.0' | E |

JH: very faint, small, irregular figure, very faint double star involved.
170x250: Moderately faint. Evenly moderately concentrated. Round. Tiny. At its edge W is a faint involved star; the galaxy makes a hook figure with four moderate brightness stars E. [7765 (LR), 7766 (LR), and 7767 (LR) were not seen.]

| ga | NGC 7769 | 23 51.1 | +20 09 | Peg | 12.8p | 2.8' | 2.8' | Sb |

HH: pretty faint, pretty small, round, much brighter middle, SW of 2.
170x250: 7769 and 7771 are a little less than a third of the field apart WNW-ESE. 7769 is a little brighter. Moderately faint to moderate brightness. Evenly concentrated overall, but it has either a very faint stellar nucleus or an involved star near center. Round. Small. No involved stars. A little ways SSW is a faint star. 7771 is moderately faint. Broadly to evenly slightly concentrated. Elongated ENE-WSW. Moderately small. No involved stars. Very nearby SW is a companion galaxy [*7770 (LR)], tiny and faint; a little ways ENE is a moderately faint star.

| ga | NGC 7771 | 23 51.4 | +20 06 | Peg | 13.1b | 3.0' | 1.4' | Sa |

HH: pretty bright, pretty large, extended E-W, brighter middle, NE of 2. [See 7769.]

| oc | NGC 7772 | 23 51.8 | +16 15 | Peg | – | 5.0' | | |

JH: cluster of scattered 10th mag. stars.
12x35, 42x60: A little arc of stars. ***170x250:*** At the N end of the arc is a little triangular concentration of seven moderately faint stars. Very small and compact.

| oc | K 12 | 23 53.0 | +61 57 | Cas | 9.0 | 2.0' | | |

42x60: Visible. ***170x250:*** A moderate brightness double star and some faint stars. Wedge-shaped, pointing WSW. Very small. Moderately dense.

| ga | NGC 7782 | 23 53.9 | +07 58 | Psc | 13.1b | 2.4' | 1.2' | Sb |

HH: pretty faint, pretty large, little extended, gradually little brighter middle, 4th of 4.
170x250: Faint. Broadly slightly concentrated. Slightly elongated N-S. Small. No involved stars. Half the field SW is a pair of galaxies [*7778 (HH), *7779 (HH)] among some stars. They're pretty similar. Both are faint. Evenly to sharply moderately concentrated. Round. Very small. No involved stars. A third of the field E of these two galaxies is another galaxy [*7781 (JH)] with a couple of faint stars involved.

| oc | HAR 21 | 23 54.3 | +61 45 | Cas | 9.0 | 4.0' | | |

42x60: Faintly visible. ***170x250:*** A very small grouping of a half-dozen faint stars and a few very faint stars. W is a NE-SW line of three stars. Not really a cluster.

| ds | LAL 192 | 23 54.4 | −27 03 | Scl | 6.5" | 6.9 | 7.5 | A2 F2 |

42x60: Split. ***170x250:*** A pretty close slightly unequal pair. Both white.

| ga | NGC 7785 | 23 55.3 | +05 54 | Psc | 12.6b | 2.4' | 1.3' | E5-6 |

HH: pretty bright, pretty small, irregularly round, pretty suddenly brighter middle, resolvable, 7th mag. star W.

170x250: Moderately bright. Evenly to sharply pretty concentrated. Round. Very small. No involved stars. Making a triangle with it are a couple of moderately bright stars SE; W is the bright plotted star.

| oc | NGC 7788 | 23 56.7 | +61 23 | Cas | 9.4 | 9.0' | | |

JH: cluster, small, pretty rich, very compressed, 10th mag. star and 13th mag. and fainter stars.
[See 7790.]

| oc | NGC 7789 | 23 57.0 | +56 43 | Cas | 6.7 | 15.0' | | |

HH: cluster, very large, very rich, very much compressed, 11th to 18th mag. stars.
12x35: A granular cloud. ***42x60:*** A cloud of very faint stars plus threshold stars. ***170x250:*** A whole lot of moderately faint and faint stars. It overflows the field with the scattered outer members; the inner concentration is half the field in size. Very dense. A nice and interesting cluster.

| ga | NGC 7793 | 23 57.8 | −32 35 | Scl | 9.6b | 10.0' | 6.5' | Sd |

Bd: like a comet.
12x35: Very faintly visible. ***42x60:*** Faintly visible. ***170x250:*** Moderate brightness. Broadly moderately concentrated. Slightly elongated E-W. Large: a quarter to a

third of the field in size. Just SW of center is a very very faint involved star; nearby N is a moderate brightness star.

| oc | NGC 7790 | 23 58.4 | +61 12 | Cas | 8.5 | 17.0' | | |

HH: cluster, pretty rich, pretty compressed.
12x35, 42x60: 7790 and 7788, visible together, are a small cloud and a large scattered cluster respectively. *170x250:* 7790 is a fair number of moderate brightness and moderately faint stars. Elongated E-W. At the W end are three relatively bright stars. A quarter of the field in size. Fairly dense. 7788 is a very sparse cluster of very widely scattered bright stars filling the field, with a very small dense concentration of faint stars in the very middle. The bright stars are in a rectangular shape NW-SE.

| vs | R Cas | 23 58.4 | +51 24 | Cas | 5.4 | 13.0 | 0.89 | M6 |

12x35: Mag. 8.5. *170x250:* Orangish.

| ds | Sigma Cas | 23 59.0 | +55 45 | Cas | 3.1" | 5.0 | 7.1 | B1 B3 |

170x250: The primary has a very very close much fainter companion. The primary is white. A nice double. [STF 3049]

| ga | IC 1525 | 23 59.3 | +46 53 | And | 13.0p | 1.9' | 1.3' | Sb |

170x250: Very faint. Broadly slightly concentrated. Round. Small. No involved stars.

| ga | NGC 7798 | 23 59.4 | +20 45 | Peg | 13.0b | 1.3' | 1.1' | S |

HH: pretty faint, small, round, suddenly brighter middle, 10th mag. star SW.
170x250: Moderately faint. Broadly to evenly moderately concentrated. Round. Small. No involved stars. Nearby SSW is a moderate brightness star; a little ways NNW is a small triangle of faint stars.

| ga | NGC 7800 | 23 59.6 | +14 48 | Peg | 13.1p | 2.3' | 1.5' | Im |

HH: faint, pretty small, irregularly extended NNE-SSW.
170x250: Very faint. Broadly very slightly concentrated. Elongated NE-SW. Small. No involved stars. It makes a small triangle with a faint star ESE and a very faint star NE.

Appendix A

Making a Sky Atlas

Although a number of very advanced sky atlases are now available in print, none is likely to be ideal for any given task. Published atlases will probably have too few or too many guide stars, too few or too many deep-sky objects plotted in them, wrong-size charts, etc. I found that with MegaStar I could design and make, specifically for my survey, a "just right" personalized atlas. My atlas consists of 108 charts, each about twenty square degrees in size, with guide stars down to magnitude 8.9. I used only the northernmost 78 charts, since I observed the sky only down to −35°. On the charts I plotted only the objects I wanted to observe. In addition I made enlargements of small, overcrowded areas ("quad charts") as well as separate large-scale charts for the Virgo Galaxy Cluster, the latter with guide stars down to magnitude 11.4. I put the charts in plastic sheet protectors in a three-ring binder, taking them out and placing them on my telescope mount's clipboard as needed.

To find an object I would use the 35 mm finder (except in the Virgo Cluster, where I used the 60 mm as the finder) to point the ensemble of telescopes at the indicated spot among the guide stars. If the object was not seen in the 35 mm, as it usually was not, I would then look in the larger telescopes. If the object was not immediately visible even in the primary telescope – a not uncommon occurrence due to inexact initial pointing – I would then scan around for it. If I did not find the object after having diligently swept the area through the primary telescope I recorded it as "not visible." I never resorted to more detailed charts containing fainter guide stars in an attempt to pinpoint the object's location more precisely – a method that in many cases enables one to see challenging objects.

MegaStar uses a certain formula to [...] duce its default star symbol sizes. This formula is designed for finder charts of [...] fields with guide stars all the way from 0 to 15th magnitudes. A different for[mula] is needed to produce appropriate star sizes for charts of much larger areas of [s]ky and with fewer star magnitude categories. The formula I settled on (throug[h tria]l and error) starts with .70 mm for the smallest stars and multiplies this by 1.[4 t]o get the next symbol size (.98). To make each larger symbol size after the s[econd], .035 is subtracted from the previous multiplier. I set the border thickness to ze[ro si]nce MegaStar otherwise enlarges and elongates all the stars.

Index	Magnitude	S[tar s]ymbol size (mm)
0–5	−2.0 – (−1.1)	×1.[0 =]5.94
6	−1.0 – (−0.1)	×1.[1 =]5.47
7	0.0 – 0.9	×1.1[5 =]1.89
8	1.0 – 1.9	×1.19[=1].23
9	2.0 – 2.9	×1.225[=1.]56
10	3.0 – 3.9	×1.260[=1.]90
11	4.0 – 4.9	×1.295=2.[3]0
12	5.0 – 5.9	×1.330=1.78
13	6.0 – 6.9	×1.365=1.34
14	7.0 – 7.9	×1.400=0.98
15	8.0 – 8.9	0.70

The font for the DSO labels needs to be as small as possible in order to minimize overlap with stars, object symbols, and other labels. I chose Verdana 7 point. (Now, with my older eyes, I might make the font a little larger.)

If you undertake an all-sky survey, you will probably need a chart map (such as the one illustrated) to keep track of which parts of the sky you have already observed as you go along. Simply mark each piece of the sky as it is observed. My chart map shows both my atlas's 108 charts and, for added reference, *Sky Atlas*'s 26 charts.

The only significant drawback to a MegaStar atlas is that the DSO labels cannot be moved around to prevent overlap with guide stars. This problem, however, is much less troublesome than it might at first appear, partly because there are so many guide stars that no one is critical, and partly because stars that are run over by labels can usually still be seen anyway. In any event, the most crowded fields can of course be enlarged.

In general it is a great pleasure to cruise around the sky with the aid of my customized atlas and my binocular finder. The MegaStar program incorporates the new Tycho star catalogue, which is so accurate that, unlike with my old *Sky Atlas 2000.0*, it is a very rare occasion when I do not see the sky through the finder exactly as it appears in the atlas.

Appendix A Making a Sky Atlas

Cygnus

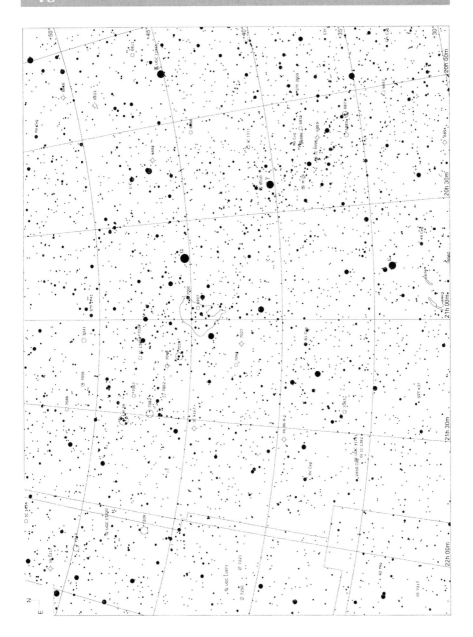

Leo

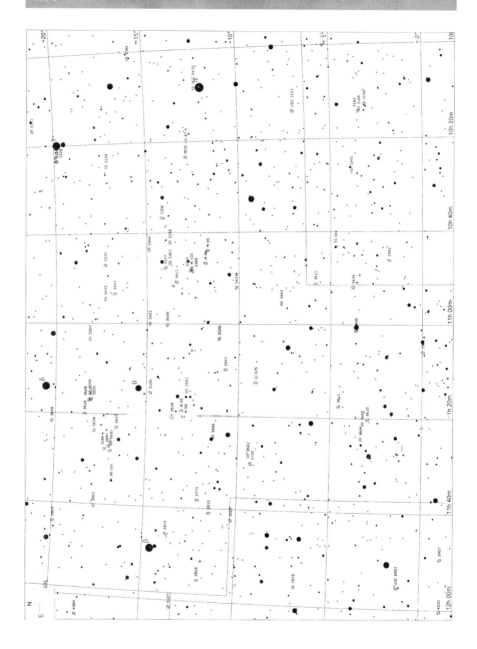

Appendix A Making a Sky Atlas 491

Sagittarius

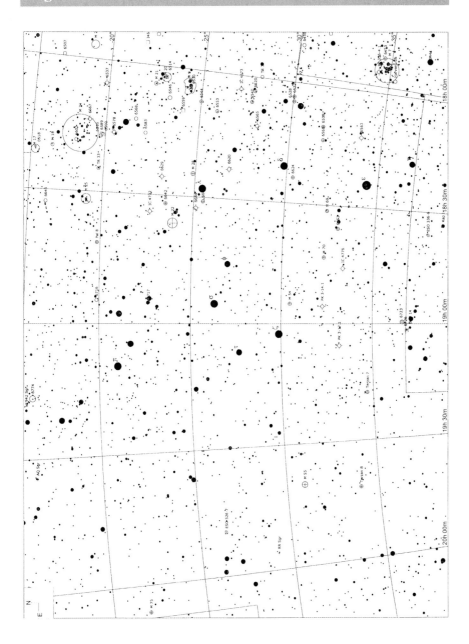

Appendix A Making a Sky Atlas

Virgo Galaxy Cluster North

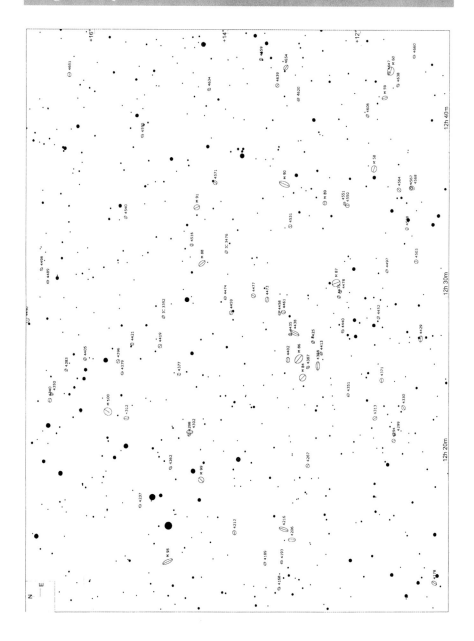

Appendix A Making a Sky Atlas

Virgo Galaxy Cluster South

Appendix A Making a Sky Atlas

Quad Chart

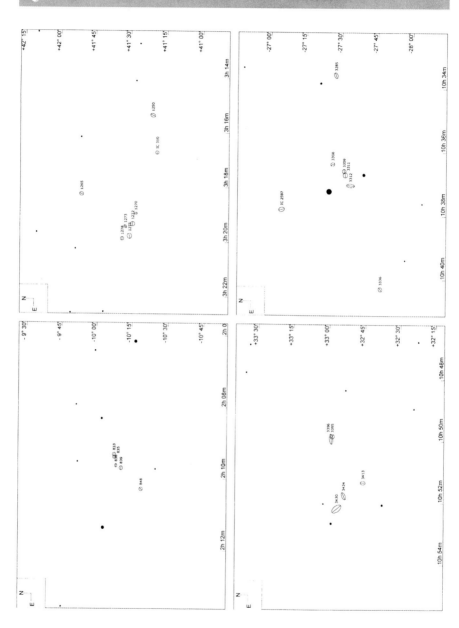

Appendix A Making a Sky Atlas 495

Chart Map

The bold/double lines and numbers indicate the corresponding *Sky Atlas* charts.

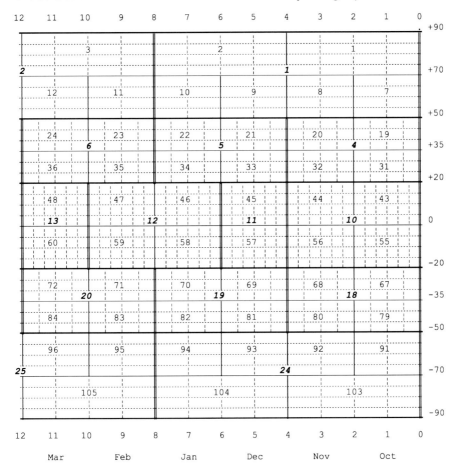

496 Appendix A Making a Sky Atlas

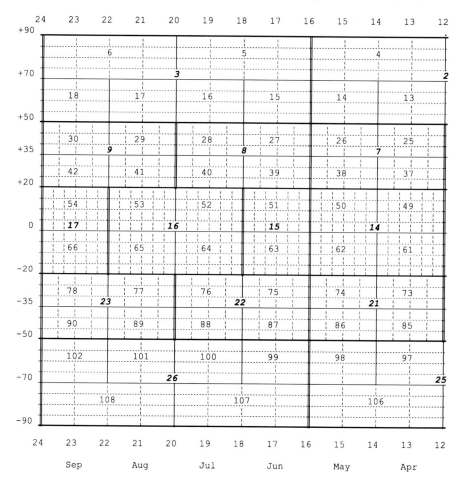

Appendix B

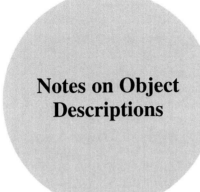

Notes on Object Descriptions

Notes on Descriptions of Galaxies

(1) General brightness is a rough measure of visual prominence, which depends on both surface brightness and size. Hence a small galaxy with a high surface brightness and a large galaxy with a low surface brightness might both be described as "moderate brightness." To put it another way, brightness comparisons among galaxies are most meaningful when the galaxies are of reasonably similar size.

(2) Unless they are confirmed by other observations or photographs, involved or very nearby stars seen in any given observation might be transient (e.g., supernovae) rather than permanent.

(3) Regarding multiple galaxies in a field: In the case of a dominant central galaxy and a few fainter ones surrounding it, directions to the additional galaxies are usually in relation to the principal galaxy. In the case of several similar galaxies in a long string or a large group, the direction to each new galaxy is usually from the last-described galaxy. In either case the described relative positions of the galaxies are obvious enough when the observation is checked against a detailed map or photograph. References to field stars within the group, however, are always in relation to the galaxy just mentioned unless otherwise stated.

(4) I neglected in my very first observation session – fortunately involving only a small fraction of all the listed galaxies – to record nearby stars as religiously as involved stars. In those cases where the omissions were most glaring, post-observation notes describing the nearby stars were appended in brackets.

(5) The names and relative directions of galaxies in pairs or small groups are given in corresponding order. Thus, "In the position is a double galaxy [*1588 (HH), 1587] with the components right next to each other E-W" means that 1588 is E and 1587 is W.

Notes on Descriptions of All Objects

(1) Small-telescope observations of bright objects in a group all appear under the brightest object. For example, 12×35 and 42×60 observations of M31, M32, and M110 are all found under M31.
(2) "Plotted" field stars are stars plotted in my atlas and on my large-scale finder charts; these are all relatively bright stars that will be found in most advanced sky atlases.
(3) A number of variable stars were too faint to be well seen when I first observed them. I had to observe these repeatedly until I caught them near the bright end of their cycles. In a handful of cases I never did catch the star at a reasonable brightness.
(4) Regarding shape: Globular clusters are almost all essentially round, so their shape is rarely mentioned. Unless the shapes of open clusters are specifically described, it is to be assumed that these objects are roughly round.
(5) As a rule, the degree of concentration of globular clusters is attenuated as compared to that of galaxies: a "sharply concentrated" globular cluster is not nearly as sharply concentrated as a sharply concentrated galaxy.
(6) It might be wondered why I use expressions such as "at its edge E/NE/SE" rather than the more normal-sounding "at its E/NE/SE edge." The reason is that the former expression is applicable with any object and with any direction, while the latter can be nonsensical, for instance in the case of a N-S edge-on galaxy with an E edge and a W edge but no NE, SE, NW, or SW edge; or in the case of a round object, which has only one edge – the circumference – not several directional edges. Whatever the shape of an object, the expression "at its edge E/NE/SE" unambiguously means a point E or NE or SE of center.
(7) Instances of glaring discrepancies between the listed data or the NGC description on the one hand and my observation on the other are too numerous and too intractable to comment on, explain, or reconcile in every case. More often than not information from all sides is given "as is" and without comment, even if one or more of the sources seems to be mistaken.
(8) The side of a non-isosceles right triangle of stars is sometimes referred to specifically as a short leg, a long leg, or a hypotenuse, rather than just a "side."
(9) The colors of stars, double stars, and variable stars are described only in views through the primary telescope. This is not because star colors are not seen through the smaller telescopes but rather because (a) the primary telescope, being so much more powerful, shows colors better, and (b) reflectors render truer colors than refractors.

Appendix C

The Visibility of Galactic Detail

It is hardly surprising that astronomical guidebook authors, being quasi-scientists themselves, have been deeply influenced by the concerns of, and the terminology employed by, professional astronomers, who exclusively use photography in their research (they don't look through eyepieces). Natural as this may be, however, at least with regard to galaxies it is in some ways unfortunate. I have examined the mammoth *Carnegie Atlas of Galaxies,* comparing the descriptions accompanying its galaxy photos with my own field observations. There is hardly any resemblance, and for good reason. Visual observers, when they observe galaxies accurately and realistically, by and large see very little of what photographs show. This is why my observations of galaxies are so unassuming.

The little story at the beginning of this book's preface, though imaginary, is based on a careful comparison I made between my observations of galaxies in Cetus and those found in a couple of leading guidebooks. In that comparison I found that the authors of these guidebooks reported seeing – in my opinion mistakenly – far more detail in galaxies than I saw during my survey.

Shown below are all the galaxies in my catalogue in which I saw some kind of irregularity, in other words something other than the usual symmetrical outline (circular, oval, or spindle) and undifferentiated light (aside from a simple brightening toward the center).[1] Most of these galaxies are relatively large and near. There are only 39 in all, out of a total of 1,860 listed galaxies – a mere 2%. Seeing detail in galaxies, therefore – at least when one observes a large number of them – is an unusual rather than a routine occurrence. And of all the various types of irregularities observed, spiral arms are the least conspicuous.

---

[1] See the Addendum.

M31	Dust lanes
247	Asymmetry
253	Unevenness, brightening, dust lane
M33	Unevenness, HII region
891	Dust lane
908	Irregular shape, unevenness
M77	Unevenness
1532	Brightening
2366	Asymmetry
2403	Asymmetry, unevenness, knot
M82	Dust lanes, unevenness
3079	Curved disk
3109	Asymmetry
3359	Unevenness, irregular shape
M108	Unevenness, dust lane
M66	Askew core, spiral arms
3628	Dust lane, irregular shape
4088	Irregular shape
4216	Dust lane
M99	Spiral arm
M106	Spiral arms
M61	Unevenness
4449	Irregular shape, graininess
4532	Irregular shape
4565	Dust lane
M104	Dust lane
4605	Asymmetrically curved disk
4618	Spiral arm
4631	Unevenness, brightening
4656	Asymmetry
4725	Unevenness, brightening
M64	Dust lane
4861	Asymmetry
M51	Spiral arms
M83	Bar, spiral arms
5248	Spiral arm
5247	Spiral arms
M101	HII regions, spiral arm
6946	Unevenness

Appendix D

Building a Binocular Telescope

Despite the manifest superiority of binocularity, standard binoculars do not compare in performance with standard monocular telescopes. The reason is simple and obvious: the large optics of serious telescopes yield magnifying and light-gathering powers that dwarf those of commonly available binoculars such as 10×50s, 20×80s, or even 25×100s. The solution to the problem of puny binoculars is in principle quite straightforward: make binoculars large enough and powerful enough for serious observing, i.e., make veritable binocular telescopes. Not only is there no good reason optically why binoculars should be limited to low power and wide fields, but they actually handle high power more comfortably than do monoculars, for at least three reasons: (1) the binocular image is brighter, counterbalancing the dimming effect of high power; (2) the greater seeing distortion at high power is more effectively "seen through" by two eyes working together; and (3) the smaller fields of high power feel less cramped due to the expansive nature of the binocular view. Only the mechanical difficulty of making two telescopes work together as one instrument makes the production of powerful binocular telescopes prohibitive.

A few amateurs (and now one company, JMI) have made large binocular telescopes consisting of two Newtonians. One observes through a Newtonian binocular by standing in front of it and facing exactly opposite to the part of the sky being studied. Such an instrument may be adequate for brief viewing sessions, but the Cassegrain design, because of the much more convenient observing position it affords – sitting and facing the sky, with the observer's heat-radiating body well away from the light path and with easy access to finders and star charts – is far better suited for a binocular telescope intended for extensive observational work, in which comfort and convenience are of the utmost importance. Mechanically, the essential requirements of a well-designed and well-built binocular telescope are that the optics of each tube assembly be mounted completely rigidly (yet, as with all telescopes, without strain) and that the two tube assemblies be precisely targetable while being firmly held in a strong housing structure.

Appendix D Building a Binocular Telescope 503

The Telescope, various Angles

504 Appendix D Building a Binocular Telescope

Appendix D Building a Binocular Telescope

Appendix D Building a Binocular Telescope

Appendix D Building a Binocular Telescope

Right Tube

510 Appendix D Building a Binocular Telescope

Left Tube

Appendix D Building a Binocular Telescope 511

Enclosure

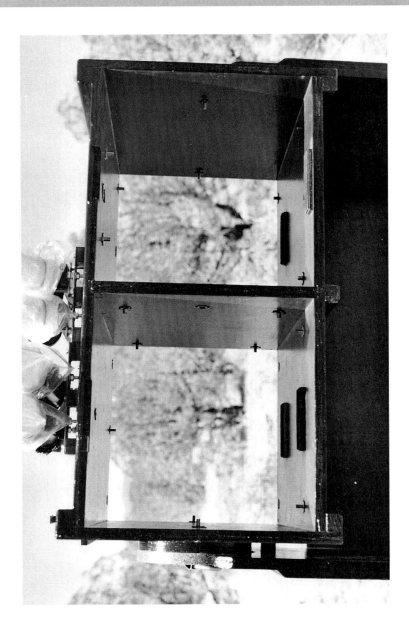

Cradle

Appendix D Building a Binocular Telescope 513

Pier

514   Appendix D Building a Binocular Telescope

Metal Weights in Pier

Appendix D Building a Binocular Telescope

There is no way that the two tube assemblies of a high-power binocular telescope of any design can be expected to maintain perfect alignment for a very long time. At magnifications of over 150×, even the slightest physical shifts – and these are unavoidable – are optically significant. This is why the two telescopes of a powerful binocular must be *independently targetable*. Otherwise the internal collimation of one or both telescopes would have to be constantly altered – inconveniently and to the detriment of image quality – in order to compensate for simple misalignment between the two telescopes. Thus the aim of the design of a powerful binocular telescope cannot be permanent alignment, like that of a small handheld binocular, but rather alignment that (a) stays precisely fixed for reasonably long periods of time (on the order of one night or at least several hours), and (b) can very quickly and easily be readjusted when necessary.

Tube Interior

Tube Interior

Appendix D Building a Binocular Telescope

Primary Mirror Cell

Primary Mirror Cell

Appendix D Building a Binocular Telescope 519

Primary Mirror Cell

Diagonal Assembly

A binocular telescope is naturally both more challenging and more costly to make than a monocular telescope. Yet I built mine out of common materials using standard tools, the most exotic being a router. The accompanying photographs are intended merely to give the prospective builder some ideas. The most important thing is to research and think out one's project carefully. For novice builders (such as I was) a preliminary study of basic carpentry is a must. Good detailed guides on telescope design and construction are listed in the Acknowledgments. Binocular builders should be aware that binoculars intended for terrestrial as well as astronomical use must be made to produce correct images, since the combination of two mirror images in a binocular results in weirdly distorted depth perception. Fortunately this is not an issue in astronomy, since there is no depth perception at all in the view of the night sky. In my setup the 35 mm binocular finder is equipped with premium Amici (i.e., correct-image) prisms to match the orientation of normal sky charts, while the 60 mm and the primary telescope use simpler standard prisms that produce mirror-reversed but slightly superior images.

522　　Appendix D Building a Binocular Telescope

Appendix D Building a Binocular Telescope 523

524 Appendix D Building a Binocular Telescope

Appendix D Building a Binocular Telescope

The Cassegrain Binocular Telescope (CBT)

I originally attempted to market and sell my telescope as a commercial product. Although this particular undertaking did not succeed, the list of the telescope's features that I initially compiled still pertains.

Optics

The main optics consist of 10" f/4-f/16 classical Cassegrain mirror sets. The secondaries are 3", and the central obstructions are 3.25" in diameter. Each tube assembly includes two 1.3" diagonal mirrors and one standard prism diagonal to redirect the light beam. All mirrors are coated with 96% enhanced aluminum.

Optics Mountings

All mirrors are attached at their backs with silicone adhesive. The heavy primary mirrors are further supported at three points along their edges by long bolts protruding from the mirror cells and wrapped in electrical tape. The primary and secondary mirror cells are attached to the tubes and the spider mounts, respectively, with powerful push-pull bolts. (For all practical purposes the internal collimation, once set, is permanent.)

Tubes

Each square tube of the CBT consists of 1/4" plywood walls over a frame of four 1" tube-length runners and eight baffles – 3/4" (2), 1/4" (5), and 1/2" (front end). The oversize tubes provide a minimum of 1" clearance between the walls and the light path. Ventilation slots at the corners of the baffles allow the free flow of air along the edges of the tubes.

Baffle Tubes

Standard front and rear Cassegrain baffle tubes prevent off-axis skylight from reaching the focal plane directly by passing around the secondary mirrors and through the primaries' perforations. The insides of the rear baffle tubes are lined with coarse sandpaper painted flat black to suppress glancing light.

Appendix D Building a Binocular Telescope   527

Focusing and Interocular Adjustment

Fine focus is achieved by sliding the eyepieces up and down in the eyepiece holders of the prism diagonals and locking them with the diagonals' set screws. Coarse focus adjustment (rarely necessary) is done either by moving the secondary mirror cells forward or backward along the optical axis or by moving the tubes that hold the diagonals in or out. The interocular spacing is adjusted by sliding the diagonal assemblies laterally on blocks attached to the backs of the primary mirror cells.

Inter-Tube Alignment

Each tube assembly is held firmly in its enclosure compartment by a number of bolts extending in from the enclosure walls and making contact with steel mending plates on the tube assemblies. The bolts, all with smoothed ends, act together with the plates as bearings and adjustment mechanisms. A wax lubricant is applied to the bolt ends and the plates when the telescope is in use. For adjustment purposes the left tube moves up and down and the right tube moves left and right about pivots at the rear of the enclosure.

Pier and Mount

The CBT's hollow pier is filled with metal weights (I use 330 pounds), making the otherwise top-heavy, 174-pound instrument extremely stable and incapable of being knocked over in the field by normal high winds. The high mounting of the CBT gives it a couple of unexpected advantages over Newtonians, whether binocular or monocular: (1) the optics stay much cleaner than do optics residing practically on the ground, and (2) the high pivot points, combined with the compact telescope design and the extremely smooth altitude and azimuth motions, make maneuvering the telescope a breeze. The bearings are classical Dobsonian: Ebony Star formica on Teflon.

Alignment Procedure

The way to align the CBT at the start of an observing session is as follows: Bring Polaris (or the nearest equivalent in the southern hemisphere) close to the center of each telescope's field. Determine the relative vertical positions of the two images by relaxing the eyes and blinking one of them quickly and repeatedly. Move the left telescope up or down as necessary until the two stars are level with each other; this sets the vertical alignment. Now put Polaris precisely at the right edge of the field in the left telescope. Move the right telescope left or right until Polaris is in the same position in this telescope. This sets the horizontal alignment; the telescopes are now aligned in both axes. (Each of the two alignment methods – blinking and edge – is uniquely appropriate for its axis.) If there is any subsequent tendency of the images to unmerge, it will most probably be the vertical alignment that needs refinement, since the eyes are vastly more sensitive to vertical misalignment than to horizontal misalignment. Simply tweak the adjustment bolts of the left telescope to move it very slightly up or down.

Addendum

Further Observations and Notes

As explained in the introduction, I did quite a bit of observing with the 10-inch binocular before I began the final set of observations between 2004 and 2008 that comprise this book. Some of the pre-project observations contain information that was missed later. Where this discrepancy was noticed during the initial compilation of the catalogue, the additional information was incorporated into the catalogue's descriptions. A few further cases of this kind, initially overlooked, are given below. Galaxies NGC 520, M65, NGC 4038, and NGC 4490 should be added to the list of galaxies with irregularities in Appendix C.

PK 119-6.1 (p. 53): 170×250: Moderate brightness. No central star. The tiniest disk; almost but not quite stellar. Strong blinking effect. Nearby NW is a little NNW-SSE arc of three stars, the middle one being the brightest, along with a couple more fainter stars. Not conspicuous, but interesting.

NGC 520 (p. 71): 170×250: The galaxy is a little bit brighter toward the NW end. It's a little bit rectangular--not a classic spindle.

NGC 772 (p. 82): 170×250: SSW is a very very small companion galaxy [770], smaller than the core of the large galaxy. Faint, but not a very low surface brightness; just tiny. Round.

NGC 3623/M65 (p. 248): 170×250: The E edge is sharper, so that may be where a dark lane is.

NGC 4038 (p. 272): 170×250: There's some indistinct unevenness in the light of 4038.

NGC 4244 (p. 285): 170×250: It doesn't taper toward the ends like 4565; it's more rectilinear. The sides are straight; the galaxy practically doesn't bulge in the middle.

NGC 4490 (p. 304): 170×250: 4490's halo turns N at its W end, hooking toward 4485.

NGC 5796 (p. 377): 170×250: A quarter of the field S is a very faint elongated galaxy [5793 (Le)].

IC 4670 (p. 409): [There is no I4670 in MegaStar, and no deep-sky object is plotted in the exact position. PK7+1.1, a planetary 3.3' N, is presumably the object seen.]

NGC 6618/M17 (p. 419): 170×250: The bar (shaped like a fat cigar) is quite bright. The hook encloses a very dark nebula, contains two bright stars, and gets fainter as it goes around S, W, and N. Near where the hook comes out of the main body is the first bright star; about half a circle around is the second bright star. The hook gets more tenuous from there, ending at a pair of moderate brightness stars at the far W end of the object. Scattered around the hook area are a few faint stars. A very interesting object.

NGC 6741 (p. 429): 170×250: Moderately bright. Essentially stellar; recognized by its blinking effect, which is moderate; otherwise it looks much like the surrounding moderate brightness stars. It makes a tiny sharp triangle with an extremely faint star W and a faint star farther NW; it makes a large quadrilateral with stars NNW and W.

NGC 7619 (p. 477): 170×250: A sixth of the field SSW of 7619 a tiny faint galaxy [7617 (dA)].

Index

A
AGC 194, 72
AGC 426, 108
AGC 1367, 259
AGC 1656, 332
Aldebaran, 123, 126, 147
Alpha 2 Lib, 376
Alpha And, 50
Alpha Aql, 438
Alpha Aqr, 462
Alpha Ari, 84
Alpha Aur, 136
Alpha Boo, 366
Alpha Cas, 59
Alpha Cep, 453
Alpha Cet, 102
Alpha CMa, 161
Alpha CMi, 179
Alpha Col, 143
Alpha CrB, 385
Alpha CVn, 330
Alpha Cyg, 447
Alpha For, 105
Alpha Gem, 176
Alpha Her, 400
Alpha Hya, 205
Alpha Leo, 220
Alpha Lep, 139

Alpha Lyn, 203
Alpha Lyr, 424
Alpha Oph, 404
Alpha Ori, 147
Alpha Peg, 473
Alpha Per, 109
Alpha PsA, 472
Alpha Sco, 393
Alpha Ser, 397
Alpha Tau, 126
Alpha Tri, 80
Alpha UMa, 241
Alpha UMi, 92
Alpha Vir, 345
Altair, 438
59 And, 86
Andromeda Galaxy, 60
ANON 0718–34, 172
Antares, 28, 393
AQ And, 53
5 Aql, 426
57 Aql, 439
12 Aqr, 451
41 Aqr, 464
94 Aqr, 477
107 Aqr, 482
AQ Sgr, 435
Arcturus, 366

1 Ari, 78
14 Aur, 136
41 Aur, 152

B

B 92, 417
B 93, 417
Barnard's Galaxy, 437
Beehive Cluster, 192
Beta 1 Cap, 444
Beta And, 67
Beta Aqr, 455
Beta Ari, 81
Beta Aur, 147
Beta Boo, 378
Beta Cas, 50
Beta Cep, 454
Beta Cet, 60
Beta CMa, 155
Beta CMi, 174
Beta Crv, 310
Beta Cyg, 434
Beta Dra, 403
Beta Eri, 134
Beta Gem, 180
Beta Her, 393
Beta Leo, 262
Beta Lep, 138
Beta Lib, 382
Beta Lyr, 427
Beta Mon, 156
Beta Oph, 406
Beta Ori, 135
Beta Peg, 473
Beta Per, 104
Beta Sco, 389
Beta Tau, 138
Beta Tri, 85
Beta UMa, 241
Beta UMi, 376
Betelgeuse, 28, 147
BL Ori, 155
Black Eye Galaxy, 330
Blinking Planetary, 437
39 Boo, 376
44 Boo, 378
Brocchi's Cluster, 434
BSO 11, 390
BSO 12, 391
BU 106, 375
BU 126, 401
Bubble Nebula, 478
Butterfly Cluster, 405

C

California Nebula, 119
1 Cam, 125
Capella, 136
3 Cen, 356
66 Cet, 86
CGCG 9–18, 226
CGCG 14–74, 311
CGCG 91–55, 201
CGCG 94–42, 226
CGCG 108–72, 389
CGCG 158–23, 276
CGCG 160–40, 331
CGCG 265–8, 204
CGCG 287–61, 187
CGCG 334–15, 244
CGCG 385–129, 72
CGCG 476–90, 480
CGCG 503–27, 78
Chi Cyg, 438
Chi Tau, 123
Christmas Tree Cluster, 160
Coathanger, 434
2 Com, 274
24 Com, 310
Coma Galaxy Cluster, 332
Coma Star Cluster, 294
Cone Nebula, 160
CR 15, 74
CR 26, 92
CR 29, 94
CR 36, 105
CR 39, 109
CR 65, 139
CR 70, 142
CR 105, 158
CR 145, 174
CR 146, 174
CR 155, 178
CR 156, 178
CR 168, 182
CR 256, 294
CR 331, 403
CR 336, 404
CR 337, 404
CR 357, 410
CR 378, 420
CR 387, 424
CR 388, 425
CR 399, 434
Crab Nebula, 140
Crescent Nebula, 443
49 Cyg, 447
61 Cyg, 452

Index

D
DA 5, 137
Delta And, 58
Delta Aql, 434
Delta Aqr, 472
Delta Boo, 381
Delta Cap, 458
Delta Cas, 71
Delta CMa, 168
Delta Crv, 303
Delta Cyg, 438
Delta Dra, 432
Delta Her, 400
Delta Leo, 244
Delta Oph, 390
Delta Ori, 139
Delta Per, 115
Delta Sco, 389
Delta Ser, 385
Delta Sgr, 419
Delta UMa, 282
Delta Vir, 329
Deneb, 447
Double Cluster, 87
Double Double, 425
17 Dra, 394
41 Dra, 411
Draco dwarf, 401
Dumbbell Nebula, 440
DUN 49, 175
DUN 78, 206

E
Eagle Nebula, 418
Epsilon 1 Lyr, 425
Epsilon 2 Lyr, 425
Epsilon Aur, 132
Epsilon Boo, 374
Epsilon Cas, 80
Epsilon CMa, 165
Epsilon Crv, 279
Epsilon Cyg, 448
Epsilon Dra, 438
Epsilon Equ, 450
Epsilon Gem, 161
Epsilon Hya, 193
Epsilon Leo, 210
Epsilon Lep, 133
Epsilon Mon, 155
Epsilon Oph, 391
Epsilon Ori, 142
Epsilon Peg, 458
Epsilon Per, 119
Epsilon Sco, 397
Epsilon Sgr, 420
Epsilon UMa, 327
Epsilon Vir, 335
32 Eri, 118
55 Eri, 127
Eskimo Nebula, 175
ESO 351–30, 65
ESO 356–4, 96
ESO 358–63, 116
ESO 371–16, 194
ESO 376–9, 231
ESO 377–29, 245
ESO 381–12, 320
ESO 383–76, 354
ESO 383–81, 354
ESO 384–2, 356
ESO 384–19, 359
ESO 384–37, 362
ESO 384–53, 363
ESO 385–30, 369
ESO 401–25, 450
ESO 404–12, 460
ESO 423–24, 140
ESO 428–11, 170
ESO 434–28, 209
ESO 436–27, 227
ESO 442–26, 326
ESO 443–24, 334
ESO 443–29, 334
ESO 443–43, 335
ESO 443–66, 336
ESO 445–2, 352
ESO 445–35, 355
ESO 462–15, 445
ESO 462–31, 446
ESO 478–6, 84
ESO 480–20, 102
ESO 481–17, 106
ESO 486–17, 133
ESO 486–19, 133
ESO 489–28, 153
ESO 490–37, 161
ESO 492–2, 169
ESO 494–26, 185
ESO 495–1, 188
ESO 495–21, 191
ESO 499–9, 212
ESO 499–23, 215
ESO 501–51, 230
ESO 501–53, 230
ESO 501–59, 231
ESO 506–33, 315
ESO 507–21, 325

534 Index

ESO 507–25, 325
ESO 507–45, 329
ESO 507–46, 329
ESO 512–18, 374
ESO 512–19, 374
ESO 514–3, 382
ESO 526–7, 438
ESO 541–4, 65
ESO 548–40, 111
ESO 548–44, 112
ESO 548–81, 115
ESO 561–2, 182
ESO 563–31, 195
ESO 566–4, 209
ESO 569–24, 240
ESO 570–2, 242
ESO 570–19, 248
ESO 576–51, 345
ESO 597–6, 446
ESO 597–41, 449
ESO 605–16, 480
Eta Aur, 134
Eta Boo, 357
Eta Cas, 62
Eta Cep, 448
Eta Cet, 67
Eta CMa, 173
Eta Dra, 392
Eta Gem, 153
Eta Leo, 219
Eta Oph, 399
Eta Ori, 137
Eta Peg, 470
Eta Ser, 419
Eta Tau, 117
Eta UMa, 354

F
Fomalhaut, 472
FOR 1, 96
FOR 2, 96
FOR 4, 96
FOR 5, 96
FOR 6, 96
Fornax dwarf, 96
Fornax Galaxy Cluster, 112

G
Gamma And, 84
Gamma Aql, 438
Gamma Ari, 80
Gamma Boo, 371

Gamma Cas, 64
Gamma Cep, 481
Gamma Cet, 98
Gamma Crv, 283
Gamma Cyg, 444
Gamma Del, 448
Gamma Dra, 409
Gamma Eri, 119
Gamma Gem, 159
Gamma Hya, 341
Gamma Leo, 224
Gamma Lyr, 429
Gamma Ori, 138
Gamma Peg, 51
Gamma Per, 103
Gamma Sge, 440
Gamma Sgr, 414
Gamma UMa, 264
Gamma UMi, 383
Gamma Vir, 317
38 Gem, 164

H
H 7, 389
H 19, 176, 392
H 24, 482
H 25, 401
H 27, 178
H 39, 392
H 56, 464
H 69, 650
H 96, 255
H 97, 375
H 101, 356
H 121, 392
Haffner 18, 181
HAR 1, 105
HAR 2, 182
HAR 15, 403
HAR 20, 439
HAR 21, 484
Helix Nebula, 467
95 Her, 411
100 Her, 414
Hercules Cluster, 395
Herschel, William and John, 16
Hickson 42C, 217
HJ 3447, 74
HJ 3506, 93
HJ 3555, 105
HJ 3752, 137
HJ 3928, 166
HJ 3945, 171

HJ 4455, 256
HJ 4839, 390
Hubble's Variable Nebula, 159
17 Hya, 196
54 Hya, 375
Hyades, 123

I

IC 10, 53
IC 93, 69
IC 166, 79
IC 167, 79
IC 171, 81
IC 180, 82
IC 206, 85
IC 207, 85
IC 219, 88
IC 223, 88
IC 239, 94
IC 267, 101
IC 284, 104
IC 289, 104
IC 310, 106
IC 334, 116
IC 335, 112
IC 342, 117
IC 344, 115
IC 348, 116
IC 351, 117
IC 356, 120
IC 361, 122
IC 381, 128
IC 382, 126
IC 391, 131
IC 396, 131
IC 418, 138
IC 421, 139
IC 438, 146
IC 455, 176
IC 456, 165
IC 458, 169
IC 464, 169
IC 465, 169
IC 467, 176
IC 512, 198
IC 520, 195
IC 529, 202
IC 546, 207
IC 575, 213
IC 600, 222
IC 676, 244
IC 694, 254
IC 749, 269

IC 750, 269
IC 764, 279
IC 829, 326
IC 844, 335
IC 879, 342
IC 982, 364
IC 983, 364
IC 1014, 369
IC 1017, 369
IC 1029, 371
IC 1066, 376
IC 1067, 376
IC 1081, 377
IC 1257, 402
IC 1265, 404
IC 1276, 415
IC 1295, 428
IC 1311, 442
IC 1369, 453
IC 1392, 456
IC 1396, 457
IC 1434, 463
IC 1438, 464
IC 1439, 464
IC 1442, 464
IC 1447, 467
IC 1454, 470
IC 1492, 479
IC 1496, 479
IC 1525, 485
IC 1558, 56
IC 1559, 57
IC 1590, 64
IC 1613, 66
IC 1637, 67
IC 1657, 68
IC 1687, 70
IC 1727, 76
IC 1747, 82
IC 1783, 85
IC 1788, 87
IC 1830, 95
IC 1848, 100
IC 1864, 100
IC 1870, 101
IC 1907, 108
IC 1953, 111
IC 1993, 117
IC 2003, 119
IC 2104, 131
IC 2120, 136
IC 2132, 139
IC 2149, 147
IC 2157, 150

IC 2158, 150
IC 2163, 154
IC 2165, 155
IC 2166, 156
IC 2174, 169
IC 2209, 183
IC 2233, 186
IC 2311, 188
IC 2469, 204
IC 2482, 205
IC 2507, 209
IC 2511, 211
IC 2514, 211
IC 2522, 214
IC 2523, 214
IC 2531, 216
IC 2533, 217
IC 2537, 219
IC 2560, 222
IC 2574, 227
IC 2580, 227
IC 2584, 227
IC 2597, 231
IC 2627, 243
IC 2764, 253
IC 2910, 255
IC 2913, 255
IC 2995, 275
IC 3010, 277
IC 3015, 278
IC 3253, 292
IC 3267, 292
IC 3392, 301
IC 3470, 306
IC 3476, 307
IC 3568, 308
IC 3583, 313
IC 4051, 332
IC 4180, 337
IC 4182, 336
IC 4197, 336
IC 4214, 341
IC 4237, 345
IC 4296, 350
IC 4299, 350
IC 4310, 352
IC 4329, 354
IC 4329A, 355
IC 4351, 359
IC 4366, 363
IC 4453, 372
IC 4460, 372
IC 4538, 383
IC 4593, 390
IC 4634, 398

IC 4665, 406
IC 4670, 409
IC 4673, 412
IC 4725, 422
IC 4732, 423
IC 4756, 424
IC 4776, 426
IC 4846, 432
IC 4996, 443
IC 4997, 444
IC 5020, 446
IC 5041, 448
IC 5117, 456
IC 5156, 461
IC 5157, 461
IC 5217, 465
IC 5271, 472
IC 5283, 473
Iota 1 Cnc, 193
Iota Aur, 131
Iota Cas, 91
Iota Dra, 384
Iota Ori, 141
Iota UMa, 197

J
J 320, 133
J 900, 155

K
K 2, 63
K 4, 94
K 5, 106
K 8, 145
K 12, 483
K 14, 55
K 16, 60
K 19, 474
K 21, 483
k Pup, 178
Kappa Boo, 365
Kappa Her, 390
Kappa Oph, 397
Kappa Ori, 145
Kemble's Cascade, 120

L
8 Lac, 468
Lagoon Nebula, 412
LAL 192, 484
LAL 194, 143
Lambda Aql, 430

Lambda Ori, 140
Lambda Sgr, 421
Lambda Tau, 119
Lambda UMa, 222
54 Leo, 239
90 Leo, 256
Leo I, 220
Leo II, 244
Leo III, 216
Little Dumbbell Nebula, 75
19 Lyn, 172
38 Lyn, 202

M
M1, 140
M2, 32, 456
M3, 353
M4, 392
M5, 383
M6, 405
M7, 408
M8, 32, 412
M9, 401
M10, 397
M11, 33, 427
M12, 396, 397
M13, 32, 395
M14, 405
M15, 455
M16, 418
M17, 32, 419
M18, 419
M19, 398
M20, 411
M21, 413
M22, 423
M23, 409
M24, 417
M25, 422
M26, 426
M27, 75, 440
M28, 420
M29, 445, 454
M30, 457
M31, 31, 59, 60, 215
M32, 59, 60
M33, 73
M34, 97
M35, 151
M36, 142
M37, 33, 146
M38, 138
M39, 456
M41, 162
M42, 32, 141, 412
M43, 141, 142
M44, 192
M45, 117
M46, 177, 179
M47, 177
M48, 186
M49, 303, 305
M50, 166
M51, 31, 348
M52, 478
M53, 338, 341
M54, 428, 441
M55, 435
M56, 432
M57, 428
M58, 313
M59, 317, 320
M60, 305, 320
M61, 290
M62, 398
M63, 340
M64, 330
M65, 248
M66, 31, 248
M67, 194
M68, 314
M69, 422, 425
M70, 425
M71, 439
M72, 449
M73, 450
M74, 74
M75, 441
M76, 75
M77, 98
M78, 144
M79, 137
M80, 391
M81, 214
M82, 31, 215
M83, 31, 350
M84, 294, 297, 303, 305
M85, 295, 305
M86, 294, 297, 305
M87, 304, 305
M88, 306
M89, 311
M90, 313
M91, 310
M92, 400
M93, 180
M94, 325
M95, 232, 233
M96, 233

M97, 245
M98, 282
M99, 31, 286
M100, 291
M101, 361
M102, 379
M103, 73
M104, 31, 315
M105, 234
M106, 31, 286
M107, 394
M108, 243
M109, 267
M110, 59, 60
MAC1119+5803, 247
MAC1122-0735, 250
MAC1131-0940A, 255
MAC1131-0940B, 255
MAC1304-3025, 335
MAC1323+3130, 344
MCG+0-47-1, 421
MCG+1-14-37, 138
MCG+1-17-3, 158
MCG+1-48-1, 427
MCG+2-27-4, 225
MCG+4-42-10, 408
MCG+8-22-9, 261
MCG+8-26-38, 370
MCG+9-19-213, 263
MCG+9-24-22, 372
MCG+10-19-63, 343
MCG+11-14-3A, 243
MCG+11-14-3B, 243
MCG+12-9-48A, 205
MCG-1-1-52, 51
MCG-1-1-57, 52
MCG-1-1-64, 52
MCG-1-2-34, 56
MCG-1-2-38, 56
MCG-1-3-18, 62
MCG-1-3-85, 66
MCG-1-3-88, 66
MCG-1-10-23, 115
MCG-1-13-30, 130
MCG-1-25-44, 211
MCG-1-26-13, 219
MCG-1-26-21, 219
MCG-1-29-13, 250
MCG-1-29-15, 250
MCG-1-32-38, 318
MCG-1-32-39, 318
MCG-1-33-1, 321
MCG-1-33-3, 322
MCG-1-35-10, 353

MCG-1-59-27, 479
MCG-1-60-15, 480
MCG-2-1-14, 50
MCG-2-8-33, 101
MCG-2-12-39, 125
MCG-2-12-41, 125
MCG-2-13-9, 129
MCG-2-15-12, 146
MCG-2-22-17, 190
MCG-2-24-7, 202
MCG-2-30-16, 259
MCG-2-33-15, 324
MCG-2-33-37, 326
MCG-2-33-82, 333
MCG-2-34-48, 346
MCG-2-34-54, 347
MCG-2-35-10, 351
MCG-2-35-11, 651
MCG-2-38-15, 375
MCG-2-41-1, 391
MCG-2-53-23, 451
MCG-3-4-39, 69
MCG-3-10-45, 117
MCG-3-13-4, 128
MCG-3-15-21, 145
MCG-3-16-18, 152
MCG-3-26-30, 220
MCG-3-28-17, 238
MCG-3-34-14, 338
MCG-3-34-61, 343
MCG-3-52-2, 444
MCG-4-16-6, 156
ME 2-1, 384
MEL 15, 92
MEL 20, 109
MEL 25, 123
MEL 71, 178
MEL 72, 178
MEL 111, 294
MKN 205, 288
MKN 509, 448
Mu Dra, 399
Mu Gem, 155
Mu Her, 406
Mu Lep, 135
Mu Lib, 375
Mu Peg, 471
Mu UMa, 225

N
N Hya, 255
n Pup, 176
NGC 1, 50

Index

NGC 2, 50
NGC 14, 50
NGC 16, 50
NGC 23, 51
NGC 24, 51
NGC 40, 51
NGC 45, 51
NGC 50, 52
NGC 57, 52
NGC 59, 52
NGC 62, 52
NGC 63, 52
NGC 80, 53
NGC 83, 53
NGC 95, 53
NGC 103, 53
NGC 108, 53
NGC 125, 54
NGC 126, 54
NGC 127, 54
NGC 128, 54
NGC 129, 54
NGC 130, 54
NGC 131, 54
NGC 133, 55
NGC 134, 54
NGC 136, 55
NGC 145, 55
NGC 146, 55
NGC 147, 55
NGC 148, 56
NGC 150, 56
NGC 151, 56
NGC 157, 56
NGC 160, 57
NGC 169, 57
NGC 175, 57
NGC 178, 57
NGC 182, 57
NGC 185, 57
NGC 188, 61
NGC 189, 58
NGC 193, 58
NGC 194, 58
NGC 198, 58
NGC 199, 58
NGC 200, 58
NGC 203, 58
NGC 204, 58
NGC 205, 59
NGC 210, 59
NGC 214, 59
NGC 221, 59
NGC 224, 60

NGC 225, 60
NGC 227, 59
NGC 234, 60
NGC 245, 61
NGC 246, 61
NGC 247, 61
NGC 253, 62
NGC 254, 61
NGC 255, 62
NGC 266, 62
NGC 270, 63
NGC 271, 63
NGC 272, 63
NGC 273, 63
NGC 274, 63
NGC 275, 63
NGC 278, 64
NGC 281, 64
NGC 288, 64
NGC 289, 64
NGC 309, 64
NGC 311, 65
NGC 315, 65
NGC 318, 65
NGC 336, 65
NGC 337, 65
NGC 337A, 65
NGC 355, 65
NGC 357, 65
NGC 358, 66
NGC 375, 66
NGC 379, 66
NGC 380, 66
NGC 381, 67
NGC 382, 66
NGC 383, 66
NGC 384, 66
NGC 385, 66
NGC 386, 66
NGC 388, 66
NGC 404, 67
NGC 407, 67
NGC 410, 67
NGC 414, 67
NGC 418, 67
NGC 420, 67
NGC 428, 68
NGC 436, 68
NGC 439, 68
NGC 441, 68
NGC 448, 68
NGC 450, 68
NGC 457, 69
NGC 467, 69

NGC 470, 69
NGC 474, 69
NGC 488, 70
NGC 491, 69
NGC 493, 70
NGC 495, 70
NGC 496, 70
NGC 499, 70
NGC 501, 70
NGC 503, 70
NGC 504, 70
NGC 507, 70
NGC 508, 70
NGC 514, 70
NGC 520, 71
NGC 521, 71
NGC 524, 71
NGC 529, 71
NGC 531, 72
NGC 533, 71
NGC 535, 72
NGC 536, 72
NGC 541, 71
NGC 542, 72
NGC 543, 72
NGC 545, 71, 72
NGC 547, 72
NGC 559, 72
NGC 578, 72
NGC 581, 73
NGC 584, 72
NGC 586, 72
NGC 596, 73
NGC 598, /93
NGC 600, 73
NGC 604, 73
NGC 609, 74
NGC 613, 73
NGC 615, 73
NGC 628, 74
NGC 636, 74
NGC 637, 75
NGC 639, 74
NGC 642, 874
NGC 650, 75
NGC 654, 76
NGC 658, 75
NGC 659, 76
NGC 660, 75
NGC 661, 76
NGC 663, 76
NGC 665, 76
NGC 672, 77
NGC 673, 77

NGC 675, 77
NGC 676, 77
NGC 677, 77
NGC 678, 78
NGC 680, 78
NGC 681, 77
NGC 687, 78
NGC 691, 78
NGC 693, 78
NGC 694, 79
NGC 697, 79
NGC 701, 79
NGC 703, 79
NGC 704, 79
NGC 705, 79
NGC 706, 79
NGC 708, 79
NGC 709, 79
NGC 710, 79
NGC 714, 80
NGC 717, 80
NGC 718, 80
NGC 720, 80
NGC 723, 80
NGC 731, 81
NGC 736, 81
NGC 738, 81
NGC 740, 81
NGC 741, 81
NGC 742, 81
NGC 744, 82
NGC 750, 81
NGC 751, 81
NGC 752, 82
NGC 753, 82
NGC 755, 81
NGC 770, 82
NGC 772, 82
NGC 776, 82
NGC 777, 83
NGC 778, 83
NGC 779, 82
NGC 783, 83
NGC 784, 83
NGC 785, 83
NGC 788, 83
NGC 799, 83
NGC 800, 83
NGC 803, 83
NGC 812, 84
NGC 818, 84
NGC 821, 84
NGC 828, 86
NGC 833, 84

Index

NGC 835, 85
NGC 838, 85
NGC 839, 85
NGC 846, 86
NGC 848, 86
NGC 864, 86
NGC 869, 87
NGC 873, 87
NGC 876, 87
NGC 877, 87
NGC 881, 87
NGC 883, 88
NGC 884, 89
NGC 887, 88
NGC 890, 88
NGC 891, 31, 89
NGC 895, 88
NGC 897, 88
NGC 899, 88
NGC 907, 89
NGC 908, 89
NGC 910, 89
NGC 912, 89
NGC 918, 90
NGC 922, 89
NGC 925, 90
NGC 936, 90
NGC 941, 90
NGC 942, 91
NGC 943, 91
NGC 945, 91
NGC 947, 91
NGC 948, 91
NGC 949, 92
NGC 955, 91
NGC 956, 92
NGC 957, 92
NGC 958, 91
NGC 959, 92
NGC 968, 93
NGC 969, 93
NGC 972, 93
NGC 974, 93
NGC 978, 93
NGC 988, 93
NGC 989, 92
NGC 991, 93
NGC 1003, 95
NGC 1012, 95
NGC 1015, 94
NGC 1016, 95
NGC 1022, 95
NGC 1023, 96
NGC 1023A, 96

NGC 1024, 95
NGC 1027, 97
NGC 1029, 95
NGC 1032, 96
NGC 1034, 94
NGC 1035, 96
NGC 1039, 97
NGC 1041, 96
NGC 1042, 97
NGC 1045, 97
NGC 1049, 96
NGC 1052, 97
NGC 1055, 97
NGC 1057, 98
NGC 1058, 98
NGC 1060, 98
NGC 1061, 98
NGC 1066, 98
NGC 1068, 98
NGC 1070, 98
NGC 1073, 99
NGC 1076, 98
NGC 1079, 99
NGC 1084, 99
NGC 1085, 100
NGC 1087, 100
NGC 1090, 100
NGC 1091, 99
NGC 1092, 99
NGC 1097, 99
NGC 1097A, 99
NGC 1098, 99
NGC 1099, 99
NGC 1100, 99
NGC 1122, 100
NGC 1134, 100
NGC 1137, 101
NGC 1140, 101
NGC 1156, 101
NGC 1160, 102
NGC 1161, 102
NGC 1169, 102
NGC 1171, 103
NGC 1172, 102
NGC 1179, 102
NGC 1186, 103
NGC 1187, 102
NGC 1189, 103
NGC 1190, 103
NGC 1191, 103
NGC 1192, 103
NGC 1193, 103
NGC 1199, 103
NGC 1201, 103

NGC 1209, 104
NGC 1220, 105
NGC 1222, 104
NGC 1232, 104
NGC 1241, 105
NGC 1242, 105
NGC 1245, 106
NGC 1250, 106
NGC 1253, 106
NGC 1253A, 106
NGC 1255, 105
NGC 1265, 107
NGC 1267, 108
NGC 1268, 108
NGC 1270, 108
NGC 1272, 107
NGC 1273, 108
NGC 1274, 108
NGC 1275, 108
NGC 1277, 108
NGC 1278, 108
NGC 1281, 108
NGC 1284, 107
NGC 1288, 107
NGC 1292, 107
NGC 1297, 107
NGC 1300, 108
NGC 1302, 108
NGC 1309, 108
NGC 1319, 109
NGC 1325, 109
NGC 1331, 109
NGC 1332, 109
NGC 1333, 110
NGC 1337, 109
NGC 1339, 110
NGC 1342, 111
NGC 1344, 110
NGC 1350, 110
NGC 1351, 110
NGC 1353, 111
NGC 1355, 112
NGC 1357, 111
NGC 1358, 111
NGC 1359, 112
NGC 1360, 111
NGC 1361, 112
NGC 1365, 31
NGC 1366, 112
NGC 1371, 112
NGC 1376, 113
NGC 1380, 112
NGC 1380A, 113
NGC 1383, 113

NGC 1385, 113
NGC 1391, 113
NGC 1393, 113
NGC 1394, 114
NGC 1395, 113
NGC 1398, 113
NGC 1400, 114
NGC 1401, 114
NGC 1402, 114
NGC 1406, 114
NGC 1407, 114
NGC 1415, 114
NGC 1417, 115
NGC 1418, 115
NGC 1421, 115
NGC 1425, 115
NGC 1426, 115
NGC 1438, 116
NGC 1439, 116
NGC 1440, 116
NGC 1444, 118
NGC 1452, 116
NGC 1453, 117
NGC 1461, 118
NGC 1481, 118
NGC 1482, 118
NGC 1491, 119
NGC 1496, 119
NGC 1499, 119
NGC 1501, 120
NGC 1502, 120
NGC 1507, 120
NGC 1513, 121
NGC 1514, 121
NGC 1516, 121
NGC 1518, 120
NGC 1521, 121
NGC 1528, 122
NGC 1530, 123
NGC 1531, 121
NGC 1532, 121
NGC 1535, 122
NGC 1537, 122
NGC 1545, 122
NGC 1550, 122
NGC 1555, 123
NGC 1560, 125
NGC 1569, 124
NGC 1573, 126
NGC 1575, 123
NGC 1579, 124
NGC 1582, 124
NGC 1587, 124
NGC 1588, 124

Index

NGC 1589, 124
NGC 1600, 125
NGC 1601, 125
NGC 1603, 125
NGC 1605, 125
NGC 1618, 126
NGC 1620, 126
NGC 1622, 126
NGC 1624, 127
NGC 1625, 126
NGC 1636, 127
NGC 1637, 127
NGC 1638, 127
NGC 1640, 127
NGC 1645, 128
NGC 1647, 128
NGC 1650, 128
NGC 1653, 128
NGC 1659, 129
NGC 1662, 129
NGC 1664, 130
NGC 1667, 129
NGC 1679, 130
NGC 1682, 130
NGC 1684, 130
NGC 1691, 130
NGC 1700, 131
NGC 1720, 132
NGC 1721, 131
NGC 1723, 131
NGC 1725, 132
NGC 1726, 132
NGC 1728, 131
NGC 1729, 132
NGC 1730, 132
NGC 1744, 132
NGC 1746, 133
NGC 1778, 134
NGC 1779, 133
NGC 1784, 133
NGC 1788, 134
NGC 1798, 135
NGC 1800, 134
NGC 1807, 134
NGC 1817, 135
NGC 1832, 135
NGC 1857, 136
NGC 1879, 136
NGC 1883, 138
NGC 1888, 137
NGC 1889, 137
NGC 1893, 137
NGC 1904, 137
NGC 1907, 138

NGC 1912, 138
NGC 1931, 139
NGC 1952, 140
NGC 1954, 139
NGC 1957, 139
NGC 1960, 142
NGC 1961, 144
NGC 1964, 140
NGC 1976, 141
NGC 1977, 141
NGC 1979, 140
NGC 1980, 141
NGC 1981, 140
NGC 1982, 142
NGC 1985, 143
NGC 1999, 142
NGC 2017, 143
NGC 2022, 144
NGC 2023, 144
NGC 2024, 144
NGC 2068, 144
NGC 2071, 145
NGC 2089, 145
NGC 2090, 145
NGC 2099, 146
NGC 2106, 146
NGC 2110, 146
NGC 2112, 147
NGC 2126, 149
NGC 2129, 149
NGC 2139, 149
NGC 2141, 150
NGC 2146, 154
NGC 2149, 150
NGC 2158, 150
NGC 2168, 151
NGC 2169, 151
NGC 2170, 150
NGC 2174, 151
NGC 2175, 152
NGC 2179, 151
NGC 2182, 151
NGC 2183, 152
NGC 2185, 152
NGC 2186, 152
NGC 2188, 152
NGC 2192, 153
NGC 2194, 153
NGC 2196, 153
NGC 2204, 153
NGC 2206, 153
NGC 2207, 154
NGC 2215, 154
NGC 2217, 155

NGC 2223, 155
NGC 2227, 156
NGC 2232, 156
NGC 2236, 156
NGC 2243, 157
NGC 2244, 157
NGC 2245, 157
NGC 2247, 157
NGC 2250, 157
NGC 2251, 158
NGC 2252, 158
NGC 2254, 158
NGC 2258, 163
NGC 2259, 159
NGC 2261, 159
NGC 2262, 159
NGC 2263, 159
NGC 2264, 160
NGC 2266, 160
NGC 2267, 160
NGC 2268, 170
NGC 2269, 160
NGC 2271, 160
NGC 2272, 160
NGC 2273, 163
NGC 2273B, 162
NGC 2274, 162
NGC 2275, 162
NGC 2276, 174
NGC 2280, 174
NGC 2281, 163
NGC 2282, 163
NGC 2283, 163
NGC 2286, 163
NGC 2287, 163
NGC 2288, 164
NGC 2289, 163
NGC 2290, 163
NGC 2291, 164
NGC 2292, 162
NGC 2293, 163
NGC 2294, 164
NGC 2295, 163
NGC 2300, 176
NGC 2301, 164
NGC 2302, 164
NGC 2304, 164
NGC 2309, 165
NGC 2311, 165
NGC 2314, 169
NGC 2320, 166
NGC 2322, 167
NGC 2323, 166
NGC 2324, 166

NGC 2325, 165
NGC 2327, 166
NGC 2329, 168
NGC 2331, 167
NGC 2335, 167
NGC 2336, 174
NGC 2337, 169
NGC 2339, 167
NGC 2340, 169
NGC 2341, 168
NGC 2342, 168
NGC 2343, 168
NGC 2344, 170
NGC 2345, 168
NGC 2346, 169
NGC 2347, 169
NGC 2353, 170
NGC 2354, 170
NGC 2355, 171
NGC 2359, 171
NGC 2360, 171
NGC 2362, 171
NGC 2363, 175
NGC 2366, 175
NGC 2367, 172
NGC 2368, 172
NGC 2371, 173
NGC 2374, 173
NGC 2380, 172
NGC 2383, 173
NGC 2384, 173
NGC 2392, 175
NGC 2395, 174
NGC 2396, 174
NGC 2401, 175
NGC 2403, 177
NGC 2414, 176
NGC 2415, 177
NGC 2418, 177
NGC 2419, 178
NGC 2420, 178
NGC 2421, 177
NGC 2422, 177
NGC 2423, 178
NGC 2432, 179
NGC 2437, 179
NGC 2438, 179
NGC 2439, 179
NGC 2440, 180
NGC 2441, 181
NGC 2447, 180
NGC 2452, 180
NGC 2453, 180
NGC 2455, 181

NGC 2460, 183
NGC 2467, 181
NGC 2479, 182
NGC 2482, 181
NGC 2483, 182
NGC 2485, 182
NGC 2486, 183
NGC 2487, 183
NGC 2489, 182
NGC 2493, 183
NGC 2495, 183
NGC 2500, 184
NGC 2506, 183
NGC 2507, 184
NGC 2509, 183
NGC 2510, 184
NGC 2511, 184
NGC 2513, 184
NGC 2517, 184
NGC 2523, 187
NGC 2525, 185
NGC 2527, 184
NGC 2532, 185
NGC 2533, 185
NGC 2537, 186
NGC 2539, 185
NGC 2541, 187
NGC 2543, 186
NGC 2545, 186
NGC 2548, 186
NGC 2549, 188
NGC 2551, 190
NGC 2552, 189
NGC 2554, 187
NGC 2555, 187
NGC 2559, 187
NGC 2562, 189
NGC 2563, 189
NGC 2566, 188
NGC 2567, 188
NGC 2571, 188
NGC 2577, 189
NGC 2580, 189
NGC 2587, 189
NGC 2588, 189
NGC 2591, 192
NGC 2595, 190
NGC 2599, 190
NGC 2604, 190
NGC 2608, 191
NGC 2610, 191
NGC 2613, 191
NGC 2618, 191
NGC 2619, 192

NGC 2627, 191
NGC 2632, 192
NGC 2633, 194
NGC 2634, 194
NGC 2634A, 194
NGC 2635, 192
NGC 2639, 193
NGC 2646, 194
NGC 2648, 192
NGC 2649, 193
NGC 2654, 194
NGC 2655, 196
NGC 2658, 192
NGC 2663, 193
NGC 2665, 193
NGC 2672, 194
NGC 2673, 194
NGC 2681, 195
NGC 2682, 194
NGC 2683, 195
NGC 2685, 196
NGC 2693, 196
NGC 2694, 197
NGC 2695, 195
NGC 2697, 195
NGC 2698, 196
NGC 2699, 196
NGC 2701, 197
NGC 2708, 196
NGC 2709, 196
NGC 2712, 198
NGC 2713, 197
NGC 2715, 199
NGC 2716, 197
NGC 2718, 197
NGC 2721, 197
NGC 2732, 200
NGC 2742, 198
NGC 2748, 200
NGC 2749, 198
NGC 2750, 198
NGC 2751, 198
NGC 2752, 198
NGC 2756, 199
NGC 2763, 198
NGC 2765, 199
NGC 2768, 200
NGC 2770, 199
NGC 2775, 199
NGC 2776, 200
NGC 2781, 199
NGC 2782, 200
NGC 2784, 200
NGC 2787, 202

NGC 2789, 201
NGC 2798, 201
NGC 2799, 201
NGC 2804, 201
NGC 2805, 203
NGC 2806, 201
NGC 2809, 201
NGC 2810, 204
NGC 2811, 201
NGC 2814, 203
NGC 2815, 201
NGC 2820, 203
NGC 2825, 202
NGC 2830, 202
NGC 2831, 202
NGC 2832, 202
NGC 2834, 202
NGC 2835, 201
NGC 2841, 203
NGC 2848, 202
NGC 2851, 203
NGC 2854, 204
NGC 2855, 203
NGC 2856, 204
NGC 2857, 204
NGC 2859, 204
NGC 2865, 204
NGC 2872, 204
NGC 2874, 205
NGC 2880, 205
NGC 2884, 205
NGC 2889, 205
NGC 2902, 206
NGC 2903, 206
NGC 2907, 206
NGC 2911, 207
NGC 2914, 207
NGC 2916, 207
NGC 2920, 207
NGC 2921, 207
NGC 2924, 207
NGC 2935, 207
NGC 2942, 208
NGC 2945, 208
NGC 2947, 207
NGC 2950, 208
NGC 2962, 208
NGC 2964, 209
NGC 2967, 208
NGC 2968, 209
NGC 2970, 209
NGC 2974, 208
NGC 2976, 211
NGC 2983, 209

NGC 2985, 212
NGC 2986, 209
NGC 2990, 210
NGC 2992, 210
NGC 2993, 210
NGC 2997, 210
NGC 2998, 211
NGC 3001, 210
NGC 3003, 211
NGC 3016, 212
NGC 3018, 212
NGC 3019, 212
NGC 3020, 212
NGC 3021, 212
NGC 3022, 211
NGC 3023, 211
NGC 3024, 212
NGC 3027, 215
NGC 3031, 214
NGC 3032, 213
NGC 3034, 215
NGC 3038, 212
NGC 3041, 213
NGC 3044, 213
NGC 3049, 214
NGC 3051, 213
NGC 3052, 213
NGC 3054, 214
NGC 3055, 214
NGC 3056, 214
NGC 3067, 215
NGC 3073, 218
NGC 3077, 218
NGC 3078, 216
NGC 3079, 218
NGC 3081, 216
NGC 3084, 216
NGC 3087, 216
NGC 3089, 216
NGC 3091, 217
NGC 3094, 217
NGC 3095, 217
NGC 3096, 217
NGC 3098, 218
NGC 3100, 217
NGC 3108, 218
NGC 3109, 218
NGC 3110, 219
NGC 3115, 219
NGC 3124, 219
NGC 3137, 220
NGC 3145, 220
NGC 3147, 222
NGC 3152, 221

Index

NGC 3156, 220
NGC 3158, 221
NGC 3160, 221
NGC 3162, 221
NGC 3165, 221
NGC 3166, 221
NGC 3169, 221
NGC 3175, 221
NGC 3177, 222
NGC 3182, 223
NGC 3183, 224
NGC 3184, 223
NGC 3185, 222
NGC 3187, 223
NGC 3190, 223
NGC 3193, 223
NGC 3198, 224
NGC 3200, 223
NGC 3203, 224
NGC 3206, 225
NGC 3222, 225
NGC 3223, 224
NGC 3224, 224
NGC 3226, 225
NGC 3227, 225
NGC 3239, 226
NGC 3241, 226
NGC 3242, 226
NGC 3245, 227
NGC 3246, 227
NGC 3254, 227
NGC 3259, 228
NGC 3264, 228
NGC 3274, 228
NGC 3277, 228
NGC 3281, 228
NGC 3285, 228
NGC 3287, 228
NGC 3294, 229
NGC 3300, 229
NGC 3301, 230
NGC 3307, 230
NGC 3308, 229
NGC 3309, 229
NGC 3310, 231
NGC 3311, 230
NGC 3312, 230
NGC 3313, 230
NGC 3314A, 229
NGC 3316, 230
NGC 3319, 231
NGC 3320, 231
NGC 3329, 232
NGC 3336, 231

NGC 3338, 232
NGC 3344, 232
NGC 3346, 232
NGC 3348, 234
NGC 3351, 232
NGC 3359, 233
NGC 3365, 233
NGC 3367, 233
NGC 3368, 233
NGC 3370, 233
NGC 3377, 234
NGC 3379, 234
NGC 3381, 235
NGC 3384, 235
NGC 3389, 235
NGC 3390, 234
NGC 3392, 236
NGC 3393, 235
NGC 3394, 236
NGC 3395, 235
NGC 3396, 235
NGC 3403, 238
NGC 3411, 236
NGC 3412, 236
NGC 3413, 236
NGC 3414, 237
NGC 3418, 237
NGC 3421, 236
NGC 3422, 236
NGC 3423, 236
NGC 3424, 237
NGC 3430, 237
NGC 3432, 237
NGC 3433, 237
NGC 3434, 237
NGC 3437, 238
NGC 3445, 239
NGC 3447, 238
NGC 3447A, 238
NGC 3448, 239
NGC 3449, 238
NGC 3450, 234
NGC 3454, 239
NGC 3455, 239
NGC 3456, 238
NGC 3458, 239
NGC 3462, 239
NGC 3464, 239
NGC 3471, 240
NGC 3483, 240
NGC 3485, 240
NGC 3486, 240
NGC 3489, 240
NGC 3495, 240

548   Index

NGC 3504, 241
NGC 3506, 241
NGC 3507, 241
NGC 3510, 242
NGC 3511, 241
NGC 3512, 242
NGC 3513, 242
NGC 3516, 242
NGC 3521, 242
NGC 3528, 243
NGC 3529, 243
NGC 3547, 243
NGC 3549, 243
NGC 3556, 243
NGC 3562, 244
NGC 3571, 244
NGC 3577, 245
NGC 3583, 244
NGC 3585, 244
NGC 3587, 245
NGC 3593, 245
NGC 3595, 245
NGC 3596, 245
NGC 3599, 245
NGC 3600, 246
NGC 3605, 246
NGC 3607, 246
NGC 3608, 246
NGC 3610, 247
NGC 3611, 246
NGC 3613, 247
NGC 3614, 247
NGC 3614A, 247
NGC 3619, 248
NGC 3621, 247
NGC 3623, 248
NGC 3626, 248
NGC 3627, 248
NGC 3628, 31, 249
NGC 3629, 249
NGC 3630, 249
NGC 3631, 249
NGC 3634, 249
NGC 3635, 249
NGC 3640, 249
NGC 3641, 249
NGC 3642, 250
NGC 3646, 250
NGC 3649, 250
NGC 3652, 250
NGC 3655, 250
NGC 3657, 251
NGC 3658, 251
NGC 3659, 250

NGC 3664, 251
NGC 3665, 251
NGC 3666, 251
NGC 3668, 252
NGC 3669, 252
NGC 3672, 251
NGC 3673, 251
NGC 3674, 252
NGC 3675, 252
NGC 3681, 252
NGC 3683, 253
NGC 3683A, 255
NGC 3684, 253
NGC 3686, 253
NGC 3687, 253
NGC 3689, 253
NGC 3690, 254
NGC 3691, 253
NGC 3692, 254
NGC 3693, 254
NGC 3705, 254
NGC 3715, 254
NGC 3717, 255
NGC 3718, 255
NGC 3726, 255
NGC 3729, 256
NGC 3732, 256
NGC 3733, 256
NGC 3735, 256
NGC 3738, 256
NGC 3756, 247
NGC 3759, 247
NGC 3769, 257
NGC 3769A, 257
NGC 3773, 257
NGC 3780, 258
NGC 3782, 257
NGC 3786, 258
NGC 3788, 258
NGC 3789, 257
NGC 3793, 258
NGC 3798, 258
NGC 3801, 258
NGC 3802, 258
NGC 3810, 258
NGC 3811, 259
NGC 3813, 259
NGC 3817, 259
NGC 3818, 259
NGC 3819, 259
NGC 3822, 259
NGC 3825, 259
NGC 3835, 260
NGC 3842, 259

Index

NGC 3865, 260
NGC 3866, 260
NGC 3872, 260
NGC 3877, 260
NGC 3885, 260
NGC 3887, 260
NGC 3888, 261
NGC 3889, 261
NGC 3891, 261
NGC 3892, 261
NGC 3893, 261
NGC 3894, 261
NGC 3895, 261
NGC 3896, 261
NGC 3898, 262
NGC 3900, 262
NGC 3904, 262
NGC 3907, 262
NGC 3912, 262
NGC 3913, 262
NGC 3916, 263
NGC 3917, 263
NGC 3921, 263
NGC 3923, 263
NGC 3928, 263
NGC 3930, 263
NGC 3936, 263
NGC 3938, 264
NGC 3941, 264
NGC 3945, 264
NGC 3949, 264
NGC 3953, 264
NGC 3955, 265
NGC 3956, 265
NGC 3957, 265
NGC 3958, 265
NGC 3959, 266
NGC 3962, 265
NGC 3963, 265
NGC 3967, 265
NGC 3968, 266
NGC 3972, 266
NGC 3973, 266
NGC 3976, 266
NGC 3977, 266
NGC 3981, 266
NGC 3982, 267
NGC 3985, 267
NGC 3987, 267
NGC 3989, 268
NGC 3990, 268
NGC 3992, 267
NGC 3993, 268
NGC 3997, 268
NGC 3998, 267
NGC 3999, 268
NGC 4000, 268
NGC 4005, 268
NGC 4008, 268
NGC 4010, 269
NGC 4011, 268
NGC 4013, 268
NGC 4015, 268
NGC 4016, 269
NGC 4017, 269
NGC 4022, 268
NGC 4023, 268
NGC 4024, 268
NGC 4026, 269
NGC 4027, 269
NGC 4027A, 269
NGC 4030, 271
NGC 4032, 271
NGC 4033, 272
NGC 4035, 271
NGC 4036, 272
NGC 4037, 272
NGC 4038, 272
NGC 4039, 272
NGC 4041, 272
NGC 4045, 273
NGC 4045A, 273
NGC 4047, 273
NGC 4050, 273
NGC 4051, 273
NGC 4060, 274
NGC 4061, 274
NGC 4062, 273
NGC 4064, 274
NGC 4065, 274
NGC 4066, 274
NGC 4068, 273
NGC 4070, 274
NGC 4072, 274
NGC 4073, 274
NGC 4074, 274
NGC 4076, 274
NGC 4077, 274
NGC 4079, 274
NGC 4085, 275
NGC 4087, 275
NGC 4088, 275
NGC 4094, 275
NGC 4096, 275
NGC 4100, 276
NGC 4102, 276
NGC 4104, 276
NGC 4105, 276

NGC 4106, 276
NGC 4108, 276
NGC 4108B, 276
NGC 4109, 277
NGC 4111, 277
NGC 4116, 277
NGC 4117, 277
NGC 4121, 277
NGC 4123, 277
NGC 4124, 277
NGC 4125, 277
NGC 4128, 278
NGC 4129, 278
NGC 4133, 278
NGC 4136, 278
NGC 4138, 17, 278
NGC 4139, 274
NGC 4143, 17, 278
NGC 4144, 279
NGC 4145, 279
NGC 4147, 279
NGC 4150, 280
NGC 4151, 279
NGC 4152, 280
NGC 4156, 279
NGC 4157, 280
NGC 4158, 280
NGC 4162, 280
NGC 4164, 281
NGC 4165, 281
NGC 4168, 281
NGC 4169, 280
NGC 4170, 280
NGC 4174, 280
NGC 4175, 280
NGC 4178, 281
NGC 4179, 281
NGC 4183, 281
NGC 4185, 281
NGC 4189, 281
NGC 4192, 282
NGC 4193, 282
NGC 4194, 282
NGC 4203, 282
NGC 4204, 282
NGC 4206, 282
NGC 4212, 283
NGC 4214, 283
NGC 4215, 283
NGC 4216, 283
NGC 4217, 283
NGC 4220, 284
NGC 4224, 284
NGC 4226, 283

NGC 4233, 284
NGC 4235, 284
NGC 4236, 284
NGC 4237, 284
NGC 4241, 285
NGC 4242, 285
NGC 4244, 285
NGC 4245, 285
NGC 4248, 285
NGC 4250, 285
NGC 4251, 286
NGC 4254, 286
NGC 4256, 286
NGC 4258, 286
NGC 4259, 288
NGC 4260, 286
NGC 4261, 286
NGC 4262, 287
NGC 4263, 287
NGC 4264, 287
NGC 4266, 288
NGC 4267, 287
NGC 4268, 287, 288
NGC 4270, 287, 288
NGC 4273, 288
NGC 4274, 287
NGC 4277, 288
NGC 4278, 288
NGC 4281, 288
NGC 4283, 288
NGC 4284, 289
NGC 4286, 288
NGC 4290, 289
NGC 4291, 288
NGC 4292, 289
NGC 4293, 289
NGC 4294, 289
NGC 4298, 289
NGC 4299, 289
NGC 4302, 289, 290
NGC 4303, 290
NGC 4303A, 290
NGC 4304, 290
NGC 4307, 290
NGC 4310, 290
NGC 4312, 290
NGC 4313, 291
NGC 4314, 291
NGC 4319, 288, 290
NGC 4321, 291
NGC 4322, 291
NGC 4324, 291
NGC 4326, 292
NGC 4329, 291

Index

NGC 4330, 291
NGC 4332, 291
NGC 4333, 292
NGC 4339, 292
NGC 4340, 292, 293
NGC 4341, 292
NGC 4342, 292
NGC 4343, 292
NGC 4346, 292
NGC 4350, 293
NGC 4351, 293
NGC 4357, 293
NGC 4361, 293
NGC 4365, 293
NGC 4366, 293
NGC 4369, 294
NGC 4371, 294
NGC 4374, 294
NGC 4377, 294
NGC 4378, 295
NGC 4379, 294
NGC 4380, 295
NGC 4382, 295
NGC 4383, 295
NGC 4385, 295
NGC 4386, 294
NGC 4387, 296
NGC 4388, 296
NGC 4389, 295
NGC 4392, 295
NGC 4393, 296
NGC 4394, 295, 296
NGC 4395, 296
NGC 4396, 296
NGC 4402, 296
NGC 4403, 297
NGC 4404, 297
NGC 4405, 297
NGC 4406, 297
NGC 4410A, 298
NGC 4410B, 298
NGC 4411A, 298
NGC 4411B, 298
NGC 4412, 297
NGC 4413, 297
NGC 4414, 297
NGC 4415, 298
NGC 4416, 298
NGC 4417, 298
NGC 4419, 298
NGC 4420, 298
NGC 4421, 298
NGC 4424, 299
NGC 4425, 299
NGC 4429, 299
NGC 4430, 299
NGC 4431, 300
NGC 4432, 299
NGC 4434, 299
NGC 4435, 294, 299
NGC 4436, 300
NGC 4438, 294, 299, 300
NGC 4440, 300
NGC 4442, 300
NGC 4448, 300
NGC 4449, 300
NGC 4450, 301
NGC 4452, 301
NGC 4454, 301
NGC 4455, 301
NGC 4457, 302
NGC 4458, 302
NGC 4459, 302
NGC 4460, 301
NGC 4461, 302
NGC 4462, 302
NGC 4464, 302
NGC 4465, 303
NGC 4467, 303
NGC 4468, 303
NGC 4469, 303
NGC 4470, 303
NGC 4472, 303
NGC 4473, 294, 303
NGC 4474, 303
NGC 4476, 303
NGC 4477, 294, 304
NGC 4478, 304
NGC 4479, 304
NGC 4480, 304
NGC 4483, 304
NGC 4485, 304
NGC 4486, 305
NGC 4487, 305
NGC 4488, 305
NGC 4489, 305
NGC 4490, 304
NGC 4492, 305
NGC 4494, 305
NGC 4496, 306
NGC 4496B, 306
NGC 4497, 306
NGC 4498, 306
NGC 4500, 306
NGC 4501, 306
NGC 4503, 306
NGC 4504, 307
NGC 4512, 307

NGC 4516, 308
NGC 4517, 307
NGC 4517A, 307
NGC 4519, 308
NGC 4521, 307
NGC 4522, 308
NGC 4525, 308
NGC 4526, 309
NGC 4527, 309, 310
NGC 4528, 309
NGC 4531, 309
NGC 4532, 309
NGC 4534, 309
NGC 4535, 310
NGC 4536, 310
NGC 4539, 310
NGC 4540, 310
NGC 4545, 310
NGC 4546, 311
NGC 4548, 310
NGC 4550, 311
NGC 4551, 311
NGC 4552, 311
NGC 4555, 311
NGC 4559, 312
NGC 4561, 312
NGC 4564, 312
NGC 4565, 31, 312
NGC 4567, 312
NGC 4568, 312
NGC 4569, 313
NGC 4570, 313
NGC 4571, 313
NGC 4578, 313
NGC 4579, 313
NGC 4580, 314
NGC 4586, 314
NGC 4589, 313
NGC 4590, 314
NGC 4592, 314
NGC 4593, 314
NGC 4594, 315
NGC 4595, 314
NGC 4596, 315
NGC 4597, 315
NGC 4602, 315
NGC 4604, 316
NGC 4605, 315
NGC 4606, 316
NGC 4607, 316
NGC 4608, 316
NGC 4612, 316
NGC 4618, 316
NGC 4620, 317

NGC 4621, 317
NGC 4623, 318
NGC 4625, 317
NGC 4626, 318
NGC 4627, 317
NGC 4628, 318
NGC 4630, 318
NGC 4631, 317
NGC 4632, 318
NGC 4633, 319
NGC 4634, 319
NGC 4635, 319
NGC 4636, 319
NGC 4637, 319
NGC 4638, 319
NGC 4639, 319
NGC 4643, 319
NGC 4647, 319
NGC 4648, 317
NGC 4649, 320
NGC 4651, 320
NGC 4653, 320
NGC 4654, 320
NGC 4656, 320
NGC 4657, 320, 321
NGC 4658, 321
NGC 4659, 321
NGC 4660, 321
NGC 4663, 321
NGC 4665, 321
NGC 4666, 321
NGC 4668, 321
NGC 4670, 322
NGC 4673, 322
NGC 4682, 322
NGC 4684, 322
NGC 4688, 322
NGC 4689, 322
NGC 4691, 323
NGC 4694, 323
NGC 4697, 323
NGC 4698, 323
NGC 4699, 323
NGC 4700, 323
NGC 4701, 324
NGC 4710, 324
NGC 4712, 325
NGC 4713, 324
NGC 4725, 324
NGC 4731, 325
NGC 4733, 325
NGC 4736, 325
NGC 4742, 325
NGC 4747, 326

Index

NGC 4750, 324
NGC 4753, 326
NGC 4754, 326
NGC 4756, 326
NGC 4760, 327
NGC 4762, 326
NGC 4763, 327
NGC 4771, 327
NGC 4772, 327
NGC 4775, 327
NGC 4779, 327
NGC 4781, 328
NGC 4782, 328
NGC 4783, 328
NGC 4784, 328
NGC 4786, 328
NGC 4787, 328
NGC 4789, 328
NGC 4789A, 328
NGC 4790, 329
NGC 4791, 329
NGC 4793, 328
NGC 4794, 328
NGC 4795, 329
NGC 4796, 329
NGC 4800, 329
NGC 4802, 330
NGC 4808, 330
NGC 4814, 329
NGC 4818, 330
NGC 4820, 331
NGC 4825, 331
NGC 4826, 330
NGC 4830, 331
NGC 4839, 331
NGC 4842A, 331
NGC 4845, 331
NGC 4855, 332
NGC 4856, 332
NGC 4861, 331
NGC 4864, 332
NGC 4866, 332
NGC 4868, 331
NGC 4869, 332
NGC 4874, 332
NGC 4877, 333
NGC 4878, 333
NGC 4880, 333
NGC 4888, 333
NGC 4889, 332
NGC 4890, 333
NGC 4897, 334
NGC 4898, 332
NGC 4899, 334

NGC 4900, 333
NGC 4902, 334
NGC 4904, 334
NGC 4908, 332
NGC 4911, 334
NGC 4914, 333
NGC 4915, 335
NGC 4921, 334
NGC 4923, 334
NGC 4927, 334
NGC 4933, 335
NGC 4933B, 335
NGC 4936, 335
NGC 4939, 335
NGC 4941, 335
NGC 4951, 336
NGC 4955, 336
NGC 4958, 336
NGC 4965, 336
NGC 4970, 336
NGC 4981, 337
NGC 4984, 337
NGC 4995, 337
NGC 4999, 337
NGC 5005, 337
NGC 5012, 338
NGC 5015, 338
NGC 5017, 338
NGC 5018, 339
NGC 5020, 338
NGC 5022, 339
NGC 5023, 338
NGC 5024, 338
NGC 5030, 339
NGC 5033, 339
NGC 5035, 339, 340
NGC 5037, 339
NGC 5038, 340
NGC 5042, 340
NGC 5044, 340
NGC 5046, 340
NGC 5047, 340
NGC 5049, 340
NGC 5053, 341
NGC 5054, 341
NGC 5055, 340
NGC 5061, 341
NGC 5066, 341
NGC 5068, 342
NGC 5072, 342
NGC 5073, 342
NGC 5076, 342
NGC 5077, 342
NGC 5078, 17, 342

NGC 5079, 342
NGC 5084, 342
NGC 5085, 343
NGC 5087, 343
NGC 5088, 343
NGC 5101, 343
NGC 5105, 343
NGC 5111, 344
NGC 5112, 344
NGC 5119, 344
NGC 5124, 345
NGC 5125, 344
NGC 5126, 345
NGC 5127, 344
NGC 5128, 31
NGC 5129, 344
NGC 5132, 345
NGC 5133, 345
NGC 5134, 345
NGC 5135, 346
NGC 5140, 346
NGC 5146, 346
NGC 5147, 346
NGC 5150, 346
NGC 5152, 346
NGC 5153, 346, 347
NGC 5161, 347
NGC 5169, 347
NGC 5170, 348
NGC 5171, 347
NGC 5172, 347
NGC 5173, 347
NGC 5174, 348
NGC 5176, 347
NGC 5177, 347
NGC 5179, 347
NGC 5180, 347
NGC 5188, 349
NGC 5193, 349
NGC 5193A, 349
NGC 5194, 348
NGC 5195, 348, 349
NGC 5198, 349
NGC 5204, 348
NGC 5205, 349
NGC 5211, 350
NGC 5216, 349
NGC 5218, 349
NGC 5220, 350
NGC 5230, 350
NGC 5236, 350
NGC 5247, 31, 351
NGC 5248, 31, 351
NGC 5253, 352

NGC 5254, 352
NGC 5257, 352
NGC 5258, 352
NGC 5264, 352
NGC 5272, 353
NGC 5273, 353
NGC 5276, 353
NGC 5292, 354
NGC 5296, 353
NGC 5297, 353
NGC 5298, 354
NGC 5300, 354
NGC 5302, 354
NGC 5306, 355
NGC 5308, 353
NGC 5311, 355
NGC 5313, 355
NGC 5320, 355
NGC 5322, 355
NGC 5324, 356
NGC 5326, 355
NGC 5328, 356
NGC 5330, 356
NGC 5334, 356
NGC 5349, 356
NGC 5350, 356, 357
NGC 5351, 356
NGC 5353, 357
NGC 5354, 357
NGC 5355, 357
NGC 5357, 358
NGC 5358, 357
NGC 5362, 357
NGC 5363, 358
NGC 5364, 358
NGC 5371, 358
NGC 5375, 359
NGC 5376, 357
NGC 5377, 359
NGC 5379, 358
NGC 5383, 359
NGC 5389, 358
NGC 5394, 359
NGC 5395, 359
NGC 5397, 360
NGC 5398, 360
NGC 5406, 360
NGC 5419, 362
NGC 5422, 360
NGC 5423, 362
NGC 5424, 362
NGC 5426, 361, 362
NGC 5427, 362
NGC 5430, 360

Index 555

NGC 5434, 362
NGC 5436, 362
NGC 5437, 362
NGC 5438, 362
NGC 5440, 361
NGC 5443, 360
NGC 5444, 362
NGC 5445, 362
NGC 5447, 361
NGC 5448, 361
NGC 5457, 361
NGC 5461, 361
NGC 5462, 361
NGC 5466, 363
NGC 5468, 363
NGC 5471, 361
NGC 5472, 363
NGC 5473, 362
NGC 5474, 363
NGC 5476, 364
NGC 5480, 363
NGC 5481, 363
NGC 5484, 364
NGC 5485, 364
NGC 5488, 364
NGC 5490, 364
NGC 5490C, 364
NGC 5493, 364
NGC 5494, 365
NGC 5496, 365
NGC 5506, 365
NGC 5507, 365
NGC 5523, 365
NGC 5527, 365
NGC 5529, 365
NGC 5531, 366
NGC 5532, 366
NGC 5533, 366
NGC 5534, 366
NGC 5556, 367
NGC 5557, 366
NGC 5560, 366
NGC 5566, 367
NGC 5569, 367
NGC 5574, 367
NGC 5576, 367
NGC 5577, 367
NGC 5582, 367
NGC 5584, 368
NGC 5585, 366
NGC 5595, 368
NGC 5597, 368
NGC 5600, 368
NGC 5605, 369

NGC 5613, 368
NGC 5614, 368
NGC 5615, 368
NGC 5629, 369
NGC 5631, 369
NGC 5633, 369
NGC 5634, 370
NGC 5636, 370
NGC 5638, 17, 370
NGC 5641, 369
NGC 5645, 370
NGC 5653, 370
NGC 5656, 370
NGC 5660, 370
NGC 5665, 371
NGC 5667, 370
NGC 5668, 372
NGC 5669, 371
NGC 5673, 371
NGC 5676, 371
NGC 5678, 371
NGC 5687, 372
NGC 5689, 372
NGC 5690, 372
NGC 5691, 372
NGC 5694, 373
NGC 5701, 373
NGC 5707, 372
NGC 5713, 373
NGC 5719, 373
NGC 5728, 373
NGC 5735, 374
NGC 5739, 374
NGC 5740, 374
NGC 5741, 375
NGC 5742, 375
NGC 5746, 374
NGC 5750, 375
NGC 5756, 375
NGC 5757, 375
NGC 5768, 376
NGC 5770, 376
NGC 5774, 376, 377
NGC 5775, 377
NGC 5791, 377
NGC 5792, 377
NGC 5793, 377
NGC 5796, 377
NGC 5806, 377
NGC 5812, 378
NGC 5813, 378
NGC 5814, 378
NGC 5824, 378
NGC 5831, 378

NGC 5832, 377
NGC 5838, 378
NGC 5839, 379
NGC 5845, 379
NGC 5846, 379
NGC 5846A, 379
NGC 5850, 379
NGC 5854, 379
NGC 5861, 380
NGC 5864, 380
NGC 5865, 380
NGC 5866, 379
NGC 5869, 380
NGC 5874, 379
NGC 5875, 380
NGC 5878, 381
NGC 5879, 380
NGC 5885, 381
NGC 5892, 381
NGC 5897, 382
NGC 5898, 382
NGC 5899, 381
NGC 5903, 382
NGC 5904, 383
NGC 5905, 381
NGC 5907, 381
NGC 5908, 381, 382
NGC 5915, 383
NGC 5916, 383
NGC 5919, 383
NGC 5921, 383
NGC 5928, 384
NGC 5929, 384
NGC 5930, 384
NGC 5936, 384
NGC 5937, 384
NGC 5949, 384
NGC 5953, 385
NGC 5954, 385
NGC 5956, 385
NGC 5957, 386
NGC 5962, 386
NGC 5963, 385
NGC 5964, 386
NGC 5965, 385
NGC 5966, 386
NGC 5968, 387
NGC 5970, 386
NGC 5981, 386
NGC 5982, 30, 386
NGC 5984, 387
NGC 5985, 30, 386, 387
NGC 5987, 387
NGC 5994, 387

NGC 5996, 387
NGC 6000, 388
NGC 6004, 388
NGC 6012, 388
NGC 6014, 388
NGC 6015, 388
NGC 6026, 389
NGC 6058, 389
NGC 6070, 390
NGC 6085, 390
NGC 6086, 390
NGC 6093, 391
NGC 6106, 391
NGC 6118, 392
NGC 6121, 392
NGC 6127, 391
NGC 6140, 391
NGC 6144, 393
NGC 6155, 393
NGC 6166, 393
NGC 6171, 394
NGC 6173, 393
NGC 6174, 393
NGC 6181, 394
NGC 6205, 395
NGC 6207, 395
NGC 6210, 395
NGC 6217, 394
NGC 6218, 396
NGC 6223, 395
NGC 6229, 396
NGC 6235, 397
NGC 6236, 396
NGC 6239, 396
NGC 6240, 397
NGC 6254, 397
NGC 6266, 398
NGC 6269, 397
NGC 6273, 398
NGC 6284, 398
NGC 6287, 399
NGC 6293, 399
NGC 6304, 399
NGC 6309, 399
NGC 6316, 400
NGC 6325, 401
NGC 6333, 401
NGC 6340, 399
NGC 6341, 400
NGC 6342, 401
NGC 6355, 402
NGC 6356, 402
NGC 6357, 402
NGC 6366, 403

Index   557

NGC 6368, 402
NGC 6369, 403
NGC 6383, 404
NGC 6384, 403
NGC 6389, 404
NGC 6395, 402
NGC 6401, 405
NGC 6402, 405
NGC 6404, 405
NGC 6405, 405
NGC 6411, 404
NGC 6412, 403
NGC 6416, 406
NGC 6425, 406
NGC 6426, 406
NGC 6434, 404
NGC 6439, 407
NGC 6440, 407
NGC 6445, 407
NGC 6451, 407
NGC 6453, 407
NGC 6469, 408
NGC 6475, 408
NGC 6482, 408
NGC 6484, 408
NGC 6486, 408
NGC 6487, 408
NGC 6490, 409
NGC 6494, 409
NGC 6495, 409
NGC 6500, 409
NGC 6501, 409
NGC 6503, 407
NGC 6507, 410
NGC 6509, 410
NGC 6514, 411
NGC 6517, 411
NGC 6520, 412
NGC 6522, 412
NGC 6523, 412
NGC 6528, 413
NGC 6530, 413
NGC 6531, 413
NGC 6535, 413
NGC 6537, 413
NGC 6539, 413
NGC 6540, 414
NGC 6543, 409
NGC 6544, 414
NGC 6546, 414
NGC 6548, 414
NGC 6549, 414
NGC 6553, 415
NGC 6555, 414

NGC 6558, 415
NGC 6559, 415
NGC 6563, 416
NGC 6565, 415
NGC 6567, 417
NGC 6568, 416
NGC 6569, 416
NGC 6572, 416
NGC 6574, 416
NGC 6577, 416
NGC 6578, 417
NGC 6580, 416
NGC 6583, 417
NGC 6589, 417
NGC 6590, 417
NGC 6595, 418
NGC 6603, 418
NGC 6604, 418
NGC 6605, 417
NGC 6611, 418
NGC 6613, 419
NGC 6618, 419
NGC 6620, 419
NGC 6624, 420
NGC 6625, 420
NGC 6626, 420
NGC 6629, 421
NGC 6631, 421
NGC 6632, 420
NGC 6633, 421
NGC 6637, 422
NGC 6638, 421
NGC 6642, 422
NGC 6643, 419
NGC 6644, 422
NGC 6645, 422
NGC 6649, 422
NGC 6652, 423
NGC 6654, 420
NGC 6656, 423
NGC 6661, 423
NGC 6664, 424
NGC 6674, 424
NGC 6681, 425
NGC 6682, 424
NGC 6683, 425
NGC 6690, 423
NGC 6694, 426
NGC 6701, 425
NGC 6702, 426
NGC 6703, 426
NGC 6704, 427
NGC 6705, 427
NGC 6709, 427

NGC 6712, 428
NGC 6715, 428
NGC 6716, 428
NGC 6717, 429
NGC 6720, 428
NGC 6738, 429
NGC 6741, 429
NGC 6749, 430
NGC 6751, 430
NGC 6755, 431
NGC 6756, 431
NGC 6760, 431
NGC 6764, 431
NGC 6765, 431
NGC 6772, 432
NGC 6774, 432
NGC 6778, 433
NGC 6779, 432
NGC 6781, 433
NGC 6790, 434
NGC 6791, 433
NGC 6792, 433
NGC 6800, 434
NGC 6802, 434
NGC 6803, 435
NGC 6804, 435
NGC 6807, 435
NGC 6809, 435
NGC 6811, 435
NGC 6814, 436
NGC 6815, 436
NGC 6818, 437
NGC 6819, 436
NGC 6822, 437
NGC 6823, 437
NGC 6824, 437
NGC 6826, 437
NGC 6830, 439
NGC 6833, 438
NGC 6834, 439
NGC 6837, 439
NGC 6838, 439
NGC 6842, 439
NGC 6852, 440
NGC 6853, 440
NGC 6857, 441
NGC 6864, 441
NGC 6866, 441
NGC 6869, 440
NGC 6871, 441
NGC 6874, 441
NGC 6879, 442
NGC 6881, 442
NGC 6883, 442

NGC 6884, 442
NGC 6885, 442
NGC 6886, 443
NGC 6888, 442
NGC 6891, 443
NGC 6894, 443
NGC 6900, 444
NGC 6903, 445
NGC 6905, 444
NGC 6906, 445
NGC 6907, 445
NGC 6910, 445
NGC 6913, 445
NGC 6923, 446
NGC 6925, 447
NGC 6926, 446
NGC 6927, 446
NGC 6928, 446
NGC 6929, 446
NGC 6930, 446
NGC 6934, 446
NGC 6939, 446
NGC 6940, 447
NGC 6946, 447
NGC 6951, 447
NGC 6956, 448
NGC 6960, 448
NGC 6961, 448
NGC 6962, 448
NGC 6964, 448
NGC 6965, 448
NGC 6967, 449
NGC 6979, 449
NGC 6981, 449
NGC 6992, 449
NGC 6994, 450
NGC 6997, 450
NGC 7000, 450
NGC 7006, 451
NGC 7008, 451
NGC 7009, 451
NGC 7013, 451
NGC 7023, 450
NGC 7026, 452
NGC 7027, 452
NGC 7031, 452
NGC 7039, 453
NGC 7042, 453
NGC 7043, 453
NGC 7044, 453
NGC 7048, 453
NGC 7058, 454
NGC 7062, 454
NGC 7063, 454

Index

NGC 7065A, 454
NGC 7067, 454
NGC 7078, 455
NGC 7080, 455
NGC 7082, 455
NGC 7086, 455
NGC 7089, 456
NGC 7092, 456
NGC 7099, 457
NGC 7127, 458
NGC 7128, 458
NGC 7129, 457
NGC 7130, 459
NGC 7133, 457
NGC 7135, 459
NGC 7137, 458
NGC 7139, 458
NGC 7142, 458
NGC 7154, 459
NGC 7156, 459
NGC 7160, 459
NGC 7167, 460
NGC 7171, 460
NGC 7172, 460
NGC 7173, 460
NGC 7174, 460
NGC 7176, 460
NGC 7177, 460
NGC 7180, 461
NGC 7183, 461
NGC 7184, 461
NGC 7185, 461
NGC 7209, 462
NGC 7217, 462
NGC 7218, 462
NGC 7221, 463
NGC 7223, 462
NGC 7225, 463
NGC 7226, 463
NGC 7229, 464
NGC 7235, 463
NGC 7243, 464
NGC 7245, 464
NGC 7250, 465
NGC 7251, 465
NGC 7252, 465
NGC 7261, 465
NGC 7265, 465
NGC 7267, 466
NGC 7280, 466
NGC 7281, 466
NGC 7284, 467
NGC 7285, 467
NGC 7292, 466

NGC 7293, 467
NGC 7296, 466
NGC 7302, 467
NGC 7309, 468
NGC 7313, 468
NGC 7314, 468
NGC 7318, 468
NGC 7319, 468
NGC 7320, 468
NGC 7330, 468
NGC 7331, 468
NGC 7332, 469
NGC 7335, 469
NGC 7337, 469
NGC 7339, 469
NGC 7340, 469
NGC 7351, 469
NGC 7354, 469
NGC 7361, 470
NGC 7371, 470
NGC 7377, 470
NGC 7380, 470
NGC 7383, 471
NGC 7385, 471
NGC 7386, 471
NGC 7387, 471
NGC 7389, 471
NGC 7391, 471
NGC 7392, 471
NGC 7416, 472
NGC 7419, 471
NGC 7443, 472
NGC 7444, 472
NGC 7448, 472
NGC 7454, 473
NGC 7457, 473
NGC 7463, 473
NGC 7464, 473
NGC 7465, 473
NGC 7469, 473
NGC 7479, 474
NGC 7490, 474
NGC 7492, 474
NGC 7497, 474
NGC 7507, 475
NGC 7510, 474
NGC 7513, 475
NGC 7515, 475
NGC 7530, 475
NGC 7532, 475
NGC 7537, 476
NGC 7538, 475
NGC 7541, 475
NGC 7547, 476

NGC 7549, 476
NGC 7550, 476
NGC 7554, 476
NGC 7556, 476
NGC 7557, 476
NGC 7558, 476
NGC 7562, 476
NGC 7585, 476
NGC 7596, 476
NGC 7600, 477
NGC 7606, 477
NGC 7611, 477
NGC 7617, 477
NGC 7619, 477
NGC 7623, 477
NGC 7625, 477
NGC 7626, 477
NGC 7631, 477
NGC 7635, 478
NGC 7640, 478
NGC 7653, 478
NGC 7654, 478
NGC 7662, 478
NGC 7673, 479
NGC 7678, 479
NGC 7686, 479
NGC 7711, 480
NGC 7714, 480
NGC 7715, 480
NGC 7716, 480
NGC 7720, 480
NGC 7721, 481
NGC 7723, 481
NGC 7725, 481
NGC 7727, 481
NGC 7741, 481
NGC 7742, 481
NGC 7743, 482
NGC 7752, 482
NGC 7753, 482
NGC 7755, 482
NGC 7756, 482
NGC 7757, 482
NGC 7762, 483
NGC 7765, 483
NGC 7766, 483
NGC 7767, 483
NGC 7768, 483
NGC 7769, 483
NGC 7770, 483
NGC 7771, 483
NGC 7772, 483
NGC 7778, 484
NGC 7779, 484

NGC 7781, 484
NGC 7782, 484
NGC 7785, 484
NGC 7788, 484
NGC 7789, 484
NGC 7790, 485
NGC 7793, 484
NGC 7798, 485
NGC 7800, 485
NGC 7814, 50
NGC 7817, 50
North America Nebula, 450
Nu Dra, 403
Nu Hya, 235
Nu Oph, 410
Nu UMa, 247

O

Omega For, 93
Omicron 2 CMa, 166
Omicron Cep, 476
Omicron Cet, 88
Omicron Oph, 401
Omicron UMa, 190
36 Oph, 400
70 Oph, 413
Orion Nebula, 141
Owl Nebula, 245

P

PAL 2, 128
PAL 5, 382
PAL 6, 406
PAL 8, 424
PAL 10, 433
PAL 11, 438
PAL 12, 458
Pegasus dwarf, 479
Perseus Galaxy Cluster, 108
PGC 17965, 145
PGC 20274, 169
PGC 30367, 225
PGC 30392, 225
Phi 2 Cnc, 190
Phi Sgr, 426
Pi 3 Ori, 129
Pi Boo, 373
Pi Her, 400
Pi Hya, 363
Pi Sco, 389
Pi Sgr, 431
PK 3+2.1, 405

Index   561

PK 3–14.1, 429
PK 3–17.1, 430
PK 7+1.1, 409
PK 38+12.1, 418
PK 51+9.1, 427
PK 52–2.2, 435
PK 64+5.1, 435
PK 86–8.1, 456
PK 107–2.1, 472
PK 111–2.1, 478
PK 119–6.1, 53
PK 164+31.1, 183
PK 171–25.1, 118
PK 190–17.1, 133
PK 194+2.1, 155
PK 342+27.1, 384
PK 356–4.1, 408
PK 359–0.1, 406
Pleiades, 117, 123
Polaris, 92, 392
Pollux, 180
Procyon, 179
35 Psc, 52
65 Psc, 62
Psi 1 Psc, 66
Psi Cyg, 440
Psi Dra, 405
Psi UMa, 243
2 Pup, 180
5 Pup, 181
PZ 4, 389
PZ 6, 410

R
R And, 53
R Aql, 430
R Cas, 485
R CMi, 168
R For, 91
R Hya, 348
R Leo, 211
R Lep, 132
R Scl, 72
R Ser, 388
R Tri, 94
Regulus, 220
Rho Her, 402
Rho Oph, 392
Rho Per, 103
Rho Pup, 185
Rigel, 135
Ring Nebula, 428
Ring-Tail Galaxy, 272

Rosette Nebula, 157
RR Sco, 397
RR Sgr, 440
RS Cyg, 443
RT Cap, 443
RV Cyg, 457
RV Mon, 165
RY Dra, 330
RZ Peg, 462

S
S Cep, 456
12 Sco, 390
S CrB, 383
S Scl, 52
S Sct, 427
Saturn Nebula, 451
Sculptor dwarf, 65
Scutum Star Cloud, 424
59 Ser, 421
Seven Sisters, 117
35 Sex, 232
SHJ 145, 303
SHJ 151, 339
SHJ 243, 400
SHJ 345, 466
Siamese Twins, 312
Sigma Cas, 485
Sigma CMa, 165
Sigma CrB, 391
Sigma Lib, 378
Sigma Ori, 143
Sigma Sco, 392
Sigma Sgr, 429
Sirius, 161
Small Sagittarius Star Cloud, 417
Sombrero Galaxy, 315
Spica, 345
SS Vir, 295
ST 2, 86
ST 23, 106
ST 24, 58
ST Cam, 130
Stephan's Quintet, 468
STF 7, 110
STF 12, 52
STF 14, 139
STF 35, 403
STF 38, 426
STF 43, 434
STF 60, 62
STF 61, 62
STF 79, 65

STF 88, 66
STF 93, 92
STF 100, 68
STF 147, 75
STF 162, 78
STF 174, 78
STF 180, 80
STF 205, 84
STF 222, 86
STF 227, 86
STF 231, 86
STF 262, 91
STF 299, 98
STF 331, 101
STF 389, 110
STF 401, 110
STF 464, 118
STF 470, 118
STF 471, 119
STF 528, 123
STF 550, 125
STF 559, 125
STF 570, 126
STF 590, 127
STF 630, 133
STF 653, 136
STF 668, 135
STF 701, 137
STF 716, 139
STF 730, 139
STF 738, 140
STF 752, 141
STF 761, 143
STF 762, 143
STF 774, 143
STF 845, 152
STF 872, 153
STF 900, 155
STF 919, 156
STF 958, 163
STF 982, 164
STF 1009, 167
STF 1062, 172
STF 1104, 176
STF 1110, 176
STF 1138, 180
STF 1146, 181
STF 1177, 185
STF 1196, 186
STF 1223, 190
STF 1245, 191
STF 1268, 193
STF 1270, 193

STF 1273, 193
STF 1295, 196
STF 1311, 198
STF 1334, 202
STF 1415, 222
STF 1424, 224
STF 1466, 232
STF 1474, 234
STF 1487, 239
STF 1520, 246
STF 1524, 247
STF 1552, 256
STF 1559, 257
STF 1596, 274
STF 1657, 310
STF 1669, 316
STF 1692, 330
STF 1694, 324
STF 1695, 330
STF 1744, 344
STF 1777, 353
STF 1788, 357
STF 1821, 365
STF 1835, 368
STF 1864, 373
STF 1877, 374
STF 1884, 375
STF 1888, 376
STF 1890, 376
STF 1909, 378
STF 1954, 385
STF 1962, 387
STF 1965, 387
STF 1998, 389
STF 1999, 389
STF 2010, 390
STF 2032, 391
STF 2078, 394
STF 2130, 399
STF 2140, 400
STF 2161, 402
STF 2241, 405
STF 2264, 411
STF 2272, 413
STF 2280, 414
STF 2308, 411
STF 2316, 421
STF 2375, 426
STF 2379, 426
STF 2382, 425
STF 2383, 425
STF 2417, 429
STF 2452, 428

Index 563

STF 2486, 432
STF 2578, 438
STF 2579, 438
STF 2594, 439
STF 2603, 438
STF 2605, 440
STF 2609, 440
STF 2644, 443
STF 2671, 444
STF 2716, 447
STF 2727, 448
STF 2735, 449
STF 2737, 450
STF 2741, 450
STF 2745, 451
STF 2758, 452
STF 2762, 452
STF 2769, 452
STF 2806, 454
STF 2840, 459
STF 2863, 461
STF 2883, 463
STF 2903, 465
STF 2909, 467
STF 2922, 468
STF 2935, 470
STF 2978, 474
STF 2998, 477
STF 3001, 476
STF 3049, 485
STF 3053, 49
STF 3127, 400
STT 437, 454
Swan Nebula, 419

T
118 Tau, 139
T Cep, 452
Tau Cet, 76
Tau Scl, 74
Tau Sco, 394
Tau Sgr, 430
TER 7, 432
TER 8, 436
Theta 1 Tau, 124
Theta 2 Tau, 124
Theta Aql, 442
Theta Aur, 147
Theta Leo, 245
Theta Oph, 401
Theta Ser, 429
Theta UMa, 206

TOM 5, 118
TR 1, 74
TR 2, 94
TR 5, 158
TR 6, 174
TR 7, 174
TR 27, 404
TR 28, 404
TR 31, 410
TR 32, 418
TR 33, 420
TR 34, 424
TR 35, 425
Trapezium, 141
6 Tri, 86
Trifid Nebula, 412
TT Cyg, 436
TU Gem, 152
TX Psc, 482

U
U Cyg, 444
U Hya, 230
U Ori, 147
UGC 863, 70
UGC 1183, 75
UGC 1281, 78
UGC 1347, 80
UGC 1866, 89
UGC 1886, 90
UGC 1933, 90
UGC 2069, 94
UGC 2296, 100
UGC 2729, 109
UGC 2765, 111
UGC 2800, 114
UGC 3069, 126
UGC 3108, 127
UGC 3253, 136
UGC 3504, 159
UGC 3511, 161
UGC 3574, 164
UGC 3580, 165
UGC 3685, 168
UGC 3691, 167
UGC 3696, 168
UGC 3714, 170
UGC 3730, 170
UGC 3804, 172
UGC 3828, 173
UGC 4028, 181
UGC 4151, 184

UGC 4289, 187
UGC 4305, 188
UGC 4375, 189
UGC 4841, 201
UGC 4883, 202
UGC 4904, 201
UGC 5139, 208
UGC 5364, 216
UGC 5373, 217
UGC 5459, 219
UGC 5470, 220
UGC 5522, 221
UGC 5612, 226
UGC 6132, 242
UGC 6253, 244
UGC 6446, 252
UGC 6510, 255
UGC 6628, 258
UGC 6793, 262
UGC 6903, 266
UGC 6917, 267
UGC 6930, 267
UGC 6983, 269
UGC 7490, 293
UGC 7577, 300
UGC 7690, 307
UGC 7698, 308
UGC 7699, 308
UGC 8041, 329
UGC 8201, 336
UGC 8320, 339
UGC 8614, 351
UGC 8658, 352
UGC 9024, 364
UGC 9215, 368
UGC 9288, 369
UGC 9310, 370
UGC 9749, 380
UGC 10144, 389
UGC 10497, 395
UGC 10502, 395
UGC 10528, 396
UGC 10803, 400
UGC 10822, 401
UGC 11337, 425
UGC 11453, 434
UGC 11466, 436
UGC 11762, 455
UGC 11775, 456
UGC 11781, 456
UGC 11920, 462
UGC 11973, 464
UGC 12433, 475

UGC 12590, 478
UGC 12613, 479
UGC 12632, 479
UGC 12704, 480
UGC 12914, 49
UGC 12915, 49
UGCA 21, 77
UGCA 32, 90
UGCA 34, 94
UGCA 67, 109
UGCA 90, 123
UGCA 95, 129
UGCA 103, 135
UGCA 104, 135
UGCA 106, 135
UGCA 108, 136
UGCA 114, 146
UGCA 127, 154
UGCA 128, 154
UGCA 137, 187
UGCA 150, 199
UGCA 167, 206
UGCA 168, 206
UGCA 180, 209
UGCA 200, 219
UGCA 205, 220
UGCA 212, 229
UGCA 241, 258
UGCA 247, 261
UGCA 250, 264
UGCA 266, 272
UGCA 270, 275
UGCA 282, 304
UGCA 289, 311
UGCA 322, 335
UGCA 324, 336
UGCA 330, 337
UGCA 348, 342
UGCA 353, 345
UGCA 378, 360
UGCA 400, 380
UGCA 408, 384
UGCA 444, 49
Ursa Minor dwarf, 380
UU Aur, 158
UV Cam, 120
UX Dra, 433

V
V Aql, 430
V CrB, 388
V Hya, 237

Index 565

54 Vir, 339
84 Vir, 353
V Oph, 393
V389 Cyg, 452
V460 Cyg, 457
V623 Cas, 105
V1942 Sgr, 433
Vega, 424
Veil Nebula, 449
Virgo Cluster, 305
VY UMa, 233

W
W CMa, 167
W Ori, 133
Whirlpool Galaxy, 348
WLM, 49
WNC 2, 137
WZ Cas, 49

X
X Cnc, 196
Xi Boo, 376
Xi Cep, 461
Xi Gem, 161
Xi Lup, 389
Xi Pup, 181
Xi Sco, 389

Y
Y CVn, 322
Y Hya, 212
Y Tau, 144

Z
Z Psc, 69
Zeta 1 Ant, 206
Zeta Aql, 430
Zeta Aqr, 467
Zeta Cep, 463
Zeta CMa, 154
Zeta Cnc, 186
Zeta CrB, 387
Zeta Cyg, 453
Zeta Dra, 399
Zeta Her, 395
Zeta Hya, 196
Zeta Leo, 222
Zeta Lyr, 426
Zeta Oph, 394
Zeta Ori, 143
Zeta Peg, 469
Zeta Per, 118
Zeta Psc, 68
Zeta Sgr, 430
Zeta Tau, 143
Zeta UMa, 344
Zeta Vir, 350

Printed by Publishers' Graphics LLC
BT20121030.19.18.25